National Conference on Progress in Mathematics Towards Industrial Applications (PMTIA 2019)

Journal of Physics: Conference Series Volume 1377

Chennai, India
27 - 28 September 2019

ISBN: 978-1-7138-1133-6
ISSN: 1742-6588

Printed from e-media with permission by:

Curran Associates, Inc.
57 Morehouse Lane
Red Hook, NY 12571

Some format issues inherent in the e-media version may also appear in this print version.

Institute of Physics
Dirac House, Temple Back
Bristol BS1 6BE UK

Phone: 44 1 17 929 7481
Fax: 44 1 17 920 0979

techtracking@iop.org

Additional copies of this publication are available from:

Curran Associates, Inc.
57 Morehouse Lane
Red Hook, NY 12571 USA
Phone: 845-758-0400
Fax: 845-758-2633
Email: curran@proceedings.com
Web: www.proceedings.com

TABLE OF CONTENTS

Journal of Physics: Conference Series **1377** (2019) 011001 doi:10.1088/1742-6596/1377/1/011001

SRM Institute of Science & Technology
Ramapuram Campus

Dr. Shakeela Sathish,
Head of the Department
Department of Mathematics

SRM Institute of Science and Technology is the number one private institution in the country, which has set standards in experimental education and knowledge creation across various fields. It has inspired students with an international standards of education, world class infrastructure and industry ready curriculum. A continuous emphasis on modern teaching methods, global links for research, student exchange and knowledge sharing and industry-institute interactions have consistently placed SRM, well ahead in every field. SRM Institute of Science and Technology has been categorized as grade 'A' by Ministry of Human Resource Development (MHRD), Government of India. SRM is accredited by NAAC with 'A++' grade in the year 2018. SRM Institute of Science and Technology Ramapuram established in 2004, is located at the heart of the city. This charming enclave provides the facilities and infrastructre, necessary for Dental, Health Sciences, Engineering and Technology and Science and Humanities.

We are pleased to inform you that, Department of Mathematics is organisising a National conference on "Progress in Mathematics towards Industrial Applications", which is aiming to promote interaction between academy and industry, leading to innovation in both fields and providing unique opportunity to discuss the latest ideas, problems and methodologies, and contributing to the advancement of science and technology. It also encourages industrial sectors to propose challenging problems where Mathematicians can provide insights and fresh perspectives.

The purpose of this conference is to bring together leading academic scientists, experts from industry and young researchers and provide a platform in which significant advances in Mathematics are presented & widespread discussions are held, to promote the use of innovative Mathematics in industry. This conference will provide a unique opportunity to exchange ideas,

Journal of Physics: Conference Series **1377** (2019) 011001 doi:10.1088/1742-6596/1377/1/011001

problems and methodologies, bridging the gap between Mathematics and industry and contributing to the advancement of science and technology to benefit the society. The conference will have keynote addresses by renowned experts and Research paper presentations.

Dr. Shakeela Sathish

Convener
PMTIA– 2019

Peer review statement

All papers published in this volume of *Journal of Physics: Conference Series* have been peer reviewed through processes administered by the proceedings Editors. Reviews were conducted by expert referees to the professional and scientific standards expected of a proceedings journal published by IOP Publishing.

Bi-conditional Cordial Labeling for Extended Duplicate Graph of Certain Classes of Graphs

E.Nandagopal[1] and V.Maheswari[2]

[1]Sriram Engineering College, Perumalapattu, Tiruvallur- 602024.
[2]Vels *Institute* of Science Technology and Advanced Studies (VISTAS), Chennai–6000117.

mail: maheswari.sbs@velsuniv.ac.in, enandhu77@gmail.com.

Abstract –In this article we admit Bi- Conditional for Duplicate graph of Quadrilateral Snake $EDG(QS_m)m \geq 2$, ***Double Quadrilateral SnakeGraphs*** $EDG(DQS_m)m \geq 2$, ***and Triangular ladder graph.***$EDG(TQL_m)m \geq 2$.

AMS Subject Classification: 05C78
Keywords: graph labeling, Extended Duplicate, labeling, Bi conditional labeling.

1. Introduction:

Murali, Thirusangu and Madura Meenakshi introduced new graph labeling called Bi-conditional cordial labeling [3]. Let G (N, L) be a graph. If there exist a mapping $g: N \rightarrow \{0,1\}$ such that the induced function $g^*(L) = \{0,1\}$ given by $g^*(uv) = \begin{cases} 1: \text{if } g(u) = g(v) \\ 0: \text{if } g(u) \neq g(v) \end{cases}$ for every uv∈ L(G), if the number of nodes labeled '0' and the number of nodes labeled '1' differ by at most '1' and also the number of lines labeled '0' and the number of lines labeled '1' differ by at most 1. Murali, Thirusangu, Balamurugan proved Combination Cordial Labeling for Flower and Corona Graphs [7]. Selvam, Thirusangu, they are proved Z_3-vertex and Z_3-edge magic total labeling for the duplicate graph of quadrilateral snake graph. [4]. Nandagopal, Maheswari and Vijayakumar existence on Signed product, sum difference, and Sum Divisor Cordial labeling for the Duplicate graphs of Quadrilateral snake Graph [5]. Ulaganathan, Selvam, and Vijayakumar proved the duplicate of Bistar, Double star and Triangular ladder graphs existence on Signed Product labeling [8].

Definition 1.1

Let G (N, L) be a simple graph. A duplicate graph of G is DG = (N_1, L_1), where the nodes set $N_1 = N \cup N'$ and $N \cap N' = \emptyset$ and $g: N \rightarrow N'$ is bijective (for $n \in N$, we write $g(n) = n'$ for convenience) and the lines set L_1 of DG is defined as: The line xy is in L if and only if both xy′ and x′y are lines in L_1.

Definition: 1.2

Let $DG = (N_1, L_1)$ be a duplicate graph of the path graph $G(N, L)$. we add an line between any one nodes from N to any other node in 'N'. For convenience, let us take $n_2 \in N$ and $n_2' \in N'$ and thus line(n_2, n'_2) is formed. This graph is called the Extended Duplicate Graph of the path P_k and it's denoted by $EDG(p_k)$.

Remarks:

1.3 The Extended Duplicate graph of Quadrilateral snake graph is denoted by $EDG(QS_m)m \geq 2$, and it has 6m+2 nodes and 8m lines. Where m is the connected path of the Quadrilateral graphs

1.4 The Extended Duplicate graph of Double Quadrilateral snake graph is denoted by

$EDG(DQS_m) m \geq 2$, and it has 10m+2 nodes and 14m lines. Where m is the connected path of the Double Quadrilateral graphs.

1.5 The Extended Duplicate graph of Triangular ladder graph is denoted by $EDG(TL_m) m \geq 2$, and it has 4m+4 nodes and 8m+2 lines. Where m is the number of rungs in the ladder.

2. Bi-conditional cordial labeling $EDG(QS_m)$

Algorithm 2.1: Allocation of labeled nodes

Let $N \rightarrow \{n_1, n_2 \dots \dots n_{3m+1}.n'_1, n'_2, \dots \dots .n'_{3m+1}\}$
$L \rightarrow \{l_1, l_2, l_3 \dots \dots, l_{4m}.l'_1, l'_2, l'_3 \dots \dots \dots .l'_{4m}\}$
For $1 \leq k \leq m$
$n_{3k-2} \leftarrow 1$ $n_{3k-1} \leftarrow 0$. $n_{3k} \leftarrow 0$
$n'_{3k+1} \leftarrow 0, n'_{3k-1} \leftarrow 1$
For $k = 3m + 1$
$n_k \leftarrow 1$
For $1 \leq k < m$
$n'_{3k+3} \leftarrow 1$
Fix $n'_1 \leftarrow 1, n'_3 \leftarrow 0$

Theorem 2.2: The extended duplicate graph of Quadrilateral Snake Graph $EDG(QS_m)$ $m \geq 2$, Bi-conditional cordial labeling admits.

Proof:

Let $N \rightarrow \{n_1, n_2, \dots \dots, n_{3m+1}.n'_1, n'_2, \dots \dots .n'_{3m+1}\}$, $L \rightarrow \{l_1, l_2, l_3 \dots \dots, l_{4m}.l'_1, l'_2, l'_3 \dots \dots \dots .l'_{4m}\}$ be the set of nodes and lines of the duplicate graph of Quadrilateral graph $EDG(QS_m)$. using the algorithm 2.1, each of the $3m + 1$nodes receive label 0 and 1 respectively. Using the induced function g^* defined by $g^*(uv) = \begin{cases} 1: \text{if } g(u) = g(v) \\ 0: \text{if } g(u) \neq g(v) \end{cases}$, the m lines namely $l_1, l_5, l_9, l_{13} \dots . l_{4m-3}$ receive label 1, the m lines namely $l_4, l_8, l_{12}, l_{16} \dots . l_{4m}$ label 1, the 1 line namely l_3 receive label 1. $m-1$ lines namely $l'_5, l'_9, l'_{13}, l'_{17}, \dots \dots \dots \dots l'_{4m-3}$. receive label 1. $m-1$ lines namely $l'_8, l'_{12}, l'_{16}, \dots \dots \dots \dots \dots \dots l'_{4m}$.receive label 1. One line namely l'_2 receive label 1.
Thus 4m lines receive label 1 and the m lines namely $l_2, l_6, l_{10}, l_{14}, \dots \dots \dots \dots l_{4m-2}$. receive label 0, the $m-1$ lines namely $l_7, l_{11}, l_{15}, l_{19}, \dots \dots \dots \dots l_{4m-1}$ receive label 0, $m-1$ lines namely $l'_6, l'_{10}, l'_{14}, \dots \dots \dots \dots \dots \dots l'_{4m-2}$ receive label 0, m lines namely $l'_3, l'_7, l'_{11}, \dots \dots \dots \dots l'_{4m-1}$. receive label 0. 2 lines namely l'_1 and l'_4 receive label 0. Thus the $4m$ lines receive label0. Hence the duplicate graph of the Quadrilateral snake $EDG(QS_m)$ $m \geq 2$, Bi-Conditional Cordial labeling are admits.

2.3. Illustration:

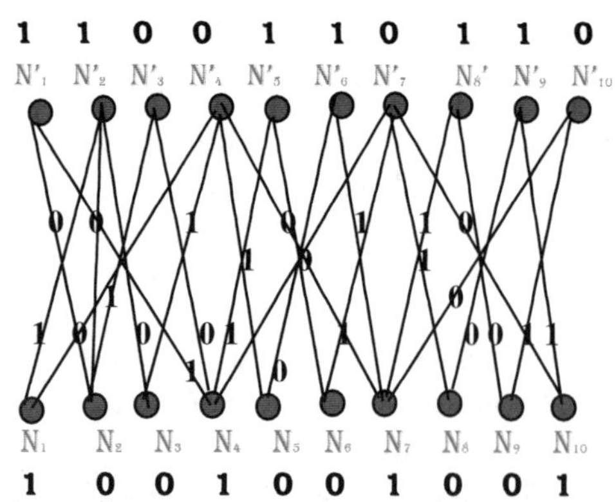

Bi-conditional cordial labeling $EDG(QS_3)$

3. Bi-conditional cordial labeling $EDG(DQS_m)$

Algorithm 3.1: Allocation of labeled nodes

Case1: When m is Even
$For 1 \leq k \leq m$
$fix\, n_1 \to 1,\ n_1' \to 0$
do
{
$if\ 1 \leq k \leq \frac{m}{2}$
{
$g(n_{10k-8}) = 0.g\ (n_{10k-5}) = 1\ g(n_{10k-7}) = 0.g(n_{10k-4}) = 1\ g(n_{10k-6}) = 0.g(n_{10k-1}) = 1$
$g(n_{10k-3}) = 0.g(n_{10k}) = \ 1\ g(n_{10k-2}) = 0.$
}
End the process.
do
{
$if\ 1 \leq k < \frac{m}{2}$
{
$g(n'_{10k-8}) = 1g(n'_{10k-5}) = 0\ g(n'_{10k-7}) = 0g(n'_{10k-4}) = 0\ g(n'_{10k-6}) = 1g(n'_{10k-1}) = 1\ g(n'_{10k-3}) = 1g(n'_{10k}) = \ 0$
$g(n'_{10k-2}) = 1g(n'_{10k-5}) = \ 0$
}
End the process.
$For\ k = 5m + 1$
$n_k \to 1. n_k' \to 0$

Case2: When m is Odd
$For\ \ 1 \leq k \leq m$
do
{
$if\ 1 \leq k \leq \frac{m+1}{2}$
{
$g(n_{10k-8}) = 1.g(n_{10k-7}) = 0\ g(n_{10k-5}) = 1.g(n_{10k-6}) = 0\ g(n_{10k-9}) = \ 0.g(n_{10k-4}) = 0$
$g(n'_{10k-8}) = 1.g(n'_{10k-7}) = 1\ g(n'_{10k-5}) = 0.g(n'_{10k-6}) = 1\ g(n'_{10k-9}) = \ 0.g(n'_{10k-4}) = 1$
}
End the process.
do.
{
$if\ 1 \leq k\ < \frac{m+1}{2}$
{
$g(n_{10k-3}) = 1g(n_{10-2}) = 0\ g(n_{10k-1}) = 0\,g(n_{10k}) = 1\ \ g(n'_{10k-3}) = 1.g(n'_{10-2}) = 0\ g(n'_{10k-1}) = 1g(n'_{10k}) = 0$
}
End the process.

Theorem 3.2: The extended duplicate graph of Double Quadrilateral Snake Graph $EDG(DQS_m) m \geq 2$, Bi-conditional cordial labeling admits.

Proof:
When m is Even
Let $N \rightarrow \{n_1, n_2 \dots \dots n_{5m+1}, n_1', n_2' \dots \dots n'_{5m+1}\} E \rightarrow \{l_1, l_2 \dots \dots, l_{7m}, l_1', l_2', \dots \dots l_{7m}'\}$ be the set of nodes and lines of the duplicate graph of Double Quadrilateral graph $EDG(DQS_m)$. Using **case1**, each of the 7m lines receive label 1 and 0 respectively. Using the algorithm 3.1, the induced function

$$g^*(uv) = \begin{cases} 1: \text{if } g(u) = g(v) \\ 0: \text{if } g(u) \neq g(v) \end{cases}$$

The 7m lines receive label 1 as below
For i = 2 to m
{
do
{
$g^*(l_{12}) = g^*(l_{26}) = \dots = g^*(l_{7m-2}) \rightarrow 1/* \; upto \frac{m}{2} lines */$
$g^*(l_1) = g^*(l_{15}) = \dots = g^*(l_{7m-1}) \rightarrow 1/* \; upto \frac{m}{2} lines */$
$g^*(l_8), g^*(l_9) = g^*(l_{22}), g^*(l_{23}) = \dots = g^*(l_{7m-6}), g^*(l_{7m-5}) \rightarrow 1/* \; upto \; m \; lines */$
$g^*(l_5), g^*(l_6) = g^*(l_{19}), g^*(l_{20}) = \dots = g^*(l_{7m-9}), g^*(l_{7m-3}) \rightarrow 1/* \; upto \; m \; lines */$
$g^*(l_1'), g^*(l_2') = g^*(l_{15}'), g^*(l_{16}') = \dots = g^*(l_{7m-13}), g^*(l_{7m-12}) \rightarrow 1/* \; upto \; m \; lines */$
$g^*(l_8'), g^*(l_9') = g^*(l_{22}'), g^*(l_{23}') = \dots = g^*(l_{7m-6}), g^*(l_{7m-5}) \rightarrow 1/* \; upto \; m \; lines */$
$g^*(l_{11}'), g^*(l_{12}') = g^*(l_{25}'), g^*(l_{26}') = \dots = g^*(l_{7m-3}), g^*(l_{7m-2}) \rightarrow 1/* \; upto \; m \; lines */$
$g^*(l_{13}') = g^*(l_{27}') = \dots = g^*(l_{7m-1}) \rightarrow 1/* \; upto \; \frac{m}{2} \; lines */$
$g^*(l_6') = g^*(l_{20}') = \dots = g^*(l_{7m-3}) \rightarrow 1/* \; upto \; \frac{m}{2} \; lines */$
}
End the process.
Hence **7m lines received label 1,**
Similarly, the remaining 7m lines label 0 as below
{
do
{
$g^*(l_7) = g^*(l_{21}) = \dots = g^*(l_{7m-7}) \rightarrow 0/* \; upto \frac{m}{2} lines */$
$g^*(l_2), g^*(l_3), g^*(l_4), = g^*(l_{16}), g^*(l_{17}) g^*(l_{18}), \dots =$
$g^*(l_{7m-12}), g^*(l_{7m-11}) g^*(l_{7m-10}), \rightarrow 0/* \; upto \; \frac{3m}{2} \; lines */$
$g^*(l_{10}), g^*(l_{11}) = g^*(l_{24}), g^*(l_{25}) = \dots = g^*(l_{7m-4}), g^*(l_{7m-3}) \rightarrow 0/* \; upto \; m \; lines */$
$g^*(l_{13}), g^*(l_{14}) = g^*(l_{27}), g^*(l_{28}) = \dots = g^*(l_{7m-6}), g^*(l_{7m-7}) \rightarrow 0/* \; upto \; m \; lines */$
$g^*(l_3'), g^*(l_4'), g^*(l_5') = g^*(l_{17}'), g^*(l_{18}'), g^*(l_{19}') = \cdots$
$= g^*(l_{7m-11}), g^*(l_{7m-10}), g^*(l_{7m-9}) \quad \rightarrow 0/* \; upto \; \frac{3m}{2} \; lines */$
$g^*(l_7') = g^*(l_{14}') = \dots = g^*(l_{7m-7}) \rightarrow 0/* \; upto \; \frac{m}{2} \; lines */$
$g^*(l_{10}'), g^*(l_{14}') = g^*(l_{24}'), g^*(l_{28}') = \dots = g^*(l_{7m-4}), g^*(l_{7m}) \rightarrow 0/* \; upto \; m \; lines */$
}
End the process.

Hence 7mlines received label 0.

When m is Odd

Let $N \rightarrow \{n_1, n_2 \dots\dots n_{5m+1}, n'_1, n'_2, \dots\dots. n'_{5m+1}\} E \rightarrow \{l_1, l_2 \dots\dots, l_{7m}. l'_1, l'_2, \dots\dots. l'_{7m}\}$ be the set of nodes and lines of the duplicate graph of Double Quadrilateral graph $EDG(\,QS_m\,)$. Using **case2,** each of the 7m lines receive label 1 and 0 respectively. Using the induced function g^* defined by

$$g^*(uv) = \begin{cases} 1: \text{if } g(u) = g(v) \\ 0: \text{if } g(u) \neq g(v) \end{cases}$$

For i = 3 to m

{

do

{

$g^*(l_1), g^*(l_2) = g^*(l_{15}), g^*(l_{16}) =, \dots = g^*(l_{7m-6}), g^*(l_{7m-5}) \rightarrow 0/*\, upto\ m+1\ lines\ */$

$g^*(l_3), g^*(l_6) = g^*(l_{17}), g^*(l_{20}) =, \dots = g^*(l_{7m-4}), g^*(l_{7m-1}) \rightarrow 0/*\, upto\ m+1\ lines\ */$

$g^*(l_8) = g^*(l_{22}) = \dots = g^*(l_{7m-13}) \rightarrow 0/*\, upto \frac{m-1}{2} lines\ */$

$g^*(l_{14}) = g^*(l_{28}) = \dots = g^*(l_{7m-7}) \rightarrow 0/*\, upto \frac{m-1}{2} lines\ */$

$g^*(l'_1), g^*(l'_{15}) = g^*(l'_{29}) = \dots = g^*(l'_{7m-6}) \rightarrow 0/*\, upto\ \frac{m+1}{2}\ lines\ */$

$g^*(l'_4), g^*(l'_{18}) = g^*(l'_{32}) = \dots = g^*(l'_{7m-3}) \rightarrow 0/*\, upto\ \frac{m+1}{2}\ lines\ */$

$g^*(l'_6), g^*(l'_{20}) = g^*(l'_{34}) = \dots = g^*(l'_{7m-1}) \rightarrow 0/*\, upto\ \frac{m+1}{2}\ lines\ */$

$g^*(l'_9), g^*(l'_{10}) = g^*(l'_{23}), g^*(l'_{24}) = \dots = g^*(l'_{7m-12}), g^*(l'_{7m-11}) \rightarrow 0/*\, upto\ m-1\ lines\ */$

$g^*(l'_{13}) = g^*(l'_{27}) = \dots = g^*(l'_{7m-8}) \rightarrow 0/*\, upto\ \frac{m-1}{2}\ lines\ */$

$g^*(l'_{11}), g^*(l'_{12}) = g^*(l'_{25}), g^*(l'_{26}) = \dots = g^*(l'_{7m-10}), g^*(l'_{7m-11}) \rightarrow 0/*\, upto\ m-1\ lines\ */$

}

End the process.

Hence 7mlines receive label 0,

Similarly, the remaining 7m lines label 1 as below

{

do

{

$g^*(l_4), g^*(l_5) = g^*(l_{18}), g^*(l_{19}) =, \dots = g^*(l_{7m-3}), g^*(l_{7m-2}) \rightarrow 1/*\, upto\ m+1\ lines\ */$

$g^*(l_9), g^*(l_{10}) = g^*(l_{23}), g^*(l_{24}) =, \dots = g^*(l_{7m-12}), g^*(l_{7m-11}) \rightarrow 1/*\, upto\ m-1\ lines\ */$

$g^*(l_7) = g^*(l_{21}) = \dots = g^*(l_{7m}) \rightarrow 1/*\, upto \frac{m+1}{2} lines\ */$

$g^*(l_{11}) = g^*(l_{25}) = \dots = g^*(l_{7m-10}) \rightarrow 1/*\, upto \frac{m-1}{2} lines\ */$

$g^*(l_{12}), g^*(l_{13}) = g^*(l_{26}), g^*(l_{27}) =, \dots = g^*(l_{7m-9}), g^*(l_{7m-8}) \rightarrow 1/*\, upto\ m-1\ lines\ */$

$g^*(l'_2), g^*(l'_3) = g^*(l'_{16}), g^*(l'_{17}) = \dots = g^*(l'_{7m-5}), g^*(l'_{7m-4}) \rightarrow 1/*\, upto\ m+1\ lines\ */$

$g^*(l'_8) = g^*(l'_{22}) = \cdots = g^*(l'_{7m-13}) \rightarrow \frac{1}{*\, upto} \frac{m-1}{2}\ lines\ */$

$g^*(l'_7), g^*(l'_{21}) = g^*(l'_{35}) = \dots = g^*(l'_{7m}) \rightarrow 1/*\, upto\ \frac{m+1}{2}\ lines\ */$

$g^*(l_5'), g^*(l_{19}') = g^*(l_{33}') =\ldots= g^*(l'_{7m-2}) \rightarrow 1/*\ upto\ \frac{m+1}{2}\ lines\ */$

$g^*(l_{14}'), g^*(l_{28}') = g^*(l_{42}') =\ldots= g^*(l'_{7m-7}) \rightarrow 1/*\ upto\ \frac{m-1}{2}\ lines\ */$

}

End the process.

Thus 7m lines receive label 1.

Hence the duplicate graph of the Double Quadrilateral snake $EDG(\ QS_m\ )\ m \geq 2$, Bi-Conditional Cordial labeling admits.

3.3. Illustration:

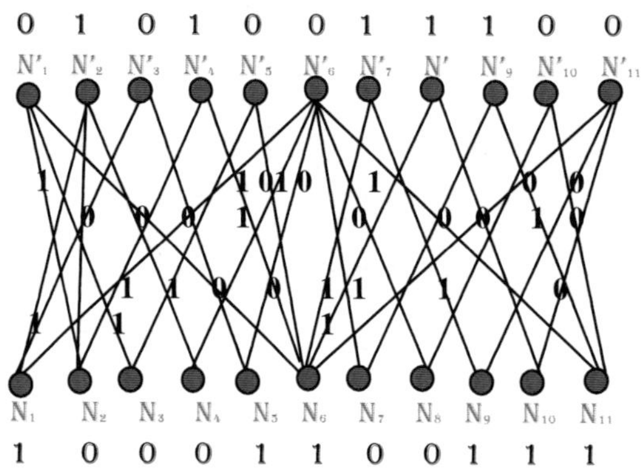

Bi-conditional cordial labeling $EDG(\ DQS_2)$

4. Bi-conditional cordial labeling $DG(TL_m)$

Algorithm 4.1: Allocation of labeled nodes

Let $N \rightarrow \{n_1, n_2 \ldots.., n_{2m+2}. n_1', n_2', n_3', \ldots \ldots. n_{2m+2}'\}$
$L \rightarrow \{l_1, l_2, l_3 \ldots.., l_{4m+1}. l_1', l_2', l_3' \ldots \ldots. l_{4m+1}'\}$
For $1 \leq k \leq m+1$
$n_{2k-1} \leftarrow 1\ n_{2k} \leftarrow 0.$
For $1 \leq k \leq m$
Fix $n'_1 \leftarrow 1. n'_2 \leftarrow 0.$
$n'_{2k+1} \leftarrow 0, n'_{2k+2} \leftarrow 1.$

Theorem 4.2: The extended duplicate graph of Triangular ladder Graph $EDG(TQL_m)m \geq 2$. Bi-conditional cordial labeling admits.

Proof:

Let $N \rightarrow \{n_1, n_2 \ldots.., n_{2m+2}. n_1', n_2', n_3', \ldots \ldots. n_{2m+2}'\}$ $L \rightarrow \{l_1, l_2, l_3, \ldots.., l_{4m+1}. l_1', l_2', l_{3,}' \ldots \ldots. l_{4m+1}'\}$ be the set of nodes and lines of the duplicate graph of Triangular ladder graph $DG(\ TLS_m\ )$. using the algorithm 4.1, each of the $2m+2$ nodes receive label 0 and 1 respectively. Using the induced function g^* defined by $g^*(uv) = \begin{cases} 1: \text{if } g(u) = g(v) \\ 0: \text{if } g(u) \neq g(v) \end{cases}$ the m lines namely $l_2, l_6, l_{10}\ \ , \ldots \ldots. l_{4m-2}$. receive label 0, the m lines namely $l_4, l_8, l_{12}\ \ , \ldots \ldots. l_{4m}$. receive label 0, the 1 line namely l_1 receive label 0. $m-1$ lines namely $l'_6, l'_{10}, l'_{14}, \ldots \ldots \ldots \ldots \ldots \ldots l'_{4m-2.}$ receive label 0. $m-1$ lines namely $l'_8, l'_{12}, l'_{16}, \ldots \ldots \ldots. \ldots \ldots \ldots \ldots l'_{4m}$. receive label 0, and 2 lines namely l'_1 and l'_3receive label 0.

Thus $4m + 1$ lines receive label 0.and the m lines namely $l'_7, l'_{11}, \dots \dots \dots \dots . \dots l'_{4m-1}$. receive label 1, the m lines namely $l_5, l_9, l_{13}, \dots \dots \dots \dots . \dots . l_{4m+1}$ receive label 1, $m-1$ lines namely $l_7, l_{11}, l_{15}, \dots \dots \dots \dots . \dots \dots l_{4m-1}$. receive label 1, m lines namely $l'_5, l'_9, l'_{13}, \dots \dots \dots \dots . \dots l'_{4m+1}$. receive label 1. 2 lines namely $l'_2$and l'_4 receive label 1. **Thus the $4m + 1$ lines receive label 1.**
Hence the duplicate graph of the Triangular ladder Snake Graph EDG(TL_m)$m \geq 2$, is Bi-Conditional Cordial labeling admits.

4.3. Illustration:

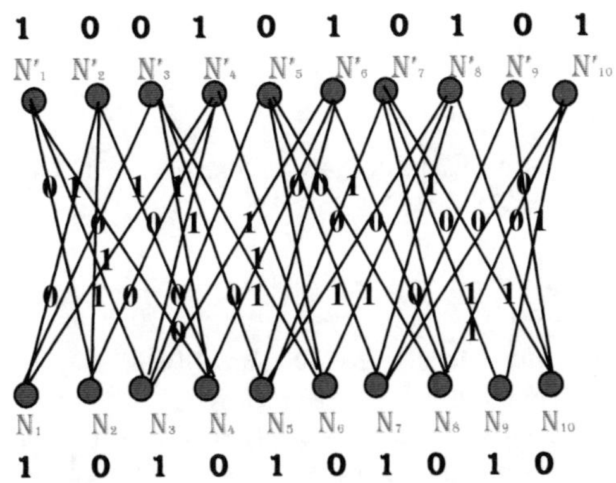

Bi-conditional cordial labeling $EDG(\,TL_4)$

5. Conclusion

We have proved the existence of the Bi-Conditional cordial labeling in Extended Duplicate graph of the Quadrilateral Snake $EDG(QS_m)m \geq 2$, Double Quadrilateral $EDG(DQS_m)m \geq 2$, and Triangular Ladder Snake Graph. $EDG(TL_m)m \geq 2$.

References

[1] Gallian A, *A Dynamic Survey of Graph Labeling*, The Electronic Journal of Combinatorics, #DS6, (2014).

[2] Sampath kumar E, *"On duplicate graphs"*, Journal of the Indian Math. Soc. **37** , No.285 – 293, (1973).

[3] Murali B.J, Thirusangu K, Madura Meenakshi R, Biconditional cordial Labeling of Cycles, International Journal of Applied Engineering Research (IJAER), 10(80) (2015), 188-191.

[4] Selvam B, Thirusangu K Z_3 *-vertex magic total labeling and Z3 -edge magic total labeling for the extended duplicate graph of quadrilateral snake* National Conference on Mathematical Techniques and its Applications – (2018).

[5] Nanda gopal E, Maheswari V, Vijaya kumar P *"Special Type labeling for the Extended Duplicate Graph of Quadrilateral Snake Graph"*, International Journal of Information and Computing Science ISSN NO: 0972-1347 Vol. **2**, No.59-68, (2019).

[6] Thirusangu K and Ezhilarasi R *"On adjacent vertex distinguishing total coloring of quadrilateral snake"* ISSN 2347-1921 Volume **13** Number 01 Journal of Advances in Mathematics.-(2017).

[7] Murali B J, Thirusangu K, Balamurugan B J *Combination Cordial Labeling of Flower Graphs and Corona Graph*, International Journal of Pure and Applied Mathematics Volume **117** No. 12 297-307, (2017)

[8] Ulaganathan P P, Selvam B, Vijayakumar P *On Signed Product Cordial labeling in duplicate graphs of Bistar, Double Star and Triangular Ladder Graph'*. International Journal of Mathematics Trends and Technology –(2016).

Journal of Physics: Conference Series **1377** (2019) 012002 doi:10.1088/1742-6596/1377/1/012002

Bifurcation and dynamical behaviour of a fractional order Lorenz-Chen-Lu like chaotic system with discretization

A. George Maria Selvam[1] and D. Abraham Vianny[2]

[1, 2] Department of Mathematics, Sacred Heart College (Autonomous), Tirupattur – 635 601, Vellore Dist., Tamil Nadu, India

E-mail : agmshc@gmail.com

Abstract. In this article, the dynamical behaviour of a discrete fractional order Lorenz-Chen-Lu like chaotic system is investigated. A discretization process is used to obtain a discrete version. The system has three constants, three fixed points and generates typical chaotic attractors. Numerical verifications are presented for the analysis of dynamical system. The chaotic attractors, bifurcation diagrams and phase portraits are provided for distinctive values of parameters. The sensitivity to initial conditions of the parameter values for the fixed points is also discussed.

1. Introduction

A fractional order system is a dynamical system which can be formulated by fractional differential equations containing derivatives of non-integer order. Its applications are found in several fields of engineering and science such as biology, diffusion processes, viscoelastic materials, magnetic fields, electro chemistry and finance systems. In modern years fractional order systems have attracted many researchers due of their applications [6, 8, 10].

Chaos theory as a new division of mathematics and physics has given birth to a new technique of exploring the universe and is a key tool to be aware of the behavior of the processes in the world. In current years, chaotic behaviors have been useful in special areas of engineering and science such as physics, economics, medicine, biology, electronics, mechanics, ecology and etc., [6, 10].

In 1950s, the meteorologist E.N.Lorenz acquired the LGP-30 computer having internal memory16KB, using computer, in 1963, he discovered the chaotic motion on a chaotic attractor [1, 2]. The model studied by E.N.Lorenz arising in weather prediction was described by autonomous system of differential equations containing nonlinear terms. It is famous for having chaotic solutions for few parameters and initial values represented by

$$x' = a(y - x),\ y' = cx - y - xz\ ,\ z' = xy - bz.$$

These equations contain three state variables $x\ , y\ , z$. For typical values of $a = 10, b = 8/3$ and $c = 28$, the famous butterfly attractor evolves. On the other hand, the Chen's system [3, 4, 9] is described by the following set of equations

$$x' = a(y - x),\ \ y' = (c - a)x + cy - xz\ ,\ z' = xy - bz$$

which has a set of chaotic parameters $a = 35,\ b = 3, c = 28$. The Chen's system is a dual of the Lorenz system but evolves more sophisticated attractors. The Lu system [5, 14] is described by the following set of equations

$$x' = a(y - x),\ \ y' = cy - xz\ ,\ z' = xy - bz$$

which has a set of chaotic parameters $a = 36,\ b = 3, c = 20$.

The article is organized as follows. In section 2, we present the basic definitions of the fractional derivatives and integrals. In section 3, we describe the chaotic system and explain the discretization process of fractional order system. In section 4, we analyze the equilibrium points with stability. Further we present illustrative examples and numerical simulations. In section 5, we present the bifurcation diagrams and chaotic attractors in x, y, z-space. Section 6, deals with the sensitivity of the system to initial values.

2. Fractional Derivatives and Integrals

We present some important definitions of the fractional derivatives and integrals [6].

Definition 1. The Riemann-Liouville fractional derivative is expressed as follows:

$$D^{\alpha}f(\tau) = \frac{1}{\Gamma(n-\alpha)}\left(\frac{d}{d\tau}\right)^{n}\int_{0}^{\tau}\frac{f(\tau)}{(\tau-\xi)^{\alpha-n+1}}d\xi.$$

Definition 2. The Caputo fractional derivative is defined by

$$D^{\alpha}f(\tau) = \frac{1}{\Gamma(n-\alpha)}\int_{0}^{\tau}\frac{f^{n}(\xi)}{(\tau-\xi)^{\alpha-n+1}}d\xi,$$

where Γ is the gamma function. The Riemann-Liouville and Caputo fractional derivatives are widely used in the application of real life problems [6, 8, 10].

3. Fractional Order Lorenz-Chen-Lu Like Chaotic System and Discretization

In some known literatures, authors have analyzed the Yang system similar to Lorenz like system [12],

$$\begin{aligned} x'(t) &= a(y(t) - x(t)) \\ y'(t) &= cx(t) - x(t)z(t) \\ z'(t) &= x(t)y(t) - bz(t) \end{aligned} \tag{1}$$

These equations contain three state variables x, y, z and three state parameters a, b, c. They think that found the system to have two interesting chaotic attractors (called as Yang–Chen attractor) when $(a, b, c) = \left(10, \frac{8}{3}, 16\right)$ and $(a, b, c) = (35,3,35)$ respectively.

Here we are concerned with the fractional order chaotic system given by

$$\begin{aligned} D^{\alpha}x(t) &= a(y(t) - x(t)) \\ D^{\alpha}y(t) &= cx(t) - x(t)z(t) \\ D^{\alpha}z(t) &= x(t)y(t) - b\,z(t) \end{aligned} \tag{2}$$

where α is the fractional order. Now we are interested in discretizing fractional order chaotic system given in the form.

$$\begin{aligned} D^{\alpha}x(t) &= a\left(y\left(r\left[\frac{t}{r}\right]\right) - x\left(r\left[\frac{t}{r}\right]\right)\right) \\ D^{\alpha}y(t) &= cx\left(r\left[\frac{t}{r}\right]\right) - x\left(r\left[\frac{t}{r}\right]\right)z\left(r\left[\frac{t}{r}\right]\right) \\ D^{\alpha}z(t) &= x\left(r\left[\frac{t}{r}\right]\right)y\left(r\left[\frac{t}{r}\right]\right) - b\,z\left(r\left[\frac{t}{r}\right]\right) \end{aligned} \tag{3}$$

with initial condition $x(0) = x_0, y(0) = y_0, z(0) = z_0$.

Applying the discretization process [7, 11, 13] and letting $t \rightarrow (n+1)r$, we end up with the discretized system as given below

$$\begin{aligned} x_{n+1} &= x_n + s(a(y_n - x_n)), \\ y_{n+1} &= y_n + s(cx_n - x_n z_n), \\ z_{n+1} &= z_n + s(x_n y_n - b\,z_n), \end{aligned} \tag{4}$$

where $s = \frac{r^{\alpha}}{\Gamma(1+\alpha)}$.

4. Equilibrium Points and Numerical Simulations

Now we analyze the stability of the system (4) which has the following three equilibrium points.

(i) $F_0 = (0,0,0)$,

(ii) $F_1 = \left(\sqrt{bc}, \sqrt{bc}, c\right)$

(iii) $F_2 = \left(-\sqrt{bc}, -\sqrt{bc}, c\right)$

Linearization of the system (4) about F_0 yield the characteristic equation is

$$P(\lambda) = \lambda^3 + \lambda^2[s(a+b) - 3] + \lambda\,[3 - 2s(a+b) + s^2a(b-c)] + [s(a+b) - s^2a(b-c) - s^3abc - 1]$$

Let

$$a_1 = s(a+b) - 3,\ \ a_2 = 3 - 2s(a+b) + s^2a(b-c),$$
$$a_3 = s(a+b) - s^2a(b-c) - s^3abc - 1,$$

From the Jury test [15], if $P(1) > 0, P(-1) < 0$ and $P(0) = a_3 < 0, |b_3| > b_1, c_2 > |c_3|$, where $b_3 = 1 - a_3^2$, $b_2 = a_1 - a_3a_2$, $b_1 = a_2 - a_3a_1$, $c_3 = b_3^2 - b_1^2$ and $c_2 = b_3b_2 - b_1b_2$ then the roots of $P(\lambda)$ satisfy $|\lambda_j| < 1$ and thus F_0 is asymptotically stable. Suppose $P(1) < 0$ then F_0 is unstable.

Linearization of the system (4) about F_1 or F_2 yield the characteristic equation is

$$P(\lambda) = \lambda^3 + \lambda^2[s(a+b) - 3] + \lambda\,[3 - 2s(a+b) + s^2b(a+c)] + [2s^3abc - s^2b(a+c) + s(a+b) - 1]$$

Let

$$a_{11} = s(a+b) - 3,\ \ a_{22} = 3 - 2s(a+b) + s^2b(a+c),$$
$$a_{33} = 2s^3abc - s^2b(a+c) + s(a+b) - 1.$$

From the Jury test [15], if $P(1) > 0, P(-1) < 0$ and $P(0) = a_{33} < 0, |b_{33}| > b_{11}, c_{22} > |c_{33}|$, where $b_{33} = 1 - a_{33}^2$, $b_{22} = a_{11} - a_{33}a_{22}$, $b_{11} = a_{22} - a_{33}a_{11}$, $c_{33} = b_{33}^2 - b_{11}^2$ and $c_{22} = b_{33}b_{22} - b_{11}b_{22}$ then the roots of $P(\lambda)$ satisfy $|\lambda_j| < 1$ and thus F_1 or F_2 is asymptotically stable. Suppose $P(1) < 0$ then F_1 or F_2 is unstable.

We discuss the time plots, phase portraits of 3D chaotic attractor in xyz – space and 2D chaotic attractor xy, xz, yz – planes of the system (4) with initial values $(x_0, y_0, z_0) = (0.1, 0.1, 0.1)$.

Example 1. Taking $a = 0.3$, $b = 0.1$, $c = 0.8$, $r = 0.1$, $\alpha = 0.9$ and applying Jury conditions, we get $P(1) = 0.00010765 > 0$, $P(-1) = -7.7942 < 0$ and $a_{33} = -0.9494 < 0$. Then the equilibrium point F_1 is asymptotically stable. Figure 1 (a – e) are the time plots, phase portraits of 2D and 3D chaotic attractors.

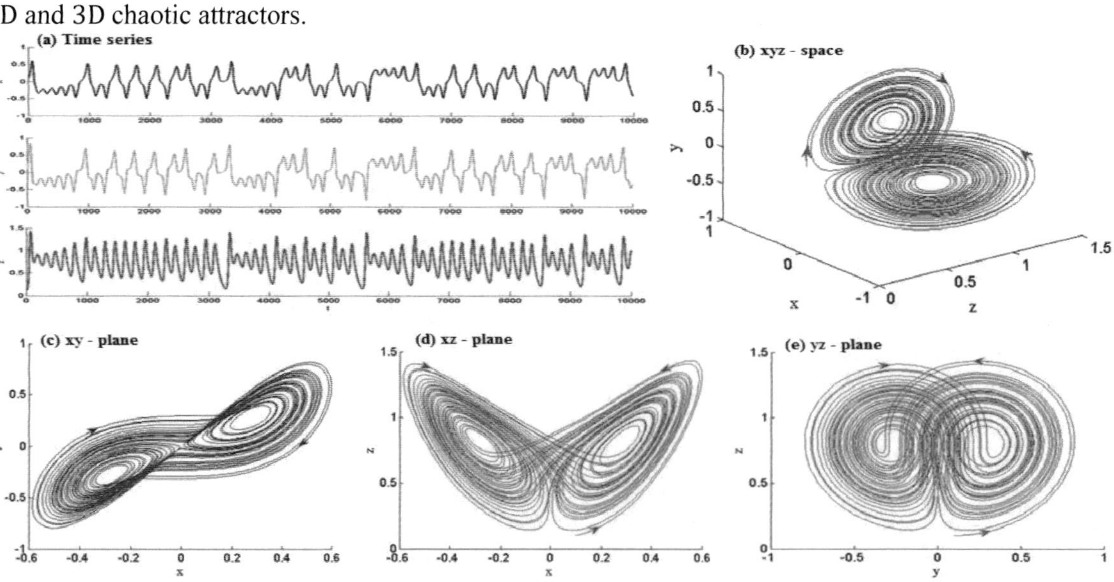

Figure 1. Time plots and phase portraits of the Lorenz like chaotic system

Example 2. Taking $a = 0.3$, $b = 0.1$, $c = 0.9$, $r = 0.1$, $\alpha = 0.9$ and applying Jury conditions, we get $P(1) = 0.00012111 > 0$, $P(-1) = -7.7946 < 0$ and $a_{33} = -0.9496 < 0$. Then the equilibrium point F_1 is asymptotically stable. Figure 2(a − e) are the time plots, phase portraits of 2D and 3D chaotic attractors.

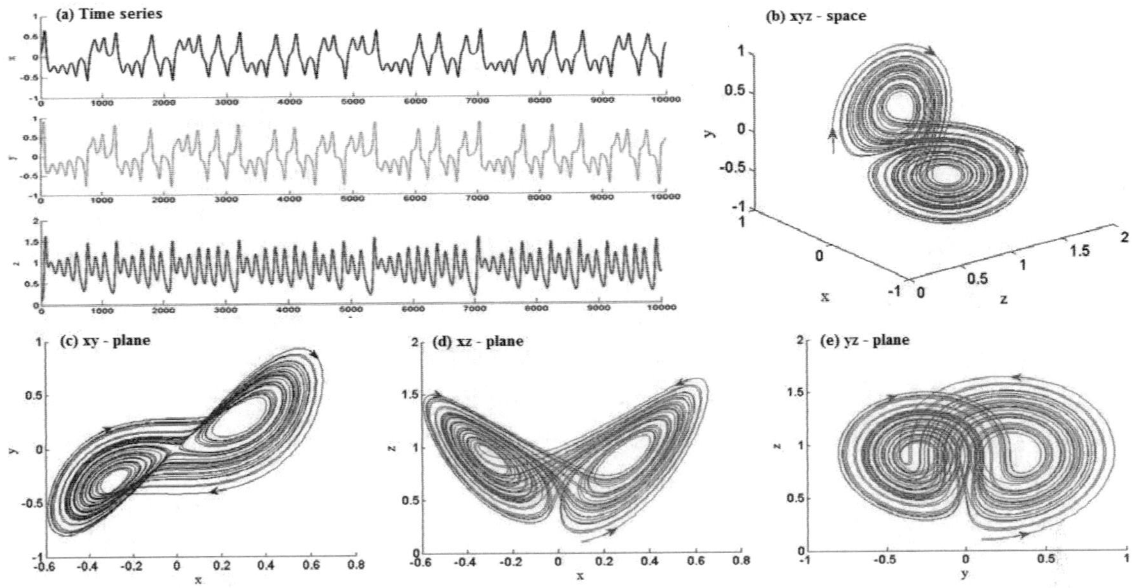

Figure 2. Time plots and phase portraits of the Lorenz like chaotic system

Example 3. Taking $a = 0.3$, $b = 0.2$, $c = 1.4$, $r = 0.1$, $\alpha = 0.9$ and applying Jury conditions, we get $P(1) = 0.0003769 > 0$, $P(-1) = -7.7495 < 0$ and $a_{33} = -0.9400 < 0$. Then the equilibrium point F_1 is asymptotically stable. Figure 3(a − e) are the time plots, phase portraits of 2D and 3D chaotic attractors.

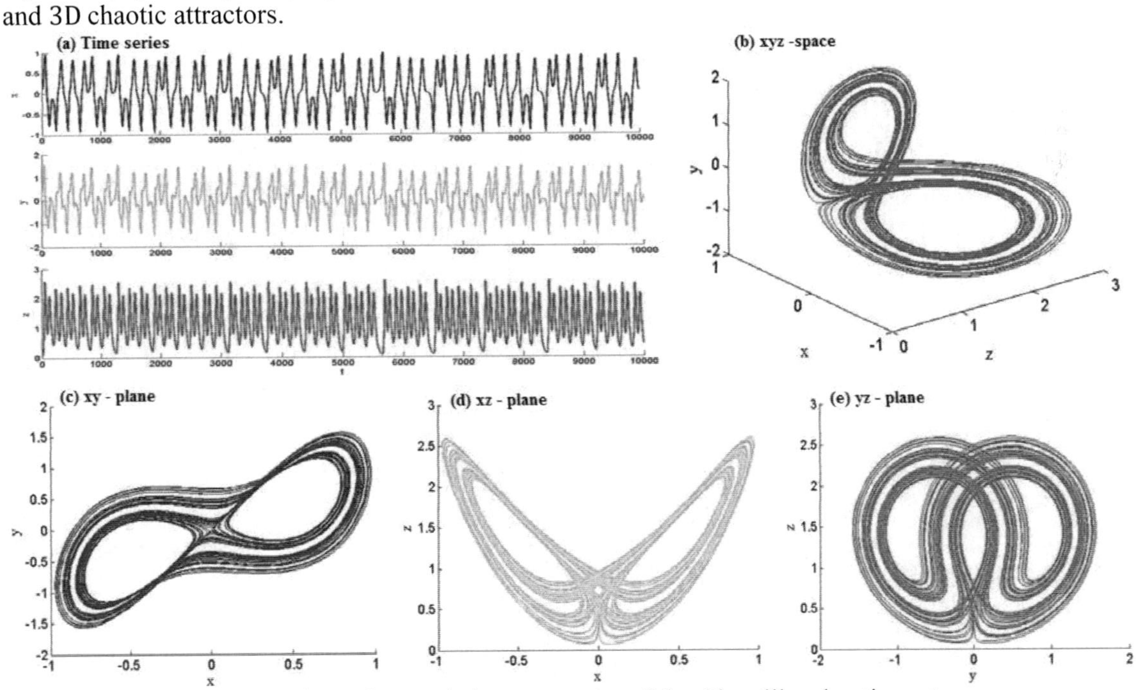

Figure 3. Time plots and phase portraits of the Chen like chaotic system

Example 4. Taking $a = 0.3,\ b = 0.07,\ c = 4.0,\ r = 0.1,\ \alpha = 0.9$ and applying Jury conditions, we get $P(1) = 0.00037679 > 0,\quad P(-1) = -7.8162 < 0$ and $a_{33} = -0.9563 < 0$. Then the equilibrium point F_1 is asymptotically stable. Figure 4(a − e) are the time plots, phase portraits of 2D and 3D chaotic attractors.

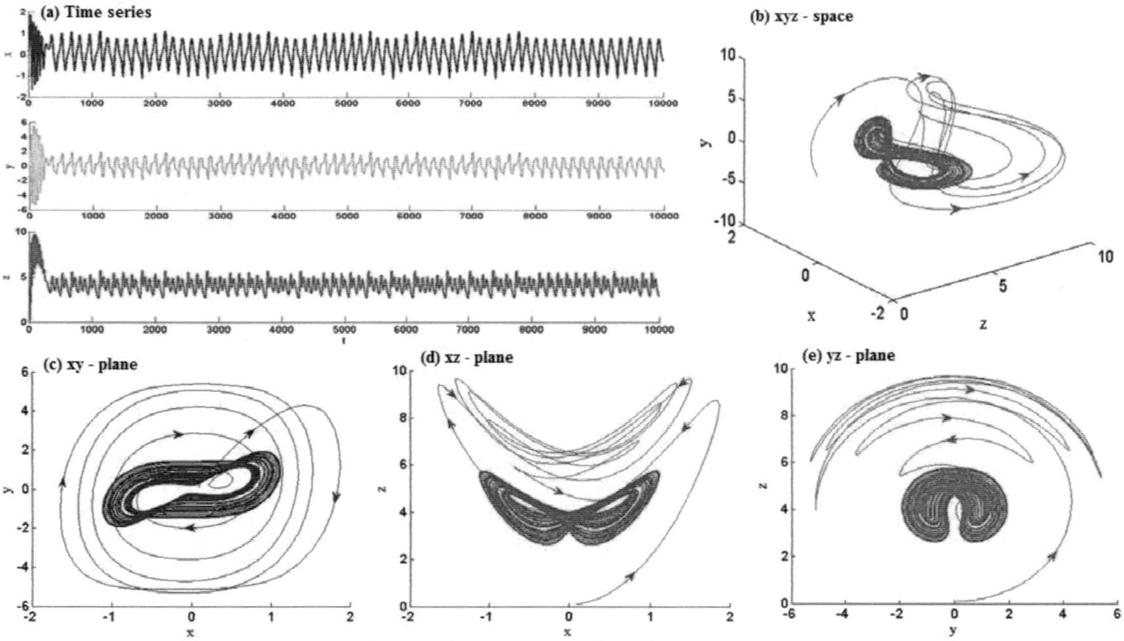

Figure 4. Time plots and phase portraits of the Lu like chaotic system

Example 5. Taking $a = 0.9,\ b = 0.07,\ c = 3.5,\ r = 0.1,\ \alpha = 0.9$ and applying Jury conditions, we get $P(1) = 0.00098908 > 0,\quad P(-1) = -7.5017 < 0$ and $a_{33} = -0.8773 < 0$. Then the equilibrium point F_1 is asymptotically stable. Figure 5(a − e) are the time plots, phase portraits of 2D and 3D chaotic attractors.

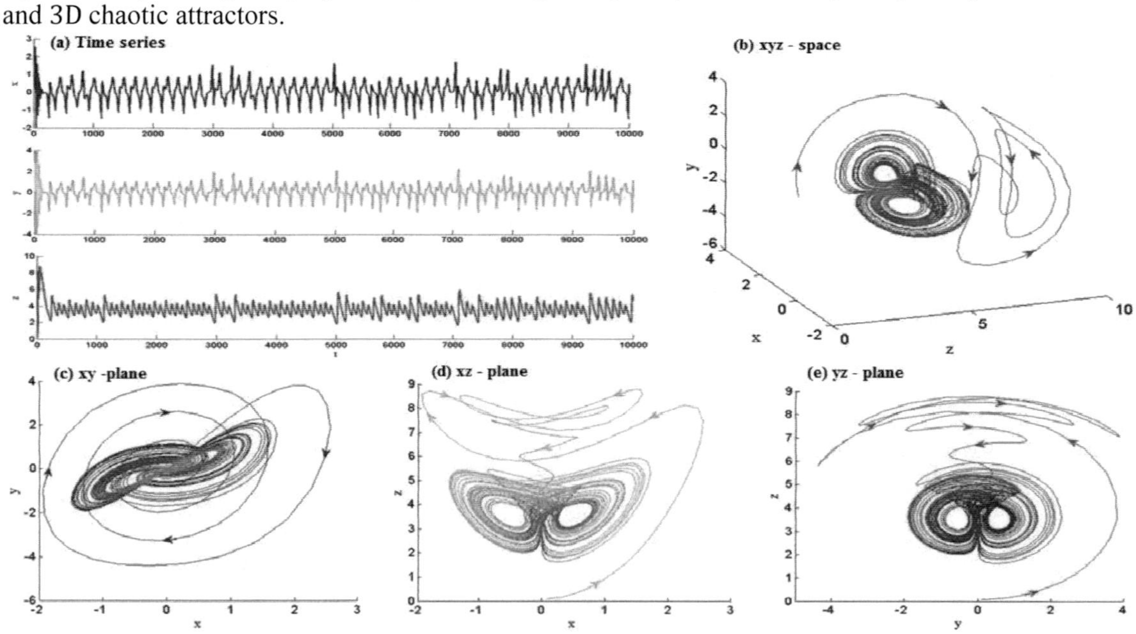

Figure 5. Time plots and phase portraits of the Chen like chaotic system

Example 6. Taking $a = 0.28$, $b = 0.07$, $c = 3.5$, $r = 0.1$, $\alpha = 0.9$ and applying Jury conditions, we get $P(1) = 0.00030771 > 0$, $P(-1) = -7.8255 < 0$ and $a_{33} = -0.9584 < 0$. Then the equilibrium point F_1 is asymptotically stable. Figure 6(a − e) are the time plots, phase portraits of 2D and 3D chaotic attractors.

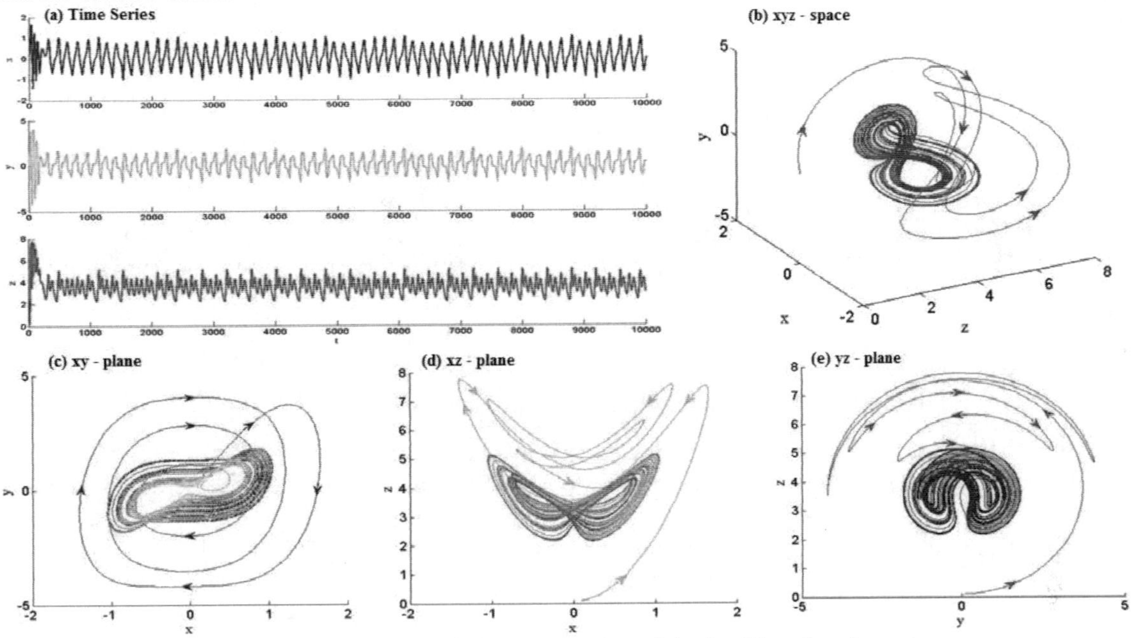

Figure 6. Time plots and phase portraits of the Lu like chaotic system

5. Bifurcation and Chaotic Attractor

In dynamical systems, it is observed that a slight smooth change in the bifurcation parameter values of a system results in a sudden qualitative or topological change in its behaviour and thus bifurcation occurs. Generally, at bifurcation point, it is noticed that the local stability properties of equilibria, other invariant sets or periodic orbits are set to alter [13, 15]. Now, we analyze the bifurcation for the variables x, y, z and the parameter values $b = 0.5, c = 3, r = 0.1$, $\alpha = 0.9$ fixing and varying $a \in [0, 3]$ (see figure.7). We present phase portraits for x, y, z - space corresponding to Figure. 7 with varying $a \in [0, 3]$, (see Figure.8). It is quite interesting to discuss the phase portraits of various a corresponding to the bifurcation diagram in figure.7 presented in figure.8. For $a \in (0.25, 0.31)$ the fixed point is stable as in Figure.8(a,b,c), but it moves from stable to unstable through the occurrence of limit cycle for $a = 0.32$ in Figure.8(d). For $a = 1.2$ is the Lu like chaotic attractor as in Figure.8(e). For $a \in (1.5, 3.0)$ is the Chen like chaotic attractor as in Figure.8(f,g,h).

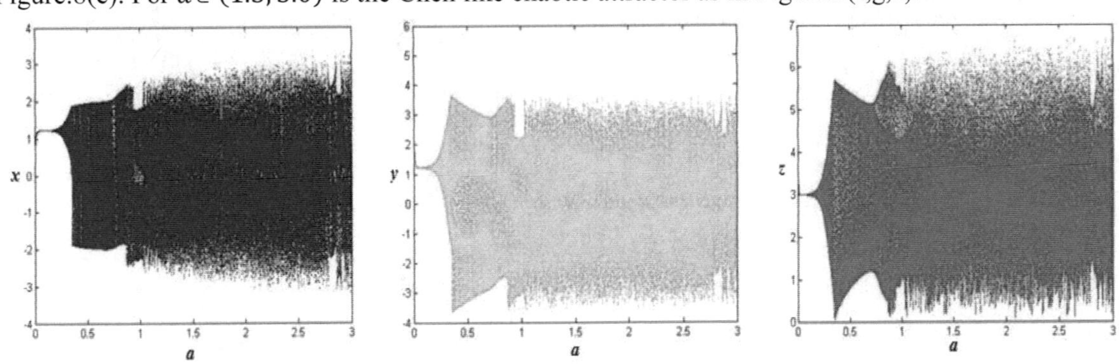

Figure 7. Bifurcation diagrams with the initial values ($x_0 = 0.1$, $y_0 = 0.1$, $z_0 = 0.1$)

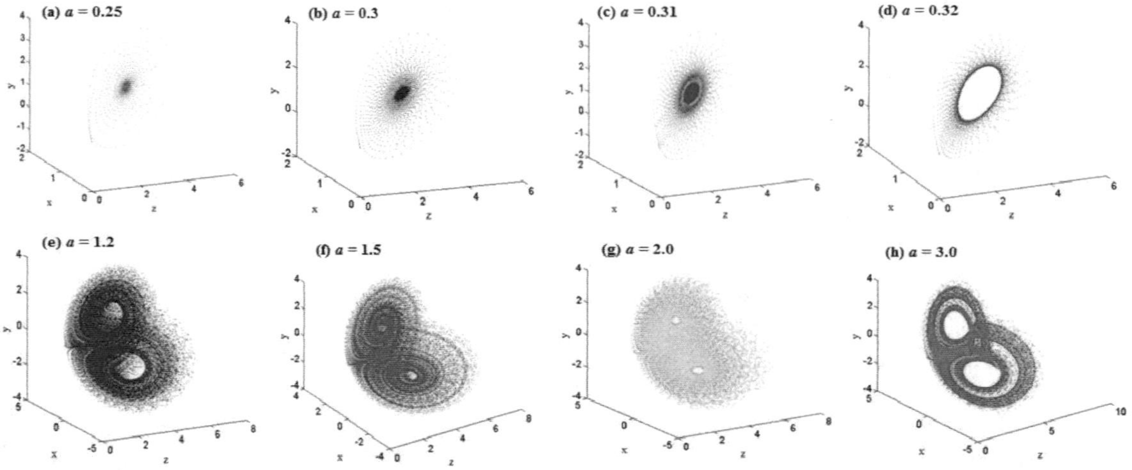

Figure 8. Phase portraits for x, y, z-space corresponding to Figure 7.

The bifurcation diagrams for the parameter values $a = 0.5$, $c = 3.0$, $r = 0.1$, $\alpha = 0.9$ fixing and varying $b \in [0, 1.5]$, (see Figure.9). We present phase diagrams for x, y, z - space corresponding to Figure 9. For $b \in (0.1, 0.2)$ is the Lu like chaotic attractor as in Figure.10(a-e). For $b = 0.78$ the fixed point is unstable through the occurrence of limit cycle as in Figure.10(f), but it moves from unstable to stable for $a \in (0.8, 0.9)$ as in Figure.10(g,h).

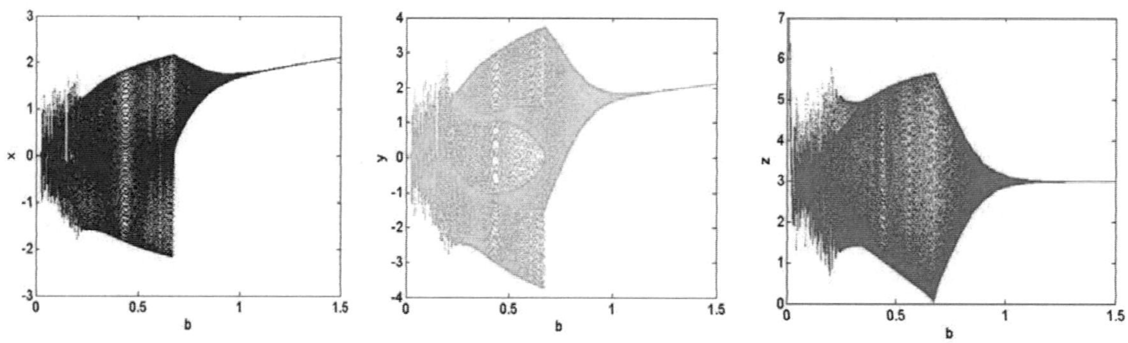

Figure 9. Bifurcation diagrams with the initial values ($x_0 = 0.1,\ y_0 = 0.1,\ z_0 = 0.1$)

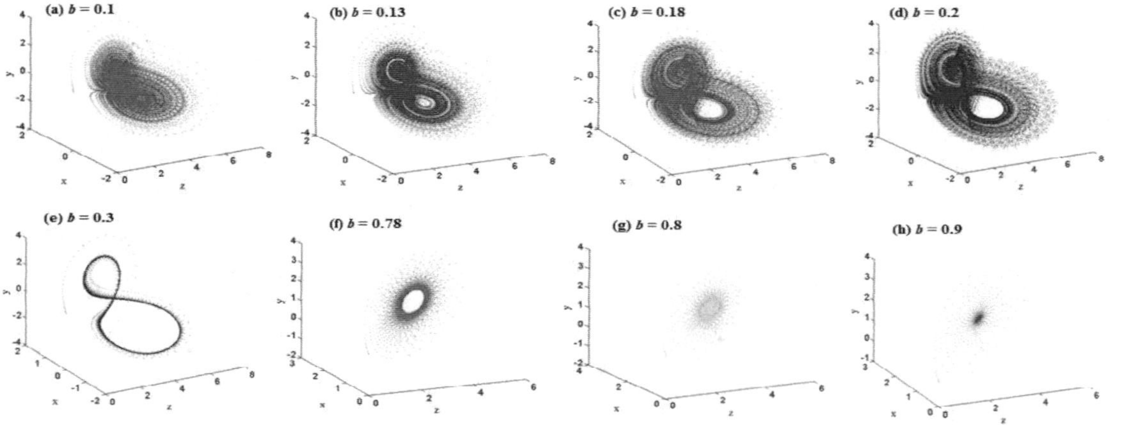

Figure 10. Phase portraits for x, y, z-space corresponding to Figure 9.

The bifurcation diagrams for the parameter values $a = 0.3$, $b = 0.08$, $r = 0.1$, $\alpha = 0.9$ are fixing and varying $c \in [0, 3]$ (see figure.11). It is quite interesting to discuss the phase portraits of various c corresponding to the bifurcation diagram in figure.11 presented in figure.12. For $c = 0.3$ is the fixed point is stable as in Figure.12(a). For $c = 0.5$ is the Lorenz like chaotic attractor as in Figure.12(b). For $c \in (1.0, 1.5)$ is the Chen like chaotic attractor as in Figure.12(c,d). For $c \in (1.7, 3.0)$ is the Lu like chaotic attractor as in Figure.12(e-h).

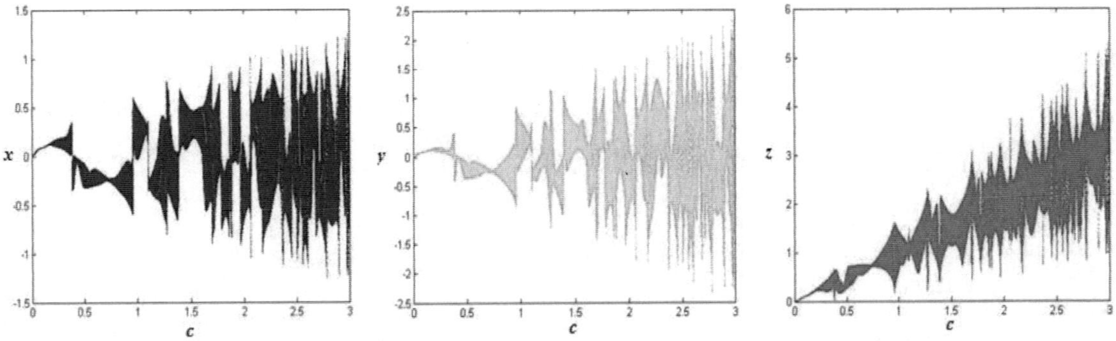

Figure 11. Bifurcation diagrams with the initial values ($x_0 = 0.1$, $y_0 = 0.1$, $z_0 = 0.1$)

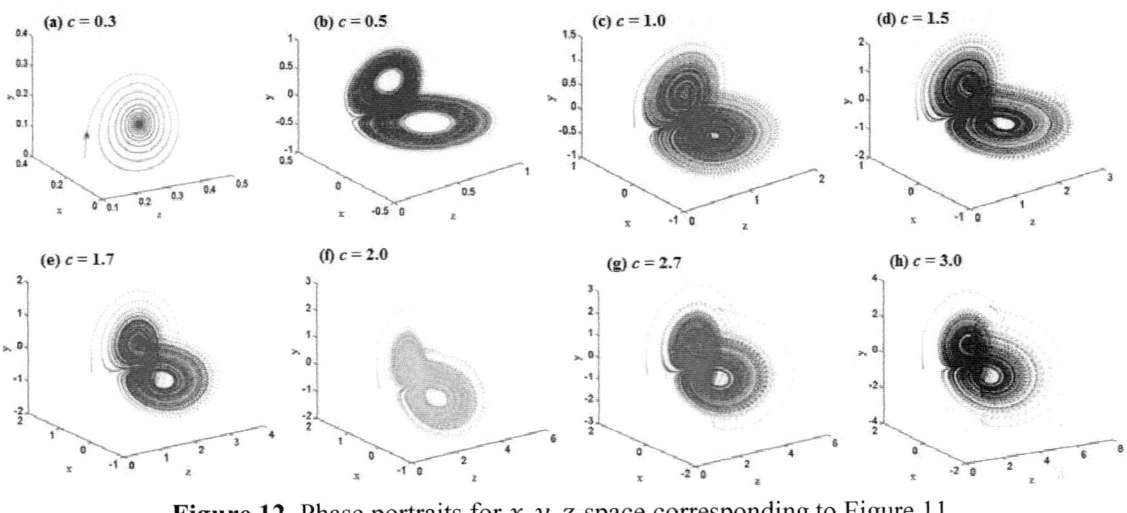

Figure 12. Phase portraits for x, y, z-space corresponding to Figure 11.

6. Sensitivity to Initial Values

In this section, we discuss the sensitivity of Lorenz-Chen-Lu like chaotic system (4) to initial values. We consider time plots for 3-sets of initial conditions $(x_0, y_0, z_0), (x_0 + 0.0001, y_0, z_0)$; $(x_0, y_0, z_0), (x_0, y_0 + 0.0001, z_0)$ and $(x_0, y_0, z_0), (x_0, y_0, z_0 + 0.0001)$ respectively. Sensitivity to initial values implies that a small change in the initial values leads to fluctuations which may be negligible initially but builds up rapidly. Thus, small change in initial values of the current path may lead to significantly different future behavior [15]. Case (i). We consider the parameter values $a = 0.3$, $b = 0.1$, $c = 0.8$, $r = 0.1$, $\alpha = 0.9$ and analyze the sensitivity to initial values on 3D Lorenz like chaotic system with $(x_0 = 0.1, y_0 = 0.1, z_0 = 0.1)$, $(x_0 = 0.1, y_0 = 0.1001, z_0 = 0.1)$ and $(x_0 = 0.1, y_0 = 0.1, z_0 = 0.1001)$, see figure-13. Case (ii). We take the values $a = 0.9$, $b = 0.07$, $c = 3.5$, $r = 0.1$, $\alpha = 0.9$ and analyze the sensitivity to initial values on 3D Chen like chaotic system with

$(x_0 = 0.1, y_0 = 0.1,\ z_0 = 0.1)$, $(x_0 = 0.1,\ y_0 = 0.1001,\ z_0 = 0.1)$ and $(x_0 = 0.1,\ y_0 = 0.1,\ z_0 = 0.1001)$, see figure-14.

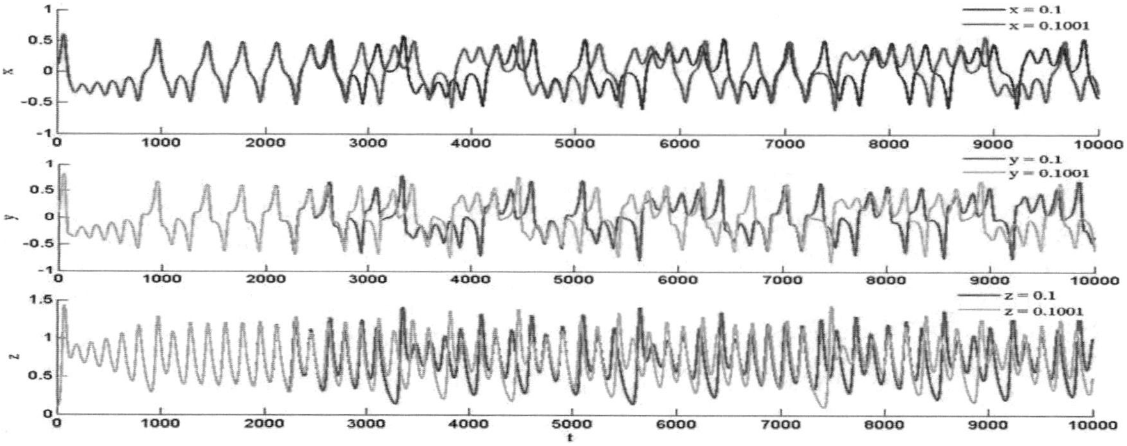

Figure 13. Sensitivity to initial values on 3D Lorenz like chaotic system

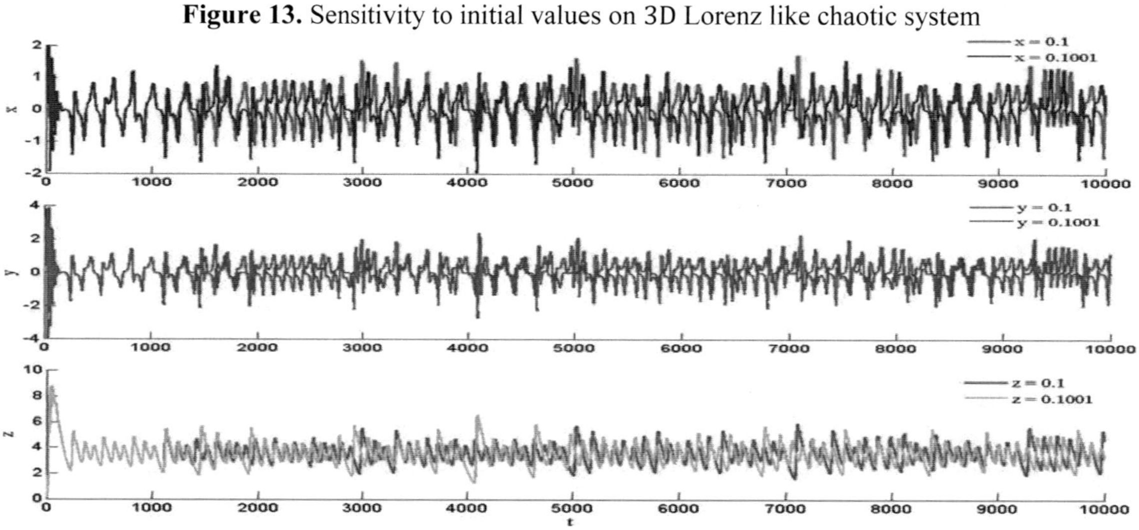

Figure 14. Sensitivity to initial values on 3D Chen like chaotic system

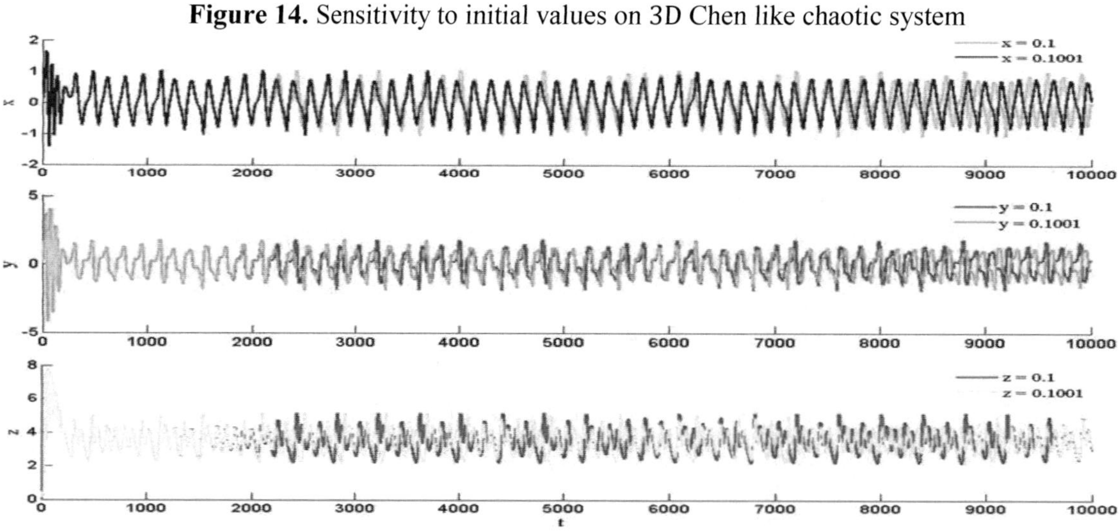

Figure 15. Sensitivity to initial values on 3D Lu like chaotic system

Case (iii). we consider the parameter values $a = 0.01$, $b = 0.5$, $h = 0.1$, $\alpha = 0.9$ and analyze the sensitivity to initial values on 3D Lu like chaotic system with $(x_0 = 0.1, y_0 = 0.1, z_0 = 0.1)$, $(x_0 = 0.1, y_0 = 0.1001, z_0 = 0.1)$ and $(x_0 = 0.1, y_0 = 0.1, z_0 = 0.1001)$, see figure-15.

7. Conclusion

This article investigated the dynamics of a discrete fractional order Lorenz-Chen-Lu like chaotic system with order in $(0, 1)$ which is developed from the system presented in (4) by using a discretization method. We established the existence of three equilibrium points and Jury condition is applied to analyze the stability of the system (4). Numerically, we have illustrated the dynamical behavior of the system (4) by means of chaotic attractors, bifurcation diagrams and phase diagrams for different set of parameter values and we also discussed the sensitivity analysis of the system.

References

[1] Edward Lorenz N 1963 *Deterministic Non-periodic Flow* Journal of the Atmospheric Sciences 20 pp 130 – 141

[2] Etienne Ghys 2013 The Lorenz Attractor, a Paradigm for Chaos Chaos 154 *Springer Basel*

[3] Chen G and Ueta T 1999 *Yet another chaotic attractor* International Journal of Bifurcation and Chaos **9** 7 pp 1465 – 1466

[4] Ueta T and Chen G 2000 *Bifurcation analysis of Chen's equation* International Journal of Bifurcation and Chaos **10** 8 pp 19 – 31

[5] Jinhu Lu and Guanrong Chen 2002 *A New Chaotic Attractor Coined* International Journal of Bifurcation and Chaos **12** 3 pp 659 – 661

[6] Keith B. Oldham and Jerome Spanier 1974 The Fractional Calculus Theory and Applications of Differentiation and Integration to Arbitrary Order *Dover Publications* INC

[7] Ravi P Agarwal, Ahmed MA El-Sayed and Sanaa M Salman 2013 *Fractional-order Chuas system: discretization bifurcation and chaos* Advances in Difference Equations, 2013:320.

[8] Miller K.S and Ross B 1993 An Introduction to the Fractional Calculus and Fractional Differential Equations *John Wiley & Sons* New York NY USA

[9] Leonov G A and Kuznetsov N V 2015 *On differences and similarities in the analysis of Lorenz, Chen and Lu systems* Applied Mathematics and Computation 256 pp 334 – 343

[10] Ivo Petras 2010 Fractional order Nonlinear Systems - Modeling Analysis and Simulation Higher Education Press *Springer International Edition*

[11] George Maria Selvam A, Paul Loganathan M, Janagaraj R and Abraham Vianny D 2018 *Chaotic behavior in discrete Chen system with fractional order* International Journal of Technical Innovation in Modern Engineering & Science **4** 11 pp 310 – 315

[12] Haijum Wang and Xianyi Li 2015 *True and False Chaotic Attractors in a 3-D Lorenz-type System* Annual Review of Chaos Theory, Bifurcations and Dynamical Systems **5** pp 33–41

[13] George Maria Selvam A and Abraham Vianny D 2019 *Dynamic analysis of a discrete fractional order 3D chaotic system* American International Journal of Research in Science Technology Engineering & Mathematics pp 243 – 251

[14] Jinhu Lu, Tianshou Zhou and Suochun Zhang 2002 *Chaos synchronization between linearly coupled chaotic systems* Chaos, Solutions & Fractals **14** pp 529 – 541

[15] George Maria Selvam A and Abraham Vianny D 2019 *Bifurcation and Chaotic Behavior of a Discrete Fractional Order Lorenz System* AIP Conference Proceedings 2112 pp 020052-1–020052-13

Existence and uniqueness of solutions for boundary value problem of fractional order difference equations

A. George Maria Selvam[1] and D. Abraham Vianny[2]

[1, 2] Department of Mathematics, Sacred Heart College (Autonomous), Tirupattur – 635 601, Vellore Dist., Tamil Nadu, India

E-mail : agmshc@gmail.com

Abstract. In this article, we investigate the existence and uniqueness of solutions for boundary value problem (BVP) of fractional order difference equations (FODE) of the form

$$\Delta^{\upsilon} w(s) = -f(s+\upsilon-1, w(s+\upsilon-1)),$$
$$w(\upsilon-2) = \psi(w), \; w(\upsilon+k) = \varphi(w),$$

where $s \in [0,k]_{N_0}$, $f : [\upsilon-2, \; \upsilon-1, \dots, \upsilon+k]_{N_{\upsilon-2}} \times R \to [0,+\infty]$ is a continuous function $\psi, \varphi : C\left([\upsilon-2, \upsilon+k\,]_{N_{\upsilon-2}}\right) \to R$ are given functions and $1 < \upsilon \leq 2$. Existence and uniqueness of solutions are established by the Brouwer fixed point theorem and contraction mapping principle.

2010 Mathematics Subject Classification: 26A33, 39A12, 47H10
Keywords: Discrete fractional order, existence, fixed point and uniqueness

1. Introduction

Mathematicians recently have shown great interest in FODE due to the development in discrete fractional calculus and theory of fractional calculus [1, 4, 5]. In Podlubney [2], the applications of the fractional calculus are provided in the field of control theory, electro chemistry, capacitor theory, diffusion analysis and electrical circuits etc., F.M.Atici et.el [7] discussed a transform method and initial value problem in discrete fractional calculus. Recently, F.M.Atici et.el [8] developed basic theory of discrete fractional calculus for summation by parts and Leibniz rule. Some authors have reported the uniqueness and existence of solutions for BVP's.

F.M.Atici et.el [10] considered a two-point BVP for the finite FODE of the type,

$$\Delta^{\upsilon} w(s) = -f(s+\upsilon-1, w(s+\upsilon-1)),$$
$$w(\upsilon-2) = 0, \; w(\upsilon+k+1) = 0,$$

where $s \in 1,2,3,\dots,k+1$, $f : [\upsilon, \upsilon+k]_{N_{\upsilon-1}} \times R \to R$ is a continuous function $k \geq 2$ and $1 < \upsilon \leq 2$. They formulated the Green function and discussed the existence of positive solutions for the BVP's of finite FODE.

Goodrich in [5] investigated a discrete fractional BVP of the type,

$$\Delta^{\upsilon} w(s) = -f(s+\upsilon-1, w(s+\upsilon-1)),$$
$$w(\upsilon-2) = g(w), \; w(\upsilon+k) = 0,$$

where $s \in 1,2,3,\dots,k$, $f : [\upsilon-2, \; \upsilon-1, \dots, \upsilon+k-1]_{N_{\upsilon-2}} \times R \to R$ is a continuous function, $1 < \upsilon \leq 2$ and $g : C\left([\upsilon-2, \upsilon+k\,]_{Z_{\upsilon-3}}, R\right)$ are the given functions. Goodrich established the uniqueness of solutions to this problem by the Brouwer fixed point theorem and contraction mapping principle and also positive solutions established by the Guo-Krasnosel'skii fixed point theorem.

Goodrich in [6] established positive solutions to integer-order and nonlocal fractional difference equations.

$$\Delta^{\upsilon} w(s) = -f(s+\upsilon-1, w(s+\upsilon-1)),$$
$$w(\upsilon-2) = \psi(w), \; w(\upsilon+k) = \varphi(w), \tag{1}$$

where $s \in [0,k]_{N_0}$, $f : [\upsilon-2, \upsilon-1, \dots, \upsilon+k]_{N_{\upsilon-2}} \times R \to [0,+\infty]$ is a continuous function, $1 < \upsilon \leq 2$ and $\psi, \varphi : C\left([\upsilon-2, \upsilon+k\,]_{N_{\upsilon-2}}\right) \to R$ are given functions. Goodrich established the existence of positive solutions to this problem by the Guo-Krasnosel'skii fixed point theorem.
In this article, we investigate the existence and uniqueness of solutions for BVP of FODE as in (1) by applying the Brouwer fixed point theorem and contraction mapping principle.
The plan of this article is as follows. In Section 2, we provide basic definition and some necessary lemma's. In Section 3, we present the existence and uniqueness of solutions by using the Brouwer theorem and contraction mapping principle.

2. Preliminaries

In this segment, some essential definitions and theory of fractional calculus are provided.
Definition. [4, 5] The generalized falling function is defined by

$$s^{\upsilon} = \frac{\Gamma(s+1)}{\Gamma(s+1-\upsilon)}.$$

Definition. [4, 5] Let $\upsilon > 0$. The $\upsilon-$ th fractional sum is defined by

$$\Delta^{-\upsilon} f(s) = \Delta^{-\upsilon} f(s,a) := \frac{1}{\Gamma(\upsilon)} \sum_{\xi=a}^{s-\upsilon} (s-\xi-1)^{(\upsilon-1)} f(\xi)$$

for $s \in \{a+\upsilon, \; a+\upsilon+1, \dots\} = N_{a+\upsilon}$. Let $\upsilon > 0$. The $\upsilon-$ th fractional difference is defined by $\Delta^{\upsilon} f(s) := \Delta^{N} \Delta^{\upsilon-N} f(s)$ where $s \in N_{a+\upsilon}$.

Lemma 2.1. [4, 5] Let s and υ be any numbers $s^{(\upsilon)}$ and $s^{(\upsilon-1)}$ is defined then

$$\Delta s^{(\upsilon)} = \upsilon s^{(\upsilon-1)} .$$

Lemma 2.2. [4, 5] $s \leq r$, then $s^{(\mu)} \leq r^{(\mu)}$ for any $\mu > 0$.

Lemma 2.3. [5] Let $0 \leq N-1 < \upsilon < N$. Then

$$\Delta^{-\upsilon} \Delta^{\upsilon} w(s) = w(s) + c_1 s^{(\upsilon-1)} + c_2 s^{(\upsilon-2)} + \dots + c_N s^{(\upsilon-N)}$$

for some $c_i \in R$ with $1 \leq i \leq N$.

Lemma 2.4. [5] For s and ξ, for with both $(s-\xi-1)^{(\upsilon)}$ and $(s-\xi-2)^{(\upsilon)}$ are defined, we have

$$\Delta_{\xi}\left[(s-\xi-1)^{(\upsilon)}\right] = -\upsilon\,(s-\xi-2)^{(\upsilon-1)}.$$

Lemma 2.5. [5] The function $\frac{1}{\Gamma(\upsilon-1)}\left[s^{(\upsilon-2)} - \frac{1}{k+2}\, s^{(\upsilon-1)}\right]$ is strictly decreasing in s, for $s \in [\upsilon-2, \upsilon+k\,]_{N_{\upsilon-2}}$. In addition

$$\min_{s \in [\upsilon-2, \upsilon+k\,]_{N_{\upsilon-2}}} \left[\frac{1}{\Gamma(\upsilon-1)}\left(s^{(\upsilon-2)} - \frac{1}{k+2}\, s^{(\upsilon-1)}\right)\right] = 0$$

and

$$\max_{s \in [\upsilon-2, \upsilon+k\,]_{N_{\upsilon-2}}} \left[\frac{1}{\Gamma(\upsilon-1)}\left(s^{(\upsilon-2)} - \frac{1}{k+2}\, s^{(\upsilon-1)}\right)\right] = 1.$$

Lemma 2.6. [6] The $1 < \upsilon \leq 2$, $h : [\upsilon - 1, \upsilon + k]_{N_{\upsilon-1}} \to R$ and $\psi, \varphi : R^{k+3} \to R$. Function w is a solution for BVP of FODE in (1) which is given by

$$\Delta^{\upsilon} w(s) = -h(s + \upsilon - 1), \tag{2}$$
$$w(\upsilon - 2) = \psi(w), \ w(\upsilon + k) = \varphi(w), \tag{3}$$

where $w(s)$ iff $s \in [0, k]_{N_0}$, for $s \in [\upsilon - 2, \upsilon + k]_{N_{\upsilon-2}}$ has

$$w(s) = -\frac{1}{\Gamma(\upsilon)} \sum_{\xi=0}^{s-\upsilon} (s - \xi - 1)^{(\upsilon-1)} h(\xi + \upsilon - 1) + \psi(w)\left[\frac{s^{(\upsilon-2)}}{\Gamma(\upsilon - 1)} - \frac{s^{(\upsilon-1)}}{(k + 2)\Gamma(\upsilon - 1)}\right]$$
$$+ \varphi(w)\left[\frac{s^{(\upsilon-1)}}{(\upsilon + k)^{(\upsilon-1)}}\right]$$
$$+ \frac{s^{(\upsilon-1)}}{\Gamma(\upsilon)(\upsilon + k)^{(\upsilon-1)}} \sum_{\xi=0}^{k} (\upsilon + k - \xi - 1)^{(\upsilon-1)} h(\xi + \upsilon - 1) \tag{4}$$

3. Existence and Uniqueness of Solutions

We see that BVP of FODE (1) can be written as an equivalent summation. We prove that (1) has atleast one solution. Let w be a solution of (1) iff w is the fixed point of the operator $S : R^{k+3} \to R^{k+3}$, has

$$(Sw)(s) = -\frac{1}{\Gamma(\upsilon)} \sum_{\xi=0}^{s-\upsilon} (s - \xi - 1)^{(\upsilon-1)} f(s + \upsilon - 1, w(s + \upsilon - 1))$$
$$+ \psi(w)\left[\frac{s^{(\upsilon-2)}}{\Gamma(\upsilon - 1)} - \frac{s^{(\upsilon-1)}}{(k + 2)\Gamma(\upsilon - 1)}\right] + \varphi(w)\left[\frac{s^{(\upsilon-1)}}{(\upsilon + k)^{(\upsilon-1)}}\right]$$
$$+ \frac{s^{(\upsilon-1)}}{\Gamma(\upsilon)(\upsilon + k)^{(\upsilon-1)}} \sum_{\xi=0}^{k} (\upsilon + k - \xi - 1)^{(\upsilon-1)} f(s + \upsilon - 1, w(s + \upsilon - 1)). \tag{5}$$

for $s \in [\upsilon - 2, \upsilon + k]_{N_{\upsilon-2}}$.

Theorem 3.1. We define $\|w\| = \max_{s \in [\upsilon-2, \upsilon+k]_{N_{\upsilon-2}}} |w(s)|$. Suppose that $f(s, w)$, $\psi(w)$ and $\varphi(w)$ are Lipschitz in w. There exists $\omega, \delta, \gamma > 0$ such that

$$|f(s, w_1) - f(s, w_2)| \leq \omega \|w_1 - w_2\|,$$
$$|\psi(w_1) - \psi(w_2)| \leq \delta \|w_1 - w_2\|,$$
$$|\varphi(w_1) - \varphi(w_2)| \leq \gamma \|w_1 - w_2\|,$$

for any function w_1, w_2 defined on $[\upsilon - 2, \upsilon + k]_{N_{\upsilon-2}}$, then if the condition

$$2\omega \prod_{j=1}^{k} \left(\frac{\upsilon + j}{j}\right) + (\delta + \gamma) < 1 \tag{6}$$

holds, then the BVP of FODE (1) has a unique solution.

Proof. The function w_1 and w_2 are given and we show that S is a contraction mapping.

$$
\begin{aligned}
\|Sw_1 - Sw_2\| \le\ & \omega \|w_1 - w_2\| \max_{s\in[\upsilon-2,\upsilon+k]_{N_{\upsilon-2}}} \left[\frac{1}{\Gamma(\upsilon)}\sum_{\xi=0}^{s-\upsilon}(s-\xi-1)^{(\upsilon-1)}\right] \\
& + \delta\|w_1 - w_2\| \max_{s\in[\upsilon-2,\upsilon+k]_{N_{\upsilon-2}}} \left[\frac{s^{(\upsilon-2)}}{\Gamma(\upsilon-1)} - \frac{s^{(\upsilon-1)}}{(k+2)\Gamma(\upsilon-1)}\right] \\
& + \gamma\|w_1 - w_2\| \max_{s\in[\upsilon-2,\upsilon+k]_{N_{\upsilon-2}}} \left[\frac{s^{(\upsilon-1)}}{(\upsilon+k)^{(\upsilon-1)}}\right] \\
& + \omega\|w_1 - w_2\| \max_{s\in[\upsilon-2,\upsilon+k]_{N_{\upsilon-2}}} \left[\frac{s^{(\upsilon-1)}}{\Gamma(\upsilon)(\upsilon+k)^{(\upsilon-1)}}\sum_{\xi=0}^{k}(\upsilon+k-\xi-1)^{(\upsilon-1)}\right]
\end{aligned} \tag{7}
$$

Applying the Lemma (2.4), we simplify the RHS of (7), we get

$$
\begin{aligned}
\frac{1}{\Gamma(\upsilon)}\sum_{\xi=0}^{s-\upsilon}(s-\xi-1)^{(\upsilon-1)} &= \frac{1}{\Gamma(\upsilon)}\left[-\frac{1}{\upsilon}(s-\xi)^{(\upsilon)}\right]_{\xi=0}^{s-\upsilon+1} \\
&= \frac{1}{\Gamma(\upsilon+1)}\left[s^{(\upsilon)}\right] \\
&= \frac{1}{\Gamma(\upsilon+1)}\frac{\Gamma(s+1)}{\Gamma(s+1-\upsilon)} \\
&\le \frac{\Gamma(\upsilon+k+1)}{\Gamma(\upsilon+1)\Gamma(k+1-\upsilon)}
\end{aligned}
$$

$$
\frac{1}{\Gamma(\upsilon)}\sum_{\xi=0}^{s-\upsilon}(s-\xi-1)^{(\upsilon-1)} = \prod_{j=1}^{k}\left(\frac{\upsilon+j}{j}\right) \tag{8}
$$

Applying the Lemma (2.2), we simplify the RHS of (7), we get

$$
\frac{s^{(\upsilon-1)}}{(\upsilon+k)^{(\upsilon-1)}} \le \frac{(\upsilon+k)^{(\upsilon-1)}}{(\upsilon+k)^{(\upsilon-1)}} = 1 \tag{9}
$$

Applying the Lemma (2.4), we simplify the RHS of (7), we get

$$
\begin{aligned}
\frac{s^{(\upsilon-1)}}{\Gamma(\upsilon)(\upsilon+k)^{(\upsilon-1)}}\sum_{\xi=0}^{k}(\upsilon+k-\xi-1)^{(\upsilon-1)} &\le \frac{1}{\Gamma(\upsilon)}\sum_{\xi=0}^{k}(\upsilon+k-\xi-1)^{(\upsilon-1)} \\
&= \frac{1}{\Gamma(\upsilon)}\left[-\frac{1}{\upsilon}(\upsilon+k-\xi)^{(\upsilon)}\right]_{\xi=0}^{k+1} \\
&= \frac{\Gamma(\upsilon+k+1)}{\Gamma(\upsilon+1)\Gamma(k+1)}
\end{aligned}
$$

$$
\frac{s^{(\upsilon-1)}}{\Gamma(\upsilon)(\upsilon+k)^{(\upsilon-1)}}\sum_{\xi=0}^{k}(\upsilon+k-\xi-1)^{(\upsilon-1)} = \prod_{j=1}^{k}\left(\frac{\upsilon+j}{j}\right) \tag{10}
$$

Applying the Lemma (2.5), we simplify the RHS of (7), we get

$$
\left[\frac{s^{(\upsilon-2)}}{\Gamma(\upsilon-1)} - \frac{s^{(\upsilon-1)}}{(k+2)\Gamma(\upsilon-1)}\right] \le \left[\frac{(\upsilon-2)^{(\upsilon-2)}}{\Gamma(\upsilon-1)} - \frac{(\upsilon-2)^{(\upsilon-1)}}{(k+2)\Gamma(\upsilon-1)}\right]
$$

$$\leq \left[\frac{(\upsilon-2)^{(\upsilon-2)}}{\Gamma(\upsilon-1)} - \frac{0}{(k+2)\Gamma(\upsilon-1)}\right]$$
$$= \frac{\Gamma(\upsilon-1)}{\Gamma(\upsilon-1)}$$

$$\left[\frac{s^{(\upsilon-2)}}{\Gamma(\upsilon-1)} - \frac{s^{(\upsilon-1)}}{(k+2)\Gamma(\upsilon-1)}\right] = 1 \tag{11}$$

Substituting (8), (9), (10) and (11) in (7), we get

$$\|Sw_1 - Sw_2\| \leq \omega \|w_1 - w_2\| \left[\prod_{j=1}^{k}\left(\frac{\upsilon+j}{j}\right)\right] + \delta \|w_1 - w_2\|(1) + \gamma \|w_1 - w_2\|(1)$$
$$+ \omega \|w_1 - w_2\| \left[\prod_{j=1}^{k}\left(\frac{\upsilon+j}{j}\right)\right]$$

$$\|Sw_1 - Sw_2\| \leq 2\omega \|w_1 - w_2\| \left[\prod_{j=1}^{k}\left(\frac{\upsilon+j}{j}\right)\right] + \delta \|w_1 - w_2\| + \gamma \|w_1 - w_2\|$$

$$\|Sw_1 - Sw_2\| \leq \|w_1 - w_2\| \left[2\omega \prod_{j=1}^{k}\left(\frac{\upsilon+j}{j}\right) + (\delta+\gamma)\right] \|w_1 - w_2\|. \tag{12}$$

Then condition (6) holds. Hence the BVP of FODE (1) has a unique solution.

We apply the Brouwer fixed point theorem

Theorem 3.2. Assume $A_1 > 0$ is a constant and there exists, such that $f(s,w)$ satisfies the inequality

$$\max_{(s,w)\in[\upsilon-2,\upsilon+k]_{N_{\upsilon-2}}\times[-A_1,A_1]} |f(s,w)| \leq \frac{A_1}{\left[2\left(\frac{\Gamma(\upsilon+k+1)}{\Gamma(\upsilon+1)\Gamma(k+1)}+1\right)\right]} \tag{13}$$

and $\psi(w)$ and $\varphi(w)$ are functions satisfying the inequlity

$$\max_{w\in[-A_1,A_1]} |\psi(w)| \leq \frac{A_1}{\left[2\left(\frac{\Gamma(\upsilon+k+1)}{\Gamma(\upsilon+1)\Gamma(k+1)}+1\right)\right]}, \tag{14}$$

$$\max_{w\in[-A_1,A_1]} |\varphi(w)| \leq \frac{A_1}{\left[2\left(\frac{\Gamma(\upsilon+k+1)}{\Gamma(\upsilon+1)\Gamma(k+1)}+1\right)\right]}. \tag{15}$$

Then the BVP of FODE (1) has alteast one solution w_0 satisfying $|w_0(s)| \leq A_1$ for all $s \in [\upsilon-2, \upsilon+k]_{N_{\upsilon-2}}$.

Proof. We consider the Banach space $B := \{w \in R^{k+3} : \|w\| \leq A_1\}$. S is a continuous operator as defined in (5). We show that $S : B \to B$, whenever $\|Sw\| \leq A_1$. We apply the Brouwer theorem to give the conclusion.

For f, ψ and φ are given functions then the inequalities (13), (14) and (15) holds.

$$\Phi = \frac{A_1}{\left[2\left(\frac{\Gamma(\upsilon+k+1)}{\Gamma(\upsilon+1)\Gamma(k+1)}+1\right)\right]}, \tag{16}$$

where Φ is positive constant, we get

$$\begin{aligned}\|Sw\| \leq & \max_{s\in[\upsilon-2,\upsilon+k]_{N_{\upsilon-2}}} \frac{1}{\Gamma(\upsilon)}\sum_{\xi=0}^{s-\upsilon}(s-\xi-1)^{(\upsilon-1)}\left|f(\xi+\upsilon-1,w(\xi+\upsilon-1))\right| \\ & + \max_{s\in[\upsilon-2,\upsilon+k]_{N_{\upsilon-2}}} |\psi(w)|\left[\frac{s^{(\upsilon-2)}}{\Gamma(\upsilon-1)}-\frac{s^{(\upsilon-1)}}{(k+2)\Gamma(\upsilon-1)}\right] \\ & + \max_{s\in[\upsilon-2,\upsilon+k]_{N_{\upsilon-2}}} |\varphi(w)|\left[\frac{s^{(\upsilon-1)}}{(\upsilon+k)^{(\upsilon-1)}}\right] \\ & + \max_{s\in[\upsilon-2,\upsilon+k]_{N_{\upsilon-2}}} \left[\frac{s^{(\upsilon-1)}}{\Gamma(\upsilon)(\upsilon+k)^{(\upsilon-1)}}\sum_{\xi=0}^{k}(\upsilon+k-\xi-1)^{(\upsilon-1)}\right] \\ & \times\left|f(\xi+\upsilon-1,w(\xi+\upsilon-1))\right|\end{aligned}$$

$$\begin{aligned}\|Sw\| \leq \Phi & \max_{s\in[\upsilon-2,\upsilon+k]_{N_{\upsilon-2}}} \frac{1}{\Gamma(\upsilon)}\sum_{\xi=0}^{s-\upsilon}(s-\xi-1)^{(\upsilon-1)} \\ & + \Phi \max_{s\in[\upsilon-2,\upsilon+k]_{N_{\upsilon-2}}} \left[\frac{s^{(\upsilon-2)}}{\Gamma(\upsilon-1)}-\frac{s^{(\upsilon-1)}}{(k+2)\Gamma(\upsilon-1)}\right] \\ & + \Phi \max_{s\in[\upsilon-2,\upsilon+k]_{N_{\upsilon-2}}} \left[\frac{s^{(\upsilon-1)}}{(\upsilon+k)^{(\upsilon-1)}}\right] \\ & + \Phi \max_{s\in[\upsilon-2,\upsilon+k]_{N_{\upsilon-2}}} \left[\frac{s^{(\upsilon-1)}}{\Gamma(\upsilon)(\upsilon+k)^{(\upsilon-1)}}\sum_{\xi=0}^{k}(\upsilon+k-\xi-1)^{(\upsilon-1)}\right]. \qquad (17)\end{aligned}$$

Applying the Lemma (2.4), we simplify the RHS of (17), we get

$$\begin{aligned}\frac{1}{\Gamma(\upsilon)}\sum_{\xi=0}^{s-\upsilon}(s-\xi-1)^{(\upsilon-1)} + \frac{s^{(\upsilon-1)}}{\Gamma(\upsilon)(\upsilon+k)^{(\upsilon-1)}}\sum_{\xi=0}^{k}(\upsilon+k-\xi-1)^{(\upsilon-1)} \\ = \frac{\Gamma(\upsilon+k+1)}{\Gamma(\upsilon+1)\Gamma(k+1)}+\frac{\Gamma(\upsilon+k+1)}{\Gamma(\upsilon+1)\Gamma(k+1)} \\ = \frac{2\Gamma(\upsilon+k+1)}{\Gamma(\upsilon+1)\Gamma(k+1)}. \qquad (18)\end{aligned}$$

Applying the Lemma (2.5), we simplify the RHS of (17), we get

$$\frac{s^{(\upsilon-2)}}{\Gamma(\upsilon-1)}-\frac{s^{(\upsilon-1)}}{(k+2)\Gamma(\upsilon-1)} \leq \frac{(\upsilon-2)^{(\upsilon-2)}}{\Gamma(\upsilon-1)}-\frac{(\upsilon-2)^{(\upsilon-1)}}{(k+2)\Gamma(\upsilon-1)} = \frac{\Gamma(\upsilon-1)}{\Gamma(\upsilon-1)} = 1. \qquad (19)$$

Applying the Lemma (2.2), we simplify the RHS of (17), we get

$$\frac{s^{(\upsilon-1)}}{(\upsilon+k)^{(\upsilon-1)}} \leq \frac{(\upsilon+k)^{(\upsilon-1)}}{(\upsilon+k)^{(\upsilon-1)}} = 1. \qquad (20)$$

Substituting (18), (19) and (20) in (17), we get

$$\begin{aligned}\|Sw\| &\leq \Phi\left[\frac{2\Gamma(\upsilon+k+1)}{\Gamma(\upsilon+1)\Gamma(k+1)}\right]+\Phi+\Phi \\ \|Sw\| &\leq \Phi\left[\frac{2\Gamma(\upsilon+k+1)}{\Gamma(\upsilon+1)\Gamma(k+1)}\right]+2\Phi \\ \|Sw\| &\leq 2\Phi\left[\frac{\Gamma(\upsilon+k+1)}{\Gamma(\upsilon+1)\Gamma(k+1)}+1\right]. \qquad (21)\end{aligned}$$

By inserting (16) in (21), we get

$$\|Sw\| \leq 2\frac{A_1}{\left[2\left(\frac{\Gamma(\upsilon+k+1)}{\Gamma(\upsilon+1)\Gamma(k+1)}+1\right)\right]}\left[\frac{\Gamma(\upsilon+k+1)}{\Gamma(\upsilon+1)\Gamma(k+1)}+1\right] = A_1. \quad (22)$$

We deduce that $S : B \to B, Sw_0 = w_0$ with $w_0 \in B$ and the function w_0 is a solution (1) and w_0 satisfies $|w_0(s)| \leq A_1$ for each $s \in [\upsilon-2, \upsilon+k]_{N_{\upsilon-2}}$. The Brouwer theorem is there exist, hence the map S is a fixed point. Thus the proof ends.

4. Conclusion

In this article we have considered a discrete BVP of fractional order. The main contributions of this article, the existence and uniqueness of solutions for BVP of FODE with $1 < \upsilon \leq 2$ are investigated by the Brouwer fixed point theorem and contraction mapping principle.

References

[1] Keith B. Oldham and Jerome Spanier 1974 The Fractional Calculus Theory and Applications of Differentiation and Integration to Arbitrary Order *Dover Publications* INC
[2] Podlubny I 1999 Fractional Differential Equations, Mathematics in Science and Engineering *Academic Press* New York
[3] Agarwal R P, Meehan M and Reegan D O 2001 Fixed Point Theory and Applications *Cambridge University Press* Cambridge
[4] Michael T. Holm 2011 The Theory of Discrete Fractional Calculus: Development and Application, Ph.D Dissertation.
[5] Christopher S Goodrich and Allan C Peterson 2015 Discrete Fractional Calculus *Springer*
[6] Christopher S Goodrich 2011 *On Positive Solutions to Nonlocal Fractional and Integer-Order Difference Equations* Applicable Analysis and Discrete Mathematics 5 pp 122 – 132
[7] Ferhan M.Atici and Paul W. Eloe 2007 *A transform method in discrete fractional calculus* InternationalJournal of Differential Equations 2 pp 165 – 176
[8] Ferhan M.Atici and Paul W. Eloe 2009 *Initial value problems in discrete fractional calculus* Proceedings of American Mathematical Society 137(3) pp 981– 989.
[9] Ferhan M.Atici and Sevgi Sengul 2010 *Modeling with fractional difference equations* Journal of Mathematical Analysis and Application 369 pp 1– 9
[10] Ferhan M.Atici and Paul W. Eloe 2011 *Two-point boundary value problems for finite fractional difference equations* Journal of Difference Equations and Applications 17 (4) pp 445 – 456

EMD of Linear Benzenoid Chain Using Chemical Graph

M Raji[1] and G Jayalalitha[2]

[1] Research Scholar, Department of Mathematics, VELS Institute of Science, Technology & Advanced Studies, Chennai, Tamil Nadu, India.
[2] Professor, Department of Mathematics, VELS Institute of Science, Technology & Advanced Studies, Chennai, Tamil Nadu, India.

E-mail: [1] rajialagumurugan@gmail.com, [2] ragaji94@yahoo.com

Abstract. Let G_m be a undirected graph. Having a Molecule, atoms indicate by vertices and bonds indicate by edges. The main aim of this paper explains Extended Medium Domination on Linear Benzenoid Chain like Benzene, Naphthalene, Anthracene and their generalisation will be proved.
Key Words: Distance in Graph, $u - v$ path, Extended Medium Domination Number.
AMS Subject Classification: 05C12

1. Introduction

Chemical graph theory depends on the combination of graphs and chemistry. A Molecular Graph or Chemical Graph may be an illustration of the molecular formula of a matter in terms of graph theory. A chemical graph may be a connected graph whose vertices correspond to the atoms of the compound and edges correspond to chemical bonds [7]. Let $\mathcal{L}_{\mathfrak{h}}$ be the Linear Benzenoid Chain with $\mathfrak{h}$ hexagons [6, 8]. In this paper, Considering G_m as a finite connected simple graph.

1.1. Connected Graph [2]

A graph G_m is connected if there is at least one path between every pair of vertices.

1.2. Distance (u, v) [1, 2]

The Length of a shortest $u - v$ path in a connected graph G_m is called the distance from a vertex u to a vertex v.

1.3. $EDOM(u, v)$ [3, 8]

$edom(u, v)$ is the summation of number of $u - v$ paths of length one, two and three.

1.4. Extened Total Dominating Vertices $(ETDV(G_m))$ [3, 8]

Total number of vertices that dominate every pair of two vertices is defined as

$$(ETDV(G_m)) = \sum_{\forall u,v \in V(G_m)} edom(u, v)$$

1.5. $EMD(G_m))$ [3, 8]
Extended Medium Domination Number is defined as $EMD(G_m)) = \frac{2ETDV(G_m)}{n(n-1)}$

2. Extended Medium Domination Number

This section describes the Extended Medium Domination number for Linear Benzenoid Chain. Also It explains the Extended Medium Domination Number of Linear Benzenoid Chain $\mathcal{L}_{\mathfrak{h}}$ is

$$EMD(\mathcal{L}_{\mathfrak{h}}) = 2\sum_{k=1}^{\mathfrak{h}} \frac{25k-7}{(4k+2)(4k+1)}$$

by Induction method.

2.1. Benzene [8]

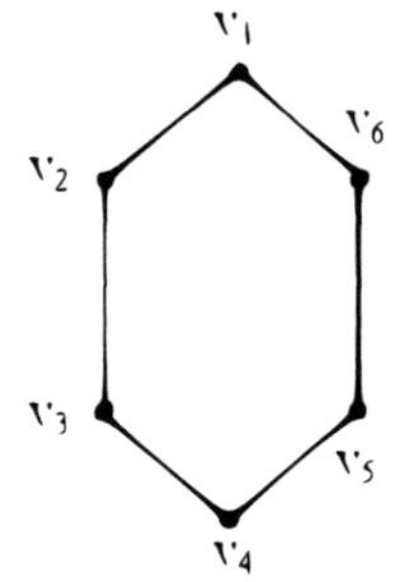

Figure 1. $\mathcal{L}_1$

$edom(v_1, v_2) = 1;$ $edom(v_1, v_3) = 1;$ $edom(v_1, v_4) = 2;$ $edom(v_1, v_5) = 1;$
$edom(v_1, v_6) = 1;$ $edom(v_2, v_3) = 1;$ $edom(v_2, v_4) = 1;$ $edom(v_2, v_5) = 2;$
$edom(v_2, v_6) = 1;$ $edom(v_3, v_4) = 1;$ $edom(v_3, v_5) = 1;$ $edom(v_3, v_6) = 2;$
$edom(v_4, v_5) = 1;$ $edom(v_4, v_6) = 1;$ $edom(v_5, v_6) = 1$

$$ETDV(\mathcal{L}_1) = 18$$

So,

$$EMD(\mathcal{L}_1) = \frac{6}{5}$$

2.2. Napathalene [5, 8]

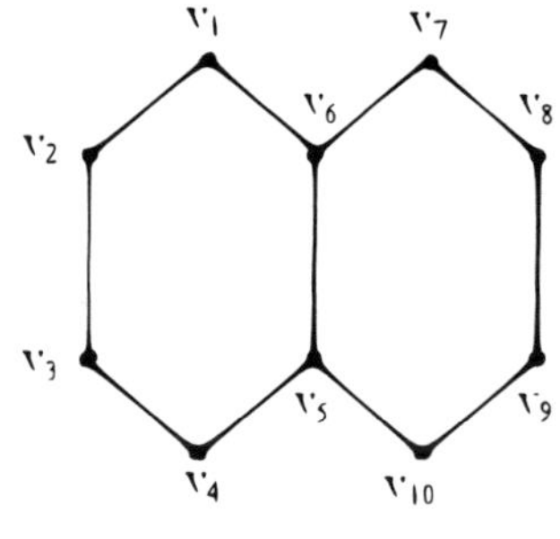

Figure 2. $\mathcal{L}_2$

$edom(v_1, v_2) = 1$; $edom(v_1, v_3) = 1$; $edom(v_1, v_4) = 2$; $edom(v_1, v_5) = 1$;
$edom(v_1, v_6) = 1$; $edom(v_1, v_7) = 1$; $edom(v_1, v_8) = 1$; $edom(v_1, v_9) = 0$;
$edom(v_1, v_{10}) = 1$; $edom(v_2, v_3) = 1$; $edom(v_2, v_4) = 1$; $edom(v_2, v_5) = 2$;
$edom(v_2, v_6) = 1$; $edom(v_2, v_7) = 1$; $edom(v_2, v_8) = 0$; $edom(v_2, v_9) = 0$;
$edom(v_2, v_{10}) = 0$; $edom(v_3, v_4) = 1$; $edom(v_3, v_5) = 1$; $edom(v_3, v_6) = 2$;
$edom(v_3, v_7) = 0$; $edom(v_3, v_8) = 0$; $edom(v_3, v_9) = 0$; $edom(v_3, v_{10}) = 1$;
$edom(v_4, v_5) = 1$; $edom(v_4, v_6) = 1$; $edom(v_4, v_7) = 0$; $edom(v_4, v_8) = 0$;
$edom(v_4, v_9) = 1$; $edom(v_4, v_{10}) = 1$; $edom(v_5, v_6) = 1$; $edom(v_5, v_7) = 1$;
$edom(v_5, v_8) = 2$; $edom(v_5, v_9) = 1$; $edom(v_5, v_{10}) = 1$; $edom(v_6, v_7) = 1$;
$edom(v_6, v_8) = 1$; $edom(v_6, v_9) = 2$; $edom(v_6, v_{10}) = 1$; $edom(v_7, v_8) = 1$;
$edom(v_7, v_9) = 1$; $edom(v_7, v_{10}) = 2$; $edom(v_8, v_9) = 1$; $edom(v_8, v_{10}) = 1$;
$edom(v_9, v_{10}) = 1$

$$ETDV(\mathcal{L}_2) = 43$$

Therefore,

$$EMD(\mathcal{L}_2) = \frac{43}{45}$$

2.3. Anthracene [4, 8]

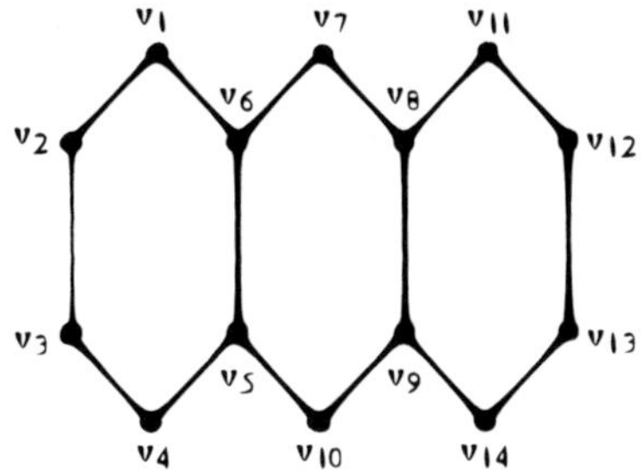

Figure 3. $\mathcal{L}_3$

$edom(v_1, v_2) = 1$; $edom(v_1, v_3) = 1$; $edom(v_1, v_4) = 2$; $edom(v_1, v_5) = 1$;
$edom(v_1, v_6) = 1$; $edom(v_1, v_7) = 1$; $edom(v_1, v_8) = 1$; $edom(v_1, v_9) = 0$;
$edom(v_1, v_{10}) = 1$; $edom(v_1, v_{11}) = 0$; $edom(v_1, v_{12}) = 0$; $edom(v_1, v_{13}) = 0$;
$edom(v_1, v_{14}) = 0$; $edom(v_2, v_3) = 1$; $edom(v_2, v_4) = 1$; $edom(v_2, v_5) = 2$;
$edom(v_2, v_6) = 1$; $edom(v_2, v_7) = 1$; $edom(v_2, v_8) = 0$; $edom(v_2, v_9) = 0$;
$edom(v_2, v_{10}) = 0$; $edom(v_2, v_{11}) = 0$; $edom(v_2, v_{12}) = 0$; $edom(v_2, v_{13}) = 0$;
$edom(v_2, v_{14}) = 0$; $edom(v_3, v_4) = 1$; $edom(v_3, v_5) = 1$; $edom(v_3, v_6) = 2$;
$edom(v_3, v_7) = 0$; $edom(v_3, v_8) = 0$; $edom(v_3, v_9) = 0$; $edom(v_3, v_{10}) = 1$;
$edom(v_3, v_{11}) = 0$; $edom(v_3, v_{12}) = 0$; $edom(v_3, v_{13}) = 0$; $edom(v_3, v_{14}) = 0$;
$edom(v_4, v_5) = 1$; $edom(v_4, v_6) = 1$; $edom(v_4, v_7) = 1$; $edom(v_4, v_8) = 0$;
$edom(v_4, v_9) = 1$; $edom(v_4, v_{10}) = 1$; $edom(v_4, v_{11}) = 0$; $edom(v_4, v_{12}) = 0$;
$edom(v_4, v_{13}) = 0$; $edom(v_4, v_{14}) = 0$; $edom(v_5, v_6) = 1$; $edom(v_5, v_7) = 1$;
$edom(v_5, v_8) = 2$; $edom(v_5, v_9) = 1$; $edom(v_5, v_{10}) = 1$; $edom(v_5, v_{11}) = 0$;
$edom(v_5, v_{12}) = 0$; $edom(v_5, v_{13}) = 0$; $edom(v_5, v_{14}) = 1$; $edom(v_6, v_7) = 1$;
$edom(v_6, v_8) = 1$; $edom(v_6, v_9) = 2$; $edom(v_6, v_{10}) = 1$; $edom(v_6, v_{11}) = 1$;
$edom(v_6, v_{12}) = 0$; $edom(v_6, v_{13}) = 0$; $edom(v_6, v_{14}) = 0$; $edom(v_7, v_8) = 1$;
$edom(v_7, v_9) = 1$; $edom(v_7, v_{10}) = 2$; $edom(v_7, v_{11}) = 1$; $edom(v_7, v_{12}) = 1$;
$edom(v_7, v_{13}) = 0$; $edom(v_7, v_{14}) = 1$; $edom(v_8, v_9) = 1$; $edom(v_8, v_{10}) = 1$;
$edom(v_8, v_{11}) = 1$; $edom(v_8, v_{12}) = 1$; $edom(v_8, v_{13}) = 2$; $edom(v_8, v_{14}) = 1$;
$edom(v_9, v_{10}) = 1$; $edom(v_9, v_{11}) = 1$; $edom(v_9, v_{12}) = 2$; $edom(v_9, v_{13}) = 1$;

Journal of Physics: Conference Series **1377** (2019) 012004 doi:10.1088/1742-6596/1377/1/012004

$edom(v_9, v_{14}) = 1$; $edom(v_{10}, v_{11}) = 1$; $edom(v_{10}, v_{12}) = 0$; $edom(v_{10}, v_{13}) = 1$;
$edom(v_{10}, v_{14}) = 1$; $edom(v_{11}, v_{12}) = 1$; $edom(v_{11}, v_{13}) = 1$; $edom(v_{11}, v_{14}) = 2$;
$edom(v_{12}, v_{13}) = 1$; $edom(v_{12}, v_{14}) = 1$; $edom(v_{13}, v_{14}) = 1$

$$ETDV(\mathcal{L}_3) = 68$$

Hence,

$$EMD(\mathcal{L}_3) = \frac{68}{91}$$

Theorem 2.1. *The Extended Medium Domination Number of Linear Benzenoid Chain $\mathcal{L}_\mathfrak{h}$ is $EMD(\mathcal{L}_\mathfrak{h}) = 2\sum_{k=1}^{\mathfrak{h}} \frac{25k-7}{(4k+2)(4k+1)}$*

Proof. Let $\mathcal{L}_\mathfrak{h}$ be Linear Benzenoid Chain with $\mathfrak{h}$ Hexagones.
To Prove: $EMD(\mathcal{L}_\mathfrak{h}) = 2\sum_{k=1}^{\mathfrak{h}} \frac{25k-7}{(4k+2)(4k+1)}$ by Induction

Step 1:
Put $\mathfrak{h} = 1$ in the equation of $\mathcal{L}_\mathfrak{h}$, we get
By Calculation, we get $ETDV(\mathcal{L}_1) = 18$.
So, Extended Medium Domination Number of Linear Benzenoid Chain $\mathcal{L}_1$ is $EMD(\mathcal{L}_1) = \frac{6}{5}$.

Step 2:
Take $\mathfrak{h} = 2$ in the equation of $\mathcal{L}_\mathfrak{h}$, we get
By Calculation, $ETDV(\mathcal{L}_2) = 43$.
Hence Extended Medium Domination Number of Linear benzenoid Chain ($\mathcal{L}_2$) is $EMD(\mathcal{L}_2) = \frac{43}{45}$.
Assume that $\mathcal{L}_{\mathfrak{h}-1}$ is Linear Benzenoid Chain with $\mathfrak{h} - 1$ Hexagons where $\mathfrak{h} \geq 2 \in N$ [8]
Then the Extended Medium Domination Number of Linear Benzenoid Chain $\mathcal{L}_{\mathfrak{h}-1}$ is $EMD(\mathcal{L}_{\mathfrak{h}-1}) = 2\sum_{i=2}^{\mathfrak{h}} \frac{25i-7}{(4i+2)(4i+1)}$.

We have to show $EMD(\mathcal{L}_\mathfrak{h}) = 2\sum_{i=1}^{\mathfrak{h}} \frac{25i-7}{(4i+2)(4i+1)}$ for $\mathcal{L}_\mathfrak{h}$, Linear Benzenoid Chain with $\mathfrak{h}$ Hexagons, where $\mathfrak{h} \geq 1 \in N$.
The Total number of vertices that dominate every pair of two vertices is $25\mathfrak{h} - 7$ for $\mathfrak{h}$ hexagons.

$$\text{Hence, } EMD(\mathcal{L}_\mathfrak{h}) = \frac{2(18)}{6(5)} + 2\sum_{i=2}^{\mathfrak{h}} \frac{25i-7}{(4i+2)(4i+1)}$$
$$= 2\left\{\frac{25(1)-7}{(4(1)+2)(4(1)+1)} + \sum_{i=2}^{\mathfrak{h}} \frac{25i-7}{(4i+2)(4i+1)}\right\}$$
$$= 2\sum_{i=1}^{\mathfrak{h}} \frac{25i-7}{(4i+2)(4i+1)}$$

By the Principle of Mathematical induction,

$$EMD(\mathcal{L}_\mathfrak{h}) = 2\sum_{k=1}^{\mathfrak{h}} \frac{25k-7}{(4k+2)(4k+1)}$$

for $\mathfrak{h}$ Hexagons where $\mathfrak{h} \geq 1 \in N$.

Journal of Physics: Conference Series **1377** (2019) 012004 doi:10.1088/1742-6596/1377/1/012004

3. Conclusion

In this Paper, the Extended Medium Domination Number for Benzene, Naphthalene and Anthracene's Molecular Graphs are obtained. Also It explained Extended Medium Domination Number of Linear Benzenoid Chain $\mathcal{L}_{\mathfrak{h}}$ is

$$EMD(\mathcal{L}_{\mathfrak{h}}) = 2\sum_{k=1}^{\mathfrak{h}} \frac{25k-7}{(4k+2)(4k+1)}$$

by Induction Method.

References

[1] Duygu Vargorn and Pinar Dundar 2011 The Medium Domination Number Of a Graph *International Journal of Pure and Applied Mathematics* **70** No. 3 pp 297-306

[2] Arumugam S 2014 Invitation to Graph Theory Scitech Publications(India) Pvt Limited ISBN 139788187328469

[3] Mahadevan G, Vijaylakshmi V and Sivagnanam C 2017 Further Results of Extended Medium Domination Number of some special Types of Graphs *International Journal of Engineering Science, Advanced Computing and Bio-Technology* **8** No. 2 pp 55-74

[4] Vijayalakshmi S, Raji M and Jayalalitha G 2018 Degree based in Molecular Graph of Organic compounds on Domination, *Journal of Advanced Research in Dynamical and Control Systems* **10** 6 pp 962-66

[5] Jayalalitha G, Raji M and Senthil S 2019 Hosoya Polynomial and Wiener Index of Molecular Graph of Naphthalene Based On Domination *International Journal of Scientific Research and Review* **8** 1 pp 1238-41

[6] Nisreen Bukhary 2010 Domination in Benzenoids, Virginia Commonwealth University, VCU Scholars Compass

[7] Nenad Trinajstic 1983 Chemical Graph Theory, Second Edition, CRC Press

[8] Venkatesan S, Raji M and Jayalalitha G 2019 Medium Domination Number in Linear Polyacene *International Journal of Innovative Technology and Exploring Engineering* **8** 9S2 pp 459-61

Difference cordial labeling and strongly multiplicative labeling for some extended duplicate graph

R. Avudainayaki[1], B. Selvam[2], P.P. Ulaganathan[3]

[1]Department of Mathematics, Sri Sairam Institute of Technology, Chennai – 600 044, India
[2,3]Department of Mathematics, S.I.V.E.T. College, Gowrivakkam, Chennai – 600 073, India

avuramesh2004@gmail.com

Abstract. The aim of this paper is to prove that the extended duplicate graph of arrow graph and splitting graph of path admits difference cordial labeling and strongly multiplicative labelling.

Key words: Arrow graph, Splitting graph of path, Duplicate graph, Extended duplicate graph, Difference cordial labeling, Strongly multiplicative labeling.

1. Introduction

E.Sampthkumar [8,9] introduced the concept of splitting graph and duplicate graph..P.Vijaya kumar et. al., have proved that duplicate graph admits many labeling [12,13,14]. K.Thirusangu et. al., have introduced the concept of extended duplicate graph [11]. Selvam et. al., have proved many result in extended duplicate graph [1,2,3,5,6,10] .In [7] Ponraj, Shatish Narayanan and Kala introduced the notions of difference cordial labeling. The strongly multiplicative labeling was introduced by Beineke and Hegde [4].

2. Preliminaries

Definition 2.1 An arrow graph A^{n}_{m} with width 'n' and length 'm' is obtained by joining a vertex 'v' with superior vertices of $P_t \times P_m$ by ' t' new edges from one end. Clearly the total number of vertices is 2m+1 and the total number of edges is 3m.

Example: Arrow graph

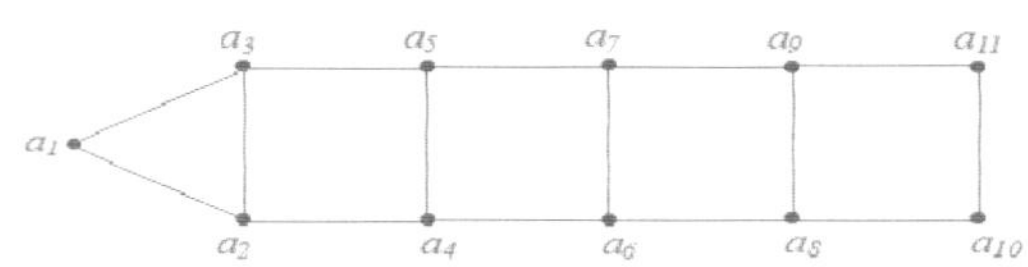

Figure 1 $A^2{}_5$

Definition 2.2 Let G (V,E) be a simple graph and the duplicate graph of G is DG (V_1, E_1) ,where the vertex set $V_1 = V \cup V'$ and $V \cap V' = \phi$ and $\varphi : V \to V'$ is bijective and the edge set E_1 of DG is defined as the edge a_ia_j is in E if and only if both $a_ia_j{}'$ and $a_i{}'a_j$ are edges in E_1.

Definition2.3 Let DG (V_1, E_1) be a duplicate graph of the arrow graph G(V,E). Extended duplicate graph of arrow graph is obtained by adding the edge $a_2\ a_2'$ to the duplicate graph. It is denoted by EDG (A_m^2), $m \geq 2$. Clearly the total number of vertices is 4m+2 and the total number of edges is 6m+1.

Example: Extended duplicate graph of Arrow graph

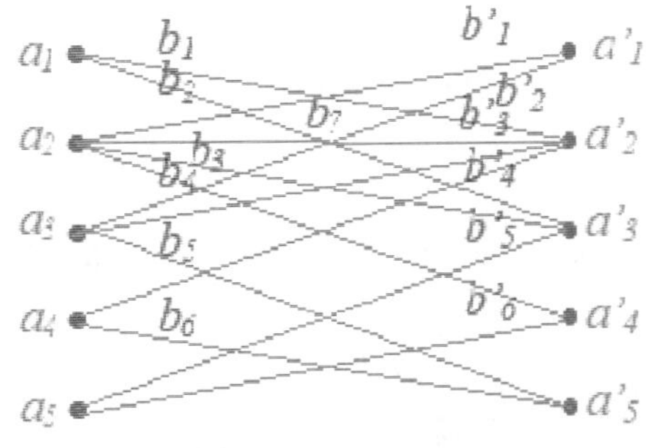

Figure 2 ***EDG(A_2^2)***

Definition 2.4 For each vertex 'a' of a graph G, take a new vertex a'. Join a' to all the vertices of G adjacent to a. The graph Spl(G) thus obtained is called splitting graph of G.

Example: Splitting graph of path P_6

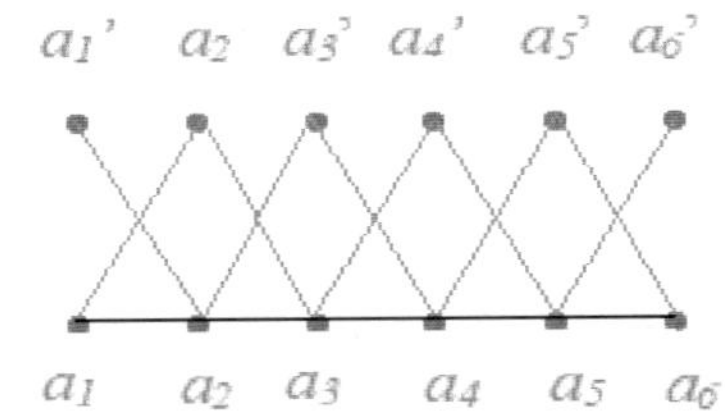

Figure 3 Spl***(P_6)***

Definition 2.5 Let DG (V_1,E_1) be a duplicate graph of splitting graph of path G(V,E). Extended duplicate graph of splitting graph of path is obtained by adding the edge $a_2\ a_2'$ to the duplicate graph. It is denoted by EDG Spl(P_m), $m \geq 2$. Clearly the total number of vertices is 4m and the total number of edges is 6m-5.

Example: Extended duplicate graph of Splitting graph of Path

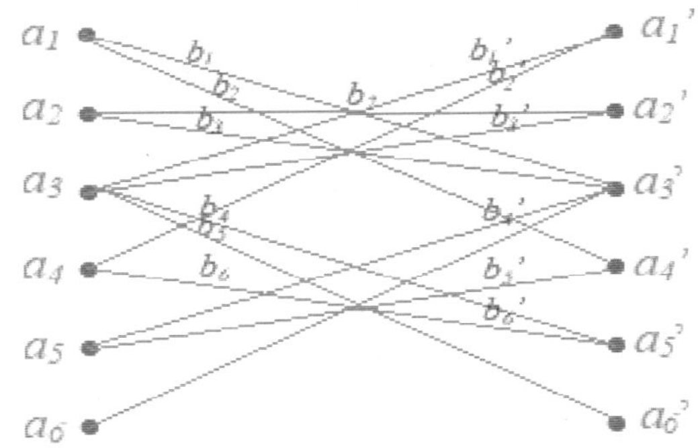

Figure 4 ***EDG(Spl(P_3))***

Definition 2.6 Let G be (p,q) graph. Let $\varphi : V(G) \rightarrow \{1,2,........p\}$. For every edge $a_i a_j$ assign the label $|\varphi(a_i) - \varphi(a_j)|$. φ is called difference cordial labeling if φ is 1 – 1 and $|b_\varphi(0) - b_\varphi(1)| \leq 1$ where $b_\varphi(0)$ and $b_\varphi(1)$ denote the number of edges labeled with 1 and not labeled with 1 respectively. A graph with a difference cordial labeling is called a difference cordial graph.

Definition 2.7 Let G(V,E) be a graph is said to be strongly multiplicative if φ: V(G)→{1,2,.....,|V|} } is bijective such that the induced map $\varphi^*(a_i a_j) = \varphi(a_i)\varphi(a_j)$ are distinct.

3. Main results

Algorithm: 3.1

Procedure [Difference cordial labeling for EDG (A_m^2), m≥ 2]

V ← $\{a_1, a_2, \ldots, a_{2m}, a_{2m+1}, a'_1, a'_2, \ldots, a'_{2m}, a'_{2m+1}\}$
E ← $\{b_1, b_2, \ldots, b_{3m}, b_{3m+1}, b'_1, b'_2, \ldots, b'_{3m}, b'_{3m}\}$
$a_1 \leftarrow 1; a_2 \leftarrow 3; a_3 \leftarrow 5$
$a'_1 \leftarrow 4; a'_2 \leftarrow 2; a'_3 \leftarrow 6$
for s = 0 to m-2 do
$a_{4+2s} \leftarrow 9+4s$
$a_{5+2s} \leftarrow 7+4s$
$a'_{4+2s} \leftarrow 8+4s$
$a'_{5+2s} \leftarrow 10+4s$
end for
end procedure

Theorem 3.1: The EDG(A_m^2) m ≥ 2 is difference cordial labeling.

Proof: Let EDG (A_m^2), m ≥ 2 be the extended duplicate graph of arrow graph. In the sequence of labeling the vertices, define a function φ : V (G) → {1,2,.... ,4m+2} as deliver in algorithm 3.1 as follows:

$\varphi(a_1) = 1$, $\varphi(a_2) = 3$, $\varphi(a_3) = 5$, $\varphi(a'_1) = 4$, $\varphi(a'_2) = 2$, $\varphi(a'_3) = 6$

For $0 \leq s \leq m-2$, $\varphi(a_{4+2s}) = 9+4s$, $\varphi(a_{5+2s}) = 7+4s$, $\varphi(a'_{4+2s}) = 8+4s$, $\varphi(a'_{5+2s}) = 10+4s$

Thus the total number of 4m+2 vertices is labeled.

For each edge $a_i a_j$ assign the label $|\varphi(a_i) - \varphi(a_j)|$

The edge functions are as follows:

$\varphi^*(b_1) = 1$, $\varphi^*(b_2) = 0$, $\varphi^*(b_3) = 0$, $\varphi^*(b'_1) = 1$, $\varphi^*(b'_2) = 1$,
$\varphi^*(b'_3) = 0$, $\varphi^*(b_{3m+1}) = 1$

For $0 \leq s \leq m-2$, $\varphi^*(b_{4+3s}) = 0$, $\varphi^*(b_{5+3s}) = 0$, $\varphi^*(b_{6+3s}) = 1$,
$\varphi^*(b'_{4+3s}) = 0$, $\varphi^*(b'_{5+3s}) = 1$, $\varphi^*(b'_{6+3s}) = 1$

All the 6m+1 edges are labeled .Hence the EDG(A_m^2),, m ≥ 2 admits difference cordial labeling.

Algorithm: 3.2

Procedure [Strongly multiplicative labeling for EDG (A_m^2), m≥ 2]

V ← $\{a_1, a_2, \ldots, a_{2m}, a_{2m+1}, a'_1, a'_2, \ldots, a'_{2m}, a'_{2m+1}\}$
E ← $\{b_1, b_2, \ldots, b_{3m}, b_{3m+1}, b'_1, b'_2, \ldots, b'_{3m}, b'_{3m}\}$
$a_1 \leftarrow 1; a_2 \leftarrow 4; a_3 \leftarrow 5$
$a'_1 \leftarrow 6; a'_2 \leftarrow 2; a'_3 \leftarrow 3$
for s = 0 to m-2 do

$a_{4+2s} \leftarrow 8+4s$
$a_{5+2s} \leftarrow 9+4s$
$a'_{4+2s} \leftarrow 7+4s$
$a'_{5+2s} \leftarrow 10+4s$
end for
end procedure

Theorem 3.2: The EDG(A^2_m), m ≥ 2 is strongly multiplicative labeling.

Proof: Let EDG (A^2_m), m ≥ 2 be the extended duplicate graph of arrow graph. In the sequence of labeling the vertices, define a function $\varphi : V(G) \rightarrow \{1,2,....,4m+2\}$ as deliver in algorithm 3.2 as follows:

$\varphi(a_1) = 1$, $\varphi(a_2) = 4$, $\varphi(a_3) = 5$, $\varphi(a'_1) = 6$, $\varphi(a'_2) = 2$, $\varphi(a'_3) = 3$

for $0 \le s \le m-2$; $\varphi(a_{4+2s}) = 8+4s$, $\varphi(a_{5+2s}) = 9+4s$, $\varphi(a'_{4+2s}) = 7+4s$, $\varphi(a'_{5+2s}) = 10+4s$

Thus the total number of 4m+2 vertices is labeled.

The edge function defined by $\varphi^*(a_i a_j) = \varphi(a_i)\,\varphi(a_j)$

$\varphi^*(b_1) = 2$, $\varphi^*(b_2) = 3$, $\varphi^*(b_3) = 12$, $\varphi^*(b_4) = 28$, $\varphi^*(b_5) = 50$, $\varphi^*(b_6) = 80$, $\varphi^*(b'_1) = 24$, $\varphi^*(b'_2) = 30$, $\varphi^*(b'_3) = 10$, $\varphi^*(b'_4) = 16$, $\varphi^*(b'_5) = 27$, $\varphi^*(b'_6) = 63$, $\varphi^*(b_{3m+1}) = 8$

For $0 \le s \le m-3$,

$\varphi^*(b_{7+3s}) = 88+76s+16s^2$, $\varphi^*(b_{8+3s}) = 126+92s+16s^2$, $\varphi^*(b_{9+3s}) = 168+104s+16s^2$

$\varphi^*(b'_{7+3s}) = 84+76s+16s^2$, $\varphi^*(b'_{8+3s}) = 130+92s+16s^2$, $\varphi^*(b'_{9+3s}) = 143+96s+16s^2$

All the 6m+1 edges are labeled and are all distinct. Hence the EDG(A^2_m), m ≥ 2 admits strongly multiplicative labeling.

Example: Difference cordial labeling and strongly multiplicative labeling for EDG of arrow graph

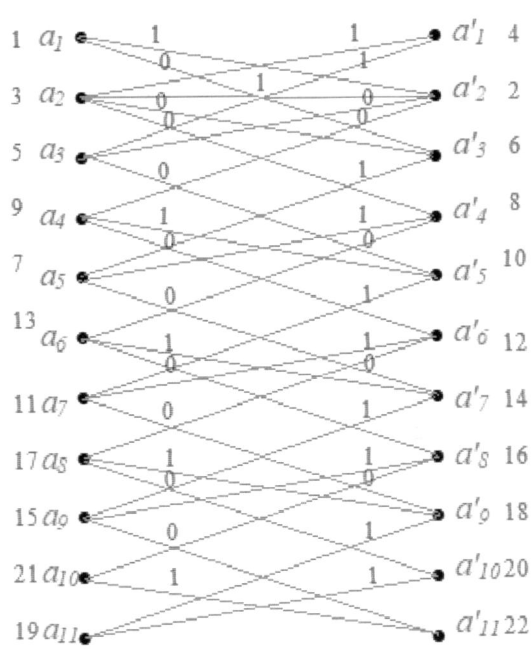

Figure 5 $EDG\left(A_5^2\right)$

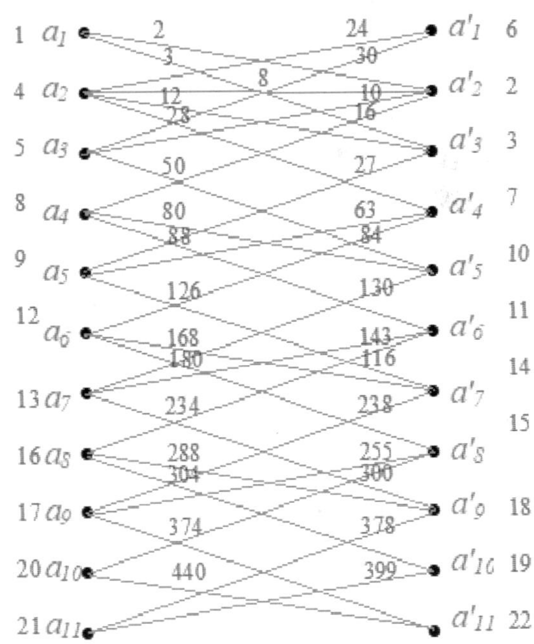

Figure 6 $EDG\left(A_5^2\right)$

Algorithm: 3.3

Procedure [Difference cordial labeling for *EDG (Spl(P$_m$))*, m ≥ 2]

V ← { $a_1, a_2, \dots, a_{2m}, a'_1, a'_2, \dots, a'_{2m}$}
E ← {$b_1, b_2, \dots, b_{3m-3}, b_{3m-2}, b'_1, b'_2, \dots, b'_{3m-3}$}
$a_1 \leftarrow$ 4m; $a_2 \leftarrow$ 4m-2;
$a'_1 \leftarrow$ 1; $a'_2 \leftarrow$ 3;
for s = 0 to [(m-2)/2] do
$a_{3+4s} \leftarrow$ 4+4s
$a_{4+4s} \leftarrow$ 2+4s
$a'_{3+4s} \leftarrow$ 4m-3-4s
$a'_{4+4s} \leftarrow$ 4m-1-4s
end for
for s = 0 to [(m-3)/2] do
$a_{5+4s} \leftarrow$ 4m-4-4s
$a_{6+4s} \leftarrow$ 4m-6-4s
$a'_{5+4s} \leftarrow$ 5+4s
$a'_{6+4s} \leftarrow$ 7+4s
end for
end procedure

Theorem 3.3: The EDG(Spl(P$_m$)), m ≥ 2 is difference cordial labeling.

Proof: Let EDG (Spl(P$_m$)) , m ≥ 2 be a extended duplicate graph of splitting path graph. In the sequence of labeling the vertices, define a function φ : V → {1,2,…..4m} as deliver in algorithm 3.3. The vertices a_1, a_2, a'_1 and a'_2 get label '4m' , '4m-2' , '1' and '3' respectively ;
for $0 \le s \le [(m-2)/3]$, the vertices a_{3+4s} get label '4+4s' ; the vertices a_{4+4s} get label '2+4s' ; the vertices a'_{3+4s} get label '4m-3-4s' ; the vertices a'_{4+4s} get label '4m-1-4s' ;
for $0 \le s \le [(m-3)/3]$, the vertices a_{5+4s} get label '4m-4-4s' ; the vertices a_{6+4s} get label '4m-6-4s' ; the vertices a'_{5+4s} get label '5+4s' and the vertices a'_{6+4s} get label '7+4s' ; Thus the total number of 4m vertices is labeled.
The induced function φ *: E → {0,1} such that
$\varphi^*(a_i a_j) = |\varphi(a_i) - \varphi(a_j)|$, where $a_i, a_j \in V$
The induced function yields the label '0' for the edges b_1, b'_1, b_{3m-2} ; the label '1' for the edges b_2, b_3, $b'_2$and b'_3 ;
for $0 \le s \le [(m-3)/3]$, the edges b_{4+6s} get label '1' ; the edges b_{5+6s} get label '0'; the edges b_{6+6s} get label '0'; the edges b'_{4+6s} get label '1' ; the edges b'_{5+6s} get label '0' ; the edges b'_{6+6s} get label '0' ;
for $0 \le s \le [(m-4)/3]$,the edges b_{7+6s} get label '0' ; the edges b_{8+6s} get label '1'; the edges b_{9+6s} get label '1'; the edges b'_{7+6s} get label '0' ; the edges b'_{8+6s} get label '1' ; the edges b'_{9+6s} get label '1' ;
Thus in both the cases all the 6m-5 edges are labeled. Hence the EDG (Spl(P$_m$)), m ≥ 2 , is difference cordial.

Algorithm: 3.4

Procedure [Strongly multiplicative labeling for *EDG (Spl(P$_m$))*, m ≥ 2]

V ← { $a_1, a_2, \dots, a_{2m}, a'_1, a'_2, \dots, a'_{2m}$}
E ← {$b_1, b_2, \dots, b_{3m-3}, b_{3m-2}, b'_1, b'_2, \dots, b'_{3m-3}$}
$a_1 \leftarrow$ 1; $a_2 \leftarrow$ 4;
$a'_1 \leftarrow$ 3; $a'_2 \leftarrow$ 2;
for s = 0 to m-2 do
$a_{3+2s} \leftarrow$ 7+4s

$a_{4+2s} \leftarrow 6+4s$
$a'_{3+2s} \leftarrow 5+4s$
$a'_{4+2s} \leftarrow 8+4s$

end for

end procedure

Theorem 3.4: The EDG Spl(P_m), $m \geq 2$ is strongly multiplicative labeling.

Proof: Let EDG(Spl(P_m)) , $m \geq 2$ be a extended duplicate graph of splitting path graph. In the sequence of labeling the vertices, define a function $\varphi : V \rightarrow \{1,2,\ldots,4m\}$ as deliver in algorithm 3.4.

$\varphi(a_1) = 1,\ \varphi(a_2) = 4,\ \varphi(a'_1) = 4,\ \varphi(a'_2) = 2$

for $0 \leq s \leq m-2$; $\varphi(a_{3+2s}) = 7+4s$, $\varphi(a_{4+2s}) = 6+4s$, $\varphi(a'_{3+2s}) = 5+4s$, $\varphi(a'_{4+2s}) = 8+4s$

Thus the total number of 4m vertices is labeled.

The edge function is defined by $\varphi^*(a_i a_j) = \varphi(a_i)\ \varphi(a_j)$.

The edge functions are as follows:

$\varphi^*(b_1) = 5$, $\varphi^*(b_2) = 8$, $\varphi^*(b_3) = 15$, $\varphi^*(b'_1) = 28$, $\varphi^*(b'_2) = 24$, $\varphi^*(b'_3) = 14$, $\varphi^*(b_{3m+1}) = 6$;

For $0 \leq s \leq m-3$;

$\varphi^*(b_{4+3s}) = 63+64s+16s^2$, $\varphi^*(b_{5+3s}) = 84+76s+16s^2$, $\varphi^*(b_{6+3s}) = 54+60s+16s^2$

$\varphi^*(b'_{4+3s}) = 55+64s+16s^2$, $\varphi^*(b'_{5+3s}) = 50+60s+16s^2$, $\varphi^*(b'_{6+3s}) = 88+76s+16s^2$

All the 6m-5 edges are labeled and are different.

Thus EDG (Spl(P_m)), $m \geq 2$ admits strongly multiplicative labeling.

Example: Difference cordial labeling and strongly multiplicative labeling for the EDG of splitting graph of path

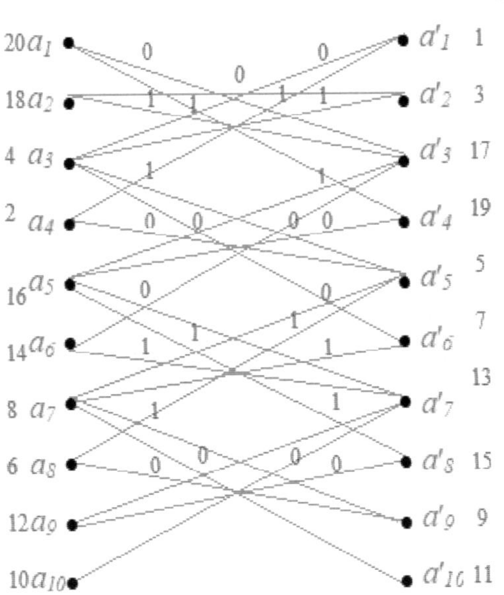

Figure 7 ***EDG(Spl(P₅))***

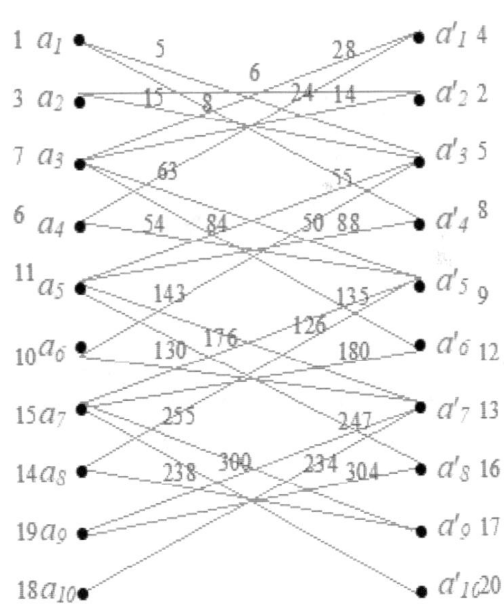

Figure 8 ***EDG(Spl(P₅))***

4. Conclusion

Here we investigated algorithms and proved that the EDG(A^2_m),m ≥ 2 and EDG(Spl(P_m)), m ≥2 admits difference cordial labeling and strongly multiplicative labeling.

References

[1] Anitha K, Selvam B and Thirusangu K 2017 Some graceful labeling for the extended duplicate graph of kite graph *Global Journal of Pure and Applied Mathematics* **13(**1) pp 112–115.

[2] Avudainayaki R, Selvam B and Ulaganathan P P (2018) E-cordial and product e-cordial labeling for the extended duplicate graph of splitting graph of path *Journal of Physics: Conference Series*.

[3] Avudainayaki R, Selvam B and Ulaganathan P P (2017) 3-Equitable and even sum cordial labeling for the extended duplicate graph of arrow graph *Global Journal of Pure and Applied Mathematics* **13(**1) pp 108–111.

[4] Beineke L W and Hegde S M (2001) Strongly Multiplicative graphs *Discuss Math. Graph Theory* **21** pp 63–75.

[5] Indira P, Selvam B and Thirusangu K (2017) Product cordial and prime cordial labeling for the extended duplicate graph of quadrilateral snake graph *Global Journal of Pure and Applied Mathematics* **13**(1) pp 162–165.

[6] Indira P, Selvam B and Thirusangu K (2018) Z_3 -vertex magic total labeling and Z_3 -edge magic total labeling for the extended duplicate graph of quadrilateral snake *Journal of Physics: Conference Series*.

[7] Ponraj R, Sathish Narayanan S and Kala R (2013) Difference cordial labeling graphs *Global J. Math. Sciences: Theory and Practical* **3** pp 192–201.

[8] Sampathkumar E and Walikar H B (2015) On splitting graph of a graph *J. Karnatak Univ. Sci.* **25** and **26** (Combined) (1980-81), 13-16,19, # DS6 (2015) pp 189–198.

[9] Sampathkumar E (1973) On duplicate graphs *Journal of the Indian Math. Soc.* **37** pp 285–293.

[10] Selvam B, Anitha K and Thirusangu K (2016) Cordial and Product Cordial Labeling for the Extended Duplicate Graph of Kite Graph *International Journal of Mathematics And its Applications* **4**(3–B) pp 61–68.

[11] Thirusangu K, Selvam B and Ulaganathan P P (2010) Cordial labeling in extended duplicate twig graphs *International Journal of computer, mathematical sciences and applications* **4**(3-4) pp 319–328.

[12] Vijaya kumar P, Ulaganathan P P and Thirusangu K (2017) Various Harmonious Labeling in Some Duplicate Graphs *International Journal of Pure and Applied Mathematics* **117**(5) pp 145–161.

[13] Vijaya kumar P, Ulaganathan P P and Thirusangu K (2017) Mean Labeling in Some Duplicate Graphs *International Journal of Pure and Applied Mathematics* **117**(11) pp 201–210.

[14] Vijaya kumar P, Ulaganathan P P and Thirusangu K (2018) V-Cordial Labelling in Some Duplicate Graphs *International Journal of Mathematics Trends and Technology (IJMTT)* pp 150–155.

3-Total edge sum cordial and Integer edge cordial labeling for the extended duplicate graph of triangular snake

P Indira[1], B Selvam[2], KThirusangu[3]

[1] Research Scholar, S.I.V.E.T College, University of madras, Chennai, India.
[1] Department of Mathematics, St.Thomas College of Arts and Science,Chennai,India.
[2,3] Department of Mathematics, S.I.V.E.T. College, Gowrivakkam, Chennai-600 073, India.
Email: indu-sel@yahoo.in[1], kbmselvam@gmail.com[2], kthirusangu@gmail.com[3]

indu-sel@yahoo.in

Abstract. In this paper, we have investigate the extended duplicate graph of triangular snake graph is 3-total edge sum cordial labeling and Integer edge cordial labeling.

Keywords : Triangular snake Graph, 3-Total edge sum cordial , Integer edge cordial.

AMS Subject Classification : 05C78.

1. Introduction.

The notion of 3-total edge sum cordial labeling was introduced by Tenguria Abha and Verma Rinku [4]. The idea of Integer edge cordial graph was orginated by P.Maya and T.Nicholas[5] and they have proved that some standard graphs such as path, cycle, wheel, helm, closed helm, star graph $K_{1,n}$ are integer edge cordial . E.Sampthkumar [1] introduced the concept of duplicate graph. Thirusangu et al.[2] have introduced the notion of extended duplicate graph.

2. Preliminaries

The following definitions are useful for understanding its results of this research paper.

Definition :2.1 Triangular snake

A triangular snake TS_m is obtained from a path $u_1, u_2, \ldots, u_{m+1}$ by connecting u_i and u_{i+1} to a new vertex v_i , for $1 \le i \le m$, where 'm' is the number of edges of the path.

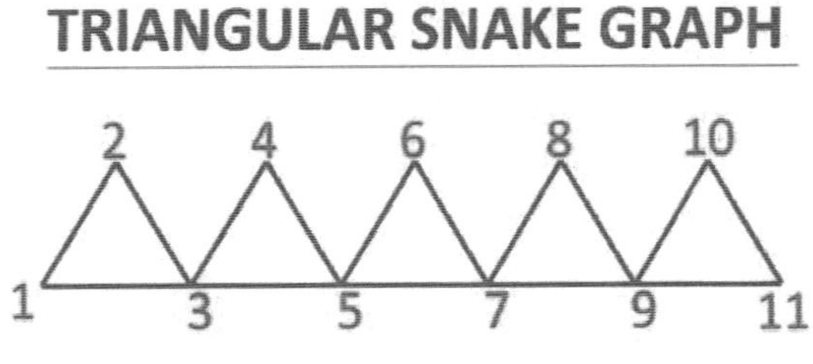

Definition : 2.2 Duplicate graph
A Simple graph G with vertex set V and edge set E. The duplicate graph of G is DG = (V_1, E_1) where the vertex set V_1=V∪V′ and V∩V′ = ϕ and h : V →V′ is bijective . The edge set E_1 of DG is defined as the edge ab∈E iff both edges ab' and $a'b$ are in E_1.
Definition : 2.3 Extended duplicate graph of triangular snake
Let DG = (V_1, E_1) be a duplicate graph of the triangular snake graph G(V, E). Extended duplicate graph of triangular snake is attained by accumulating the edge $v_2 v_2'$ to the duplicate graph. This is indicated by EDG (TS_m). And it has 4m+2 vertices and 6m+1 edges, 'm' is the number of edges of the path.
Definition : 2.4 3-Total edge sum cordial labeling
For Graph G the edge labeling function is defined as h : E(G) → {0,1,2} and induced vertex labeling h*: V(G) →{0,1,2} is given as if $e_1,e_2,e_3\ldots e_n$ are the edge incident on vertex v then $h^*(v) = h(e_1) +_3 h(e_2) +_3 h(e_3) +_3 \ldots +_3 h(e_n)$. Then the map h is called 3-total edge sum cordial labeling of a graph G if $|h(i) - h(j)| \leq 1$; i , j ∈ {0,1,2} where h(x) denotes the total no of vertices and edges labeled with x={0,1,2}.
Definition :2.5 Integer edge cordial labeling
A Simple and connected graph G with p vertices and q edges. Let
h: E→ $\left[\left(-\frac{q}{2}\right)\ldots\left(\frac{q}{2}\right)\right]^*$ or $\left[-\left\lfloor\frac{q}{2}\right\rfloor\ldots\left\lfloor\frac{q}{2}\right\rfloor\right]$ as q is even or odd, be an injective map which induces a vertex labeling h* such that $h^*(v_i) = \begin{cases} 0, & if\ \Sigma h(ei) \leq 0 \\ 1, & if\ \Sigma h(ei) > 0 \end{cases}$
Let $v_h(i)$ = number of vertices labeled with i where i = 0 or 1. h* is said to be integer edge cordial if $|v_{h^*}(0) - v_{h^*}(1)| \leq 1$. A graph is said to be integer edge cordial graph if it admits integer edge cordial labeling.

3. Main results

3.1. 3-Total edge sum cordial labeling (ts_m) , m≥ 1
Here, an algorithm presence itself and proves the existence of 3-total Edge sum cordial labeling for the EDG(TS_m).
Algorithm-3.1
Procedure - [3-total edge sum cordial labeling for EDG(TS_m) , m ≥ 1]

V ← $\{v_1, v_2, \ldots, v_{2m}, v_{2m+1}, v_1', v_2', \ldots, v_{2m}', v_{2m+1}'\}$
E ← $\{e_1, e_2, \ldots, e_{3m}, e_{3m+1}, e_1', e_2', \ldots, e_{3m}'\}$
e_1← 0, e_2← 2, e_3←'2' , e'_1←'1, e'_2←1, e'_3←0, e_{3m+1}←0
for i=0 to [(m-2)/3] do
  e'_{4+9i} ← 1
  e_{4+9i} ←1
  e'_{5+9i} ← 2
  e'_{6+9i} ←0
end for
for i=0 to [(m-2)/3] do
  for j=0 to 1 do
    e_{5+j+9i} ← 2
  end for.
end for
for i=0 to [(m-3)/3] do
  e_{8+9i} ← 1
  e'_{9+9i} ←1

```
end for
for i=0 to [(m-3)/3] do
    for j=0 to 1 do
            e7+2j+9i ← 2
            e'7+j+9i ←0
    end for
end for
for i= 0 to [(m-4)/3] do
            e10+9i ← 1
            e11+9i ←2
            e12+9i ← 0
            e'12+9i ←1
end for
for i=0 to [(m-4)/3] do
    for j=0 to 1 do
            e'10+j+9 ← 0
    end for
end for
end procedure
```

Theorem 3.1 : The EDG(TSm), m≥1 admits 3-total edge sum cordial labeling.

Proof: Let TS_m be the Triangular snake graph and $EDG(TS_m)$ be the extended duplicate graph of triangular snake.

Define the set of vertices and edges are

$$V(G) \leftarrow \{v_1, v_2, \ldots, v_{2m}, v_{2m+1}, v_1', v_2', \ldots, v_{2m}', v_{2m+1}'\}$$

$$E(G) \leftarrow \{e_1, e_2, \ldots, e_{3m}, e_{3m+1}, e_1', e_2', \ldots, e_{3m}'\}$$

Define the function h: E(G) → {0,1,2} as given in algorithm 3.1

The edges e_1, e'_3 and e_{3m+1} receive label 0 ; the edges e'_1 and e'_2 receive label 1 ; the edges e_2 and e_3 receive label 2 .
For i= 0 to (m-2)/3, the edges e'_{4+9i} and e_{4+9i} receive label 1; the edges e'_{5+9i} receive label 2; the edges e'_{6+9i} receive label 0.
For i= 0 to (m-2)/3 and j=0 to 1, the edges e_{5+j+9i} receive label 2.
For i=0 to (m-3)/3 , the edges e_{8+9i} and e'_{9+9i} receive label 1.
For i= 0 to (m-3)/3 and $0 \le j \le 1$, the edges $e_{7+2j+9i}$ receive label 2; the edges e'_{7+j+9i} receive label 0.
For i=0 to (m-4)/3, the edges e_{10+9i} and e'_{12+9i} receive label 1; the edges e_{11+9i} receive label 2; the edges e_{12+9i} receive label 0.
For i = 0 to (m-4)/3 and $0 \le j \le 1$, the edges $e'_{10+j+9i}$ receive label 0.
Thus the entire 6m+1 edges are labeled.

The induced vertex labeling h* : V(G) ← {0,1,2} is defined by

$$h^*(v_i) = h(e_1) +_3 h(e_2) +_3 \ldots +_3 h(e_n), \quad \text{for all vertex } v \in V.$$

The vertices v_1 and v'_1 receive label 2 ; v_2 and v'_2 receive label 0 ; v_3 and v'_3 receive label 1.
For i=0 to (m-2)/3, the vertices v_{4+6i} receive the label 0 and the vertices v_{5+6i} receive label 2.
For i=0 to (m-3)/3, the vertices v_{6+6i} receive the label 2 and the vertices v_{7+6i} receive label 1.
For i=0 to (m-4)/3, the vertices v_{8+6i} receive the label 0 and the vertices v_{9+6i} receive label 1.
For i=0 to (m-2) and $0 \le j \le 1$, the vertices v'_{4+j+6i} receive label 1.
For i=0 to (m-3)/3 and $0 \le j \le 1$, the vertices v'_{6+j+6i} receive label 0.
For i=0 to (m-4)/3 and $0 \le j \le 1$, the vertices v'_{8+j+6i} receive label 2.

Journal of Physics: Conference Series **1377** (2019) 012006 doi:10.1088/1742-6596/1377/1/012006

Thus the entire 4m+2 vertices are labeled.
Hence the extended duplicate graph of triangular snake is 3-total edge sum cordial .
Example 1: 3-Total edge sum cordial labeling graph is shown in figure(1) & figure(2).

3 - TOTAL EDGE SUM CORDIAL LABELING FOR THE GRAPH EDG (TS_5)

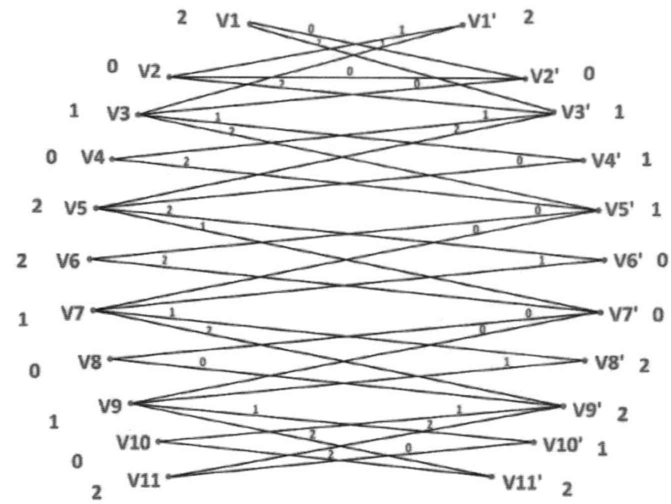

Figure 1. **EDG(TS_5)**

3 - TOTAL EDGE SUM CORDIAL LABELING FOR THE GRAPH EDG (TS_6)

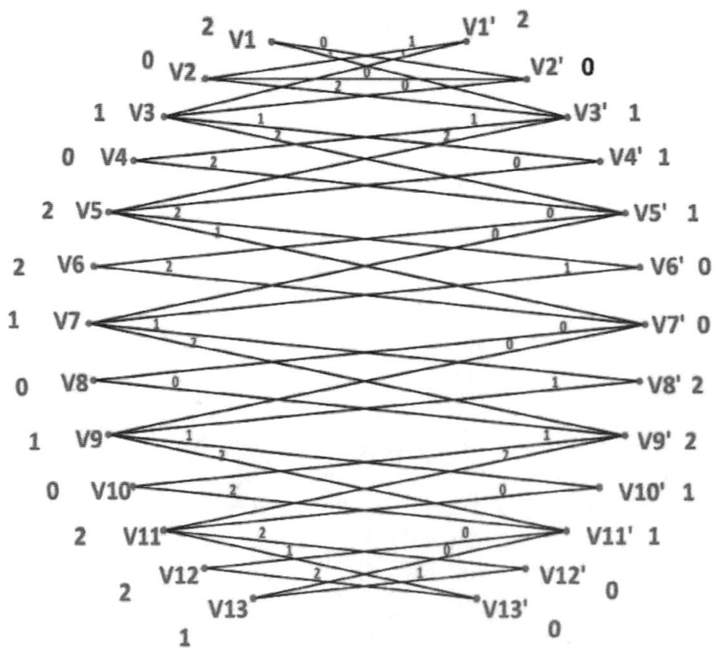

Figure 2. **EDG(TS_6)**

3.2. Integer edge cordial labeling

Algorithm-3.2

Procedure - [Integer edge cordial labeling for EDG(TS_m) , m ≥ 1]

V ← $\{v_1, v_2, \ldots, v_{2m}, v_{2m+1}, v'_1, v'_2, \ldots, v'_{2m}, v'_{2m+1}\}$

E ← $\{e_1, e_2, \ldots, e_{3m}, e_{3m+1}, e'_1, e'_2, \ldots, e'_{3m}\}$

```
        e3m+1 ← 0
for i=0 to [(m-1)/2] do
    for j=0 to 1 do
        e1+j+6i ← -1-j-6i
        e'1+j+6i ← 1+j+6i
    end for.
end for
for i=0 to [(m-2)/2] do
    for j=0 to 1 do
        e4+j+6i ← 4+j+6i
        e'4+j+6i ← -4-j-6i
    end for.
end for
for i=0 to [(m-1)/2] do
        e3+6i ← -3-6i
        e'3+6i ← 3+6i
end for
for i=0 to [(m-2)/2] do
        e6+6i ← 6+6i
        e'6+6i ← -6-6i
end for
end procedure
```

Theorem 3.2 : The extended duplicate graph of triangular snake TS_m, m ≥ 1 admits integer edge cordial labeling.

Proof: Let TS_m be the triangular snake graph and EDG(TS_m) be the extended duplicate graph of triangular snake.

Define the set of vertices and edges are

V(G)← $\{v_1, v_2, \ldots, v_{2m}, v_{2m+1}, v'_1, v'_2, \ldots, v'_{2m}, v'_{2m+1}\}$

E(G)← $\{e_1, e_2, \ldots, e_{3m}, e_{3m+1}, e'_1, e'_2, \ldots, e'_{3m}\}$

Define a function h: E → $\left[\left(-\frac{q}{2}\right) \ldots \left(\frac{q}{2}\right)\right]^*$ or $\left[-\left\lfloor\frac{q}{2}\right\rfloor \ldots \left\lfloor\frac{q}{2}\right\rfloor\right]$ as given in algorithm 3.2.

The edge e_{3m+1} receive label '0'.

For i = 0 to (m-1)/2 and $0 \le j \le 1$, the edges e_{1+j+6i} receive label '-1-j-6i' and the edges e'_{1+j+6i} receive label '1+j+6i'.

For i = 0 to (m-2)/2 and $0 \le j \le 1$, the edges e_{4+j+6i} receive label '4+j+6i and the edges e'_{4+j+6i} receive label '-4-j-6i'.

For i = 0 to (m-1)/2, the edges e_{3+6i} receive label '-3-6i' and the edges e'_{3+6i} receive label '3+6i'.

For i = 0 to (m-2)/2, the edges e_{6+6i} receive label '6+6i' and the edges e'_{6+6i} receive label '-6-6i'.

Thus the entire 6m+1 edges are labeled.

The induced vertex labeling h* such that

$$h^*(v_i) = \begin{cases} 0, & if\ \Sigma h(ei) \le 0 \\ 1, & if\ \Sigma h\ (ei) > 0 \end{cases}$$

Here $v_h*(i)$ is a number of vertices labeled with i where i = 0 or 1.
The vertices v_1 receive label 0 and v'_1 receive label 1.
For i = 0 to (m-1)/2, the vertices v_{2+4i} and v'_{3+4i} receive the label '0'; the vertices v'_{2+4i} and v_{3+4i} receive the label '1'.
For i = 0 to (m-2)/2, the vertices v_{4+4i} and v'_{5+4i} receive the label '1'; the vertices v_{5+4i} and v'_{4+4i} receive the label '0'.
Thus the entire 4m+2 vertices are labeled such that the number of vertices labeled with '1' and '0' are same
Hence the extended duplicate graph of triangular snake graph TSm(m≥1) admits integer edge cordial labeling.
Example 2: **Integer edge cordial labeling graph** is shown in figure(3) & figure(4).

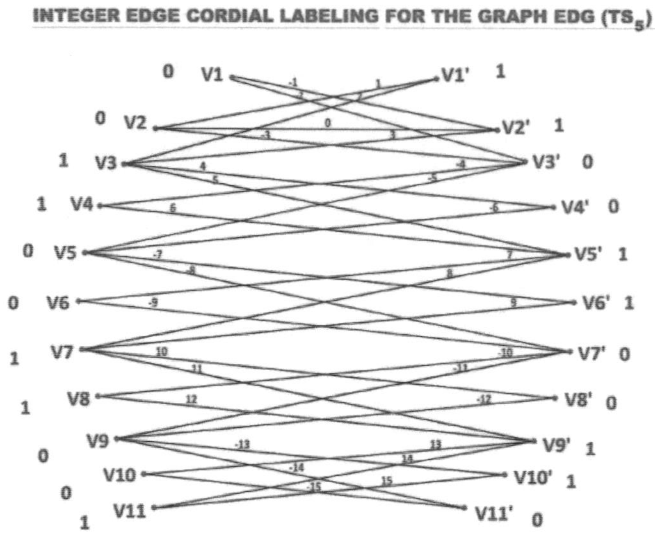

Figure 3. EDG(TS_5)

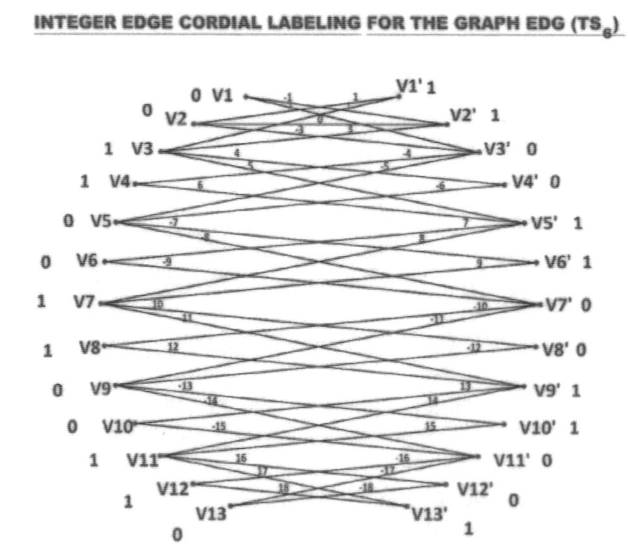

Figure 4. EDG(TS_6)

4. Conclusion

In this research paper ,we have presented and investigated algorithms and their results with 3-Total edge sum cordial labeling, Integer edge cordial labeling for the extended duplicate graph of triangular snake graph (TSm, m≥1).

References

[1] Sampathkumar E 1973 On duplicate graphs *Journal of the Indian Math. Soc.* **37** 285–293
[2] Thirusangu K, Selvam B and Ulaganathan P P 2010 Cordial labelings in extended Duplicate Graph of twig graphs *International Journal of computer, mathematical sciences and applications* **4**(3-4) 319–328
[3] Yilmaz R and cahit I 1997 E-cordial graphs *Ars combin.* **46** 251–266
[4] Tenguria Abha and Verma Rinku 2015 3-total edge sum cordial labeling of graphs, *IJCA, Foundation of computer science*, New York, USA **129**(8) 1–3
[5] Maya P and Nicholas T 2017 Integer edge cordial labeling of graphs *DJ Journal of Engineering Applied Mathematics* **3**(1) 1–12
[6] Pethanachi Selvam S and Lathamaheswari G 2014 Total 3-sum Cordial labeling *IJMA* **5**(5)
[7] Avudainayaki R, Selvam B, Ulaganathan P P 2018 E-Cordial and Product E-cordial labeling for the extended duplicate graph of splitting graph of path *Jounal of physics: conf. series* **1000** 012118
[8] Indira P, Selvam B and Thirusangu K 2018 Z_3 – Vertex magic total labeling and Z_3 – Edge magic total labeling for the extended duplicate graph of quadrilateral snake graph *Journal of Physics: Conference series* **1000**
[9] Selvam B, Thirusangu K and Indira P 2017 3-Equitable and Total 3-sum Cordial labeling for the extended duplicate graph of triangular snake *IJPAM* **113** 110–118

Edge, acyclic and star colorings on human chain graphs

K. Anitha[1], B. Selvam[2], K. Thirusangu[3]

[1]Department of Mathematics, Sri Sai Ram Engineering College, Chennai-44, India
[2,3]Department of Mathematics, S.I.V.E.T. College, Gowrivakkam, Chennai-73, India

anitha.math.sairam.edu.in

Abstract. In this paper, we determine chromatic index, acyclic chromatic and star chromatic number for the human chain and circular human chain graphs.

Index terms- Edge coloring, acyclic coloring, star coloring, human chain graph, circular human chain graph.

1. Introduction

Human chain and circular human chain graphs are simple and undirected graphs. These graphs are introduced by Anitha and Selvam[1,2] in 2019. There are many applications in bus and ring topology by using human chain graphs. The concept of acyclic and star coloring were studied by [3,4,5]. In this paper, we determine edge chromatic, acyclic chromatic and star chromatic number for the human chain and circular human chain graphs by providing algorithms.

2. Preliminaries

Definition-1 A proper vertex coloring of a graph G is to color the vertices of G in such a way that any two adjacent vertices receive distinct colors. A k-vertex coloring of G uses exactly k distinct colors. The chromatic number $\chi(G)$ is the least positive integer k for which G admits k-vertex coloring.

Definition-2 An edge coloring of a graph G is to color the edges of G in such a way that any two adjacent edges receive distinct colors. A k-edge coloring of G uses exactly k distinct colors. The chromatic index $\chi^{/}(G)$ is the least positive integer k for which G admits k-edge coloring.

Definition-3 A proper vertex coloring of a graph G is acyclic if every cycle uses at least three colors. The acyclic chromatic number of G is denoted by $\chi_a(G)$ is the least positive integer k for which G admits acyclic k-coloring.

Definition-4 A star-coloring of a graph G is a proper vertex coloring of G with the condition that no path on four vertices P_4 is 2-colored. The star chromatic number of G is denoted by $\chi_s(G)$ is the least positive integer k for each G admits star k-coloring.

Definition-5 A human chain graph is obtained from a path $u_1, u_2, \ldots, u_{2n+1}$, $n \in N$ by joining a cycle of length m (C_m) and Y-tree (Y_{m+1}), $m \geq 3$ to each u_{2i} for $1 \leq i \leq n$. The vertices of the cycle (C_m) and Y-tree (Y_{m+1}) are $v_1, v_2, \ldots, v_{(m-1)n}$ and $w_1, w_2, \ldots, w_{mn}$ respectively and the vertices of path is $u_1, u_2, \ldots, u_{2n+1}$. The human chain graph is denoted by $HC_{n,m}$.

Example-1 The human chain graph $HC_{2,3}$ is shown in figure (i)

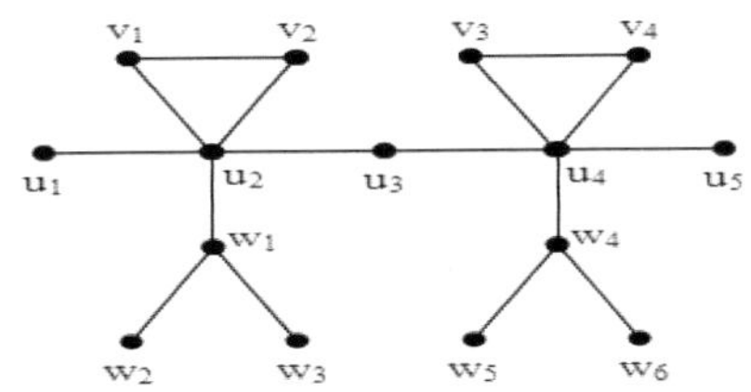

Figure (i)

Definition-6 A circular human chain graph is obtained from a circle(C_{2n}) $u_1,u_2,\ldots,u_{2n},u_1$, $n\geq 3$ by joining a cycle of length m (C_m) and Y-tree (Y_{m+1}), $m\geq3$ to each u_{2i} for $1\leq i\leq n$. The vertices of the cycle (C_m) and Y-tree (Y_{m+1}) are $v_1, v_2, \ldots, v_{(m-1)n}$ and $w_1, w_2, \ldots, w_{mn}$ respectively and the vertices of C_{2n} is $u_1, u_2, \ldots, u_{2n}$. The circular human chain graph is denoted by $CHC_{n,m}$.

Example-2 The circular human chain graph $HC_{5,5}$ is shown in figure (ii)

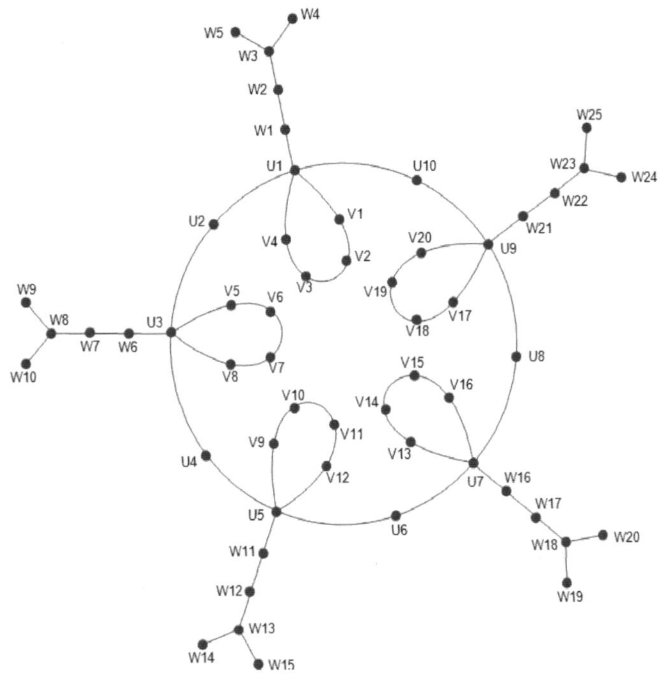

Figure (ii)

3. Main results

Theorem-1 For $n\geq 1$ and $m\geq 3$, the chromatic index of the human chain graph is 5.

Proof. Let $HC_{n,m}$ be the human chain graph with p=2mn+n+1 vertices and q=2mn+2n edges. The vertices and edges of $HC_{1,m}$ are $\{u_1, u_2, u_3, w_1, w_2, \ldots, w_m, v_1, v_2, \ldots, v_{(m-1)n}\}$ and $\{u_1u_2, u_2u_3, u_2v_1, u_2v_{(m-1)}, u_2w_1\}\cup\{v_i,v_{i+1/1\leq i\leq m-2}\}\cup\{w_i,w_{i+1/1\leq i\leq m-2}\}\cup\{w_m,w_{m-1}\}$. In $HC_{1,m}$, $\Delta(u_2)=5$, therefore $\chi'(HC_{1,m})\geq 5$. In order to color the edges of $HC_{1,m}$, $c(u_1u_2)=c_1$, $c(u_2u_3)=c_2$, $c(u_2w_1)=c_3$,

$c(u_2v_1)=c_4$, $c(u_2v_{(m-1)})=c_5$. The coloring of the remaining edges of $HC_{1,m}$ are as follows: If $m\geq3$, $c(w_mw_{m-2})=2$, $c(w_mw_{m-1})=1$; If $m>3$, for $1\leq i\leq\left\lfloor\frac{m-2}{2}\right\rfloor$, $c(v_{2i-1}v_{2i})=c_5$, $c(v_{2i}v_{2i+1})=c_4$; $c(w_{2i}w_{2i+1})=c_4$ for $1\leq i\leq\left\lfloor\frac{m}{3}\right\rfloor$; $c(w_{2i-1}w_{2i})=c_3$ for $1\leq i\leq\left\lfloor\frac{m+1}{3}\right\rfloor$; if m is odd, $c(v_{(m-1)}v_{(m-2)})=c_3$. In the way of minimum edge coloring, $\chi'(HC_{1.m})=5$. Without loss of generality, we can apply the same way of edge coloring using 5 colors to $HC_{n,m}$ for all n. Hence $\chi'(HC_{n.m})=5$.

Example-3 Edge coloring of $HC_{3,6}$ is shown in figure (iii)

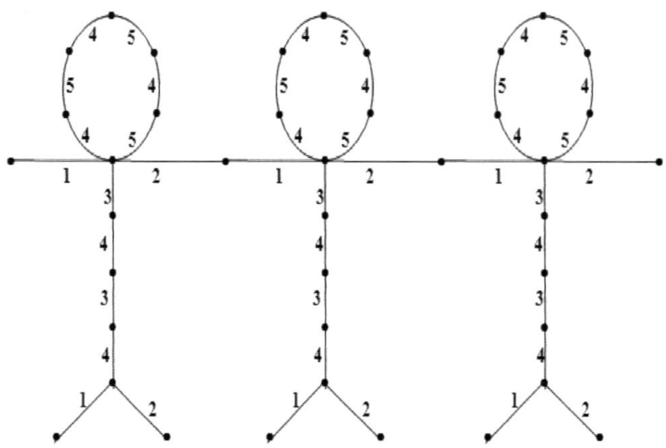

Figure (iii)

Theorem-2 For $m,n\geq 3$, the chromatic index of the circular human chain graph is 5.

Proof The circular human chain graph $CHC_{n,m}$ is obtained from the human chain graph by deleting the vertex u_{2n+1} and adding u_1 and u_{2n} by an edge. The vertices and edges of $CHC_{n,m}$ are $V(CHC_{n,m})=\{V(HC_{n,m})-u_{2n+1}\}$ and $E(CHC_{n,m})=\{E(HC_{n,m})-u_{2n}u_{2n+1}\}\cup\{u_{2n}u_1\}$. The edges of $CHC_{n,m}$ are colored as given in theorem 1 and for the remaining edge, $c(u_1u_{2n})=c_2$. In the way of minimum coloring used only 5 colors to color all the edges of $CHC_{n,m}$. Hence $\chi'(CHC_{n.m})=5$.

Example-4 Edge coloring of $HC_{3,6}$ is shown in figure (iv)

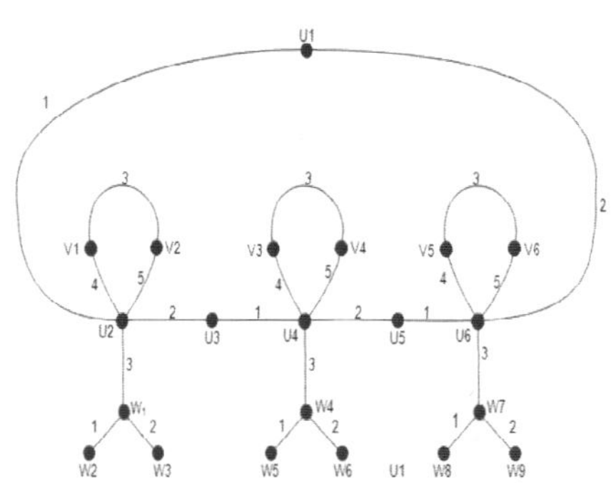

Figure (iv)

Theorem-3 For n≥ 1 and m≥ 3, $\chi_a(HC_{n,m})$ =3.

Proof. Let $HC_{n,m}$ be the human chain graph with n≥ 1 and m≥ 3. From the structure of $HC_{n,m}$, the vertices $\{u_1, u_2, \ldots, u_{2n+1}\} \cup \{w_1, w_2, \ldots, w_{mn}\}$ is acyclic and clearly the acyclic chromatic is 2. The length of each cycle in $HC_{n,m}$ is m and they do not intersect each other. By the definition of acyclic coloring, cycle is 3-colorable.

The 2mn+2n+1 vertices are colored with 3 colors as given in the following way. $c(u_1)=2$; For $1\le i\le n$ $c(u_{2i+1})=2$, $c(u_{2i})=1$; For $1\le i\le n$ $1\le j\le\left\lfloor\frac{m}{2}\right\rfloor$ $c(w_{mi-m+2j-1})=2$, $c(v_{(m-1)i-m+2j})=2$; For $1\le i\le n$ $1\le j\le\left\lfloor\frac{m-1}{2}\right\rfloor$ $c(w_{mi-m+2j})=3$, $c(v_{(m-1)i-m+2j+1})=3$; If m is even $c(w_{mi})=2$ for $1\le i\le n$ and m is odd $c(w_{mi})=3$ for $1\le i\le n$. Therefore thr human chain graph $HC_{n,m}$ contains no bi-colored cycle. Hence $\chi_a(HC_{n,m})$ =3.

Example-5 Acyclic coloring of $HC_{3,6}$ is shown in figure (v)

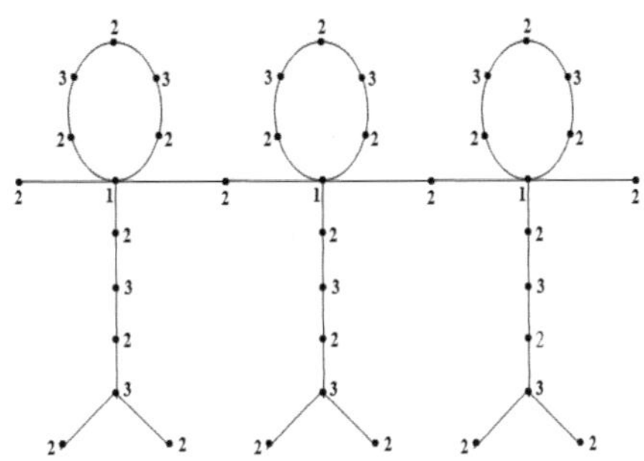

Figure (v)

Theorem-4 For m, n≥ 3, $\chi_a(CHC_{n,m})$ =3

Proof. The circular human chain graph $CHC_{n,m}$ is obtained from the human chain graph by deleting the vertex u_{2n+1} and adding the vertices u_1 and u_{2n} by an edge. In $CHC_{n,m}$, the vertices $u_1, u_2, \ldots, u_{2n}, u_1$ form a cycle of even length for n≥3 and $\{u_{2i} / 1\le i\le n\} \cup \{v_1, v_2, \ldots, v_{(m-1)n}\}$ be the vertices of cycle of length m. Clearly the acyclic chromatic number of cycle with any length is 3. For m,n≥3, the vertices $\{w_1, w_2, \ldots, w_{mn}\}$ is acyclic, therefore it is 2-acyclic colorable. The vertices $\{w_1, w_2, \ldots, w_{mn}\} \cup \{v_1, v_2, \ldots, v_{(m-1)n}\} \cup \{u_2, u_3, \ldots, u_{2n}\}$ are colored as given in theorem 3 and the color of u_1 is 3. In the way of acyclic coloring, $CHC_{n,m}$ contains no bi-colored cycle. Hence $\chi_a(CHC_{n,m})$ =3.

Example-6 Acyclic coloring of $CHC_{3,3}$is shown in figure (vi)

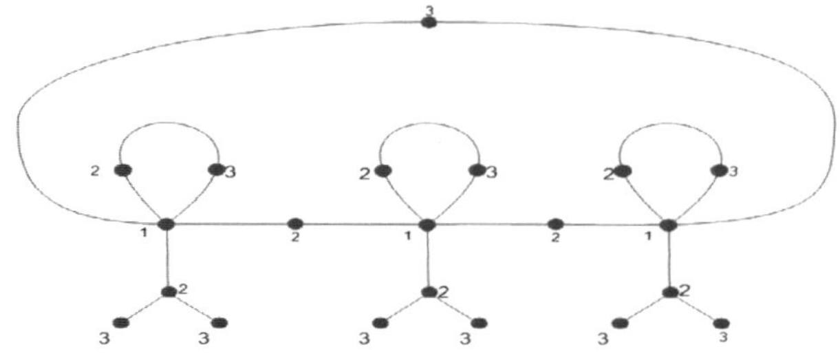

Figure (vi)

Theorem-5 For n≥ 1 and m≥ 3, $\chi_s(HC_{n,m}) = \begin{cases} 3 & if m \neq 5 \\ 4 & if m = 5 \end{cases}$

Proof. Let $HC_{n,m}$ be the human chain graph with n≥ 1 and m≥ 3. From the structure of $HC_{n,m}$, the vertices $\{u_1, u_2, \ldots, u_{2n+1}\} \cup \{w_1, w_2, \ldots, w_{mn}\}$ is acyclic, clearly the star chromatic is 3. In $HC_{n,m}$, the length of each cycle is m and do not intersect each other. By the definition of star coloring, cycle of length 5 is 4-colorable and cycle length of m (m≠5) is 3 colorable. The 2mn+2n+1 vertices are colored with 3 colors for m≠5 and 4 colors for m=5. Therefore the human chain graph $HC_{n,m}$ contains no path with bi-colored. Hence $\chi_s(HC_{n,m}) = \begin{cases} 3 & if m \neq 5 \\ 4 & if m = 5 \end{cases}$

Example-7 Star coloring of $HC_{3,6}$ is shown in figure (vii)

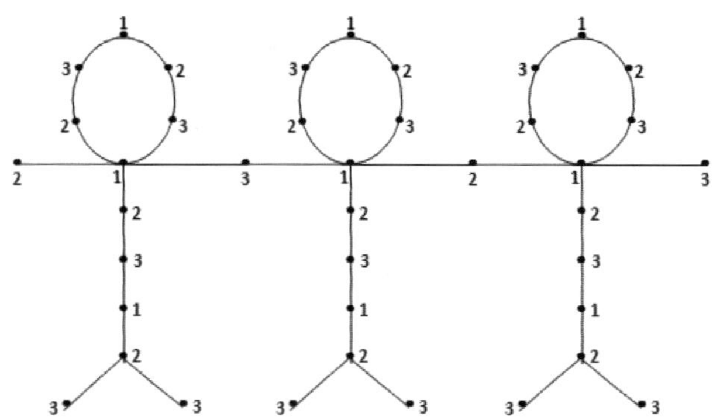

Figure (vii)

Theorem-6 For m,n≥ 3, $\chi_s((CHC_{n,m})) = \begin{cases} 3 & if m \neq 5 \\ 4 & if m = 5 \end{cases}$

Proof. Let $CHC_{n,m}$ be the circular human chain graph with m,n≥ 3. From the structure of $CHC_{n,m}$, the vertices $\{w_1 \cup w_2 \cup \ldots, \cup w_{mn}\}$ is acyclic, clearly the star chromatic number is 3.The vertices $\{u_1 \cup u_2 \cup \ldots, \cup u_{2n} \cup u_1\}$is a cycle of length greater than 5 for n≥ 3 , therefore it is also 3-colorable.

The remaining vertices $\{u_{2i}, 1 \le i \le n\} \cup \{v_1, v_2, \ldots, v_{(m-1)n}\}$ form the cycle of length m and do not intersect each other. By the definition of star coloring, cycle of length 5 is 4-colorable and cycle length of m (m≠5) is 3 colorable. Therefore the circular human graph $CHC_{n,m}$ contains no path with bi-colored.

Hence $\chi_s(CHC_{n,m}) = \begin{cases} 3 & if m \neq 5 \\ 4 & if m = 5 \end{cases}$

Example-8 Star coloring of $CH_{5,5}$ is shown in figure (viii)

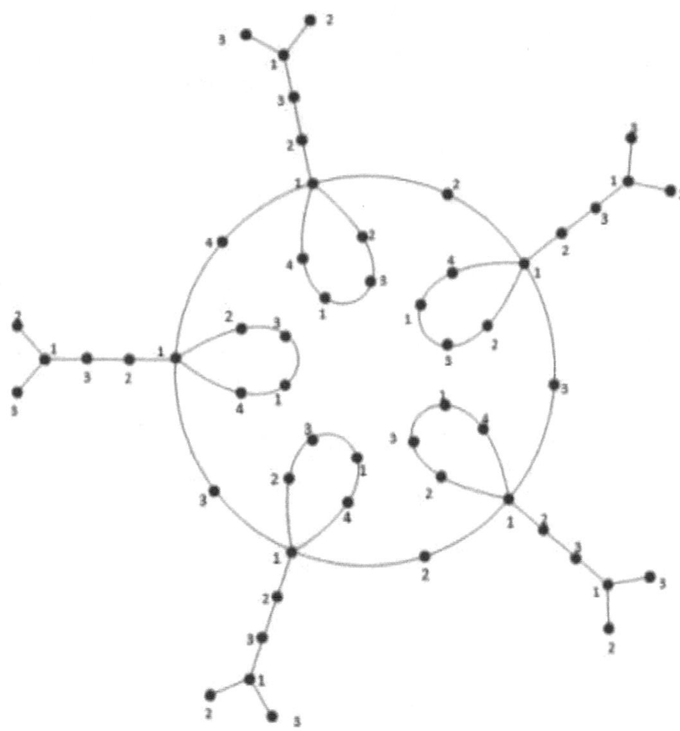

Figure (viii)

4. Conclusion

We have determined edge chromatic, acyclic chromatic and star chromatic number of human chain and circular human chain graphs.

References

[1] Anitha K and Selvam B (2018) Human chain graph *International Journal of Engineering, Science and Mathematics* **7**(8) pp 41–49.

[2] Anitha K and Selvam B (2018) Magicness of circular Human chain graph *International Journal of Scientific Research in Mathematical and Statistical Sciences* **5**(6) pp 191–195.

[3] Grunbaum B (1973) Acyclic colorings of planar graphs *Israel J. Math.* **14**(3) pp 390–412.

[4] Jethruth Emelda Mary L and Ameenal Bibi K (2017) A Study on star chromatic number of some special classes of graphs *Global Journal of Pure and Applied Mathematics* **13**(9) pp 4569–4593.

[5] Ramya N and Muthukumar (2017) On star and acyclic coloring of graphs *International Journal of Pure and Applied Mathematics* **116**(16) pp 467–470.

New class of weak open sets via R*-interior and a decomposition of continuity

R. Manoharan & L. Prasanna Venkatesh

Department of Mathematics,

Sathyabama Institute of Science and Technology, Chennai.

mano_rl@yahoo.co.in

Abstract. In this paper , some new class of sets R^*_I-S(X) , R^*_I-P(X)and R^*_I-α(X) which are the collection of all 'R^*_I-semi –open' , 'R^*_I -pre-open' and 'R^*_I-α-open set's are introduced and we analysed how the sets are related with them and examine the relationship the other similar collection of weak open sets in ideal and general topological spaces . Also we obtained an equivalent wayof representing continuous function interms of funcions which are defined using weak open sets in R*-topological spaces.

1. Introduction and Preliminaries

An ideal topological space which is simply called as an ideal space is commonly denoted by (X, τ, $\mathcal{I}$), where τ, $\mathcal{I} \subseteq$ P(X) in which τ is a topology on the non empty set X and I is an ideal [5]. We use τ(X, $\mathcal{I}$) instead of (X, τ, $\mathcal{I}$). In an ideal space, for every subset A of X,, $A^*(\mathcal{I}, \tau)$ contains the points for which every open set containing the points whose intersection with A is not a member of $\mathcal{I}$ [5]. Also, $cl^*(A) = A \cup A^*(\mathcal{I}, \tau)$ satisfies all the axioms of being a closure operator[6] for some topology that is denoted as $\tau^*(\mathcal{I}, \tau)$. The advantage of the topology τ^* is which contains the actual topology τ. Later, Manoharan *et al* [8] introduced R^*-topology for finite sets which is finer thanτ*($\mathcal{I}$ τ). The members are the 'R*-topology' are called 'R*-open sets'. Different authors introduced new class of weak open sets like 'semi-open' , 'pre-open' , 'α-open' , 'β-open' , 'α^*-set'([7],[9],[10],[1] , [3]) and some other sets like 'C-set', 't-set' and 's-se't . These concepts in topological spaces are extended to ideal space ([2],[4]) with replacing the closure operator by cl*.

1.1 *Definition* [8]: Let $A \subseteq \tau(X, \mathcal{I})$. If the condition $A^* \setminus A \in \mathcal{I}$ is satisfied, then A is said to be an "R^*-perfect set".

1.2 Definition [8]: Let A$\subseteq$ τ(X, $\mathcal{I}$). Then the largest of the R^*-open sets which are containing in A is called as R^*-interior of A .

We will use the notation A^{∇} for R^*-interior of a set A and interior of A by A^{o} .Also int*(A) is used for interior of the set A with respect to the topology τ*. As usual $\overline{A}$, denotes closure of A.

As we have three different topologies namely τ, τ^* and "R*-topology" defined on a set , we can compare the topological properties with respect to this different topologies. It is looking like doing the similar work again but for the completeness of the newly developed topology which also finer than the actual topology, we have to extend the concepts which are already defined for the actual topology on a set.

1.3 Lemma: If A $\subseteq$ τ(X, $\mathcal{I}$), then $A^{o} \subseteq int^*(A) \subseteq A^{\nabla}$.

2. "R^*_I-Semi open", "R^*_I- Pre open" and "R^*_I-α-open sets"

Many authors introduced weak open sets in "τ^*-topology" by changing corresponding closure operator. In this section using R^*-interior of a set , we introduce "R^*_I-semi-open", "R^*_I-pre-open" and "R^*_I-α-open sets" and study their relationship with some already existing sets.

2.1 Definition : Let A $\subseteq$ τ(X, $\mathcal{I}$). Then A is

(1) "R^*_I-semi-open" if $A \subseteq \overline{A^{\Delta}}$,

(2) "R^*_I-pre-open" if A $\subseteq (\overline{A})^{\Delta}$

(3) "R^*_I-α-open" if A $\subseteq (\overline{A^{o}})^{\Delta}$

2.2 Remark : Let R^*_I-S(X) (resp. R^*_I-P(X)and R^*_I-α(X)) is used to represent the set of all "R^*_I-semi-open sets" (resp. "R^*_I-pre-open sets" and "R^*_I-α-open sets"). Let $SO^*(X)$ (resp. $PO^*(X)$ and $\alpha O^*(X)$) is used to denote the set of all "semi-I-open sets" (resp. "pre-I-open sets" and "α-I-open sets"). Similarly, SO(X, τ) (resp. PO(X, τ) , αO(X, τ, t(X, τ) and α^* (X, τ)) is to represent the set of all 'semi -open sets' (resp. 'pre- open sets ,' α-open sets', 't-set' and 'α-open set').

As we have different collection of weak open sets with respect to three different topologies , we can compare them like which one contains the other one.

2.3 Proposition: Consider an ideal space τ(X, $\mathcal{I}$). The following three results are true.

(1) SO(X, τ) $\subseteq$ R^*_I-S(X),
(2) PO(X, τ)$\subseteq$ R^*_I-P(X),
(3) αO(X, τ) $\subseteq$ R^*_I-α(X).

2.4 Corolarry : In an ideal space τ(X, $\mathcal{I}$), the following three results hold.

(1) $SO^*(X)$ $\subseteq$ R^*_I-S(X),
(2) $PO_*(X)$, $\subseteq$ R^*_I-P(X),
(3) $\alpha O^*(X)$, $\subseteq$ R^*_I-α(X).

2.5 Example : From this example discussed below we can come for a conclusion that the reverse relation will not true for the Proposition 2.3. Consider τ(Y, $\mathcal{I}$) with Y = $\{c_i / i = 1,2,3,4\}\}$
τ= {∅, Y, {c_1, c_3, c_4}{c_1, c_3},{c_4}} and $\mathcal{I}$={∅, {c_3},{c_4},{c_3, c_4}}

(1) A= {{c_3}, {c_1, c_2}} which is an "R^*_I semi-open" whereas it failed to be a semi-I-open set.
(2) A= {{c_3}, {c_2, c_3},{c_3, c_4},{c_1, c_2, c_3}, {c_2, c_3, c_4}} is R^*_I pre-open but failed to be pre-I-open.

(3) A= $\{c_1, c_2, c_3\}$ which is $R^*_I\alpha$-open but not α-I-open.

Similar examples can be given for all the results in which we used just an inclusion relation alone.

2.6 Theorem : If $\mathcal{J} = \{\varnothing\}$, Then

(1) R^*_I-S(X)= $SO^*(X)$= SO(X,τ),

(2) R^*_I-P(X)= $PO^*(X)$ = PO(X,τ),

(3) R^*_I-α(X))= $\alpha O^*(X)$= αO(X,τ).

2.7 Proposition : Every "$^*_I$-α-open set" is both "R^*_I-pre-open" and "R^*_I-semi-open".

2.8 Definition : Let A⊆ τ(X, $\mathcal{J}$). Then A is said to be a

(1) "R^*_I- t-set" if $(\overline{A})^\Delta = A^o$,

(2) " R^*_I- α^*-set " if $(\overline{A^o})^\Delta = A^o$,

(3) "' R^*_I-s-set" if $\overline{A^\Delta} = A^o$.

2.9 Definition: Let S⊆ τ(X, I). Then S is an

(1) "R^*_I-B-set" if S can be represented as intersection of an open set with an "R^*_I- t-set",

(2) "R^*_I-C-set" if S can be represented as intersection of an open set with an "R^*_I- α^*-set",

(3) "R^*_I-S set" S can be represented as intersection of an open set with if an "R^*_I-s-set".

2.10 Remark : Let R^*_I-t(X) (resp. R^*_I- α^* (X), R^*_I- s (X) , R^*_I- B(X), R^*_I- C(X), R^*_I- S(X)) used to denote the set of all "R^*_I-t-sets" ("R^*_I- α^*sets" , "R^*_I-s-sets", "R^*_I-B-sets", "R^*_I-C-sets" and "R^*_I-S-sets").

2.11 Proposition: Let A⊆ τ(X, $\mathcal{J}$). Then the below mentioned results are true

(1) If A∈ R^*_I-t(X) then A∈ t(X, τ) ,

(2) If A ∈ R^*_I- α^* (X), then A∈ α^* (X, τ),

(3) If A ∈ R^*_I- t(X), then A∈ R^*_I- α^* (X).

2.12 Proposition : Let A⊆ τ(X, $\mathcal{J}$). Then the below mentioned results are true

(1) If A ∈ R^*_I- t(X), then A∈ R^*_I- B(X),

(2) If A ∈ R^*_I- α^* (X), then A∈ R^*_I- c(X),

(3) If A ∈ R^*_I- s(X), then A∈ R^*_I-S(X).

The openness of any set can be equivalently approached by combination of two sets which we newly described in this paper via R^*-interior.

2.13 Proposition : Let A⊆ τ(X, $\mathcal{J}$). Then all the four combination of sets given below are equivalent.

(1) A ∈τ .

(2) A ∈ R^*_I-P(X) and A ∈R^*_I -B(X) .

(3) A ∈ R^*_I - α(X) and A∈ R^*_I -C(X) .

(4) A ∈ R^*_I -(X) and A ∈ R^*_I -S(X) .

3. Decomposition of continuity

Some new type of continuous functions using the sets we discussed in the previous section and also we obtain a new decomposition of continuity.

3.1 Definition : A function f: τ(X, $\mathcal{I}$) → (Y,σ) is said to be R^*_I-'f'-continuous where f∈ { α ,semi, pre,B,C, S} if for every V∈σ, $f^{-1}(V)$ is an R^*_I-'g' set where g∈ { α-open ,semi-open, pre open ,B, C, S}
We will denote the collection of all R^*_I-'f'-continuous functions as R^*_IF-'f' where f∈ { α ,semi, pre,B,C, S}

3.2 Theorem: Let f: τ(X, $\mathcal{I}$) →(Y,σ) . Then all the four combination given are equivalent:
(1) f is continuous.
(2) f∈ R^*_IF-semi and f∈ R^*_IF-B
(3) f∈ R^*_IF-pre and f∈ R^*_IF-C.
(4) f∈ R^*_IF- α and f∈ R^*_IF-S .

References

[1] M.E.Abd El-Monsef, S.N.El-Deep and R.A.Mahmoud," *β-open sets and β-continuous mappings", Bull. Fac. Sci. AssiutUniv.* ,12 (1983), 77-90.

[2] Dontchev J,"*Idealization of Ganster-Reilly decomposition theorems", http://arxiv. Org/abs/ Math.GN/9901017*, 5 Jan. 1999(Internet).

[3] Hatir E, T.Noiri and S.Yuksel," *A decomposition of continuity", Acta. Math.Hungar.* 70(1996), 145-150.

[4] Hatir E and Noiri T," *On decompositions of continuity via idealization", Acta. Math.Hungar.* 96(4)(2002), 341-349.

[5] Jankovic D and Hamlett TR," *New topologies from old via ideals", Amer. Math. Monthly*, 97(4) (1990), 295-310.

[6] Kuratowski K, "*Topology*", Vol. I, Academic press, New York, 1966.

[7] Levine N,"*Semi-open sets and semi-continuity in topological spaces", Amer. Math. Monthly,* 70 (1963), 36-41.

[8] Manoharan R and Thangavelu P, "*Some sets and Topologies in Ideal Topological spaces", Chinese Journal of Mathematics*(2013). http://dx.dio.org/10.1155/2013/973608.

[9] Mashhour AS, Abd. El-Minsef M E and El-deeb SN, " *On pre-continuous and week pre-continuous mapping", Proc. Math. Phys. Soc. Egypt*, 53 (1982), 47-53.

[10] Njastad O, " *On some classes of nearly open sets", Pacific J. Math.*, 15 (1965), 961-970'

[11] Tong J," *On decompositions of continuity in topological space",Acta. Math.Hungar.* 54(1989), 51-55.

A Comparative Study of Two Dimensional Legendre & Chebyshev Wavelet with an Extended case

V. Sumathi[1], S. Hemalatha[2], B. Sripathy[3]

[1]Department of Mathematics, SriSaiRam Engineering College, Chennai-44, Tamil Nadu, India.
[1]Research Scholar, University Of Madras, Chennai.
[2]Department of Mathematics, S.D.N.B. Vaishnav College for Women, Chennai-44.
[3]Department of Mathematics, Sastra University, Thanjavur.

Email ID: sumathi.math@sairam.edu.in

Abstract.

In this paper, we present an extended case of Two Dimensional Legendre & Chebyshev wavelet to solve a system of PDE's and a Comparative Study of these two wavelets. In this article, we construct an operational matrix from two Dimensional Chebyshev and Legendre wavelets which is then used to convert PDE into some algebraic equations and hence we solve by the method of collocation. A new operational matrix was derived and it is utilized to convert PDE with BVP to a set of algebraic equations. Some examples are given to evaluate the effectiveness of the newly found approximation technique and compared with the existing result

Keywords. TDLW, TDCW, Error in TDLW & TDCW.

1. Introduction

Wavelet analysis plays a vital role in decomposing images and sounds into waves of different periods called wavelets. Wavelets are used in localizing vibrations of both sound and image signal in detail. Multiresolution analysis is the heart of wavelet analysis, which is the decomposition of a function of different levels of resolutions . In the field of mathematics, the development of wavelets are relatively very recent compared the other domains wavelets is a simple tool in mathematics with a huge variety of utmost applications. On the other hand, wavelets appeal to engineers and scientists of several backgrounds. We take the wave equation in our example. The wavelets equations is an important second order linear PDE for the observation of waves such as water , Seismic waves, sound or light waves are the examples of mechanical waves[12]. It arises in the field of acoustics, fluid dynamics and electromagnetics. In the literature, the problems based on Vibrating string of musical instruments are studied by d 'Alembert, Bernoulli, L. Euler & J.L Lagrange[1] [2]. We take Klein –Gordan equation in our example which is a relativistic energy- momentum wave equations, its solutions are of quantum scalar or pseudo scalar field. This equation explains all the spinless particles with zero, negative and positive charges .Klein-Gordan equation plays a key role in the quantum field theory. [3] [4] "Wavelets"- Very famous topic of discussions among many scientists of science, Engineering [9] [11]. Wavelets are the new basis for the representations of functions and a few considered it as a new technique to define the analysis of time frequency and the other hand a few thought wavelets was a new mathematical subject. Wavelets acts as a versatile tool with very rich content in the mathematics and high potential for applications.[6]Wavelets are really a special type of oscillatory which has compact support that plays the platform for most of the domains. The Klien-Gordan equation is a relativistic version of the schrodinger equation which describes the behaviour of spinless particles[3] [4] [5]. For full reconciliation of quantum mechanics with relativity quantum field theory is required, in which the K.G. Equations remerges as the equations obeyed by the component of all free quantum field.[7][8] In quantum field then, the solution of the free version of the original equations still plays a vital role. They are needed to build the Hilbert Space and to express quantum field by spanning sets of Hilbert space of wave functions.

We compare our both the new ideas with the exact value for some cases with proper tabulation and hence we conclude our comparative analysis of our method followed by references. We compare the results obtained in the previous two cases as we discussed in the research paper [13], [14].

Two - dimensional Legendre wavelet [9], [10]

Two-dimensional Legendre Wavelet can be explained as a product of one dimensional Legendre Wavelet as given below.

$$\psi^L_{n,m,n',m'}(t,x)=\begin{cases}\psi^L_{n,m}(t)\psi^L_{n',m'}(x)\ ,\ if\ \dfrac{\hat{n}-1}{2^k}\le t<\dfrac{\hat{n}+1}{2^k}\\ \qquad\qquad\qquad\qquad \dfrac{n'-1}{2^{k'}}\le x<\dfrac{n'+1}{2^k},\\ 0\ , \qquad\qquad\qquad Otherwise\end{cases}$$

$$\psi^L_{n,m}(t)\ =\ \sqrt{m+\frac{1}{2}}\ 2^{\frac{k}{2}}\,p_m(2^k t-\hat{n})\ ,$$

$$\psi^L_{n',m'}(x)=\ \sqrt{m'+\frac{1}{2}}\ 2^{\frac{k'}{2}}\,p_{m'}(2^{k'}x-\hat{n})$$

$and\ m=0,1,2,....,M-1,\ m'=0,1,2,.....,M'-1,\ \hat{n}=2n-1,$

$\hat{n}'=2n'-1,\ n=1,2,3,....,2^{k-1},\ n'=1,2,3,...,2^{k'-1},\ in\ which,\ p_m\ and\ p_{m'}$ are Legendre functions of order m and m' defined in the interval [0,1] and the two-dimensional Legendre Wavelets is a orthonormal set in $\Omega=[0,1)X[0,1)$.

Operational matrices [9],[10]

Legendre wavelets operational matrix of differentiation with respect to variable t.

Let $\psi^L(t,x)$ be two-dimensional Legendre wavelet vector defined and its derivative matrix as follows:

$$\frac{\partial}{\partial t}\psi^L(t,x)=D^L_t\psi^L(t,x)$$

$D^L_t=D^L\otimes I$ is the matrix of order $2^{k-1}2^{k'-1}MM'$ Where I is the identity matrix.

Legendre wavelet operational matrix of differentiation with respect to variable x.

Let $\psi^L(t,x)$ be two-dimensional Legendre Wavelet vector defined in Equation then derivative matrix as follows:

$$\frac{\partial}{\partial x}\psi^L(t,x)\ =D^L_x\psi^L(t,x)$$

Where

$D^L_x=I\otimes D^L\ is\ a\,2^{k-1}2^{k'-1}MM'\times 2^{k-1}2^{k'-1}MM'$ matrix , also I is the identity matrix.

Extended case of Two-Dimensional Chebyshev and Legendre wavelet Operational Matrices [9], [10], [11]

Extended case of TDLW:

$ForK=3,\ n=1,\ m=0,1,2$

$$u(x,t) = (f_{1010}\ f_{1011}\ f_{1012}\ f_{1110}\ f_{1111}\ f_{1112}\ f_{1210}\ f_{1211}\ f_{1212}) \begin{pmatrix} \frac{1}{2} \\ \frac{\sqrt{3}}{2}(2x-5) \\ \frac{\sqrt{5}}{2}(12x^2-60x+74) \\ \frac{\sqrt{3}}{2}(2t-5) \\ \frac{3}{2}(2t-5)(2x-5) \\ \frac{\sqrt{15}}{2}(2t-5)(12x^2-60x+74) \\ \frac{\sqrt{5}}{2}(12t^2-60t+74) \\ \frac{\sqrt{15}}{2}(12t^2-60t+74)(2x-5) \\ \frac{5}{2}(12t^2-60t+74)(12x^2-60x+74) \end{pmatrix}$$

Extended case of TDCW

$For K = 3,\ n = 1,\ m = 0,1,2$

$$u(x,t) = (f_{1010}\ f_{1011}\ f_{1012}\ f_{1110}\ f_{1111}\ f_{1112}\ f_{1210}\ f_{1211}\ f_{1212}) \begin{pmatrix} \frac{8}{\pi} \\ \frac{8\sqrt{2}}{\pi}(8x-1) \\ \frac{8\sqrt{2}}{\pi}(128x^2-32x+1) \\ \frac{8\sqrt{2}}{\pi}(8t-1) \\ \frac{16}{\pi}(8x-1)(8t-1) \\ \frac{16}{\pi}(128x^2-32x+1) \\ \frac{8\sqrt{2}}{\pi}(128t^2-32t+1) \\ \frac{16}{\pi}(128t^2-32t+1)(8x-1) \\ \frac{16}{\pi}(128x^2-32x+1)(128t^2-32t+1) \end{pmatrix}$$

Journal of Physics: Conference Series **1377** (2019) 012009 doi:10.1088/1742-6596/1377/1/012009

2. Numerical examples

Example: 1

Consider the 2-Dimensional PDE. [13]

$$\frac{\partial^2 u(x,t)}{\partial x^2} = \frac{\partial^2 u(x,t)}{\partial t^2}$$

$$u(0,t) = \cos t \;\&\; \frac{\partial u(0,t)}{\partial x} = 0$$

$$and\ u(x,0) = \cos x \;\&\; \frac{\partial u(x,0)}{\partial t} = 0$$

The exact solution is $u(x,t) = \cos x \cos t$.

We apply the Two-Dimensional Legendre wavelet (TDLW) and Two-Dimensional Chebyshev wavelet (TDCW) [13] [14] we obtain Nine algebraic equations and solving those equations with the Collocation points and solving them using Matlab, we arrive Nine unknowns. We multiply those nine values with the formulated Operational matrices of TDLW and TDCW and get a polynomial containing x and t. Here we compare both the wavelets

Case-(i)

$K = 0,\ \hat{n} = 3, M = 3, M' = 3$ *then* $u(x,t)$ *is a* 9×1 *vector* $(TDLW)$

$K = 1, n = 1, m = 0,1,2$, *then we get* $u(x,t)$ *is a* 9×1 *vector* $(TDCW)$.

Therefore the results are furnished below.

X	t	Exact Solution	Approximate Solutions		Error in TDLW	Error in TDCW
			TDLW	TDCW		
0.1	0.5	0.87319	0.49678	0.999903	0.37641	-0.126713
0.2	0.5	0.86008	0.49678	0.999903	0.3633	-0.139823
0.3	0.5	0.83838	0.49678	0.999903	0.3416	-0.161523
0.4	0.5	0.80830	0.49678	0.999903	0.31152	-0.191603
0.5	0.5	0.77015	0.49678	0.999903	0.27337	-0.229753
0.6	0.5	0.72430	0.49678	0.999903	0.22752	-0.275603
0.7	0.5	0.67121	0.49678	0.999903	0.17443	-0.328693
0.8	0.5	0.61141	0.49678	0.999903	0.11463	-0.388493
0.9	0.5	0.54551	0.49678	0.999903	0.04873	-0.454393

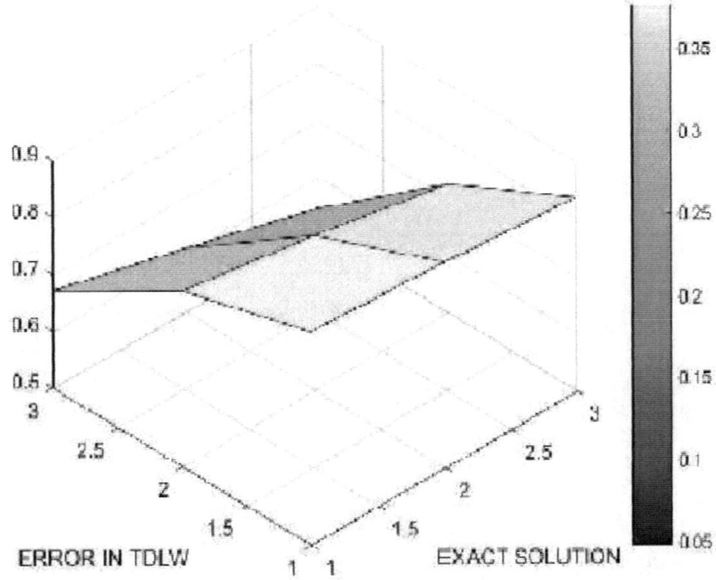
ERROR IN TDLW
EXACT SOLUTION

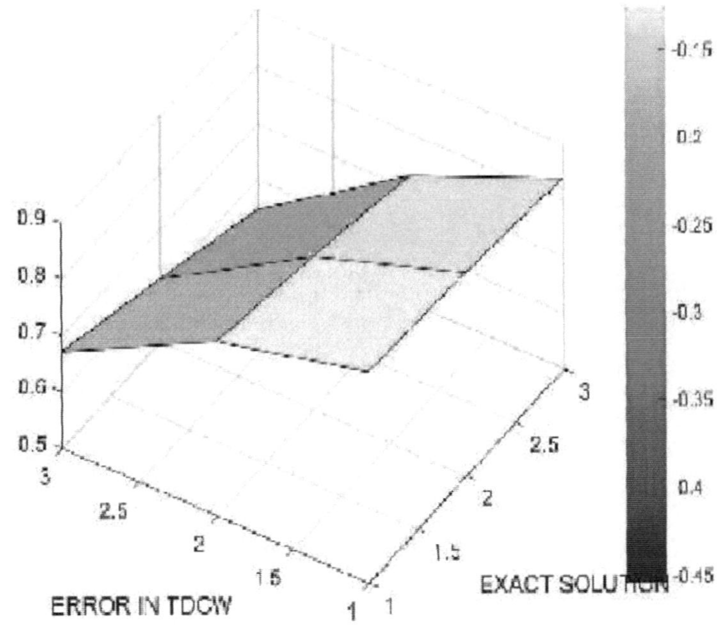
ERROR IN TDCW
EXACT SOLU

Journal of Physics: Conference Series **1377** (2019) 012009 doi:10.1088/1742-6596/1377/1/012009

Case (ii):

$u(x,t)$ *which is of a* 9×1 *matrix for* $K=1, M=3, \hat{n}=3, M'=3$.(*TDLW*)

For $n=1,\ k=2,\ m=0,1,2$ *we obtain a* 9×1 *matrix* (*TDCW*)

Thus we obtain

X	t	Exact Solution	Approximate Solutions		Error in TDLW	Error in TDCW
			TDLW	TDCW		
0.1	0.5	0.87319	0.99929	0.99993	-0.1261	-0.12674
0.2	0.5	0.86008	0.99929	0.99993	-0.13921	-0.13985
0.3	0.5	0.83838	0.99929	0.99993	-0.16091	-0.16155
0.4	0.5	0.80830	0.99929	0.99993	-0.19099	-0.19163
0.5	0.5	0.77015	0.99929	0.99993	-0.22914	-0.22978
0.6	0.5	0.72430	0.99929	0.99993	-0.27499	-0.27563
0.7	0.5	0.67121	0.99929	0.99993	-0.32808	-0.32872
0.8	0.5	0.61141	0.99929	0.99993	-0.38788	-0.38852
0.9	0.5	0.54551	0.99929	0.99993	-0.45378	-0.45442

Case (iii): *For* $n=1, K=3,\ m=0,1,2$ (*TDCW*) *Extended case*

X	t	Exact Solutions	Approximate Solutions	Error in TDCW
			TDCW	
0.1	0.5	0.87319	1.82622	-0.95303
0.2	0.5	0.86008	4.70723	-3.84715
0.3	0.5	0.83838	7.58824	-6.74986
0.4	0.5	0.80830	10.46925	-9.66095
0.5	0.5	0.77015	13.35026	-12.58011
0.6	0.5	0.72430	16.23127	-15.50697
0.7	0.5	0.67121	19.13052	-18.45931
0.8	0.5	0.61141	19.44683	-18.83542
0.9	0.5	0.54551	24.87431	-24.3288

Journal of Physics: Conference Series **1377** (2019) 012009 doi:10.1088/1742-6596/1377/1/012009

Example: 2 Consider the Klein-Gordon Equation [14]

$u(x,0) = 1 + \sin x,\ u_t(x,0) = 0$ *and the boundary*

$u_{tt} - u_{xx} = u;\ 0 < x < 1$ *subject to the initial conditions*

conditions $u(0,t) = 0$, $u(1,t)$ 0 , $t > 0$ *with the exact solution is,* $u(x,t) = \sin x + \cosh t$

Case (i):

X	t	Exact Solution	Approximate Solution		Error in TDLW	Error in TDCW
			TDLW	TDCW		
0.1	0.5	1.22745	-1.3992	1.49014	2.62665	-0.26269
0.2	0.5	1.32629	-1.4906	1.83512	2.81689	-0.50883
0.3	0.5	1.42314	-1.4101	1.95103	2.83324	-0.52789
0.4	0.5	1.51701	-1.3294	1.72664	2.84641	-0.20963
0.5	0.5	1.60705	-1.2488	2.14078	2.85585	-0.53373
0.6	0.5	1.69226	-1.1682	2.55493	2.86046	-0.86267
0.7	0.5	1.77184	-1.0876	2.96907	2.85944	-1.19723
0.8	0.5	1.84498	1.0071	3.38322	0.83788	-1.53824
0.9	0.5	1.91095	-0.9264	3.79736	2.83735	-1.88641

Case (ii):

X	t	Exact Solution	Approximate Solution		Error in TDLW	Error in TDCW
			TDLW	TDCW		
0.1	0.5	1.22745	-1.93540	-0.424084	3.16285	1.651534
0.2	0.5	1.32629	-1.76920	1.174875	3.09549	0.151415
0.3	0.5	1.42314	-1.60292	2.543355	3.026064	-1.120215
0.4	0.5	1.51701	-1.43664	3.681355	2.95365	-2.164345
0.5	0.5	1.60705	-1.27037	4.588870	2.87742	-2.98182
0.6	0.5	1.69226	-1.10409	5.265912	2.796354	-3.573652
0.7	0.5	1.77184	-0.77154	5.712469	2.54338	-3.940629
0.8	0.5	1.84498	-1.33375	5.928540	3.17873	-4.08356
0.9	0.5	1.91095	-0.60526	5.914142	2.516213	-4.003192

Case (iii): (Extended Case)

X	t	Exact Solutions	Approximate Solutions	Error in TDCW
			ETDLW	
0.1	0.5	1.22745	1.982087	-0.754637
0.2	0.5	1.32629	1.968100	-0.641810
0.3	0.5	1.42314	1.695910	-0.272770
0.4	0.5	1.51701	1.940480	-0.423470
0.5	0.5	1.60705	1.926628	-0.319578
0.6	0.5	1.69226	1.912771	-0.220511
0.7	0.5	1.77184	1.898915	-0.127075
0.8	0.5	1.84498	1.885058	-0.040078
0.9	0.5	1.91095	1.871202	0.039748

Journal of Physics: Conference Series **1377** (2019) 012009 doi:10.1088/1742-6596/1377/1/012009

Example: 3

Consider another system of PDE. [13]

$u_t + u_x + u = (x-t)^2$ *with conditions*

$u(x,0) = x^2$; $u(0,t) = t^2$

The Exact solution is $u(x,t) = (x-t)^2$

Case (i):

X	t	Exact Solutions	Approximate Solutions	Error in TDCW
			TDCW	
0.50	0.50	0.0000	0.269080	-0.269080
0.50	0.25	0.0625	0.287809	-0.225309
0.50	0.85	0.1225	0.187264	-0.064764
0.25	0.50	0.0625	0.269082	-0.206582
0.25	0.25	0.0000	0.231630	-0.231630
0.25	0.85	0.3600	0.175451	0.184549
0.85	0.50	0.1225	0.287809	-0.165309
0.85	0.25	0.3600	0.006913	0.353087
0.85	0.85	0.0000	-0.105445	0.105445

Case (ii):

X	t	Exact Solutions	Approximate Solutions	Error in TDCW
			TDCW	
0.1	0.5	0.1600	0.402336	-0.242336
0.2	0.5	0.0900	0.218992	-0.128992
0.3	0.5	0.0400	0.035648	0.004352
0.4	0.5	0.0100	-0.147696	0.157696
0.5	0.5	0.0000	-0.331040	0.331040
0.6	0.5	0.0100	-0.514384	0.524384
0.7	0.5	0.0400	-0.697728	0.737728
0.8	0.5	0.0900	0.881072	-0.791072
0.9	0.5	0.1600	0.081464	0.078536

Case (iii): (Extended case)

X	t	Exact Solutions	Approximate Solutions	Error in TDCW
			TDCW	
0.1250	0.1250	0.00000	2.697400	-2.697400
0.1250	0.2133	0.00779	2.697400	-2.689610
0.1250	0.0366	0.00718	2.697400	-2.690220
0.2133	0.1250	0.00779	0.742600	-0.734810
0.2133	0.2133	0.00000	0.742600	-0.742600
0.2133	0.0366	0.03122	0.742600	-0.711380
0.0366	0.1250	0.00781	0.024750	-0.016940
0.0366	0.2133	0.03122	0.024750	0.006470
0.0366	0.0366	0.00000	0.024750	-0.024750

3. Conclusion
We have the TDCW and TDLW with some examples. However, as we extend the case, we are not getting the accurate value and the refinement is very less. We have noticed in ODE if we increase the scale we will get a better refinement, but it is not the case here in PDE., we have compared the error of our proposed method by comparing it with exact value and gave a graph with Matlab coding. Our examples give the proof of our comparative technique among the TDCW,TDLW and the exact values.

References
[1] Evans L (1998) Partial differential equations American Mathematical society providence.
[2] Linear Wavelets Equations. Eqworld: The Work of Mathematical Equations.
[3] Gordan , Walter (1926) Der Comptone ffekt nachder schrodinger schen Theorie Z.Physik. 40:117 doi: 10.1007/BFO/390840
[4] Greiner, W (2000) Relativistic Quantum Mechanics Wave Equations (3rd. edition) Springer Verlag. ISBN 3-5406-74578.
[5] Hariharan G (2011) Haar Wavelet method for solving the Klein-Gordan and the Sine-Gordan Equqtions International journal of nonlinear Science II(2) pp 180⌋189.
[6] Charles K. Chui (1992) Wavelets analysis and its applications: An introduction to wavelets: San Diego USA.
[7] A Chebyshev series Approximations for Linear second order partial differential equations with complicated conditions.
[8] A New Legendre wavelets decomposition method for solving PDE'S –Malaya Journal of Mathematik 1 (1) (2014) 72-81, Naima Ablaoui –Lahmar, Omar Belhamiti, and Sidi Mohammed Bahri.
[9] Applications of wavelet collocation method for hyperbolic partial differential equations via Matrices –Somveer Singh , Vijay kumar Patel, Vineet Kumar Singh – Applied mathematics and computations 320 (2018) 404-424.
[10] Two-Dimensional wavelets collocation method for Electromagnetic waves in Dielectric Media-Vijay Kumar Patel, Somveer Sing , Vineet kumar Singh – Journal of computational and Applied Mathematics September 29, 2016.
[11] Sripathy B, Vijayaraju P and Hariharan G (2015) Chebyshev wavelet based approximation to some Non-Linear differential equations arising in engineering International Journal of Mathematical Analysis.
[12] Sripathy B and Sumathi V (2016) Wavelets based solution for certain Non-Linear boundary value problems Global Journal of pure and applied mathematics (GJPAM) 12(2).
[13] Sumathi V, Thangavelu K and Sripathy B (2018) Applications of Two dimensional Legendre wavelets collocation method for solving PDE's International Journal of Pure and Applied Mathematics.
[14] Sumathi V, Hemalatha S and Sripathy B (2019) Numerical solutions of PDE using Two Dimensional Chebyshev wavelets collocation method International Journal of Innovative Technology and Exploring Engineering (IJITEE) 8(4).

OTSMO Labeling on Generalized Petersen Graphs $GP(n, 2)$

V. Shendra Shainy[1], **G. Margaret Joan Jebarani**[2] **and V. Balaji**[3]
[1,3] PG and Research Department of Mathematics.
Sacred Heart College, Tirupattur - 635601, India
[2]Department of Mathematics, (Retd Head)
Auxilium College, Vellore, India

E-mail: pulibala70@gmail.com

Abstract. In this paper a discussion of OTSMO Labeling (One Third Sum Modulo 3, introduced by the Researcher) on Generalized Petersen graphs $GP(n, 2)$ is done.
Keywords: OTSMO Labeling, Generalized Petersen Graphs.
2010 AMS Subject classification : 05C78.

1. Introduction
The ways of the mind are too many. Having struck at a new labeling (One Third Sum Modulo 3) and a fruitful discussion on a two star [5], the Researcher continued the work with the labeling OTSMO on Generalized Petersen graphs. The Researcher's findings are presented in this paper.

2. Definitions
Definition 2.1. OTSMO Graph
Let $G(p, q)$ be a simple graph with p vertices and q edges. Let f be a bijective function from $V(G) \to \{1, 2, 3, \cdots p\}$ and f^* be the induced function on $E(G)$ such that $f^*(uv) = \left\lceil \frac{f(u) + f(v)}{3} \right\rceil \equiv s \mod 3$ with $f^*(E(G)) \to \{0, 1, 2\}$ and $|e_f(i) - e_f(j)| \leq 2,\ i, j \in \{0, 1, 2\}$ where $e_f(i)$ denotes the number of edges assuming the value i. A graph with this labeling is called 'One Third Sum Modulo 3' (OTSMO) graph.

Definition 2.2. Generalized Petersen Graph
The Generalized Petersen graphs $P(n, k)$ with $n \geq 3$ and $1 \leq k \leq \left\lfloor \frac{n}{2} \right\rfloor$ are defined to be graphs with $V(P(n, k)) = \{v_i, u_i : 1 \leq i \leq n\}$ and $E(P(n, k)) = \{v_i v_{i+1}, v_i u_i, u_i u_{i+k} : 1 \leq i \leq n,$ subscripts modulo $n\}$. $P(n, k)$ has $2n$ vertices and $3n$ edges.

3. Pre - requisites
As the OTSMO labeling is defined in terms of congruence modulo 3, the positive integers to be used in the $GP(n, 2)$ have to be categorized into 6 sets. They are n odd, $\equiv 0, \equiv 1, \equiv 2$ and n even, $\equiv 0, \equiv 1, \equiv 2$ modulo 3. As $GP(n, 2)$ are regular cubic graphs with three edges passing through all the polygonal vertices and star vertices, without the above categorization it is not possible to proceed. After spending a quality time and search, the Researchers found the need to provide three ways of allotting numbers for $GP(n, 2)$.

Notations
As mentioned in the pre - requisites, the sets of even positive integers congruent to zero, one and two are denoted by E_0, E_1, E_2 and the sets of odd positive integers congruent to zero, one and two are denoted by O_0, O_1, O_2. The edges are classified into i) polygonal edges, ii) star edges, iii) connecting edges denoted by $PE, SE,$ and CE. The last edges of PE, SE, CE are denoted by LE_1, LE_2, LE_3 respectively.

An outline of a counting scheme
As $OTSMO$ labeling depends on the difference in number of edges taking the values $0, 1, 2$, one has to know how many edges take these values respectively. They depend on n, odd or even, $\equiv 0, \equiv 1$ or $\equiv 2 \mod 3$.
Take $n = 3g + h$, $h = 0, 1, 2$ (n may be odd or even). There are three rules suggested. They are denoted by R_1, R_2 and R_3.

According to each rule, the last edge value will differ for PE and CE, following certain patterns. The SE will take values depending on the edges joining the vertices with odd suffixes, odd to even, even to even and even to odd when n is odd; and odd to odd and even to even when n is even.

After alloting the numbers to vertices through different rules, the counting is done particularly, proceeding in the same way as explained above.

4. Discussion and Findings

Description of Rule 1 (R_1)
This rule is applicable on $E_0 \cup E_1 \cup E_2$.
The polygonal vertices are denoted by $v_1, v_2, \cdots, v_n$ and the star vertices by $u_1, u_2, \cdots, u_n$. The $2n$ vertices are divided into sets of 4 vertices in the following manner $(v_1u_1u_2v_2), (v_3u_3u_4v_5), \cdots$. As $2n$ is even, after such a grouping only 2 vertices or no vertices will be left out. They are given the remaining numbers maintaining the movements, that is, the from the outer vertex v_1 to 2 inner vertices u_1, u_2 and to 2 outer vertices v_2 and v_3 and so on. The following diagram explains the movements.

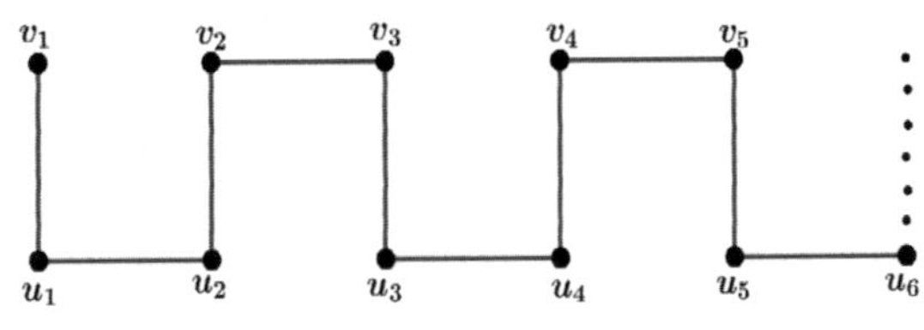

Figure 1.

Theorem 4.1. *The Generalized Petersen graph $GP(n, 2)$ is an $OTSMO$ graph with respect to $OTSMO$ labeling and the allotment of numbers according to R_1 for any n which is even.*

Proof. The rule R_1 defined on any $n \in E_1 \cup E_2 \cup E_3$ makes the $GP(n, 2)$ an $OTSMO$ graph is proved in this theorem.
The $OTSMO$ labeling with respect to R_1 is presented below.

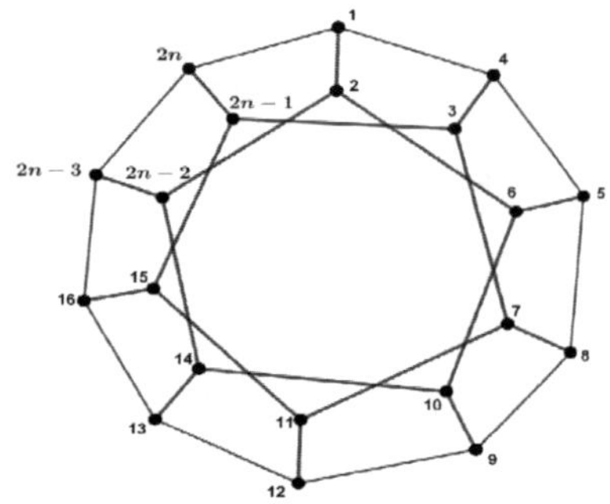

Figure 2. $GP(10, 2)$

Label for the polygonal vertices:

$f(v_{2i+1}) = 4i + 1$ for $i = 0, 1, 2, \cdots \left(\frac{n-2}{2}\right)$;

$f(v_{2j}) = 4j$ for $j = 1, 2, \cdots \frac{n}{2}$.

Label for the star vertices:

$f(u_s) = 2s$ for $s = 1, 3, 5, \cdots (n-1)$;

$f(u_{2t}) = 4t - 1$ for $t = 1, 2, \cdots \frac{n}{2}$.

Now the labeling for polygonal edges is provided.

For edges joining the polygonal vertices with odd to even suffixes:

$$f^*(v_{2i+1}, v_{2j}) = \frac{(4i+1)+4j}{3} \equiv k \mod 3$$
$$\text{for } (i, j) = (0, 1), (1, 2), \cdots, \left(\frac{n-2}{2}, \frac{n}{2}\right);$$

For edges joining the polygonal vertices with even to odd suffixes:

$$f^*(v_{2j}, v_{2i+1}) = \frac{(4i+1)+4j}{3} \equiv k \mod 3 \text{ for } (i, j) = (1, 1), (2, 2), \cdots, \left(\frac{n}{2}, \frac{n}{2}\right);$$

$$LE_1 : f^*(v_n, v_1) = \frac{2n+1}{3} \equiv k \mod 3.$$

Labeling for star edges:

In $GP(n, 2)$, as n is even, the star edges connect the vertices $u_s u_{s+2}$ (s odd) and $u_\ell u_{\ell+2}$ (ℓ even). The edges are given by,

$u_1u_3, \cdots, u_s u_{s+2}, \cdots, u_{n-1}u_n$ and $u_2u_4, u_4u_6, \cdots, u_{2n-2}u_2$.

The edge values are,

$$f^*(u_s, u_{s+2}) = \frac{4s+4}{3} \equiv k \mod 3 \text{ for } s = 1, 3, \cdots, (n-3);$$

$$f^*(u_{n-1}, u_1) = \frac{2(n-1)+2}{3} = \frac{2n}{3} \equiv k \mod 3;$$

$$f^*(u_{2t}, u_{2(t+1)}) = \frac{8t+2}{3} \equiv k \mod 3 \text{ for } t = 1, 2, \cdots, \left(\frac{n-2}{2}\right);$$

$$f^*(u_n, u_2) = \frac{2n+2}{3} \equiv k \mod 3.$$

The labeling for connecting edges

The edges are $v_1u_1, v_2u_2, v_3u_3, \cdots, v_{n-1}u_{n-1}$, that is both suffixes are odd or both are even.

$$f^*(v_{2i+1}, u_s) = \frac{4i+2s+1}{3} \equiv k \mod 3$$
$$\text{for } (i, s) = (0, 1), (1, 3), (2, 5), \cdots, \left(\frac{n-2}{2}, (n-1)\right);$$

$$f^*(v_{2j}, u_{2t}) = \frac{4j+(4t-1)}{3} \equiv k \mod 3 \text{ for } (j, t) = (1, 1), (2, 2), \cdots, \left(\frac{n}{2}, \frac{n}{2}\right);$$

Journal of Physics: Conference Series **1377** (2019) 012010 doi:10.1088/1742-6596/1377/1/012010

The counting of $e_f(0), e_f(1)$ and $e_f(2)$ with respect to R_1
Case 1:
When n is even, $n \equiv 0 \mod 3$.
Here $n = 3g + h,\ h = 0,\ n = 3g$.
There are g repeated triplets of the pattern $2, 0, 1$;
with $LE_1 = \dfrac{2n+1}{3} \equiv 1 \mod 3$.
There are g repeated triplets of the pattern $0, 1, 2$;
with $LE_3 = \dfrac{2n+2n-1}{3} = \dfrac{4n-1}{3} \equiv 2 \mod 3$, in the connected edges.
There are $\dfrac{g}{2}$ repeated triplets of the pattern $2, 1, 0$ and $\dfrac{g}{2}$ repeated triplets of the pattern $1, 0, 2$,
with $LE_2 = \dfrac{(2n-1)+3}{3} = \dfrac{2n+2}{3} \equiv 2 \mod 3$, in the star edges.
These are labeled below:
$PE : 201, 201, 201, \cdots, g$ triplets $\left(LE_1 : \dfrac{2n+1}{3}\right)$;
$SE : 210, 210, 210 \left(\dfrac{g}{2}\text{ triplets}\right), \cdots, 102, \dfrac{g}{2},\ 102, \dfrac{g}{2},\ \left(LE_2 : \dfrac{2n+2}{3}\right)$;
$CE : 012, 012, \cdots, 012, \cdots, g$ triplets, $\left(LE_3 : \dfrac{4n-1}{3}\right)$.
They are tabulated below.

	PE	SE	CE	Total
$e_f(0)$	g	g	g	$3g = n$
$e_f(1)$	g	g	g	$3g = n$
$e_f(2)$	g	g	g	$3g = n$

So, $|e_f(i) - e_f(j)| \le 2$ is true. Hence the result is proved for $n \equiv 0, n$ even.
Case 2: n even, $n \equiv 1 \mod 3$, with respect to R_1, $n = 3g + 1$.
In PE there are g repeated triplets of the pattern $201, 201, \cdots$.
The $LE_1 : \dfrac{2n+1}{3} = \dfrac{6g+2+1}{3} \equiv 0$.
In SE, the first $\dfrac{n-2}{2}$ edges follow the pattern $210, 210, \cdots$.
The last but one edge connecting odd suffix vetrtices is
$\dfrac{2n-2+2}{3} = \dfrac{6g+2}{3} \equiv 2$.
$f^*(u_{n-1}, u_1) = \dfrac{2n-2+2}{3} = \dfrac{2n}{3} \equiv 2 \mod 3$.
The next $\left(\dfrac{n-2}{2}\right)$ edges connecting even suffix vertices follow the pattern $1, 0, 2 \cdots$.
The LE_2 is $\dfrac{2n-1+3}{3} = \dfrac{6g+2-1+3}{3} \equiv 1 \mod 3$.
In CE, there are g repeated triplets of the form $012, 012, \cdots$.
The $LE_3 = \dfrac{2n+(2n-1)}{3} = \dfrac{4n-1}{3} = \dfrac{12g+4-1}{3} \equiv 0 \mod 3$.
The listing and tabulating are done below:
$PE : 201, 201 \cdots, \boxed{0}$
$SE : 210, 210, \cdots, \boxed{2} \cdots, 102, \cdots, \boxed{1}$
$CE : 012, 012, \cdots, \boxed{0}$

	PE	SE	CE	Total
$e_f(0)$	$(g+1)$	$(g-1)$	$(g+1)$	$(3g+1)=n$
$e_f(1)$	g	$(g+1)$	g	$(3g+1)=n$
$e_f(2)$	g	$(g+1)$	g	$(3g+1)=n$

The result stands verified for n even and $n \equiv 1 \mod 3$

Case 3: n **even,** $n \equiv 2 \mod 3$

$n = 3g+2$.

In PE there are g repeated triplets of the form $201, 201, \cdots$

The last two edges are 2, $LE_1 = \frac{2n+1}{3} \equiv \frac{6g+4+1}{3} \equiv 2$

In SE, as in case 2, the first $\frac{n-2}{2}$ edges are of the form $2, 1, 0 \cdots$

The last edge connecting odd suffix vertices is,

$\frac{2n-2+2}{3} = \frac{2n}{3} = \frac{2(3g+2)}{3} = \frac{6g+4}{3} \equiv 1 \mod 3.$

The next $\frac{n-2}{2}$ edges connecting even suffix vertices is of the form 1, 0, 2.

The last two edges are 2, $LE_2 = \frac{(2n+2)}{3} = \frac{6g+6}{3} \equiv 0.$

In CE there are g repeated triplets of the form $012, 012, \cdots$.

The last two edges are 0, $LE_3 = \frac{(4n-1)}{3} = \frac{12g+8-1}{3} \equiv 1.$

The listing and tabulating are done below.

$PE : 201, 201, \cdots ; 2, \boxed{2}$

$SE : 210, 210,$ upto $(n-2)$ edges $\boxed{1}$ 102, $\boxed{0}$

$CE : 012, 012, \cdots ; 0, \boxed{1}$

	PE	SE	CE	Total
$e_f(0)$	g	$(g+1)$	$(g+1)$	$(3g+2)=n$
$e_f(1)$	g	$(g+1)$	$(g+1)$	$(3g+2)=n$
$e_f(2)$	$(g+2)$	g	g	$(3g+2)=n$

The result is verified for n even and $n \equiv 2 \mod 3$.

The OTSMO labeling with R_1 on $GP(n, 2)$, n even for the vertices and the edges are provided above and an example is given below.

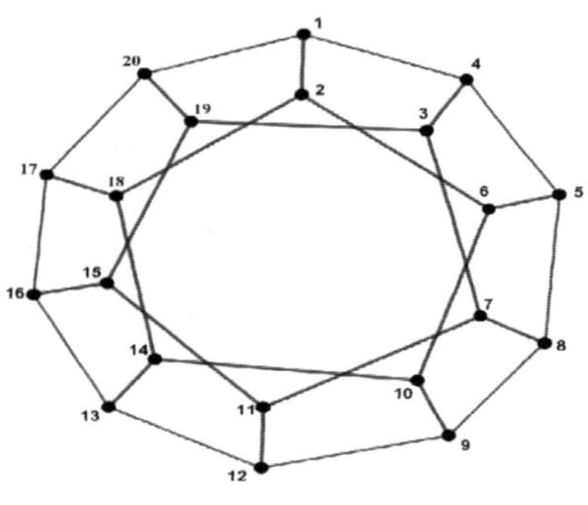

Figure 3. $GP(10, 2)$

$PE : 2, 0, 1,\ 2, 0, 1,\ 2, 0, 1,\ 0;$

$SE : 2, 1, 0,\ 2, 2, 1,\ 0, 2, 1,\ 1;$

$CE : 0, 1, 2,\ 0, 1, 2,\ 0, 1, 2,\ 0.$

Then, $e_f(0) = 10; e_f(1) = 10; e_f(2) = 10$. $|e_f(i) - e_f(j)| \leq 2; i, j = 0, 1, 2.$ are verified. □

Theorem 4.2. *The Generalized Petersen graph $GP(n,2)$ is not an OTSMO graph with respect to R_1 for n odd, $n \equiv 0 \mod 3$ and $n \equiv 2 \mod 3$ and is an OTSMO graph for n odd, $n \equiv 1 \mod 3$.*

Proof. Due to similarty of the proofs of theorems, a brief discussion is given.
Case 1: n **odd,** $n \equiv 0 \mod 3$.
$n = 3g$, g **is odd.**
$PE : 201, 201, \cdots (g-1)$ repeated triplets, $2, 0, (LE_1)$
$$LE_1 = \frac{(2n-1)+1}{3} = \frac{2n}{3} \equiv 0 \mod 3.$$
$PE : 2, 0, 1, 2, 0, 1, \cdots, 2, 0, \boxed{0}$
$SE : 2, 1, 0, 2, 1, 0, \cdots$ upto $\left(\frac{n-1}{2}\right)^{\text{th}}$ edges,
$\left(\left(\frac{n+1}{2}\right)^{\text{th}} \text{edge} = \frac{2n+3}{3} \equiv 0 \mod 3\right), 1, 0, 2,$ upto $\left(\frac{n-1}{2}\right) - 1$ edges, LE_2
$$LE_2 : \frac{(2n-3)+2}{3} \equiv \frac{2n-1}{3} \equiv 2 \mod 3$$
$SE : 2, 1, 0, 2, 1, 0, 2, \boxed{0}, 1, 0, 2, \cdots 0, 2, \boxed{2}$
$CE : 0, 1, 2, 0, 1, 2, \cdots (g-1)$ repeated triplets, $0, 1, (LE_3)$
$$LE_3 : \frac{2n + (2n-1)}{3} \equiv 2 \mod 3$$
$CE : 0, 1, 2, 0, 1, 2, \cdots 0, 1, \boxed{2}$
Two SE are shown in the following figure

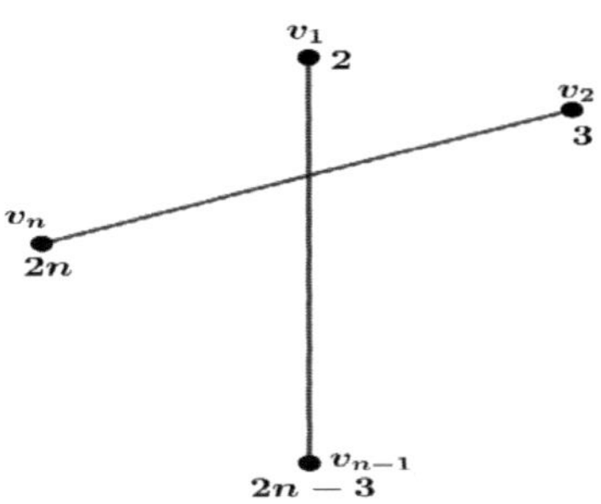

Figure 4.

Tabulating the above, we get the following.

	PE	SE	CE	Total
$e_f(0)$	$g+1$	g	g	$(3g+1)$
$e_f(1)$	$(g-1)$	$(g-1)$	g	$(3g-2)$
$e_f(2)$	g	$(g+1)$	g	$(3g+1)$

So, R_1 fails on O_0, as
$|e_f(0) - e_f(1)| = |e_f(1) - e_f(2)| = |e_f(2) - e_f(0)| = 3 \nleq 2$

Case 2: When $n \equiv 1 \mod 3$, n is odd, R_1 holds.

$h = 3g + 1$, g is even.

$$PE : LE_1 = \frac{(2n-1)+1}{3} \equiv \frac{2(3g+1)}{3} = \frac{6g+2}{3} \equiv 2;$$

$$SE : \frac{(n+1)^{\text{th}}}{2} \text{ edge: } \frac{2n+3}{3} = \frac{6g+2+3}{2} \equiv 2 \mod 3;$$

$$LE_2 : \frac{(2n-3)+2}{3} = \frac{6g+2-3+2}{3} = 1 \mod 3;$$

$$CE : LE_3 = \frac{(2n-1)+2n}{3} = \frac{4n-1}{3} = \frac{12g+4-1}{3} \equiv 0 \mod 3;$$

	PE	SE	CE	Total
$e_f(0)$	g	g	$(g+1)$	$(3g+1) = n$
$e_f(1)$	g	$(g+1)$	g	$(3g+1) = n$
$e_f(2)$	$(g+1)$	g	g	$(3g+1) = n$

It is clear that R_1 holds on O_1. That is $n \equiv 1 \mod 3, n$ is odd, $GP(n,2)$ is an OTSMO graph.

Case 3: When n is odd, $n \equiv 2 \mod 3$, R_1 fails

$n = 3g + 2, g$ odd.

$$PE : LE_1 = \frac{2n}{3} = \frac{2(3g+2)}{3} \equiv 1 \mod 3;$$

$$SE : \left(\frac{n+1}{2}\right)^{\text{th}} \text{ edge: } \frac{6g+4+3}{3} \equiv 1 \mod 3;$$

$$LE_2 : \frac{(2n-3)+2}{3} = \frac{6g+4-3+2}{3} = \frac{6g+3}{3} \equiv 0 \mod 3;$$

$$CE : LE_3 = \frac{(2n-1)+2n}{3} = \frac{12g+8-1}{3} \equiv 1 \mod 3.$$

Note:

In the star edge connecting odd suffix to even suffix $\frac{2n+3}{3} \equiv 1 \mod 3$. That is the $\left(\frac{n+1}{2}\right)^{\text{th}}$ edge.

The first even to even also takes the value 1

The $(g+1)^{\text{th}}$ triplet can contain only 2 numbers given by $1, LE_2$ i.e, $1, \boxed{0}$.

$SE : 2,1,0,\ 2,1,0,\cdots 2,1,\boxed{1},\ \boxed{1},0,2\cdots 1,\boxed{0}$.

	PE	SE	CE	Total
$e_f(0)$	g	g	$(g+1)$	$(3g+1)$
$e_f(1)$	$(g+1)$	$(g+2)$	$(g+1)$	$(3g+4)$
$e_f(2)$	$(g+1)$	g	g	$(3g+1)$

$|e_f(i) - e_f(j)| = 3 \nleq 2$.

So, R_1 fails on O_2. □

As R_1 fails on O_0 and O_2 the Researcher proceeded to find another rule R_2

Description of R_2

The polygonal vertices are allotted even numbers from 2 to $2n$ and the star vertices are allotted odd numbers from 1 to $(2n-1)$.

Theorem 4.3. *The $GP(n,2)$ is an OTSMO graph with the Rule R_2 on O_0 and E_0 but fails on O_2.*

Proof. The labeling for polygonal and star vertices with respect to R_2.

$f(v_i) = 2i$ for $i = 1,2,3\cdots n$;

$f(u_j) = 2j - 1$ for $j = 1,2,3\cdots n$.

The labeling for polygonal edges:

$$f^*(v_iv_{i+1}) = \frac{2i+2(i+1)}{3} \equiv a \mod 3;$$
$$\frac{(4i+2)}{3} \equiv a \mod 3 \text{ for } i = 1,2,3\cdots,n.$$
The labeling for star edges:
$$f^*(u_ju_{j+2}) = \frac{(2j-1)+2(j+2)-1}{3};$$
$$\frac{4j+4-1-1}{3} = \frac{4j+2}{3} \text{ for } j = 1,2,\cdots,n.$$
The labeling for connecting edges:
$$f^*(v_iu_j) = \frac{(2i)+2(j-1)}{3}$$
$$\frac{2(i+j)-1}{3} \text{ for } i,j = 1,2,\cdots n.$$
The observations are as follows and they are tabulated also.
In PE, there are g repeated triplets of the form $0,1,2\cdots$. with $LE_1 = \frac{2n+2}{3} \equiv 2 \mod 3$ (same for O_0 and E_0);
In CE, there are g repeated triplets of the form $0,1,2\cdots$. with $LE_3 = \frac{4n-1}{3} \equiv 2 \mod 3$ (same for O_0 and E_0);
In SE, the pattern is
$0,2,1,\ 0,2,1,\cdots$ with the $\left(\frac{n+1}{2}\right)^{\text{th}}$ edge value $= \frac{2n+2}{3} \equiv 2 \mod 3$ and $LE_2 = \frac{2n-2}{3} \equiv 1 \mod 3$ for $(n \in O_0)$.

	PE	SE	CE	Total
$e_f(0)$	g	g	g	$3g = n$
$e_f(1)$	g	g	g	$3g = n$
$e_f(2)$	g	g	g	$3g = n$

That is, $e_f(0) = e_f(1) = e_f(2)$

When $n \in E_0$, the $\left(\frac{n+1}{2}\right)^{\text{th}}$ edge value is 1 and PE_2 is 2 (these values for $n \in O_0$ are interchanged). Therefore $e_f(0) = e_f(1) = e_f(2)$ for $n \in E_0$.
Hence with respect to the Rule R_2, the $GP(n,2)$ when n is odd or n is even and a multiple of 3 is found to be an $OTSMO$ graph.
We now prove that R_2 fails on O_2.
$n = 3g+2$.
The LE_1 of PE is given by,
$$\frac{2n+2}{3} = \frac{2(3g+2)+2}{3} = \frac{6g+6}{3} \equiv 0 \mod 3.$$
The pattern is g repeated triplets, followed by 0 and 0.
$0,1,2,\ 0,1,2,\ \cdots\boxed{0}\,\boxed{0}$
In SE, the $\left(\frac{n+1}{2}\right)^{\text{th}}$ edge is $\frac{2n+2}{3} \equiv \frac{2(3g+2)+2}{3} = \frac{6g+6}{3} \equiv 0 \mod 3$
The $LE_2 = \frac{2n-2}{3} = \frac{2(3g+2)-2}{3} = \frac{6g+2}{3} \equiv 2 \mod 3$
The triplet of the form $0,2,1\ 0,\boxed{2},1,\ 0,\boxed{2}$
The LE_3 of CE is $\frac{4n-1}{3} = \frac{4(3g+2)-1}{3} = \frac{12g+8-1}{3} \equiv 1 \mod 3$
The pattern is g repeated triplets follwed by 0 and 2.
$0,1,2,\ 0,1,2\ \cdots 0,\boxed{2}$.

	PE	SE	CE	Total
$e_f(0)$	$(g+2)$	$(g+1)$	$(g+1)$	$(3g+4)$
$e_f(1)$	g	g	g	$(3g+1)$
$e_f(2)$	g	$(g+1)$	$(g+1)$	$(3g+1)$

For Verification

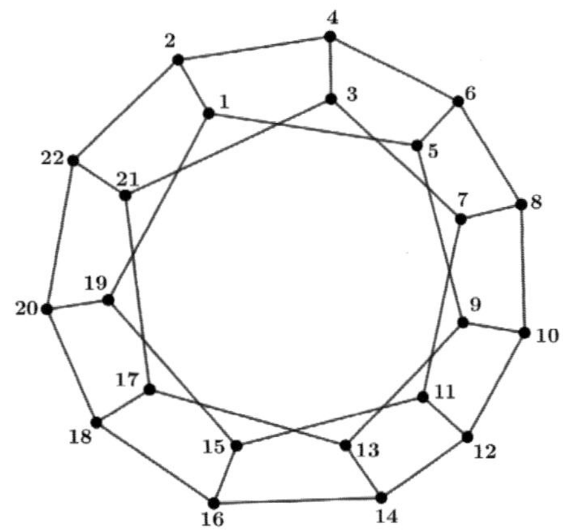

Figure 5. $GP(11,2)$

$PE: 0,1,2,\ 0,1,2,\ 0,1,2\ 0,\boxed{0};$
$SE: 0,2,1,\ 0,2,\boxed{0},\ 1,0,2,\ 1,\boxed{2};$
$CE: 0,1,2,\ 0,1,2,\ 0,1,2,\ 0,\boxed{1}.$
$|e_f(0)-e_f(1)| = |(3g+4)-(3g+1)| = 3 \nleq 2.$
R_2 fails on O_2.
Note:

(i) R_1 is valid on E_0, E_1, E_2 and O_1
(ii) R_2 is valid on O_0 and E_0
(iii) R_2 fails on O_1, O_2, E_1 and E_2

Another rule has to be introduced on O_2 □

Description of R_3
We proceed with the usuad notations. The $2n$ vertices are divided into sets of 4 vertices in the following manner. $(v_1u_1u_2v_2)$, $(v_3u_3u_4v_5)$. As $2n$ is even, after such a grouping, only 2 vertices or no vertices will be left out. The laeft out vertices they are given the remaining numbers, maintaining the following movement that is from the outer vertex v_1 to two inner vertices u_1u_2, next two outer vertices v_2v_3 and then to two inner vertices $u_3, u_4, \cdots$.

The description of R_3 is the same as R_1 with only differences in the numbers assigned to the I set $(v_1u_1u_2v_2), f(v_1) = 2n,\ f(v_2) = 2n-1,$
$f(u_1) = 1, f(u_2) = 2.$

Theorem 4.4. *The $GP(n,2)$ is an OTSMO graph with respect to R_3 when $n \in O_2 \cup E_1 \cup E_2$ and not an OTSMO graph otherwise.*

Proof. $f(v_1) = 2n, f(v_2) = 2n-1, f(u_1) = 1, f(u_2) = 2.$
The labeling for polygonal vertices:

$f(v_1) = 2n;$
$f(v_2) = 2n - 1;$
$f(v_{2i+1}) = 4i - 1 \quad \text{for } i = 1, 2, \cdots, \left(\frac{n-1}{2}\right);$
$f(v_{2j}) = 4j - 2 \quad \text{for } j = 1, 2, \cdots, \left(\frac{n-1}{2}\right);$

The labeling for star vertices:
$f(u_1) = 1;$
$f(u_2) = 2\ ;$
$f(u_{2s+1}) = 4s \quad \text{for } s = 1, 2, \cdots, \left(\frac{n-1}{2}\right);$
$f(u_{2t+2}) = 4t + 1 \quad \text{for } t = 1, 2, \cdots, \left(\frac{n-3}{2}\right);$

Labeling for the polygonal edges:
$f^*(v_1v_2) = \frac{2n + (2n-1)}{3} = \frac{4n-1}{3} \equiv k \mod 3;$
$f^*(v_2v_3) = \frac{(2n-1)+3}{3} = \frac{2n+2}{3} \equiv k \mod 3\ ;$
$f^*(v_{2i+1}v_{2j}) = \frac{(4i-1)+(4j-2)}{3};$
$f^*(v_{2i+1}v_{2j}) = \frac{4i+4j-3}{3} \equiv k \mod 3$
$\text{for } (i,j) = (1,2), (2,3), \cdots, \left(\frac{n-1}{2}, \frac{n-1}{2}\right);$
$f^*(v_nv_1) = \frac{2n-3+2n}{3} = \frac{4n-3}{3}.$

Labeling for star edges:
$f^*(u_1u_3) = \frac{1+4}{3} \equiv 2 \mod 3;$

Connecting odd suffix vertices:
$f^*(u_{2s+1}u_{2s'+3}) = \frac{4s + 4(s'+2)}{3} = \frac{4(s+s')+8}{3};$
$\text{for } (s,s') = (1,1), (2,2), \cdots, \left(\left(\frac{n-1}{2}\right) - 1, \left(\frac{n-1}{2}\right) - 1\right).$

Connecting even suffix vertices:
$f^*(u_{2t+2}u_{2t'+4}) = \frac{(4t+1) + 4(t'+1) + 1}{3} = \frac{4t + 4t' + 6}{3}$
$\text{for } (t,t') = (1,1), (2,2), \cdots, \left(\left(\frac{n-1}{2}\right) - 2, \left(\frac{n-1}{2}\right) - 2\right)$

Labeling for connecting edges:
$f^*(v_1u_1) = \frac{2n+1}{3};$
$f^*(v_2u_2) = \frac{2n+1}{3};$
$f^*(v_{2i+1}u_{2s+1}) = \frac{4i - 1 + 4s}{3} = \frac{4(i+s)-1}{3}$
$\text{for } (i,s) = (1,1), (2,2), \cdots, \left(\left(\frac{n-1}{2}\right), \left(\frac{n-1}{2}\right)\right).$
$f^*(v_{2j}v_{2t+2}) = \frac{(4j-2)+(4t+1)}{3}$
$\text{for } (j,t) = (2,1), (3,2), \cdots, \left(\left(\frac{n-1}{2}\right), \left(\frac{n-3}{2}\right)\right).$

Here $n = 3g + 2$

The edge values are listed below

$PE : 1, 0, 0\ 1, 2, 0\ 1, 2, 0\ \boxed{1}, \boxed{2};$

$: 1, 0, 0,\ 1, 2, 0,\ 1, 2, 0,\ \frac{(2n-4)+(2n-3)}{3}, \frac{2n-3+2n}{3};$

$: 1, 0, 0,\ 1, 2, 0,\ 1, 2, 0,\ \frac{4n-7}{3}, \frac{4n-3}{3};$

$SE : 2, 0, 2,\ 1, 0\ \cdots \frac{(2n-2)+2}{3}, 1, 2, 1,\ \cdots, 0\left(\frac{2n-5+1}{3}\right);$

$: 2, 0, 2,\ 1, 0, \cdots \frac{2n}{3}\ 1, 2, 1, \cdots, 0, \frac{2n-4}{3};$

$CE : \frac{2n+1}{3}, \frac{2n+1}{3}, 1,\ 2, 0, 1,\ 2, 0, 1,\ \frac{(2n-4)+(2n-5)}{3}, \frac{(2n-3)+(2n-2)}{3};$

$: \frac{2n+1}{3}, \frac{2n+1}{3}, 1,\ 2, 0, 1,\ 2, 0, 1,\ \frac{4n-9}{3}, \frac{4n-5}{3}.$

Replacing $n = 3g + 2$

$PE : 1, 0, 0,\ 1, 2, 0,\ 1, 2, 0,\ \frac{12g+5}{3}, \frac{12g+5}{3}$

$: 1, 0, 0,\ 1, 2, 0,\ 1, 2, 0,\ 2, \boxed{2}$

$SE : 2, 0, 2,\ 1, 0, \cdots \frac{6g+4}{3}\ 1, 2, 1, \cdots \frac{12g-6}{3}\ \frac{6g}{3}$

$: 2, 0, 2,\ 1, 0, \cdots \boxed{1}\ 1, 2, 1, \cdots 0, \boxed{0}$

$CE : \frac{2n+1}{3}, \frac{2n+1}{3}, 1,\ 2, 0, 1,\ 2, 0, 1,\ \frac{4n-9}{3}, \frac{4n-5}{3}$

$: \frac{6g+5}{3}, \frac{6g+5}{3}, 1,\ 2, 0, 1,\ 2, 0, 1,\ \frac{12g-1}{3}, \frac{12g+3}{3}$

$: 2, 2, 1,\ 2, 0, 1,\ 2, 0, 1,\ 2, \boxed{0}$

Listing after counting the edge values given above,

	PE	SE	CE	Total
$e_f(0)$	$(g+1)$	$(g+1)$	g	$(3g+2)$
$e_f(1)$	$(g+1)$	$(g+1)$	g	$(3g+2)$
$e_f(2)$	g	g	$(g+2)$	$(3g+2)$

Hence $|e_f(i) - e_f(j)| = 0 \leq 2$ for $i = 0, 1, 2$

R_3 holds good on O_2.

Illustration 4.5.

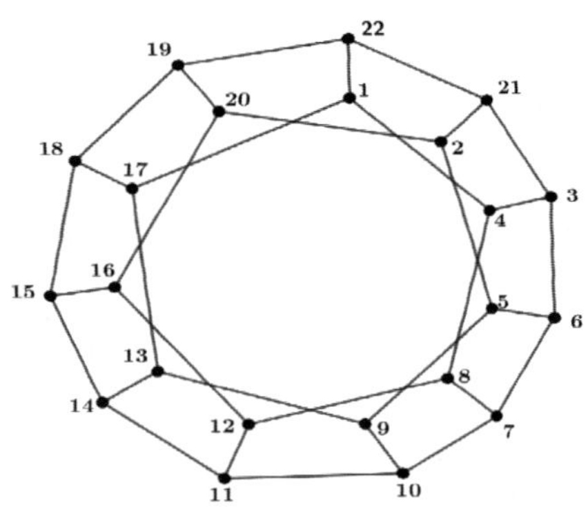

Figure 6. $GP(11, 2)$

Note: R_3 fails on O_0; R_3 is valid on O_1; R_3 is fails on E_0; R_3 is valid on E_1; R_3 is valid on E_2 □

Table on validity of rules R_i, $i = 1, 2, 3$.

	O_0	O_1	O_2	E_0	E_1	E_2
R_1	×	✓	×	✓	✓	✓
R_2	✓	×	×	✓	×	×
R_3	×	✓	✓	×	✓	✓

Conclusion

The $OTSMO$ labeling is defined in terms of congruence modulo 3 in allotting numbers on $GP(n, 2)$. While proving the invalidity of R_1 on O_0 and O_2, we have arrived at the results that $|e_f(i) - e_f(j)| = 3$ for some $i, j = 0, 1, 2$. On all other sets $|e_f(i) - e_f(j)| \leq 2$ is satisfied. These have inspired the Researchers to struck at a new labeling named as $OTSMO$ (Second) with a little modification that $|e_f(i) - e_f(j)| \leq 3$; $i, j = 0, 1, 2$. In this case, the single rule R_1 with $OTSMO$ (Second) labeling make $GP(n, 2)$ an $(OTSMO)$ (Second) graph for all values of n, that is for n odd or even, $\equiv 0, \equiv 1$ and $\equiv 2$.

The $(OTSMO)$ and $(OTSMO)$ (Second) labelings can be applied on a variety of graphs.

Acknowledgment

One of the author (Dr. V. Balaji) acknowledges University Grants Commission, SERO, Hyderabad and India for financial assistance $(No.FMRP5766/15(SERO/UGC))$.

References

[1] J. A. Gallian, *A dynamic survey of graph labeling,* The Electronic Journal of Combinatorics, (2016), #DS6.
[2] F. Harary, *Graph Theory,* Addision wesley reading, 1969.
[3] Kh. Md. Mominul Haque, Lin Xiaohui, Yang Yuansheng and Zahao pingzhong, On the prime labeling of Genereralized Petersen graphs $p(n, 1)$, utilitas Mathematica.
[4] Prime Labeling of Generalized Petersen graph, U. M. Prajpati, S. J. Gajjar, International Journal of Mathematics and soft computing, Vol. 5, No. 1 (2015), 65 -71.
[5] V. Shendra Shainy, G. Margaret Joan Jebarani and V. Balaji, *A New Labeling 'One Third Sum Modulo 3' With Verification and an Application* - Communicated.

Generalized Ulam-Hyers Stability of Complex Additive Functional Equation

P. Agilan[1] **and V. Vallinayagam** [2]
[1,2]Department of Mathematics, St.Joseph's College of Engineering, OMR, Chennai - 600 119, TamilNadu, India.
[1] **E-mail:** agilram@gmail.com
[2]**E-mail:** vngam05@yahoo.co.in

Abstract. The object of the present paper is to assess speculation of the Hyers-Ulam stability theorem for the complex additive functional equation on abelian groups and stability results have been gotten by a fixed point technique. This technique demonstrates that the stability is identified with some fixed point of an appropriate operator.

1. Introduction

An intriguing and renowned speak conferred by Stanislaw M.Ulam [27] in 1940 triggering the study of stability issues for varied purposeful equations. In his speak, Ulam mentioned variety of necessary unresolved mathematical issues. Among them, a matter regarding the stability of group homomorphisms appeared too abstract for anyone to achieve any conclusion.

"Let G be group and H be a metric group with metric $d(.,.)$. Given $\epsilon > 0$ does there exists a $\delta > 0$ such that if a function $f : G \to H$ satisfies

$$d\left(f(xy), f(x)f(y)\right) < \delta$$

for all $x, y \in G$, then there exists a homomorphism $a : G \to H$ with

$$d\left(f(x), a(x)\right) < \varepsilon$$

for all $x \in G$."

In the following year, Donald H.Hyers [14] was ready to provides a partial resolution to Ulam's question that was the primary important breakthrough and step toward a lot of solutions during this space. Since then, an oversized range of papers are printed in reference to varied generalizations of Ulam's problem and Hyers's theorem. In 1978, Themistocles M.Rassias [26] succeeded in extending the results of Hyers's theorem and his exciting result attracted variety of mathematicians who began to be stirred up to research the steadiness issues of many purposeful equations. because of influence of S.M. Ulam, D.H. Hyers within the work of Th.M. Rassias concerning the study of stability issues of purposeful equations, the steadiness development verified by Th.M. Rassias is termed because the Hyers-Ulam-Rassias stability thenceforth. For the last thirty five years several results regarding the Hyers-Ulam-Rassias stability of assorted purposeful equations are extensively investigated by variety of authors and there area unit several fascinating results regarding this problem (see [1, 6, 8, 9, 12, 13, 15, 16, 22, 23]).

The general solution and the permanence of the following additive FE

$$f\left(f(x)-f(y)\right)=f(x+y)+f(x-y)-f(x)-f(y) \tag{1}$$

$$\sum_{i=1}^{m} f\left(mx_i+\sum_{j=1,j\neq i}^{m} x_j\right)+f\left(\sum_{i=1}^{m} x_i\right)=2f\left(\sum_{i=1}^{m} mx_i\right) \tag{2}$$

$$f\left(\sum_{k=1}^{n} kx_k\right)+\sum_{k=1}^{n} f\left(-kx_k+\sum_{l=1,l\neq k}^{n} lx_l\right)+f\left(x_1-\sum_{k=2}^{n} kx_k\right)=nf(x_1)+(n-2)\sum_{k=2}^{n} kf\left(x_k\right) \tag{3}$$

$$f\left(\sum_{k=1}^{n} kx_k\right)+\sum_{l=2}^{n} f\left(\sum_{k=1,k\neq l}^{n} kx_k-lx_l\right)+f\left(x_1-\sum_{k=2}^{n} kx_k\right)=(n+1)f(x_1)+(n-3)\sum_{k=2}^{n} kf\left(x_k\right) \tag{4}$$

$$f\left(\sum_{p=1}^{n} px_p\right)+f\left(\sum_{p=1}^{n} p(-1)^{p+1}x_p\right)+f\left(x_1-\sum_{p=2}^{n} px_p\right)=3f(x_1)+\sum_{p=2}^{n} p(-1)^{p+1}f\left(x_p\right) \tag{5}$$

were discussed in (see [7, 19, 3, 4, 5]).

In this paper, the authors are to discover the solution and establish Hyers-Ulam stability concerning the complex additive FE

$$\begin{aligned}(\lambda^n+1)\,f\left(e^{\frac{in\pi}{3}}\vartheta+e^{\frac{-in\pi}{3}}\omega\right)+(\lambda^n-1)\,f\left(e^{\frac{-in\pi}{3}}\vartheta+e^{\frac{in\pi}{3}}\omega\right)\\ +e^{\frac{-in\pi}{3}}f(\vartheta-\omega)+e^{\frac{in\pi}{3}}f\left(\omega-\vartheta\right)=2\lambda^n cos\left(\frac{n\pi}{3}\right)[f(\vartheta)+f(\omega)]\end{aligned} \tag{6}$$

with $n\in Z$ on abelian group using two different approaches.

2. General Solution

We derive the general solution of FE (6) as follows:

Lemma 2.1. *Let G be a group and Y be a real vector space. Then the function $f:G\to Y$ satisfies the additive FE*

$$f(x+y)=f(x)+f(y) \tag{7}$$

for all $x,y\in G$ if $f:G\to Y$ satisfies the FE (6) for all $\vartheta,\omega\in G$.

Proof. Let f satisfies FE (7). Insert $x=y=0$ in (7), we arrive $f(0)=0$. Put $x=-y$ in (7), we attain $f(-y)=-f(y)$ for all $y\in G$. Henceforth f is an odd function. By changing y by x and y by $2x$ in (7), we arrive

$$f(2x)=2f(x)\quad\text{and}\quad f(3x)=3f(x) \tag{8}$$

for all $\vartheta\in G$. Moreover, for any a, we have

$$f(ax)=af(x) \tag{9}$$

By changing (x,y) by $\left(e^{\frac{in\pi}{3}}\vartheta,e^{\frac{-in\pi}{3}}\omega\right)$ in (7) and using (9), we get

$$f\left(e^{\frac{in\pi}{3}}\vartheta+e^{\frac{-in\pi}{3}}\omega\right)=e^{\frac{in\pi}{3}}f(\vartheta)+e^{\frac{-in\pi}{3}}f(\omega) \tag{10}$$

for all $\vartheta, \omega \in G$. Both side multiply by $(\lambda^n + 1)$ in (10), we arrive

$$(\lambda^n + 1) f\left(e^{\frac{in\pi}{3}}\vartheta + e^{\frac{-in\pi}{3}}\omega\right) = (\lambda^n + 1) e^{\frac{in\pi}{3}} f(\vartheta) + (\lambda^n + 1) e^{\frac{-in\pi}{3}} f(\omega) \tag{11}$$

Replacing (x, y) by $\left(e^{\frac{-in\pi}{3}}\vartheta, e^{\frac{in\pi}{3}}\omega\right)$ in (7) and using (9), we get

$$f\left(e^{\frac{-in\pi}{3}}\vartheta + e^{\frac{in\pi}{3}}\omega\right) = e^{\frac{-in\pi}{3}} f(\vartheta) + e^{\frac{in\pi}{3}} f(\omega) \tag{12}$$

for all $\vartheta, \omega \in G$. Both side multiply by $(\lambda^n - 1)$ in (12), we arrive

$$(\lambda^n - 1) f\left(e^{\frac{-in\pi}{3}}\vartheta + e^{\frac{in\pi}{3}}\omega\right) = (\lambda^n - 1) e^{\frac{-in\pi}{3}} f(\vartheta) + (\lambda^n - 1) e^{\frac{in\pi}{3}} f(\omega) \tag{13}$$

for all $\vartheta, \omega \in G$. Setting (x, y) by (ϑ, ω) in (7), and utilizing the property of odd f we have

$$f(\vartheta - \omega) = f(\vartheta) - f(\omega) \tag{14}$$

for all $\vartheta, \omega \in G$. Both side multiply by $e^{\frac{-in\pi}{3}}$ in (14), we arrive

$$e^{\frac{-in\pi}{3}} f(\vartheta - \omega) = e^{\frac{-in\pi}{3}} f(\vartheta) - e^{\frac{-in\pi}{3}} f(\omega) \tag{15}$$

for all $\vartheta, \omega \in G$. Setting (ϑ, ω) by $(\omega, -\vartheta)$ in (7), and utilizing the property of odd f we have

$$f(\omega - \vartheta) = f(\omega) - f(\vartheta) \tag{16}$$

for all $\vartheta, \omega \in G$. Both side multiply by $e^{\frac{in\pi}{3}}$ in (16), we arrive

$$e^{\frac{in\pi}{3}} f(\omega - \vartheta) = e^{\frac{in\pi}{3}} f(\omega) - e^{\frac{in\pi}{3}} f(\vartheta) \tag{17}$$

for all $\vartheta, \omega \in G$. Adding (11), (13), (15) and (17), we arrive (6). □

3. Stability Results: Direct Method

The mapping is defined as follows: $Df : G \to Y$ by

$$Df(\vartheta, \omega) = (\lambda^n + 1) f\left(e^{\frac{in\pi}{3}}\vartheta + e^{\frac{-in\pi}{3}}\omega\right) + (\lambda^n - 1) f\left(e^{\frac{-in\pi}{3}}\vartheta + e^{\frac{in\pi}{3}}\omega\right)$$
$$+ e^{\frac{-in\pi}{3}} f(\vartheta - \omega) + e^{\frac{in\pi}{3}} f(\omega - \vartheta) - 2\lambda^n cos\left(\frac{n\pi}{3}\right)[f(\vartheta) + f(\omega)]$$

with $n \in Z$.

Theorem 3.2. *For $j = \pm 1$ and $\mathcal{M} : G \times G \to [0, \infty)$ be a function such that*

$$\lim_{n\to\infty} \frac{\mathcal{M}\left(\mathcal{P}^{nj}\vartheta, \mathcal{P}^{nj}\vartheta\right)}{\mathcal{P}^{nj}} = 0 \tag{18}$$

for all $\vartheta, \omega \in G$. Let $f : G \to Y$ be an odd mapping satisfying

$$\|Df(\vartheta, \omega)\| \leq \mathcal{M}(\vartheta, \omega) \tag{19}$$

for all $\vartheta, \omega \in G$. Then $\exists$ a unique additive mapping $A : G \to Y$ and satisfying FE (6) $\ni$

$$\|f(\vartheta) - A(\vartheta)\| \leq \frac{1}{2\lambda^n \mathcal{P}} \sum_{k=\frac{1-j}{2}}^{\infty} \frac{\mathcal{M}(\mathcal{P}^{kj}\vartheta, \mathcal{P}^{kj}\vartheta)}{\mathcal{P}^{kj}} \tag{20}$$

for all $\vartheta \in G$. The mapping $A(\vartheta)$ is defined by

$$A(\vartheta) = \lim_{n\to\infty} \frac{f(\mathcal{P}^{nj}\vartheta)}{\mathcal{P}^{nj}} \tag{21}$$

for all $\vartheta \in G$.

Journal of Physics: Conference Series **1377** (2019) 012011 doi:10.1088/1742-6596/1377/1/012011

Proof. Consider $j = 1$, by changing (ϑ, ω) by (ϑ, ϑ) in (19), we get

$$\left\| 2\lambda^n f\left[\left(e^{\frac{in\pi}{3}} + e^{\frac{-in\pi}{3}}\right)\vartheta\right] - 4\lambda^n cos\left(\frac{n\pi}{3}\right) f(\vartheta)\right\| \leq \mathcal{M}(\vartheta, \vartheta) \tag{22}$$

for all $\vartheta \in G$. The above inequality can be written as

$$\left\| 4\lambda^n cos\left(\frac{n\pi}{3}\right) f(\vartheta) - 2\lambda^n f\left[\left(2cos\left(\frac{n\pi}{3}\right)\right)\vartheta\right]\right\| \leq \mathcal{M}(\vartheta, \vartheta) \tag{23}$$

for all $\vartheta \in G$. Bothside divide by $4\lambda^n cos\left(\frac{n\pi}{3}\right)$ in (23), we have

$$\left\| f(\vartheta) - \frac{f\left(2cos\left(\frac{n\pi}{3}\right)\vartheta\right)}{2cos\left(\frac{n\pi}{3}\right)}\right\| \leq \frac{\mathcal{M}(\vartheta, \vartheta)}{2.2\lambda^n cos\left(\frac{n\pi}{3}\right)} \tag{24}$$

for all $\vartheta \in G$. Consider $\mathcal{P} = 2cos\left(\frac{n\pi}{3}\right)$, Now replacing ϑ by $\mathcal{P}\vartheta$ and dividing by $\mathcal{P}$ in (24), we get

$$\left\| \frac{f(\mathcal{P}\vartheta)}{\mathcal{P}} - \frac{f(\mathcal{P}^2\vartheta)}{\mathcal{P}^2}\right\| \leq \frac{\mathcal{M}(\mathcal{P}\vartheta, \mathcal{P}\vartheta)}{2\lambda^n \mathcal{P}^2} \tag{25}$$

for all $\vartheta \in G$. From (24) and (25), we obtain

$$\begin{aligned} \left\| f(\vartheta) - \frac{f(\mathcal{P}^2\vartheta)}{\mathcal{P}^2}\right\| &\leq \left\| f(\vartheta) - \frac{f(\mathcal{P}\vartheta)}{\mathcal{P}}\right\| + \left\| \frac{f(\mathcal{P}\vartheta)}{\mathcal{P}} - \frac{f(\mathcal{P}^2\vartheta)}{\mathcal{P}^2}\right\| \\ &\leq \frac{1}{2\lambda^n \mathcal{P}}\left[\mathcal{M}(\vartheta, \vartheta) + \frac{\mathcal{M}(\mathcal{P}\vartheta, \mathcal{P}\vartheta)}{\mathcal{P}}\right] \end{aligned} \tag{26}$$

for all $\vartheta \in G$. In any case for n , we get

$$\begin{aligned} \left\| f(\vartheta) - \frac{f(\mathcal{P}^n\vartheta)}{\mathcal{P}^n}\right\| &\leq \frac{1}{2\lambda^n \mathcal{P}} \sum_{k=0}^{n-1} \frac{\mathcal{M}(\mathcal{P}^t\vartheta, \mathcal{P}^t\vartheta)}{\mathcal{P}^t} \\ &\leq \frac{1}{2\lambda^n \mathcal{P}} \sum_{k=0}^{\infty} \frac{\mathcal{M}(\mathcal{P}^t\vartheta, \mathcal{P}^t\vartheta)}{\mathcal{P}^t} \end{aligned} \tag{27}$$

for all $\vartheta \in G$. To demonstrate the convergence of the sequence

$$\left\{\frac{f(\mathcal{P}^n\vartheta)}{\mathcal{P}^n}\right\},$$

By changing x by $\mathcal{P}^m x$ and divide by $\mathcal{P}^m$ in (27), for any $m, n > 0$, we attain

$$\begin{aligned} \left\| \frac{f(\mathcal{P}^m\vartheta)}{\mathcal{P}^m} - \frac{f(\mathcal{P}^{n+m}\vartheta)}{\mathcal{P}^{(n+m)}}\right\| &= \frac{1}{\mathcal{P}^m}\left\| f(\mathcal{P}^m\vartheta) - \frac{f(\mathcal{P}^n \cdot \mathcal{P}^m\vartheta)}{\mathcal{P}^n}\right\| \\ &\leq \frac{1}{2\lambda^n \mathcal{P}} \sum_{k=0}^{n-1} \frac{\mathcal{M}(\mathcal{P}^{k+m}\vartheta, \mathcal{P}^{k+m}\vartheta)}{\mathcal{P}^{k+m}} \\ &\leq \frac{1}{2\lambda^n \mathcal{P}} \sum_{k=0}^{\infty} \frac{\mathcal{M}(\mathcal{P}^{k+m}\vartheta, \mathcal{P}^{k+m}\vartheta)}{\mathcal{P}^{k+m}} \\ &\rightarrow 0 \quad as \ m \rightarrow \infty \end{aligned}$$

for all $\vartheta \in G$. Therefore the sequence $\left\{ \frac{f(\mathcal{P}^n \vartheta)}{\mathcal{P}^n} \right\}$ is Cauchy. The completeness of Y allow us to consider that, $\exists$ a map $A : G \to Y$ such that

$$A(\vartheta) = \lim_{n\to\infty} \frac{f(\mathcal{P}^n \vartheta)}{\mathcal{P}^n} \quad \forall \ \vartheta \in G.$$

Assuming $n \to \infty$ in (27) one can see that (20) holds for all $\vartheta \in G$. To prove that A satisfies (6), replace (ϑ, ω) by $(\mathcal{P}^n \vartheta, \mathcal{P}^n \omega)$ and divide by $\mathcal{P}^n$ in (19), we obtain

$$\frac{1}{\mathcal{P}^n} \left\| Df\left(\vartheta \mathcal{P}^n, \omega \mathcal{P}^n\right) \right\| \leq \frac{1}{\mathcal{P}^n} \mathcal{M}(\mathcal{P}^n \vartheta, \mathcal{P}^n \omega)$$

for all $\vartheta, \omega \in G$. Letting $n \to \infty$ in the above inequality and using the definition of $A(\vartheta)$, we see that

$$\begin{aligned}(\lambda^n + 1)\, A\left(e^{\frac{in\pi}{3}} \vartheta + e^{\frac{-in\pi}{3}} \omega\right) + (\lambda^n - 1)\, A\left(e^{\frac{-in\pi}{3}} \vartheta + e^{\frac{in\pi}{3}} \omega\right) \\ + e^{\frac{-in\pi}{3}} A(\vartheta - \omega) + e^{\frac{in\pi}{3}} A\left(\omega - \vartheta\right) = 2\lambda^n cos\left(\frac{n\pi}{3}\right)\left[A(\vartheta) + A(\omega)\right]\end{aligned}$$

Therefore A fulfill FE (6) for all $\vartheta, \omega \in G$. To explore A is unique, we let $B(\vartheta)$ be an additional mapping satisfying (6) and (20), then

$$\begin{aligned}\|A(\vartheta) - B(\vartheta)\| &= \frac{1}{\mathcal{P}^n} \|A(\mathcal{P}^n \vartheta) - B(\mathcal{P}^n \vartheta)\| \\ &\leq \frac{1}{\mathcal{P}^n} \left\{\|A(\mathcal{P}^n \vartheta) - f(\mathcal{P}^n \vartheta)\| + \|f(\mathcal{P}^n \vartheta) - B(\mathcal{P}^n \vartheta)\|\right\} \\ &\leq \frac{1}{\lambda^n \mathcal{P}} \sum_{k=0}^{\infty} \frac{\mathcal{M}(\mathcal{P}^{k+n} x, \mathcal{P}^{k+n} \vartheta)}{\mathcal{P}^{(k+n)}} \\ &\to 0 \quad as \ n \to \infty\end{aligned}$$

for all $\vartheta \in G$. Hence A is unique. □

Corollary 3.3. *Let Π and s be non-negative real numbers. Let a function $f : G \to Y$ satisfies the inequality*

$$\|Df(\vartheta, \omega)\| \leq \begin{cases} \Pi, & \\ \Pi\left\{||\vartheta||^s + ||\omega||^s\right\}, & s \neq 1; \\ \Pi||\vartheta||^s||\omega||^s, & 2s \neq 1; \\ \Pi\left\{||\vartheta||^s||\omega||^s + \left\{||\vartheta||^{2s} + ||\omega||^{2s}\right\}\right\}, & 2s \neq 1; \end{cases} \tag{28}$$

for all $\vartheta, \omega \in G$. Then there exists a unique additive function $A : G \to Y$ such that

$$\|f(\vartheta) - A(\vartheta)\| \leq \begin{cases} \dfrac{\Pi}{2\lambda^n |\mathcal{P} - 1|}, \\ \dfrac{\Pi||\vartheta||^s}{\lambda^n |\mathcal{P} - \mathcal{P}^s|}, \\ \dfrac{\Pi||\vartheta||^{2s}}{2\lambda^n |\mathcal{P} - \mathcal{P}^{2s}|}, \\ \dfrac{3\ \Pi||\vartheta||^{2s}}{2\lambda^n |\mathcal{P} - \mathcal{P}^{2s}|} \end{cases} \tag{29}$$

for all $\vartheta \ \in \ G$.

4. Stability Results: Fixed Point Method

Let us consider G be a group and B a Banach space, respectively.

Theorem 4.4. *[18](The alternative of fixed point) Suppose that for a complete generalized metric space (X, d) and a strictly contractive mapping $T : X \to X$ with Lipschitz constant L. Then, for each given element $x \in X$, either*
(B_1) $d(T^n x, T^{n+1}x) = \infty \quad \forall \; n \geq 0$,
or
(B_2) there exists a natural number n_0 such that:
(i) $d(T^n x, T^{n+1}x) < \infty$ for all $n \geq n_0$;
(ii) The sequence $(T^n x)$ is convergent to a fixed point y^* of T;
(iii) y^* is the unique fixed point of T in the set $Y = \{y \in X : d(T^{n_0}x, y) < \infty\}$;
(iv) $d(y^*, y) \leq \frac{1}{1-L} \; d(y, Ty)$ for all $y \in Y$.

Theorem 4.5. *Let $f : G \to B$ be a mapping for which there exists functions $\mathcal{M}, \gamma : G \times G \to [0, \infty)$ with the condition*

$$\lim_{t\to\infty} \frac{\mathcal{M}(\mu_i^t \vartheta, \mu_i^t \omega)}{\mu_i^t} = 0, \tag{30}$$

where

$$\mu_i = \begin{cases} \mathcal{P}, & i = 0, \\ \frac{1}{\mathcal{P}}, & i = 1 \end{cases}$$

satisfying the functional inequality

$$\|Df(\vartheta, \omega)\| \leq \mathcal{M}(\vartheta, \omega) \tag{31}$$

for all $\vartheta, \omega \in G$. If $\exists$ an $L = L(i) < 1 \ni$ the function

$$\vartheta \to \gamma(\vartheta) = \frac{1}{2\lambda^n} \mathcal{M}\left(\frac{\vartheta}{\mathcal{P}}\right),$$

holds the property

$$\gamma(\vartheta) = L \; \mu_i \; \gamma\left(\frac{\vartheta}{\mu_i}\right) \tag{32}$$

$\forall \; \vartheta \in G$. Then $\exists$ a unique additive function $A : G \to B$ satisfying the FE (6) and

$$\|f(\vartheta) - A(\vartheta)\| \leq \frac{L^{1-i}}{1-L} \gamma(\vartheta) \tag{33}$$

holds for all $\vartheta \in G$.

Proof. Let us assume $X = \{p/p : V \to B, \; p(0) = 0\}$ and present the generalized metric on X,

$$d(p, q) = \inf\{K \in (0, \infty) :\| \, p(\vartheta) - q(\vartheta) \, \|\leq K\gamma(\vartheta), x \in G\}.$$

It is natural to perceive that (X, d) is complete.

The mapping $T : X \to X$ by

$$Tp(\vartheta) = \frac{1}{\mu_i} p(\mu_i \vartheta), \forall \qquad x \in G.$$

For $p, q \in X$,

$$\begin{aligned} d(p,q) \leq K \Rightarrow & \parallel p(\vartheta) - q(\vartheta) \parallel \leq K\gamma(\vartheta), \vartheta \in G. \\ & \Rightarrow \left\| \frac{1}{\mu_i} p(\mu_i \vartheta) - \frac{1}{\mu_i} q(\mu_i \vartheta) \right\| \leq \frac{1}{\mu_i} K\gamma(\mu_i \vartheta), \vartheta \in G, \\ & \Rightarrow \left\| \frac{1}{\mu_i} p(\mu_i \vartheta) - \frac{1}{\mu_i} q(\mu_i \vartheta) \right\| \leq LK\gamma(\vartheta), \vartheta \in G, \\ & \Rightarrow \parallel Tp(\vartheta) - Tq(\vartheta) \parallel \leq LK\gamma(\vartheta), \vartheta \in G, \\ & \Rightarrow d(Tp, Tq) \leq LK. \end{aligned}$$

Thus derive

$$d(Tp, Tq) \leq Ld(p, q),$$

$\forall\ p, q \in X$.

From (24), we have

$$\left\| f(\vartheta) - \frac{f(\mathcal{P}\vartheta)}{\mathcal{P}} \right\| \leq \frac{\mathcal{M}(\vartheta, \vartheta)}{2\lambda^n \mathcal{P}} \tag{34}$$

for all $\vartheta \in G$. Using (32) for the case $i = 0$, it reduces to

$$\left\| \frac{1}{\mathcal{P}} f(\mathcal{P}\vartheta) - f(\vartheta) \right\| \leq \frac{1}{\mathcal{P}} \gamma(\vartheta)$$

for all $\vartheta \in G$.

$$\text{i.e.,} \qquad d(Tf, f) \leq \frac{1}{\mathcal{P}} = L = L^{1-0} = L^{1-i} < \infty.$$

By varying again $\vartheta = \dfrac{\vartheta}{\mathcal{P}}$ in (34), we get

$$\left\| f(\vartheta) - \mathcal{P} f\left(\frac{\vartheta}{\mathcal{P}}\right) \right\| \leq \frac{1}{2\lambda^n} \mathcal{M}\left(\frac{\vartheta}{\mathcal{P}}\right).$$

for all $\vartheta \in G$. Using (32) when $i = 1$, it simplifies as

$$\left\| f(\vartheta) - \mathcal{P} f\left(\frac{\vartheta}{\mathcal{P}}\right) \right\| \leq \gamma(\vartheta)$$

for all $\vartheta \in G$.

$$\text{i.e.,} \qquad d(f, Tf) \leq 1 = L^0 = L^{1-1} = L^{1-i} < \infty.$$

From the above, it appears

$$d(f, Tf) \leq L^{1-i}.$$

$(B_2(i))$ holds.

By $(B_2(ii))$, it pursues that $\exists$ a fixed point A of T in $X \ni$

$$A(\vartheta) = \lim_{t \to \infty} \frac{f(\mu_i^t \vartheta)}{\mu_i^t}, \qquad \forall\ \vartheta \in G. \tag{35}$$

Claim that $A : G \to B$ is additive. To substitute (ϑ, ω) by $\left(\mu_i^t \vartheta, \mu_i^t \omega\right)$ in (31) and dividing by μ_i^t, it follows from (30) and (35), A satisfies (6) for all $\vartheta, \omega \in G$.

By $(B_2(iii))$, A is the unparellel fixed point of T in the set $Y = \{f \in X : d(Tf, A) < \infty\}$, adopting the alternative fixed point theorem A is the unique function $\ni$

$$\|f(\vartheta) - A(\vartheta)\| \leq K\gamma(\vartheta)$$

for all $\vartheta \in G$ and $K > 0$. Lastly by $(B_2(iv))$, we acquire that

$$d(f, A) \leq \frac{1}{1-L} d(f, Tf)$$

implying

$$d(f, A) \leq \frac{L^{1-i}}{1-L}.$$

Therefore we infer that

$$\| f(\vartheta) - A(\vartheta) \| \leq \frac{L^{1-i}}{1-L}\gamma(\vartheta).$$

for all $\vartheta \in G$. □

Corollary 4.6. *The function $f : G \to B$ be a mapping and $\exists$ real numbers Π and $s \ni$*

$$\|Df(\vartheta, \omega)\| \leq \begin{cases} \Pi, & \\ \Pi\left\{||\vartheta||^s + ||\omega||^s\right\}, & s \neq 1; \\ \Pi||\vartheta||^s||\omega||^s, & 2s \neq 1; \\ \Pi\left\{||\vartheta||^s||\omega||^s + \left\{||\vartheta||^{2s} + ||\omega||^{2s}\right\}\right\}, & 2s \neq 1; \end{cases} \tag{36}$$

$\forall\ \vartheta, \omega \in G$, then $\exists$ a unique additive function $A : G \to B \ni$

$$\|f(\vartheta) - A(\vartheta)\| \leq \begin{cases} \dfrac{\Pi}{2\lambda^n|\mathcal{P}-1|}, \\ \dfrac{\Pi||\vartheta||^s}{\lambda^n|\mathcal{P}-\mathcal{P}^s|}, \\ \dfrac{\Pi||\vartheta||^{2s}}{2\lambda^n|\mathcal{P}-\mathcal{P}^{2s}|}, \\ \dfrac{3\ \Pi||\vartheta||^{2s}}{2\lambda^n|\mathcal{P}-\mathcal{P}^{2s}|} \end{cases} \tag{37}$$

for all $\vartheta \in G$.

Proof. Let us set

$$\mathcal{M}(\vartheta, \omega) = \begin{cases} \Pi, \\ \Pi\left\{||\vartheta||^s + ||\omega||^s\right\}, \\ \Pi\ \ ||\vartheta||^s\ ||\omega||^s, \\ \Pi\left\{||\vartheta||^s||\omega||^s + (||\vartheta||^{2s} + ||\omega||^{2s})\right\} \end{cases}$$

for all $\vartheta, \omega \in G$. Now

$$\frac{\mathcal{M}(\mu_i^t\vartheta, \mu_i^t\omega)}{\mu_i^t} = \begin{cases} \frac{\Pi}{\mu_i^t}, \\ \frac{\Pi}{\mu_i^t}\left\{||\mu_i^t\vartheta||^s + ||\mu_i^t\omega||^s\right\}, \\ \frac{\Pi}{\mu_i^t}||\mu_i^t\vartheta||^s \ ||\mu_i^t\omega||^s, \\ \frac{\Pi}{\mu_i^t}\left\{||\mu_i^t\vartheta||^s \ ||\mu_i^t\omega||^s + \left\{||\mu_i^t\vartheta||^{2s} + ||\mu_i^t\omega||^{2s}\right\}\right\} \end{cases}$$

$$= \begin{cases} \to 0 \text{ as } t \to \infty, \\ \to 0 \text{ as } t \to \infty, \\ \to 0 \text{ as } t \to \infty, \\ \to 0 \text{ as } t \to \infty. \end{cases}$$

i.e., (30) is holds. In any case, we have

$$\gamma(\vartheta) = \frac{1}{2\lambda^n}\left[\mathcal{M}\left(\frac{\vartheta}{\mathcal{P}}, \frac{\vartheta}{\mathcal{P}}\right)\right].$$

Hence

$$\gamma(\vartheta) = \frac{1}{2\lambda^n}\left[\mathcal{M}\left(\frac{\vartheta}{\mathcal{P}}, \frac{\vartheta}{\mathcal{P}}\right)\right] = \begin{cases} \frac{\Pi}{2\lambda^n\mathcal{P}}, \\ \frac{\Pi}{2\lambda^n\mathcal{P}^s}||\vartheta||^s, \\ \frac{\Pi}{2\lambda^n\mathcal{P}^{2s}}||\vartheta||^{2s}, \\ \frac{3\Pi}{2\lambda^n\mathcal{P}^{2s}}||\vartheta||^{2s}. \end{cases}$$

Also,

$$\frac{1}{\mu_i}\gamma(\mu_i\vartheta) = \begin{cases} \frac{\Pi}{\mu_i \cdot 2\lambda^n\mathcal{P}}, \\ \frac{\Pi}{\mu_i \cdot 2\lambda^n\mathcal{P}^s}||\mu_i\vartheta||^s, \\ \frac{\Pi}{\mu_i \cdot 2\lambda^n\mathcal{P}^{2s}}||\mu_i\vartheta||^{2s}, \\ \frac{3\Pi}{\mu_i \cdot 2\lambda^n\mathcal{P}^{2s}}||\mu_i\vartheta||^{2s}. \end{cases}$$

$$= \begin{cases} \mu_i^{-1}\frac{\Pi}{2}\lambda^n\mathcal{P}, \\ \mu_i^{s-1}\frac{\Pi}{2\lambda^n\mathcal{P}^s}||\vartheta||^s, \\ \mu_i^{2s-1}\frac{\Pi}{2\lambda^n\mathcal{P}^{2s}}||\vartheta||^{2s}, \\ \mu_i^{2s-1}\frac{3\Pi}{2\lambda^n\mathcal{P}^{2s}}||\vartheta||^{2s}. \end{cases}$$

$$= \begin{cases} \mu_i^{-1}\gamma(\vartheta), \\ \mu_i^{s-1}\gamma(\vartheta), \\ \mu_i^{2s-1}\gamma(\vartheta), \\ \mu_i^{2s-1}\gamma(\vartheta). \end{cases}$$

we prove the subsequent cases for conditions by utilizing (33)
Case:1 $L = \mathcal{P}^{-1}$ if $i = 0$

$$\|f(\vartheta) - A(\vartheta)\| \leq \frac{L^{1-i}}{1-L}\gamma(\vartheta) = \frac{\left(\mathcal{P}^{-1}\right)^{1-0}}{1-(\mathcal{P})^{-1}} \cdot \frac{\Pi}{2\lambda^n \mathcal{P}} = \frac{\Pi}{2\lambda^n\left(\mathcal{P}-1\right)}.$$

Case:2 $L = \mathcal{P}$ if $i = 1$

$$\|f(\vartheta) - A(\vartheta)\| \leq \frac{L^{1-i}}{1-L}\gamma(\vartheta) = \frac{(\mathcal{P})^{1-1}}{1-\mathcal{P}} \cdot \frac{\Pi}{2\lambda^n \mathcal{P}} = \frac{\Pi}{2\lambda^n\left(1-\mathcal{P}\right)}.$$

Case:1 $L = \mathcal{P}^{s-1}$ if $i = 0$

$$\|f(\vartheta) - A(\vartheta)\| \leq \frac{L^{1-i}}{1-L}\gamma(\vartheta) = \frac{\lambda^n\left(\mathcal{P}^{s-1}\right)^{1-0}}{1-\mathcal{P}^{s-1}} \frac{\Pi}{\lambda^n \mathcal{P}^s}||\vartheta||^s = \frac{\mathcal{P}^s}{\mathcal{P}-\mathcal{P}^s}\frac{\Pi}{\lambda^n \mathcal{P}^s}||\vartheta||^s = \frac{\Pi||\vartheta||^s}{\lambda^n\left(\mathcal{P}-\mathcal{P}^s\right)}.$$

Case:2 $L = \frac{1}{\mathcal{P}^{s-1}}$ if $i = 1$

$$\|f(\vartheta) - A(\vartheta)\| \leq \frac{L^{1-i}}{1-L}\gamma(\vartheta) = \frac{\left(\frac{1}{\mathcal{P}^{s-1}}\right)^{1-1}}{1-\frac{1}{\mathcal{P}^{s-1}}} \frac{\Pi}{\lambda^n \mathcal{P}^s}||\vartheta||^s = \frac{\mathcal{P}^s}{\mathcal{P}^s-\mathcal{P}}\frac{\Pi}{\lambda^n \mathcal{P}^s}||\vartheta||^s = \frac{\Pi||\vartheta||^s}{\lambda^n\left(\mathcal{P}^s-\mathcal{P}\right)}.$$

Case:1 $L = \mathcal{P}^{2s-1}$ if $i = 0$

$$\|f(\vartheta) - A(\vartheta)\| \leq \frac{L^{1-i}}{1-L}\gamma(\vartheta) = \frac{\left(\mathcal{P}^{2s-1}\right)^{1-0}}{1-\mathcal{P}^{2s-1}} \frac{\Pi}{2\lambda^n \mathcal{P}^{2s}}||\vartheta||^{2s} = \frac{\mathcal{P}^{2s}}{\mathcal{P}-\mathcal{P}^{2s}}\frac{\Pi}{2\lambda^n \mathcal{P}^{2s}}||\vartheta||^{2s} = \frac{\Pi||\vartheta||^{2s}}{2\lambda^n\left(\mathcal{P}-\mathcal{P}^{2s}\right)}.$$

Case:2 $L = \frac{1}{\mathcal{P}^{2s-1}}$ if $i = 1$

$$\|f(\vartheta) - A(\vartheta)\| \leq \frac{L^{1-i}}{1-L}\gamma(\vartheta) = \frac{\left(\frac{1}{\mathcal{P}^{2s-1}}\right)^{1-1}}{1-\frac{1}{\mathcal{P}^{2s-1}}} \frac{\Pi}{2\lambda^n \mathcal{P}^{2s}}||\vartheta||^{2s} = \frac{\mathcal{P}^{2s}}{\mathcal{P}^{2s}-\mathcal{P}}\frac{\Pi}{2\lambda^n \mathcal{P}^{2s}}||\vartheta||^{2s} = \frac{\Pi||\vartheta||^{2s}}{2\lambda^n\left(\mathcal{P}^{2s}-\mathcal{P}\right)}.$$

Case:1 $L = \mathcal{P}^{2s-1}$ if $i = 0$

$$\begin{aligned}\|f(\vartheta) - A(\vartheta)\| \le \frac{L^{1-i}}{1-L}\gamma(\vartheta) &= \frac{\left(\mathcal{P}^{2s-1}\right)^{1-0}}{1-\mathcal{P}^{2s-1}} \frac{3\Pi}{2\lambda^n \mathcal{P}^{2s}} ||\vartheta||^{2s} \\ &= \frac{\mathcal{P}^{2s}}{\mathcal{P}-\mathcal{P}^{2s}} \frac{3\Pi}{2\lambda^n \mathcal{P}^{2s}} ||\vartheta||^{2s} \\ &= \frac{3\Pi||\vartheta||^{2s}}{2\lambda^n \left(\mathcal{P}-\mathcal{P}^{2s}\right)}.\end{aligned}$$

Case:2 $L = \frac{1}{\mathcal{P}^{2s-1}}$ if $i = 1$

$$\begin{aligned}\|f(\vartheta) - A(\vartheta)\| \le \frac{L^{1-i}}{1-L}\gamma(\vartheta) &= \frac{\left(\frac{1}{\mathcal{P}^{2s-1}}\right)^{1-1}}{1-\frac{1}{\mathcal{P}^{2s-1}}} \frac{3\Pi}{2\lambda^n \mathcal{P}^{2s}} ||\vartheta||^{2s} \\ &= \frac{\mathcal{P}^{2s}}{\mathcal{P}^{2s}-\mathcal{P}} \frac{3\Pi}{2\lambda^n \mathcal{P}^{2s}} ||\vartheta||^{2s} \\ &= \frac{3\Pi||\vartheta||^{2s}}{2\lambda^n \left(\mathcal{P}^{2s}-\mathcal{P}\right)}.\end{aligned}$$

□

5. Conclusion

We attained stability results in both direct and fixed point method in the sense of Hyers Ulam.

References

[1] **Aczel.J and Dhombres.J**, *Functional Equations in Several Variables,* Cambridge Univ, Press, 1989.

[2] **Aoki.T**, *On the stability of the linear transformation in Banach spaces*, J. Math. Soc. Japan, 2 (1950), 64-66.

[3] **Arunkumar .M, Agilan. P, Ramamoorthy.S** *Solution and Generalized Ulam-Hyers Stability of a n−Dimensional Additive Functional Equation in Banach Space and Banach Algebra: Direct and Fixed Point Methods* , Annals of Pure and Applied Mathematics, Vol. 15, No.1, 2017, 25-41.

[4] **Arunkumar. M, Agilan. P, Ramamoorthy. S, Mahesh Kumar.N**, *Generalised Ulam-Hyers stability of a n− dimensional additive functional equation in two different methods*, Int. J. Computer Aided Engineering and Technology (In press).

[5] **Arunkumar .M, Agilan .P**, *Solution and Ulam-Hyers Stability of a n Dimensional Additive Functional Equation in Banach Space and Banach Algebra: The Direct and Fixed Point Methods*, International Journal of Pure and Applied Mathematics, Vol 120 No. 9 2018, 93-104.

[6] **Czerwik .S**, *Functional Equations and Inequalities in Several Variables*, World Scientific, River Edge, NJ, 2002.

[7] **Dong Yun Shin, Hassan Azadi Kenary, Sahami .N**,*Hyers-Ulam-Rassias Stability of Functional Equation In NAB-Spaces*, International Journal of Pure and Applied Mathematics, Vol 95 No. 1 2014, 1-11

[8] **Eskandani .G.Z, Găvrută .P, Rassias .J.M, Zarghami .R**, *Generalized Hyers-Ulam Stability for a General Mixed Functional Equation in Quasi-β-normed Spaces*, Mediterr. J. Math. 8 (2011), 331-348.

[9] **Eskandani .G.Z, Găvrută .P**,*On the stability problem in quasi-Banach spaces*, Nonlinear Funct. Anal. Appl. (to appear).

[10] **Ebadian .A, Ghobadipour .N., Eshaghi Gordji .M**,*On the Stability of a Parametric Additive Functional Equation in Quasi-Banach Spaces*, Abstract and Applied Analysis, Vol 2012, Article ID 235359, 14 pages

[11] **Găvrută .P**, *A generalization of the Hyers-Ulam-Rassias stability of approximately additive mappings*, J. Math. Anal. Appl., 184 (1994), 431-436.

[12] **Găvrută .P**, *An answer to a question of Rassias .J.M concerning the stability of Cauchy functional equation,* Advances in Equations and Inequalities, Hadronic Math. Ser., (1999), 67-71.

[13] **Găvrută .P**, *On a problem of Isac .G and Rassias .Th. M. concerning the stability of mappings*, J. Math. Anal. Appl. 261 (2001), 543-553.

[14] **Hyers .D.H**, *On the stability of the linear functional equation,* Proc.Nat. Acad.Sci.,U.S.A.,27 (1941) 222-224.

[15] **Hyers .D.H, Isac G.,Rassias .Th.M**, *Stability of functional equations in several variables*, Birkhauser, Basel, 1998.

[16] **Jung .S.M**, *Hyers-Ulam-Rassias Stability of Functional Equations in Mathematical Analysis*, Hadronic Press, Palm Harbor, 2001.

[17] **Lee .D.O**, *Hyers-Ulam stability of an addtiive type functional equation,* J. Appl. Math. and Computing, 13 (2003) no.1-2, 471-477.

[18] **Margoils .B, Diaz .J.B**, *A fixed point theorem of the alternative for contractions on a generalized complete metric space*, Bull.Amer. Math. Soc. 126 74 (1968), 305-309.

[19] **Mihet .D, Saadati .R, Vaezpour .S. M**, *The Stability of an Additive Functional Equation in Menger Probabilistic φ-Normed Spaces*, Math. Slovaca 61 (2011), No. 5, 817–826.

[20] **Rassias .J.M**, *On approximately of approximately linear mappings by linear mappings,* J. Funct. Anal. USA, 46, (1982) 126-130.

[21] **Rassias .J.M**, *On approximately of approximately linear mappings by linear mappings,* Bull. Sc. Math, 108, (1984) 445-446.

[22] **Rassias J.M., Kim .H.M**, *Generalized Hyers Ulam stability for general additive functional equations in quasi-β-normed spaces.* J. Math. Anal. Appl., 356, (2009), 302-309.

[23] **Rassias .J.M, Jun .K.W, Kim .H.M**, *Approximate $(m,n)-$Cauchy - Jensen Additive Mappings in C^*-algebras*, Acta Mathematica Sinica, English Series, Vol. 27, No. 10, (2011), 1907-1922.

[24] **Ravi .K, Arunkumar .M**, *On a n- dimensional additive Functional Equation with fixed point Alternative*, Proceedings of International Conference on Mathematical Sciences 2007, Malaysia.

[25] **Ravi .K, Arunkumar .M and Rassias .J.M**, *On the Ulam stability for the orthogonally general Euler-Lagrange type functional equation*, International Journal of Mathematical Sciences, Autumn 2008 Vol.3, No. 08, 36-47.

[26] **Rassias .Th.M**, *On the stability of the linear mapping in Banach spaces,* Proc.Amer.Math. Soc., 72 (1978), 297-300.

[27] **Ulam .S.M**, *Problems in Modern Mathematics,* Science Editions,Wiley, NewYork, 1964 (Chapter VI, Some Questions in Analysis: 1, Stability).

Fuzzy cognitive maps and induced fuzzy cognitive maps approach to traffic flow

R Sujatha[1], G Kuppuswami[2]

[1]Department of Mathematics, SSN College of Engineering, Chennai, Tamilnadu, India.
[2]Department of Mathematics, Sri Venkateswaraa College of Technology, Kanchipuram, Tamilnadu, India.

Email: sujathar@ssnedu.in, gkuppuswamiji@gmail.com

Abstract. This paper presents a method to analyze the traffic flow pattern at a crowded junction in Chennai, one of the metropolitan cities in India, using waiting time in the signal in different time intervals with the help of Fuzzy Cognitive Map and Induced Fuzzy Cognitive Map.

Keywords: Fuzzy Cognitive Maps, Induced Fuzzy Cognitive Maps, Traffic flow

1. Introduction

Traffic congestion has been one of major issues that most metropolises are facing. It is believed that identification of congestion is the first step for selecting appropriate alleviation measures. Congestion impacts the movement of people. Different approaches can be used to minimize the traffic problem such as improved public transport, communication mechanisms expanded, proper handling of signal time. The first two approaches require many resources as money, labour, area etc. They seem very difficult to achieve in many places due to lack of resources. Control of traffic light signal timings is one of the least expensive and most effective means of reducing vehicular congestion in metropolitan road networks. The problem is examined by Fuzzy Cognitive Maps (FCMs) and Induced Fuzzy Cognitive Maps (IFCMs).

2. Basic concepts and preliminaries

Fuzzy Cognitive Maps describe and examine human being imagining or doings on particular areas by making examples. Lofti A. Zadeh in 1965 introduced the notion of fuzziness [1]. In 1976 Axelrod [2] presented cognitive maps. In 1986 Kosko [3] proposed FCMs based on the cognitive maps structure. Fuzzy cognitive maps for practical applications is discussed in [4]. Ever since, it is widely used in different areas [5-15]. In another applications, Ritha et al.[16] used IFCMs in medical for predictors of interest in cosmetic surgery.

An FCM is a directed graph with concepts such as policies, events as nodes and causalities as edges. It is a way to denote the causal relationship between concepts. If a concept C_j is in direct proportion with another concept C_i, then the directed edge is assigned a weight 1. If a concept C_j is in inverse proportion with another concept C_i, then the directed edge is assigned a weight -1. If there exists no relation between two concepts the value 0 is given. The connection or adjacency matrix of the FCM is $M = (e_{ij})$, e_{ij} being the edge weight.

The instantaneous behavior of each node is given as an input vector $A = (a_1, a_2, \cdots, a_n)$ where $a_i \in \{0,1\}$; 0 represents OFF and 1 represents ON position. The hidden pattern is the equilibrium state of the FCM. If the equilibrium state is a unique state vector, then is called fixed point. The dynamical system goes round and round when the causality flows through the edges like a cycle starting with concept C_i and ending at C_i when C_i is switched ON.

Suppose $A = (a_1, a_2, \cdots, a_n)$ is a vector which is passed into a dynamical system X. Then $A * E = (a'_1, a'_2, \dots, a'_n)$ after thresholding and updating the vector suppose we get $(b_1, b_2, \cdots, b_n)$. We denote this $(a'_1, a'_2, \cdots, a'_n) \hookrightarrow (b_1, b_2, \cdots, b_n)$. Thus the symbol $\hookrightarrow$ means the resultant vector has been threshold and updated. $(b_1, b_2, \cdots, b_n)$ is the fixed point of the dynamical system X.

3. Estimation of signal waiting time for different time intervals using FCMs

Signal co-ordination significantly improves the performance of network operations. Using traffic counts, the cycle times of the signals were calculated and offset optimization has been done. The road user cost (travel time, fuel consumption etc) can be considerably reduced by signal synchronization. Using FCMs we will estimate when the signal time is high and less.

In this work, we consider Velachery-Vijayanagar junction, the three road junction in Chennai. At this junction, Velachery Vijaya nagar three roads meet at the signal, one from Medavakkam to Vijaya nagar, the second from Taramani to Vijaya nagar and the third from Velachery By pass to Vijaya nagar. Since this area is also connected by fly-over train, daily huge no of vehicles passing through the signal to reach the railway junction. In other areas in Chennai, the waiting time in the signal is less compared to the week days, but in Velachery signal, the waiting time is high because Velachery is occupied by many malls. This junction is depicted in Figure 1. The different types of vehicles passing in all the three directions is counted for a period of time and classified for different time intervals morning 5 to 7, 7 to 9, 9 to 11, afternoon 11 to 1, 1 to 3, evening 3 to 5, 5 to 7, 7 to 9.

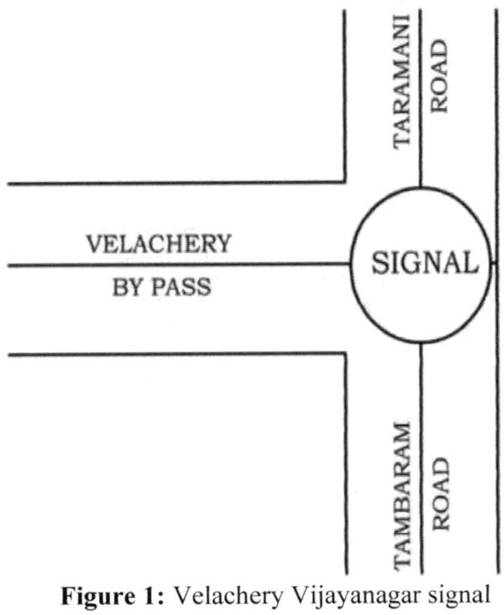

Figure 1: Velachery Vijayanagar signal

The pseudo code for the method is

* Collect the concepts (nodes) for the given problem.
* Obtain the connection square matrix M.
* Set the concept C_i (i=1, 2, 3,..., n) in ON-State
* Multiply C_i (i=1, 2, 3,..., n) with M and threshold value is calculated by assigning 1 to the first state and for the values > 0 to get C_2 .
*Multiply C_2 with M and repeat the procedure to get the fixed point.
* Continue the above process for the remaining state vector and find the hidden pattern.

The different concepts considered to study the signal waiting time are given in the table-1.

TABLE 1. CONCEPTS FOR SIGNAL WAITING TIME

Nodes	Concepts
C_1	Total no of vehicles passing in the time 5-7 A.M
C_2	Total no of vehicles passing in the time 7-9 A.M
C_3	Total no of vehicles passing in the time 9-11A.M
C_4	Total no of vehicles passing in the time 11-1P.M
C_5	Total no of vehicles passing in the time 1-3 P.M
C_6	Total no of vehicles passing in the time 3-5 P.M
C_7	Total no of vehicles passing in the time 5-7 P.M
C_8	Total no of vehicles passing in the time 7-9 P.M
C_9	Traffic is very low
C_{10}	Traffic is low
C_{11}	Traffic is medium
C_{12}	Traffic is high
C_{13}	Traffic is very high
C_{14}	Waiting time for the signal is less.
C_{15}	Waiting time for the signal is more

The connection matrix M is:

0	0	0	0	0	0	0	0	1	0	0	0	0	1	0
0	0	0	0	0	0	0	0	0	1	0	0	0	1	0
0	0	0	0	0	0	0	0	0	0	0	0	1	0	1
0	0	0	0	0	0	0	0	0	0	1	0	0	1	0
0	0	0	0	0	0	0	0	1	0	0	0	0	1	0
0	0	0	0	0	0	0	0	0	0	1	0	0	1	0
0	0	0	0	0	0	0	0	0	0	0	1	0	0	1
0	0	0	0	0	0	0	0	0	0	0	0	1	0	1
1	0	0	0	1	0	0	0	0	0	0	0	0	1	0
0	1	0	0	1	0	0	0	0	0	0	0	0	1	0
0	0	0	1	0	1	0	0	0	0	0	0	0	1	0
0	0	0	0	0	0	1	0	0	0	0	0	0	0	1
0	0	1	0	0	0	0	1	0	0	0	0	0	0	1
1	1	0	1	1	1	0	0	1	1	1	0	0	0	0
0	0	1	0	0	0	1	1	0	0	0	1	1	0	0

Case 1:

All the states are OFF state except "Total no of vehicles passing in the time 5-7 A.M" i.e., $A_1 = (1\ 0\ 0\ 0\ 0\ 0\ 0\ 0\ 0\ 0\ 0\ 0\ 0\ 0\ 0)$. The outcome vector is

$$
\begin{aligned}
A_1\mathrm{M} &= (0\ 0\ 0\ 0\ 0\ 0\ 0\ 0\ 1\ 0\ 0\ 0\ 0\ 1\ 0)\\
&\hookrightarrow (1\ 0\ 0\ 0\ 0\ 0\ 0\ 0\ 1\ 0\ 0\ 0\ 0\ 1\ 0) = A_2\\
A_2\mathrm{M} &= (2\ 1\ 0\ 1\ 2\ 1\ 0\ 0\ 2\ 1\ 1\ 0\ 0\ 2\ 0)\\
&\hookrightarrow (1\ 1\ 0\ 1\ 1\ 1\ 0\ 0\ 1\ 1\ 1\ 0\ 0\ 1\ 0) = A_3\\
A_3\mathrm{M} &= (2\ 2\ 0\ 2\ 3\ 2\ 0\ 0\ 3\ 2\ 3\ 0\ 0\ 8\ 0)\\
&\hookrightarrow (1\ 1\ 0\ 1\ 1\ 1\ 0\ 0\ 1\ 1\ 1\ 0\ 0\ 1\ 0) = A_4.
\end{aligned} \tag{1}
$$

Case 2:

All the states are OFF state except "Total no of vehicles passing in the time 9-11A.M" i.e., $A_3 = (0\ 0\ 1\ 0\ 0\ 0\ 0\ 0\ 0\ 0\ 0\ 0\ 0\ 0\ 0)$. The outcome vector is

$$
\begin{aligned}
A_3\mathrm{M} &= (0\ 0\ 0\ 0\ 0\ 0\ 0\ 0\ 0\ 0\ 0\ 0\ 1\ 0\ 1)\\
&\hookrightarrow (0\ 0\ 1\ 0\ 0\ 0\ 0\ 0\ 0\ 0\ 0\ 0\ 1\ 0\ 1) = A_4
\end{aligned}
$$

$$A_4\mathrm{M} = (0\,0\,2\,0\,0\,0\,1\,2\,0\,0\,0\,1\,2\,0\,2)$$
$$\hookrightarrow (0\,0\,1\,0\,0\,0\,1\,1\,0\,0\,0\,1\,1\,0\,1) = A_5$$
$$A_5\mathrm{M} = (0\,0\,2\,0\,0\,0\,2\,2\,0\,0\,0\,2\,3\,0\,5)$$
$$\hookrightarrow (0\,0\,1\,0\,0\,0\,1\,1\,0\,0\,0\,1\,1\,0\,1) = A_6. \quad (2)$$

Using Fuzzy Cognitive Maps from the case 1 we are getting (1 1 0 1 1 1 0 0 1 1 1 0 0 1 0) as a fixed point, i.e. total no of vehicles passing in the time 5-7 a.m, 7-9 a.m the traffic is very low, 11-1 p.m, 1-3 p.m the traffic is medium and 3-5 p.m there is less waiting time in signal. For two time intervals the traffic is low or medium. To overcome this and to correctly identify the signal waiting time Induced Fuzzy Cognitive Map is used. From the case 2 we are getting (0 0 1 0 0 0 1 1 0 0 0 1 1 0 1) is a fixed point, i.e. total no of vehicles passing in the time 9-11 a.m, total no of vehicles passing in the time 5-7 p.m, total no of vehicles passing in the time 7-9 p.m, traffic is high, traffic is very high are more waiting time in the signal.

4. Estimation of signal waiting time for different time intervals using IFCMs

Induced Fuzzy Cognitive Maps (IFCMs) is altered way of Fuzzy Cognitive Maps. The algorithmic approach for IFCM is explained in detail in [6].

Identification of the concepts, incorporating expert's opinion and formation of connection matrix is similar to FCM. We study the effect of each concept, by setting one concept at a time in ON state and the remaining in OFF state i.e., set the concept C_i (i=1, 2, 3,..., n) in ON-State. First consider C_1. Multiply C_1 with M. Similar to FCM, threshold to get the resultant vector C_1'. Now each component in C_1' is taken separately and the process of multiplication and thresholding is repeated. The vector with maximum number of one's is identified as a fixed point (say) B_1 . Multiply B_1 with M and threshold to obtain B_1'. Now each component in B_1' is taken separately and the process of multiplication and thresholding is repeated. The vector with maximum number of one's is identified as a fixed point. If this fixed point is same as B_1 , the process is terminated. At this stage we conclude that B_1 is the fixed point of C_1. This procedure is repeated for the remaining concepts and the hidden pattern is identified. Based on the hidden patterns derived the resultant flow can be identified precisely. The relation between the concepts is plotted as a graph based on the resultant flow.

The IFCM calculations are done for the FCM model discussed in section-3 and the results of are presented in the following table. In Figure 2, the IFCM of the traffic flow problem is depicted.

Using Induced Fuzzy Cognitive Maps we proved that the waiting time for the signal is less in the time interval 5-7 a.m, 7-9 a.m,11-1p.m,1-3p.m,3-5p.m. When the traffic is very low, low and medium the waiting time is signal is less. In the time interval 9-11 a.m, 5-7 p.m, 7-9 p.m signal waiting time is more. When the traffic is high and very high the signal waiting time is more. These observations clearly indicate the relationship between the time intervals, traffic volume and signal waiting time, which were the concepts of FCM.

TABLE.2. RESULTS OF IFCMS

No.	Concepts ON State	Resultant Flow
Step 1	(1 0 0 0 0 0 0 0 0 0 0 0 0 0 0)	$C_1 \rightarrow C_{14} \rightarrow C_{14}$
Step 2	(0 1 0 0 0 0 0 0 0 0 0 0 0 0 0)	$C_2 \rightarrow C_{14} \rightarrow C_{14}$
Step 3	(0 0 1 0 0 0 0 0 0 0 0 0 0 0 0)	$C_3 \rightarrow C_{15} \rightarrow C_{15}$
Step 4	(0 0 0 1 0 0 0 0 0 0 0 0 0 0 0)	$C_4 \rightarrow C_{14} \rightarrow C_{14}$
Step 5	(0 0 0 0 1 0 0 0 0 0 0 0 0 0 0)	$C_5 \rightarrow C_{14} \rightarrow C_{14}$
Step 6	(0 0 0 0 0 1 0 0 0 0 0 0 0 0 0)	$C_6 \rightarrow C_{14} \rightarrow C_{14}$
Step 7	(0 0 0 0 0 0 1 0 0 0 0 0 0 0 0)	$C_7 \rightarrow C_{15} \rightarrow C_{15}$
Step 8	(0 0 0 0 0 0 0 1 0 0 0 0 0 0 0)	$C_8 \rightarrow C_{15} \rightarrow C_{15}$
Step 9	(0 0 0 0 0 0 0 0 1 0 0 0 0 0 0)	$C_9 \rightarrow C_{14} \rightarrow C_{14}$
Step 10	(0 0 0 0 0 0 0 0 0 1 0 0 0 0 0)	$C_{10} \rightarrow C_{14} \rightarrow C_{14}$
Step 11	(0 0 0 0 0 0 0 0 0 0 1 0 0 0 0)	$C_{11} \rightarrow C_{15} \rightarrow C_{15}$
Step 12	(0 0 0 0 0 0 0 0 0 0 0 1 0 0 0)	$C_{12} \rightarrow C_{15} \rightarrow C_{15}$
Step 13	(0 0 0 0 0 0 0 0 0 0 0 0 1 0 0)	$C_{13} \rightarrow C_{15} \rightarrow C_{15}$
Step 14	(0 0 0 0 0 0 0 0 0 0 0 0 0 1 0)	$C_{14} \rightarrow C_9 \rightarrow C_{14} \rightarrow C_{14}$ $C_{14} \rightarrow C_{10} \rightarrow C_{14} \rightarrow C_{14}$ $C_{14} \rightarrow C_{11} \rightarrow C_{14} \rightarrow C_{14}$
Step 15	(0 0 0 0 0 0 0 0 0 0 0 0 0 0 1)	$C_{15 \rightarrow} C_{13} \rightarrow C_{15} \rightarrow C_{15}$

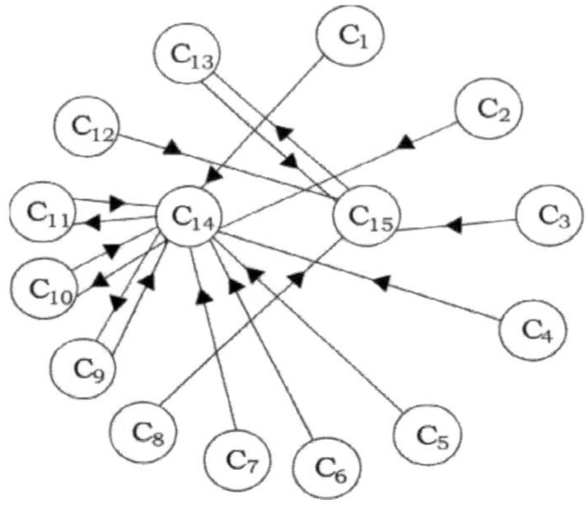

Figure 2: Induced directed graph

5. Conclusion

The concepts of Fuzzy Cognitive Maps and Induced Cognitive Maps were applied to analyse the traffic flow problem. The results indicate precisely the relation between the chosen concepts. This approach can be applied to examine the traffic flow pattern in any crowded road junction.

Acknowledgements

The authors thank Science and Engineering Research Board, Department of Science and Technology, India for providing financial assistance for carrying out this work under the project SR/S4/MS:816/12. The authors thank the Managements of SSN College of Engineering and Sri Venkateswaraa College of Technology for their support.

References

[1] Zadeh L A 1965 Fuzzy sets *Information and Control* **8** pp 338-353
[2] Axelord R 1976 The Cognitive Maps of Political Elites *Structure of Decision* Princeton University Press **72** pp 1017-18
[3] Bart Kosko 1986 Fuzzy Cognitive Maps *International Journal of Man-machine Studies* **24** pp. 65-75.
[4] Vasantha Kandasamy W B Florentin Smarandache 2003 *Fuzzy Cognitive Maps and Neutrosophic Cognitive Maps* Xiquan Phoenix
[5] Papageogiou E I Salmeron J L 2014 Methods and Algorithms for Fuzzy Cognitive Map-based modelling *Springer* pp 1-28.
[6] Dhrubajyoti G Anitha P 2015 Using fuzzy cognitive map and induced fuzzy cognitive map to analyze the real world problems *Annals of Pure and Applied Mathematics* **10** pp 153-163..
[7] Marta O Unai P 2015 Use of Fuzzy cognitive maps to study urban resilience and transformation *Environmental Innovation and Societal Tranitions* **18** pp 1-19.
[8] Salmeron L Wojciech F 2016 Dynamic optimization of Fuzzy cognitive maps for time series forecasting *Knowledge-based system* **105** pp 1-9.
[9] Ashraful Alam Md 2017 Fuzzy cognitive maps approach to identify risk factors of diabetes *Journal of Physical Sciences* **22** pp 13-21.
[10] Jose A Oswaldo T Hebert S Jose G Jorge C Danilo C 2017 Towards a Fuzzy cognitive map for opinion mining *Precedia Computer Science* **108** pp 2522-2526.
[11] Kartik P Shyamal Kumar M 2017 Risk analysis in a production system using Fuzzy cognitive map *International Journal of Mathematics in Operational Research* **11** pp 29-44.

[12] Vasiliy S Kireev Ivan S Smirno Victor S Tyunyakov 2018 Automatic Fuzzy Cognitive Map Building Online System *Procedia Computer Science* **123** pp 228-233.

[13] Kokkinos K Lakioto E Papageorgiou E Moustakas K Karayannis V 2018 Fuzzy cognitive map-based modeling of social acceptance to overcome uncertainties in establishing waste biorefinary facilities *Frontiers in Energy Research* **6** pp 1-18.

[14] Vassiliki M Peter P 2018 Re-approaching Fuzzy cognitive maps to increase the knowledge of a system *AI & Society(Springer)* **33** pp 175-188.

[15] Nabin S Waldemar K 2019 Applications of Fuzzy Cognitive Maps in Human Systems Integration *Advances in Artifical Intelligence, Software and Systems Engineering* pp 391-399.

[16] Ritha W Mary Mejrullo Merlin M 2011 Predictors of interest in cosmetic surgery-An analysis used induced fuzzy cognitive maps *Annals of Fuzzy Mathematics and Informatics* **2** pp 161-169.

EDGE BIMAGIC LABELINGS OF GRAPHS $K_{2,N}$, $K_{N,2,N}$, BISTAR, nC_4, AND DOUBLE WHEEL

Dr.S. Subbulakshmi[1], and R. Kokila[2]
[1]Dharmapuram Gnanabigai Government Arts College(Women),(Affiliated to Bharathidasan University), Mayiladuthurai-609 001, Tamilnadu, India.**Email id: drssl2006@gmail.com**
[2]Sindhi College of Arts & Science, 146, Poonamallee High Road, Numbal, Chennai-600 077, Tamilnadu, India.
Email id: kokilamanokaran@gmail.com

Abstract. Ngurahet. al. [2010] noted super magic coverings of fans and ladders. Ahmad et. al. [2011] enumerated super edge-magic labeling for the union of non-isomorphic graphs. Lopez et.al. [2012] suggested perfect super edge-magic graphs of path and cycle. Gao and Fan [2013] analyzed super edge-magic labelings of multi-star St(a1,...,an), Ali et. al. [2013] magic Labelings of type (a, b, c) of families of wheels.**In this article,** edge bimagiclabelings for bipartite complete graph, double bipartite complete graph, bistar merging with a path, one point union of finite copies of cycle having four vertices, and closed double wheel are found.

Section 1 - Introduction: Baskoro and Ngurah [2003] found super edge-magic total labelings of some classes of paths and trees. Beardon [2004] discussed magic labellings of infinite graphs, like as star and wheel. Calhoun et. al. [2005] studied totally magic labelings of comb and web graphs. Ivanco [2009] obtained supermagic regular graphs. Ivanco and Luckanicova [2002] constructed edge-magic disconnected graphs of cycles. Lee et.al.[2007] initiated edge-magic indices of (v,v + 1)-graphs. MacDougall and Wallis [2003] got strong edge-magic labeling of a cycle with a chord. McQuillan [2010] derived a technique for constructing magic labelings of 2-regular graphs. McQuillan and McQuillan [2009] observed magic labelings of triangles.

Section 2 – Necessary definitions:

Definition 2.1: A graph G(V,E) with order p and size q is edge magic if there exists a bijection $f : V \cup E \to \{1, 2,..., p + q\}$ such that $f(u) + f(v) + f(uv)$ is a constant , for all edges $uv \in E$ and it is super edge magic (SEM) if g further satisfies $g(V) = \{1, 2,..., p\}$.

Definition 2.2: A graph G(V,E) with order p and size q is edge bimagic if there exists a bijection $g : V \cup E \to \{1,2,..., p + q\}$ such that $g(u) + g(v) + g(uv)$ is a constant either c_1 or c_2 for all edges $uv \in E$, and G is super edge bimagic (SEBM) if g also satisfies $g(V) = \{1,2,..., p\}$.

Section 3 –Edge bimagiclabelings for some bipartite complete graphs and double wheel:
Definition3.1: $K_{2,n}$ is a complete connected graph obtained by joining two vertices in one side and n vertices on the other side.Its vertex set $\{v_0,v_0', v_1,v_2,\dots,v_n\}$ is and edge set is $\{v_0v_i : i = 1 \text{ to } n\}$ $\{v_0' v_i : i=1 \text{ to } n\}$.

Theorem 3.2: The bipartite complete graph $K_{2,n}$ is bi-magic.
Proof: One of the arbitrary labelings for the vertices of the given graph is as follows in figure 1.

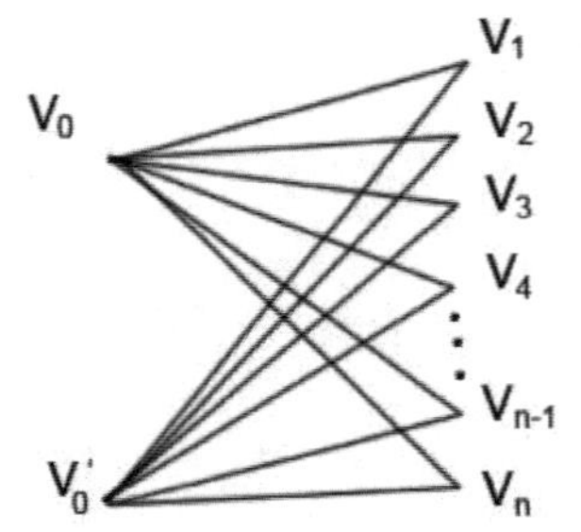

Define f : V(G){1,2,3,…,p} by
$f(v_0) = 1$ $f(v_0') = 2$
(1). n is even:
$f(v_i) = 2i+1$: i = 1 to (n/2)
$=2(i+1-n/2)$: i= (n/2)+1,…,n
(2). n is odd
$f(v_i) = 2i +2$: i= 1,2,3,… (n-1)/2
$=2(i +1-((n+1)/2))+1$: i = ((n+1)/2)+1, 2, …,n.
Define f : E(G)→{p +1, p +2,…,p+q } by
$f(v_0v_i) = 3n+4-2i$: i = 1 to n, and $f(v_0'\, vi) = 3n+3-2i$: i = 1 to n

So f satisfies the conditions for edge bi-magic labeling for vertices and edges of the given graph and so $K_{2,\, n}$ is edge bi-magic. Its bimagic numbers are $k_1 = 3n+7$, $k_2 = 2n+7$ if n is odd, and $k_1 = 3n+6$, $k_2 = 2n+7$ if n is even.

Definition 3.3: A bi-star with path is a tree whose vertex set is $\{v_1,v_2,\ldots,v_n,v_{n+k},\ldots,v_{n+k+s}\}$, and edge set is $\{v_i\, v_{i+1}$:i = 1 to(n-1)} { v_iv_{n+i} :i = 1 to k } {v_nv_{n+k+I} : i =1 to s}.

Theorem 3.4: Bistar with a path Pn (n is even) is bimagic.
Proof: One of the arbitrary labelings for the vertices of the given graph is as follows in figure 2.

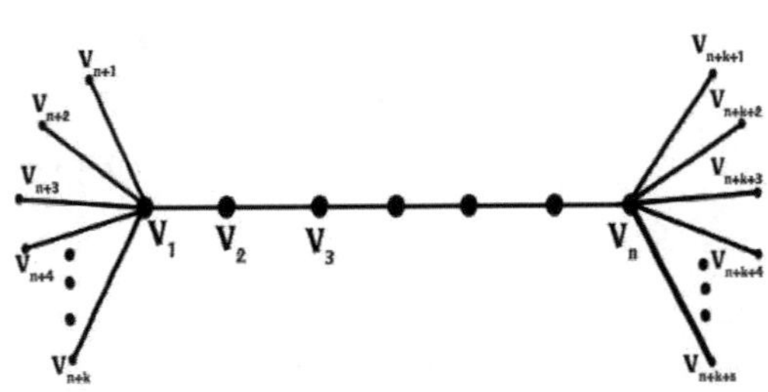

Define f : V(G){1,2,3,…,p} by
$f(v_i) = i$: i= 1 to n
Define f: E(G)→{p +1, p +2,…,p+q } by
$f(v_i\, v_{i+1})$ = 2(n+k+s)-i : i =1 to (n/2)
=2(n+k+s) – 2-(i -+ 1)=(n+k+s)+ -1 – i where i = ,,,…,(n-1).
$f(v_1v_{n+i})$ = n+2k+2s+1- i ; i =1 to k
$f(v_nv_{n+k+i})$ = n+k+2s+1- i : i =1 to s

So f satisfies the conditions for edge bi-magic labeling for vertices and edges of the given graph and so Bistar with a pathis edge bi-magic.

Definition 3.5: It is a connected graph obtained by joining any number of cycle C_4 at a common vertex whose vertex set is $\{v_0,v_1,v_2,\ldots,v_n,v_{n+1},\ldots,v_{2n}\}$ and edge set is $\{v_0v_i \,;\, i = 1$ to $2n\}$ $\{v_{2i}\, v_{2n+i} : i = 1$ to $n\}$ $\{v_{2i-1}\, v_{2n+I} : i = 1$ to $n\}$.

Theorem 3.6: One point union of finite copies of cycle having 4 vertices (nC_4) is bi-magic.
Proof: One of the arbitrary labelings for the vertices of the given graph is as follows in figure 3.

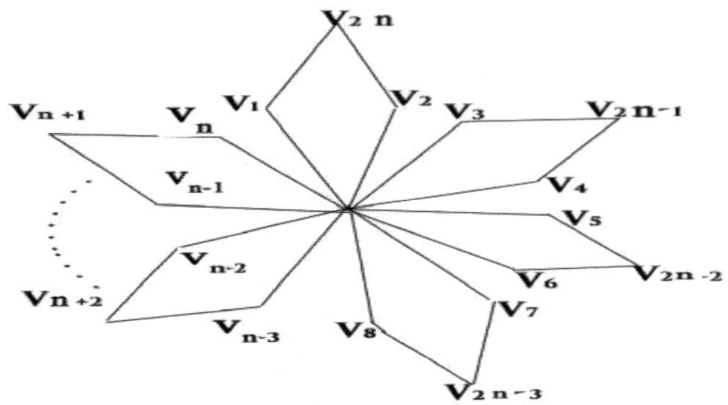

Define $f{:}V(G)\rightarrow\{1,2,3,\ldots,p\}$ by
$f(v_i) = i+1 \;: i = 0$ to $2n$.
Define f: $E(G)\rightarrow\{p+1,\ p+2,\ldots,p+q\}$ by
$f(v_{2n+i}) = 3n+2-i : i = 1$ to n;
$f(v_{2i-1}\, v_{2n+I}) = 4n+2-i : i = 1$ to n
$f(v_{2i}\, v_{2n+i}) = 5n+2-i \;: i = 1$ to n
So f satisfies the conditions for edge bi-magic labeling for vertices and edges of the given graph and so Bistar with a pathis edge bi-magic.

Definition 3.7: $K_{n,2,n}$ is a complete connected graph and the mirror image of $K_{2,n}$ and its vertex set : $\{v_0,v_0',v_1,v_2,\ldots,v_n,v_{n+1},\ldots,v_{n+k}\}$, and edge set$\{v_0vi : i = 1$ to $n\}\cup\{v_{0'}\, v_i : i=1$ to $n\}$ $\{v_0vi,\ v_{0'}\, v_i \,;\, i=n+1,\ldots,n+k\}$.

Theorem 3.8: $K_{n,2,n}$ is bi-magic.
Proof: One of the arbitrary labelings for the vertices of the given graph is as follows in figure 4.

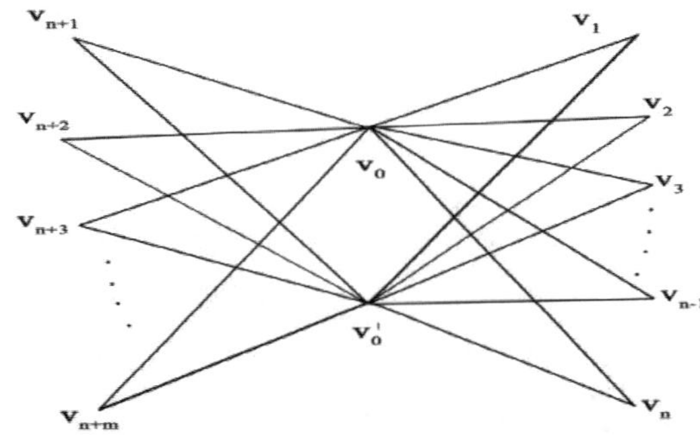

Define f : V(G){1,2,3,…,p} by
$f(v_0) = 1$ $f(v_0') = 2$; $f(v_i) = i+2$, $i = 1,2,\ldots,(n+k)$
Define f : E(G)→{p +1, p +2,…,p+q } by
$f(v_0v_1) = 3n+3k+2$; $f(v_0'v_1) = 3n+3k+1$;
$f(v_0'v_i) = 3n+3k+2- i$, $i = 2,3,\ldots,n$
$f(v_0'v_{n+i}) = 2n + 3k + 2 - i$,$i = 1,2,\ldots,k$
$f(v_0v_i) = 2n+2k+3- i$, $i = 2,3,\ldots,m$
$f(v_0v_{n+i}) = n+2k+3- i$, $i = 1,2,\ldots,n$

So f satisfies the conditions for edge bi-magic labeling for vertices and edges of the given graph and so $K_{n,2,n}$ is edge bi-magic.

Definition 3.9: The vertex set of $W_n=C_n +K_1$ is $\{v_0,v_1,\ldots,v_n\}$ and edge set is $\{v_0v_i$: i=1 to n$\}$ $\{v_iv_{i+1}$: i=1 to (n-1) ,(n+1) to (2n-1) $\}$ $\{v_nv_1\}$.

Definition 3.10: Closed helm is a connected graph obtained from a helm by joining each pendent vertex to form a cycle. The pendent edges are $\{v_iv_{n+I}$: i =1 to n$\}$.The edges of the outer cycle are $\{v_iv_{i+1}$: i=1 to (n-1) ,(n+1) to (2n-1) $\}$ $\{v_{2n}v_{2n-1}\}$

Definition 3.11: W(2, n) is closed helm whose vertex set : $\{v_0,v_1,\ldots,v_{2n}\}$ and edge set : $\{v_0v_i$: i=1 to n$\}$ $\{v_iv_{n+I}$: i=1 to n$\}$ $\{v_iv_{i+1}$: i =1 to (n-1) ,(n+1) to (2n-1) $\}$ $\{v_nv_1\}$ $\{v_{2n}v_{2n-1}\}$.The pendent edges are $\{v_iv_{n+I}$: i =1 to n$\}$.The edges of the outer cycle are $\{v_iv_{i+1}$: i =1 to (n-1) ,(n+1) to (2n-1) $\}$ $\{v_{2n}v_{2n-1}\}$.

Theorem 3.12: Closed W(2,n) (n is odd) [double wheel] is bi-magic.

Proof: One of the arbitrary labelings for the vertices of the given graph is as follows in figure

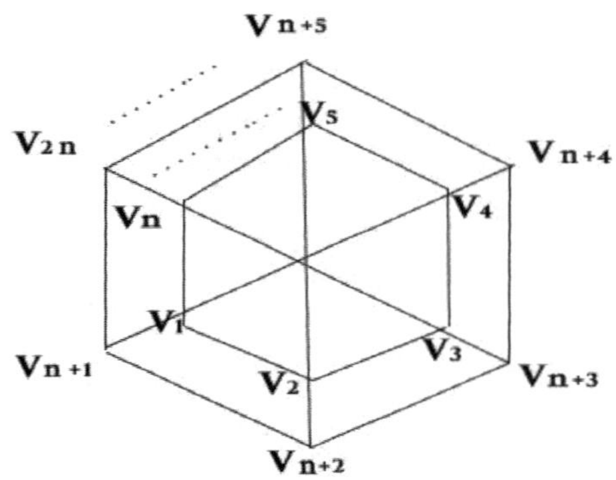

5.

Define $f:V(G)\rightarrow \{1,2,\ldots,p\}$ by

$f(v_i) = i+1, \quad i = 0 \text{ to } 2n$

$f(v_0v_i) = 6n + 2 - i, \; i = 1 \text{ to } n$

$f(v_nv_1) = 5n+1;$

$f(v_iv_{i+1}) = 5n+1- 2i, \; i = 1 \text{ to } n$

$f(v_iv_{n+i}) = 5n+2-2i, \; i = 1 \text{ to } n$

$f(v_{2n}v_{n+1}) = 3n+1;$

$f(v_{n+i}v_{n+i+1}) = 3n+1-2i, \; i = 1,2,\ldots,$

$= 3n+(i -), \; i = ,\ldots,(n-1)$

So f satisfies the conditions for edge bi-magic labeling for vertices and edges of the given graph and so W(2,n) is edge bi-magic. The bimagic numbers are $k_1 = 6n+4$, and $k_2 = 5n+4$.

1.1. A subsection

Some text.

1.1.1. A subsubsection. The paragraph text follows on from the subsubsection heading but should not be in italic.

References:

[1] A. Ahmad, S.C. Lopez, F.A. Muntaner-Batle, and M. Rius-Font, Enumerating super edge-magic labeling for the union of non-isomorphic graphs, Bull. Aust. Math. Soc., 84 (2011) no. 2, 310-321.

[2] K. Ali, M. Hussain, A. Ahmad and M. Miller, Magic Labelings of type (a,b,c) of families of wheels, Math. Computer Sci., 7 (2013) 315-319.

[3] E. Baskoro and A. A. G. Ngurah, On super edge-magic total labelings, Bull. Inst. Combin. Appl., 37 (2003) 82-87.

[4] A. F. Beardon, Magic labellings of infinite graphs, Austral. J. Combin., 30 (2004), 117-132.

[5] B. Calhoun, K. Ferland, L. Lister, and J. Polhill, Totally magic labelings of graphs, Australas. J. Combin., 32 (2005) 47-59.

[6] Z. Gao and C. Fan, Some discussion on super edge-magic labelings of St(a1,...,an), ArsCombin., 108 (2013) 187-192.

[7] J. Ivanˇco, Onsupermagic regular graphs, Math. Bohemica, 125 (2000) 99-114.

[8] J. Ivanˇco and I. Luˇckaniˇcov´a, On edge-magic disconnected graphs, SUT J. Math., 38 (2002)175-184.

[9] S. M. Lee, H. K. Ng, and Y. Wen, On the edge-magic indices of (v,v + 1)-graphs, Util. Math., 72 (2007) 97-110.

[10] S. C. Lo´pez, F. A. Muntaner-Batle, M. Rius-Font, Perfect super edge-magic graphs, Bull. Math. Soc. Sci. Math. Roumanie, 55 (103) No 2 (2012) 199-208.

[11] J. A. MacDougall and W. D. Wallis, Strong edge-magic labeling of a cycle with a chord, Australas. J. Combin., 28 (2003) 245-255.

[12] D. McQuillan, A technique for constructing magic labelings of 2-regular graphs, J. Combin. Math.Combin.Comput., 75 (2010) 129-135.

[13] D. McQuillan and J. McQuillan, Magic labelings of triangles, Discrete Mathematics, 309 (2009) 2755-2762.

[14] A. A. G. Ngurah, A. N. M. Salman, and I. W. Sudarsana, On supermagic coverings of fans and ladders, SUT J. Math., 46 (2010) 67-78.

BI-MAGIC LABELINGS OF SOME CONNECTED AND DISCONNECTED GRAPHS

Dr.S. Subbulakshmi1 , and R. Kokila 2

1DharmapuramGnanambigai Government Arts College (Women), (Affiliated to Bharathidasan University), Mayiladuthurai-609 001, Tamilnadu, India. Email id: drssl2006@gmail.com

2 Sindhi College of Arts & Science, 146, Poonamallee High Road, Numbal, Chennai-600 077, Tamilnadu, India.

Email id: kokilamanokaran@gmail.com

Abstract. Various graph labeling that generalize the idea of a magic square have been discussed. In particular ,a magic labeling on a graph with v vertices and e edges will be defined as a one to one map taking the vertices and edges onto integers 1,2,…,v+e with the property that the sum of the label on an edge and the labels of its end points is constant ,independent of the choice of edge.

Introduction

In 1970 Kotzig Rosa defined a magic valuation of a graph G(V,E) as a bijection f from V∪ E to {1,2,…,|V ∪ E | } such that for all edges xy , f(x)+f(y)+f(xy) is constant (called a magic constant).It was later called edge magic labeling. Enomoto ,Llado,Nakamigawa and Ringel called a graph G(V,E) with an edge magic total labeling that has the additional property that the vertex labels are 1 to |V| super edge magic labeling.

In 2004,Baskar Babujee introduced the notion of vertex bimagiclabeling in which there exists two constants k_1 and k_2 such that the sums involved in a specified type of magic labeling is k_1 or k_2. An edge bimagiclabeling of a graph G(V,E) with p vertices and and q edges is a bijection f from the set of vertices and edges to such that for every edge uv in E ,

f(u)+f(uv)+f(v) is one of two constants k_1 or k_2.

In this paper,we discuss the bimagiclabeling of the graphs : Helm, Bistar, Shell graph ,Rooted tree and the disconnected graph nK_3

Section 1 – Necessary definitions:

Definition1.1: An edge bi-magic total labeling of a graph G(V,E) with p vertices and q edges is a bijection f from the set of vertices and edges to such that for every edge uv in E, f(u)+f(uv)+f(v) is one of two constants k_1 or k_2 ,independent of the choice of the edge.

Definition1.2:A graph G(V,E) with order p and size q is edge bimagic if there exists a bijection g: V∪E → {1,2,…p+q}such that g(u)+g(v)+g(uv) is a constant either constant c_1 or c_2 for all edges uv in G and G is super edge bimagic if g also also satisfies g(V)={1,2,…,p}.

Section 2 – Edge bimagiclabeling of some connected graphs

Definition 2.1 The helm H_n is a connected graph whose vertex set is $\{v_0,v_1,v_2,\dots,v_n,v_{n+1},\dots,v_{2n}\}$ and edge set is $\{v_0v_i ; i = 1 \text{ to } n\} \cup \{v_i v_{i+1} : i= 1 \text{ to} (n-1)\} \cup \{v_iv_{n+i} : i = 1 \text{ to } n\} \cup \{v_n v_1\}$.It is obtained from a wheel by attaching a pendent edge at each vertex of the n- cycle .the vertex set and edge set of the n-cycle are $\{v_0,v_1,v_2,\dots,v_n,v_{n+1},\dots,v_{2n}\}$ and $\{v_0v_i ; i = 1 \text{ to } n\} \cup \{v_i v_{i+1} : i= 1 \text{ to} (n-1)\} \cup \{v_n v_1\}$.The pendent edges are $\{v_iv_{n+i} : i= 1 \text{ to } n\}$

Theorem2.2: The graph helm H_n is bi-magic

Proof: One of the arbitrary labeling of the given graph is as follows in figure 1.

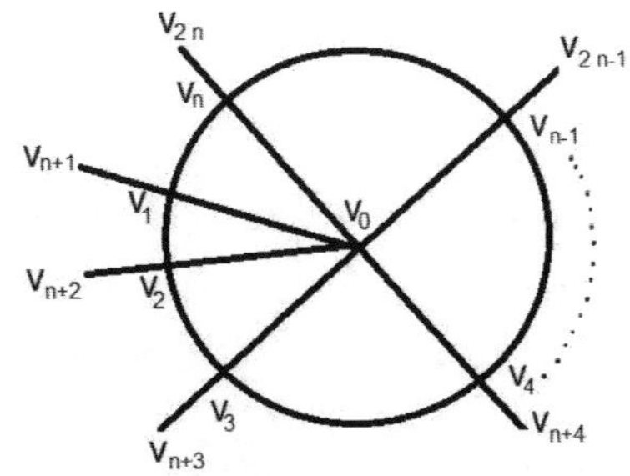

Figure 1 –One of arbitrary labelings of vertices of Helm graph Hn

Define f : V(G)→{1,2,3,…,p} by

$f(v_i) = i+1$: i= 0 to 2n

Define f : E(G) → {p+1,p+2,…,p+q} by

$f(v_0v_i) = 5n+2-i$:i = 1 to n

$f(v_i v_{i+1}) = 4n+1-2i$;i =1 to (n-1)

$f(v_iv_{n+i}) = 4n+2-2i$; i = 1 to n

$f(v_nv_1) = 4n+1$.

For all edges uv in G, f(u)+f(uv)+f(v) = f(uv) = 4n + 4 or 5n + 4. Fix $k_1 = 4n+4$; and $k_2 = 5n+4$. So $H_m$satisfies the all conditions of bi-magic graph, and helm graph H_n is bi-magic.

Definition 2.3: The graph $B_{n,n}$ is the connected graph obtained by joining the centre vertex of one of $K_{1,n}$ to the pendent vertex of the other $K_{1,n}$ .Its vertex set is $\{v_0,v_1,v_2,..,v_n,v_{n+1},.,v_{2n+2}\}$ and edge set is $\{v_1v_i$: i = 2 to (n+1)$\} \cup \{v_{n+2}\, v_i$: i = n+3,n+4,…,2n+2 $\} \cup \{v_1v_{n+3}\}$.

The vertex set and edge set of one of $K_{1,n}$ are $\{v_0,v_1,v_2,\dots,v_n,v_{n+1}\}$ and $\{v_1v_i$: i = 2 to (n+1)$\}$ and the other are $\{v_{n+2},\dots,v_{2n+2}\}$ and $\{v_{n+2}\, v_i$: i= n+3,n+4,…,2n+2 $\}$.They are joined by the edge v_1v_{n+3}.

Theorem 2.4: The graph bi-star B(n,n) is bi-magic

Proof: One of the arbitrary labeling of the given graph is as follows.

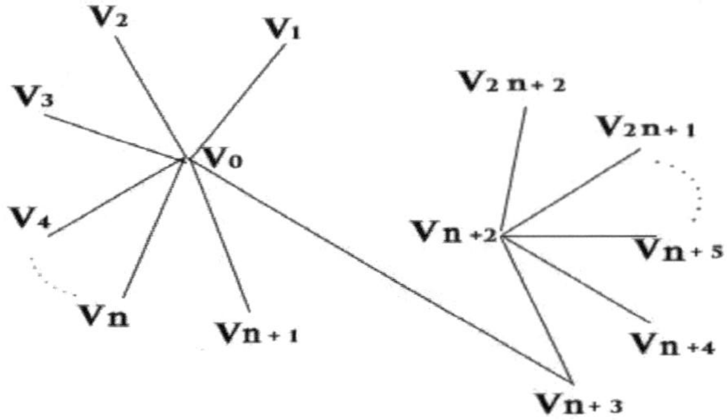

Figure 2 – One of arbitrary labeling of vertices of graph $B_{n,\,n}$.

Define f : V(G) $\rightarrow$ {1,2,3,…,p} by

$f(v_i) = i : i = 1$ to (2n+2)

Define f : E(G)$\rightarrow$ { p+1, p+2,…,p+q } by

$f(v_1v_{n+3}) = 4n+3$;

$f(v_1v_i) = 4n+4-i$; i = 2,3,…,(n+1)

$f(v_{n+2}v_i) = 4n+5-i$; i = n+3,n+4,…,(2n+2).

For all edges uv in G, f(u)+f(uv)+f(v) = f(uv) = 4n + 5 or 5n + 7. Fix k_1 = 4n+ 5; and k_2 = 5n+ 7. So $B_{n,\,n}$ fulfilled all conditions of bi-magic graph, and helm graph $B_{n,\,n}$ is bi-magic.

Definition 2.5: (Shell graph): The shell graph C(n,k) denotes a connected graph which contains a cycle C_n with k chords sharing a common end point called the apex . Its vertex set is $\{v_1,v_2,\ldots,v_n\}$ and edge set is $\{ v_i\, v_{i+1}$: i = 1 to(n-1)$\}\cup\{ v_iv_i$: i = 3 to (n-1) $\}\cup\{ v_nv_1\}$. The vertex set of cycle C_n is $\{v_1,v_2,\ldots,v_n\}$ and edge set is$\{v_iv_{i+1}$: i = 1 to(n-1)$\}\cup\{v_n\ v_1\}$. The chords are $\{v_iv_i$: i = 3 to (n-1) } and v_1 is the apex.

Theorem 2.6: The shell graph C(n,k) is bi-magic (n – even).

Proof: One of the arbitrary labeling of the given graph is as follows;

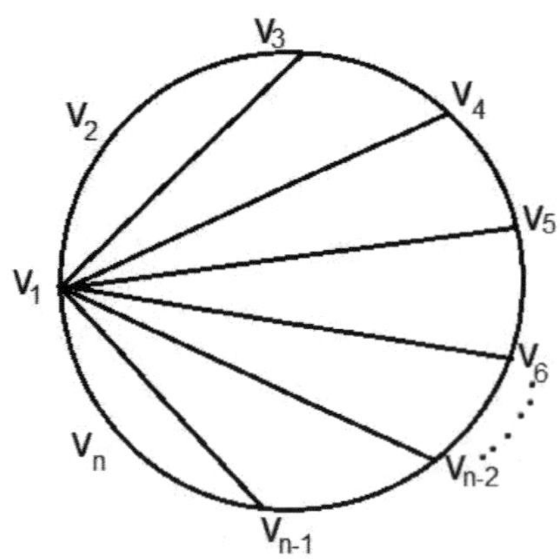

Figure 3 – One of arbitrary labeling of vertices of shell graph C(n, k).

Define f: V(G) → {1,2,3,…,p} by

$f(v_i) = i : i = 1$ to n.

Define f : E(G) → {p +1, p +2,…,p+q } by

$f(v_1v_i) = 2n+k-i+3$; i = 3 to (n-1)

$f(v_1v_n) = 2n$

$f(v_iv_{i+1}) = (2n-1)-2(i-1)$; i =1 to (n/2)

$=2n-2i+1$.

$f(v_i\, v_{i+1}) = (2n-2)-(i -(n+2)/2)$; i =(n+2)/2 to (n-1)

$=(5n-2i-2)/2$.

For all edges uv in G, f(u)+f(uv)+f(v) = f(uv) = 2n + 2 or 3n + 1. Fix k_1 = 2n+ 2; and k_2 = 3n+ 1. So C(n, k) obeyed all conditions of bi-magic graph, and shell graph C(n, k) is bi-magic.

Definition 2.7: Rooted tree is joined of many paths at a point called its root. Each path is of odd length whose vertex set = $\{v_0,v_1,\ldots,v_{k1},v_{k1+1},\ldots,v_{2k1},v_{2k1+1},\ldots,v_{nk1}\}$ and edge set $=\{v_iv_{i+1}$; $i=1,2,\ldots,k_1-1$: $i = k_1+1,\ldots,2k_1-1$; $i=2k_1+1,\ldots,3k_1-1$; ……. $i = (n-1)k_1+1,\ldots,nk_1-1\} \cup \{v_0v_j$; $j=1,k_1+1,2k_1+1,\ldots,(n-1)k_1+1\}$.

Theorem 2.8: Rooted tree is bi-magic.

Proof: One of the arbitrary labeling of the given graph is as follows;

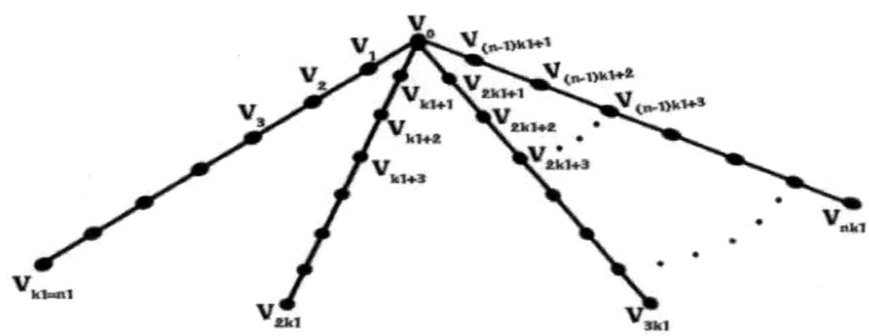

Figure 4 – One of arbitrary labeling of vertices of rooted tree

Define f:V(G) → {1,2,…,p} by

$f(v_0) = 1$;

$f(v_j)=f(v_{\ell k_1+i})$(v_lk_1+i)

$= \ell+2+n\,(i-1)$ where $l= 0,1,2,\ldots,(n-1)$, $i =1,2,3,\ldots,k_1$.

Define f:E(G) → {p+1,p+2,…,p+q} by

$f(v_0v_j)= f(v_0v_{(t-1)k_1+1})$(v_0v_t-1k_1+1)

$=2nk_1+2 - t$ where $t=1,2,\ldots,$

$f(v_{(t-1)k1+s}\, v_{(t-1)k1+s+1}) = (2nk_1-(2s-1)n-2(t-1))$; $t=1,2,\ldots,n$; $s=1,2,\ldots,\left(\frac{k-1}{2}\right)$

$f(v_{(t-1)k_1+s}v_{(t-1)k_1+s+1}) = (2nk_1+1)-2(t-1)-2\left[\left(s-\left(\frac{k+1}{2}\right)\right)\right] n$; $t=1,2,\ldots,n$; $s=\left(\frac{k+1}{2}\right),\ldots,k.$

For all edges uv in G, $f(u)+f(uv)+f(v) = f(uv) = 2\ell n_1+4; 2\ell n_1+4n_1+5$. So all conditions of bi-magic are checked in given rooted tree, and it is bi-magic.

Section 3- Disconnected graph nK_3 :

Definition3.1 :nK_3 are disconnected collection of k_3 graphs in which each k_3 is connected whose vertex set is $\{v_1,v_2,\ldots,v_n,v_{n+1},\ldots,v_{2n},v_{2n+1},\ldots,v_{3n}\}$, and edge set is $\{v_iv_{n+I} : i=1 \text{ to } n\} \cup \{v_iv_{2n+I} ; i=1 \text{ to } n\} \cup \{v_{n+i}v_{2n+I} ; i=1 \text{ to } n\}$.

Theorem 3.2: nK_3 (n is even) is bi-magic

Proof: One of the arbitrary labeling of the given graph is as follows;

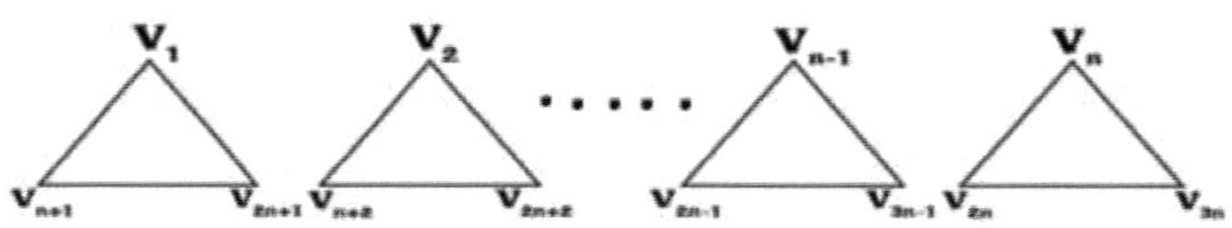

Figure 5 – One of arbitrary labeling of vertices of nk_3.

$f(v_i) = i$ (i=1 to n)

$f(v_{n+i}) = n+i$ (i=1 to n)

$f(v_{2n+i}) = 2n+i$ (i=1 to n)

Define f: $E(G) \rightarrow \{p+1,p+2,\ldots,p+q\}$ by

$f(v_iv_{n+i}) = 6n-2i+2$ (i=1 to n)

$f(v_iv_{2n+i}) = 6n-2i-10$ (i=1 to $\frac{n}{2}$)

$f(v_{n+i}v_{2n+i}) = 6n-11-2\left(i-\left(\frac{n+2}{2}\right)\right)$ $(i=\frac{n}{2}+1,\ldots,n)$

For all edges uv in G, $f(u)+f(uv)+f(v) = f(uv) = 7n + 2; 8n + 1$ So all conditions of bi-magic are satisfied in the graph nK_3 (n is even), and it is bimagic.

References:

1. J.BaskarBabujee, Bimagiclabeling in path graphs, Math .Education,38(2004),12-16

2. J.BaskarBabujee, On edge bimagiclabeling, journal of combinatorics Information and system sciences, 28 (2004), 239-244

3. J.A Gallian, A dynamic survey of graph labeling, Electronic Journal of combinatorics, 2016 #DS6

4. A.AmaraJothi ,N.G.David,J.BaskarBabujee ,super edge bimagiclabeling graphs, International journal of Math.Sci.Comput.,5 (2015) ,no.1,39-43

5. A.M.Marr, N.C.K.Phillips and W.D.Wallis, Bimagiclabeling, AKCE Journal of Graphs Combin., 6 (2009), 155-160

Journal of Physics: Conference Series **1377** (2019) 012015 doi:10.1088/1742-6596/1377/1/012015

Exponentially varying suction and injection on MHD viscous oscillatory stratified flow between parallel porous plates

L.Prasanna Venkatesh [1] and R.Manoharan

Department of Mathematics
Sathyabama Institute of Science and Technology, chennai - 600119

prasannaclassof1@gmail.com

Abstract. Start your abstract hereIn this paper we consider Magenetohydrodynamic viscous oscillatory flow of incompressible linearly density stratified fluid in a long vertical rectangular channel with injection through y – axis and suction through the plate parallel to y - axis. In the undisturbed state the density distribution is assumed to be linearly distributed. Closed form solutions are obtained using similarity transformation. The effects on axial and transverse velocity profiles due to various parameters involved in the problem are illustrated in the figures. The graphical representations of the investigation show that the stratification effects are more on transverse velocity profiles than that of axial velocity. For constant density and zero electromagnetic induction the result reduces to that of viscous oscillatory flow between parallel plates

1. Introduction

The characterisation of mingling of fluids with electrical conductivity and elctromagnetic forces is know as Magnetohydrodynamics (MHD). In 1937, Hartmann experimented the effects of magnetic field applied in transverse direction on the flow of fluid that is viscous incompressible and electrically conducting between two infinite parallel insulated stationaryplates.Motivated by the invention that plasma could be reproduced in the laboratory during 1950 the study on MHD started its developement. Due the fact that Earth has its magnetic field, crystal growth induced space weather prediction with damping of highly varying fluctuations in semiconductors melts gave way for the inventions on MHD. The problem of MHD flows are usually explained mathematically by using Navier–Stokes equations, along with its associated electrodynamics equations and Maxwell's equations. These equations are then combined into moving electrical conductor's Lorentz force and Ohm's law. Fluid flows affected by the viscosity and density differences present in the fluid is described as heterogeneous or stratified flow. These fluid flows in the field of environment and ocean reasearch draws interest in the study of stratification. Viscous fluid flowing through the boundary layer that afre unsteady and convective having variable fluid properties was considered by Vajravelu, Prasad and Chiu-On (2013). Krishnendu Bhattacharyya and Lavek (2014) analysed using fourth order R-K mehtod by applying shooting technique on boundary layer flow of magnetohydrodynamic nanofluid over an exponentially stretching permeable sheet. Azeem Shahzad, Ramzan Ali and Masood Khan (2012) derived the variable separable solution for two dimensional convective flow over a nonlinear sheet that is stretched radially.Inviscous heterogenous fluid flow due to forced oscillation was analysed in research work of many authors such as Krishna & Sharma (1969), Hendershott (1969), Rao and Rao (1971),

Sharma and Naidu (1971), K.B Naidu (1974), Prasanna Venkatesh, Ganesh and Naidu (2014) and PrasannaVenkatesh (2018). The main focus of our research problem is to consider heterogenous fluid which is viscous oscillatory flowing through two vertical plates with a porous wall induced by variable suction and injection at the porous wall that accomodates Boussinesq approximation for density. Variable separable solution for the problem is derived by using a convenient choice of stream function which is a multiple of suction or injection velocity.

2. Problem Formulation and Solution

The mathematical construction of the problem considers the flow of heterogeneous fluid which is viscous and oscillatory in a long channel of width 'h'. Both the injected velocity through the plate on y – axis as well as the suction velocity through the plate placed parallel to y – axis is varying as a function of y. The fluid flow is induced only after the start of suction and injection velocities. The distance between the plates is fixed be 'h'. The density variation is along the fluid flow which is downwards in the direction of gravity along the negative side of y – axis. The density in the motionless state is considered to be dependent on only y whereas during its motion it is assumed to be dependent of x,y & t. The following set of equation describes the mathematical model for the above said situation after applying Boussienesq's approximation in the inertial terms in the equation for creeping flow of heterogeneous fluid.

Equation of continuity

$$\frac{\partial u}{\partial x}+\frac{\partial v}{\partial y}=0(1)$$

Equation describing Incompressibility of the flow

$$\frac{\partial \rho}{\partial t}+u\frac{\partial \rho}{\partial x}+v\frac{\partial \rho}{\partial y}=0 \qquad (2)$$

Equation of conservation of motion

$$\rho_0'\frac{\partial u}{\partial t}=-\frac{\partial p}{\partial x}+\mu\left(\frac{\partial^2 u}{\partial x^2}+\frac{\partial^2 u}{\partial y^2}\right) \qquad (3)$$

$$\rho_0'\frac{\partial v}{\partial t}=-\frac{\partial p}{\partial y}+\mu\left(\frac{\partial^2 v}{\partial x^2}+\frac{\partial^2 v}{\partial y^2}\right)-\sigma_e B_0^2 v-\rho' g \qquad (4)$$

Hereμdenotes with usual notation the viscosity coefficient and ρdenotes the fluid density in general,σrepresents the conduction of electricity and $B_0 = \mu_e H_0$, $B_0$is considered to be electromagnetic induction, $\mu_e$is magnetic permeability notation and H_0 is the notation for measuring transverse magnetic field. The density expression in the motionless state is taken as

$$\rho=\rho_0(y)+\rho'(x,y,t) \qquad (5)$$

$$\rho_0(y)=\rho_0'(1-\beta y) \qquad (6)$$

ρ_0' represents constant density part. The density in motionless state denoted by$\rho_0(y)$ is a linear function of y, ρ'(y,t) denotes the density during motion of fluid, $\beta$is the stratification parameter(a constant). Brunt – Vaisala frequency N defined by the relation$N^2 = \beta g$.

$$\frac{\partial \rho'}{\partial t}=\rho_0'\beta v \qquad (7)$$

The conditions for the boundary velocities are

$$\left.\begin{array}{ll} u(0,y)=v_1 e^{\frac{-v_1}{hu_0}y}, & u(h,y)=v_2 e^{\frac{v_1}{hu_0}y} \\ v(0,y)=0, & v(h,y)=0 \end{array}\right\} \qquad (8)$$

By usual procedure of eliminating Pressure and with equation of density and then finally substituting stream function the system of equation above reduces to

$$\rho_0'\frac{\partial^2}{\partial t^2}\nabla^2\psi=\mu\frac{\partial}{\partial t}\nabla^4\psi-\left(N^2\rho_0'+\frac{\partial}{\partial t}\sigma_e B_0^{\,2}\right)\frac{\partial^2\psi}{\partial x^2} \tag{9}$$

$$\psi(x,y,t)=\Psi(x,y)\,e^{i\omega t},\quad u=u(x,y)\,e^{i\omega t},\ v=v(x,y)e^{i\omega t},\ p=p(x,y)e^{i\omega t}$$

The function $f(\eta)$ is introduced as follows:

$$\Psi=u_0e^{-\frac{v_1}{hu_0}y}f(\eta) \tag{10}$$

Where $\eta=\frac{x}{h}$ and u_o is the average entrance velocity.
Using (10) in (9)

$$\left[D^4-\frac{h^2\rho_0'}{\mu i\omega}\left(\omega^2-N^2-\frac{\sigma_e B_0^{\,2}i\omega}{\rho_0'}\right)D^2+\left(\frac{v_1^{\,4}}{u_0^{\,4}}+\frac{h^2\rho_0'\omega v_1^{\,2}}{\mu i u_0^{\,2}}\right)\right]f(\eta)=0 \tag{11}$$

The above equation clearly represents an ordinary differential equation of order 4 with constant co-efficient which are complex numbers for which the general solution is presented as follows

$$f(\eta)=c_1e^{\alpha_1\eta}+c_2e^{-\alpha_1\eta}+c_3e^{\alpha_2\eta}+c_4e^{-\alpha_2\eta} \tag{12}$$

The Conditions of the boundary are converted in terms of $f(\eta)$ are given by

$$f(0)=-1, f(1)=-(1-a)\ \text{and}\ f'(0)=0, f'(1)=0 \tag{13}$$

Where $a=1-\frac{v_1}{v_2}$, $0\le v_1\le v_2$ and u_o is the average entrance velocity.
Substituting (13) in (12)

$$c_1+c_2+c_3+c_4=-1$$

$$c_1e^{\alpha_1}+c_2e^{-\alpha_1}+c_3e^{\alpha_2}+c_4e^{-\alpha_2}=-(1-a)$$

$$c_1\alpha_1-c_2\alpha_1+c_3\alpha_2+-c_4\alpha_2=0$$

$$c_1\alpha_1e^{\alpha_1}-c_2\alpha_1e^{-\alpha_1}+c_3\alpha_2e^{\alpha_2}+-c_4\alpha_2e^{-\alpha_2}=0$$

The constant values present in the above system of equations after solvingare represented by

$$A_1=(e^{\alpha_1}-e^{\alpha_2})\left(1+\frac{\alpha_2}{\alpha_1}\right), A_2=(e^{-\alpha_2}-e^{\alpha_1})\left(1-\frac{\alpha_2}{\alpha_1}\right), A_3=e^{\alpha_1}-1+a$$

$$A_4=2\frac{\alpha_2}{\alpha_1}(e^{\alpha_2-\alpha_1}-1)-\left(\frac{\alpha_2}{\alpha_1}-1\right)(e^{\alpha_2+\alpha_1}-e^{\alpha_2-\alpha_1})$$

$$A_5=2\frac{\alpha_2}{\alpha_1}\left(1-e^{-(\alpha_2+\alpha_1)}\right)-\left(\frac{\alpha_2}{\alpha_1}+1\right)\left(e^{\alpha_1-\alpha_2}-e^{-(\alpha_2+\alpha_1)}\right)$$

$$A_6=\frac{(A_2+A_3)A_4-(A_1+A_3)A_5}{A_1A_5-A_2A_4}, A_7=-\frac{\alpha_2}{\alpha_1}\left(\frac{A_3A_4+A_3A_5}{A_1A_5-A_2A_4}\right)$$

$$c_1=A_8=\frac{(A_6+A_7)}{2} \tag{14}$$

$$c_2=A_9=\frac{(A_6-A_7)}{2} \tag{15}$$

$$c_3=A_{10}=\frac{A_3A_5}{A_1A_5-A_2A_4} \tag{16}$$

$$c_4=A_{11}=\frac{-A_3A_4}{A_1A_5-A_2A_4} \tag{17}$$

Substituting the constants

$$f(\eta)=A_8e^{\alpha_1\eta}+A_9e^{-\alpha_1\eta}+A_{10}e^{\alpha_2\eta}+A_{11}e^{-\alpha_2\eta} \tag{18}$$

$$u(x,y)=-v_1e^{\frac{-v_1}{hu_0}}\left(A_8e^{\alpha_1\eta}+A_9e^{-\alpha_1\eta}+A_{10}e^{\alpha_2\eta}+A_{11}e^{-\alpha_2\eta}\right) \tag{19}$$

$$v(x,y) = -\frac{1}{h}u_0 e^{\frac{-v_1}{hu_0}}\left(\alpha_1\left(A_8 e^{\alpha_1\eta} - A_9 e^{-\alpha_1\eta}\right) + \alpha_2\left(A_{10}e^{\alpha_2\eta} - A_{11}e^{-\alpha_2\eta}\right)\right) \tag{20}$$

$$\rho = \rho_0'(1-\beta y) + \frac{\rho_0'\beta i}{\omega}\left(1 - e^{i\omega\omega}\right)v(x,y) \tag{21}$$

3. Results and Discussion

For Figure 1 to Figure 6 the numerical values of v (x-direction), the velocity along the fluid flow direction have been depicted for different values of x (y – direction), y and various other parameters that are present in the problem. Figures 1 shows the variation in the velocity along the fluid flow direction for different values of x = 0 to 1 and y = 0.3 while values of ρ0=1.2, μ=1.5, σe=1.5, N=1.5, u0=1.2, v1=0.6, v2=1.8 and ωt (ωt=0 to π). Figure 2 represents the Transverse velocity profiles for ωt=0, x = 0 to 10 and for y=0 to 1. Figure 3 depicts the variation in Transverse velocity due to the change in magnetic parameter B0 from 0 to 10. Figure 4 represents velocity along the fluid flow direction for different β values from 2 to 10. It is noted from these graphical representations that velocity along the fluid flow decreases with respect to time, depth, magnetic parameter and stratification. Figure 5 and Figure 6 shows the variations in velocity along flow direction for different μ (0 to 1) and ρ0 values (5 to 25) respectively. The fluid are often realigned quickly around the centre of channel which indicated that the effects of stratification are more around the centre of the channel. Figure 7 and 8 represents Axial velocity for varying ωt and y values respectively. Axial velocity profiles varies inversely with time and depth. Axial velocity profiles for varying values of N (2 to 20) is shown by Figures 9. Figures 10 and 11 represents the changes shown in the Density funcction due to varying ωt and y values respectively. Figure 12 explains velocity in flow direction for different σe values (4 to 20).

4. Conclusion

The pictorial representations of Transverse and axial velocity profiles lead to the following observations. velocity along the flow direction decreases with time, depth, MHD parameter and parameter for stratification. It increases with viscosity whereas it decreases with constant density ρ_0. Axial velocity as well as density decreases with time and depth but increases with stratification parameter. The stratification effects are more at the centre than that of boundary of the channel. The stratification effects are more for transverse velocity than that of axial velocity. Density distribution is dominated by Transverse velocity profiles.

References

[1] A referenceAzeem Shahzad, Ramzan Ali, and Masood Khan. 2012. "On the Exact Solution for Axisymmetric Flow and Heat Transfer over a Nonlinear Radially Stretching Sheet". Chin. Phys. Lett. 29 (8): 1-4.

[2] D.V. Krishna, and L.V.K.V. Sharma. 1969. "Motion of an axisymmetric body in a rotating stratified fluid confined between two parallel planes". J.Fluid Mech. 38 (4): 833-842.

[3] K. Vajravelu, K.V. Prasad, and Chiu-On Ng. 2013. "Unsteady convective boundary layer flow of a viscous fluid at a vertical surface with variable fluid properties". Nonlinear Analysis: Real World Applications, Elsevier, 14 (1): 455-464.

[4] K.B.Naidu. 1974. "Stratified Viscous Flow between two Oscillating Cylinders". Indian Journal of Pure and Applied Mathematics, Indian National Science Academy. 5 (12): 1127-1136.

[5] Krishnendu Bhattacharyya and G. C. Layek. 2014. "Magnetohydrodynamic Boundary Layer Flow of Nanofluid over an Exponentially Stretching Permeable Sheet" Hindawi Publishing Corporation, Physics Research International 2014: 1-12.

[6] L. Prasanna Venkatesh. 2018. Magnetohydrodynamic Viscous Oscillatory Flow in a Narrow Channel Due to Suction or Injection through the Porous Wall, Journal of Advanced Research in

Dynamic and Control System, 3(3): 1354-1361:1943-023X.
[7] L. Prasanna Venkatesh, Ganesh S. and Naidu K.B. 2014, Viscous oscillatory stratified flow between parallel plates with a porous wall due to suction and injection through the porous wall, International Journal of Applied Engineering Research. 9(26) : 9018-9022 : 0973-4562
[8] L. Prasanna Venkatesh, S. Ganesh and K.B. Naidu. 2014. "MHD Flow of Viscous Oscillatory Stratified Fluid between Two Vertical Parallel Plates with a Porous Wall due to Suction and Injection through the Porous Wall". Pensee Journals. 76 (1): 304-316.
[9] L.V.K.V. Sharma, and K.B. Naidu. (1971). "Source in a rotating stratified fluid". Acta Mech. 13 (1-2): 21-29.
[10] M.C. Hendershot. 1969. "Impulsively started oscillations in a rotating stratified fluid". J. Fluid Mech. 38 (3): 513-527.
[11] V.S. Rao, and G.V.P. Rao. 1971. "On waves generated in rotating stratified liquids by travelling foreign effects". J. Fluid Mech. 46 (3): 447 - 464.

Figure 1. Transverse Velocity for different ωt values

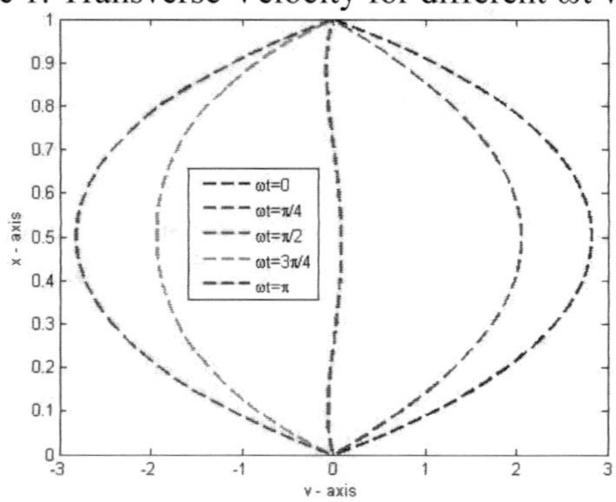

Figure 2. Transverse Velocity for varying y values

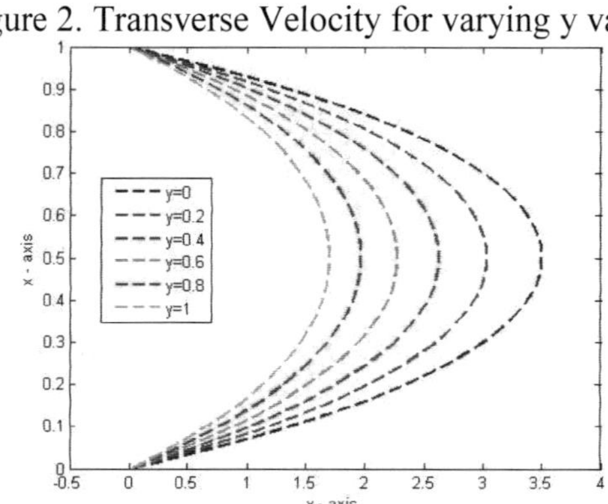

Figure 3. Transverse Velocity for varying B_0 values

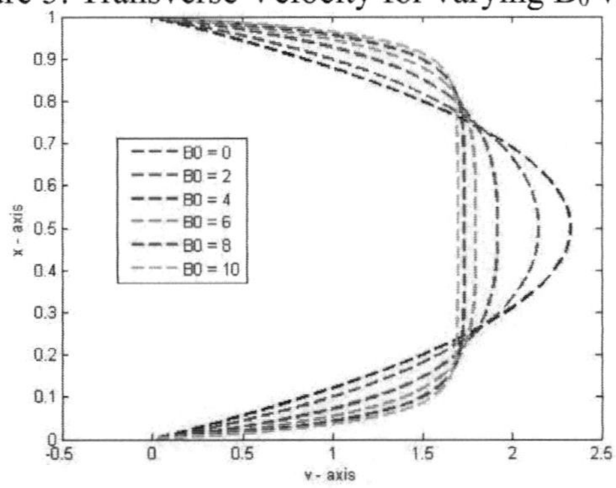

Figure 4. Transverse Velocity for varying N values

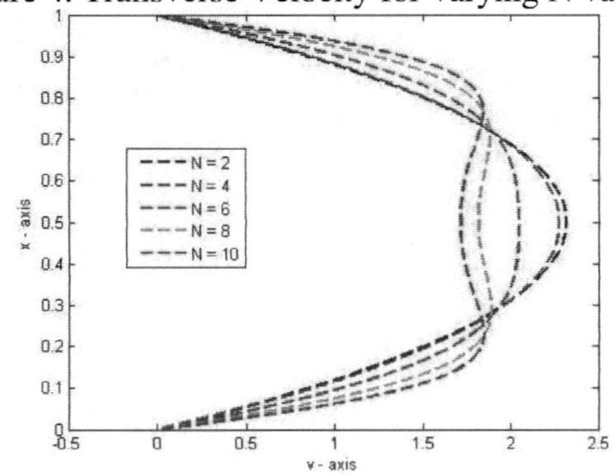

Figure 5. Transverse Velocity for varying μ values

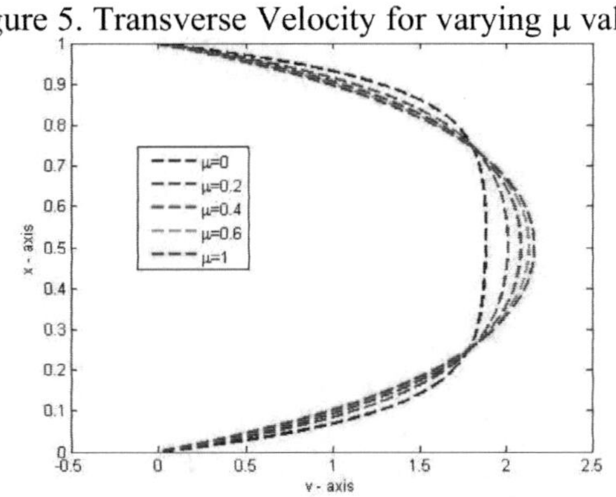

Figure 6. Transverse Velocity for different ρ_0 values

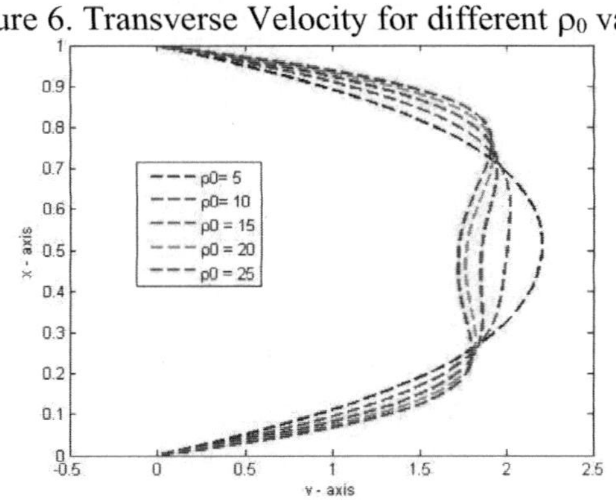

Journal of Physics: Conference Series **1377** (2019) 012015 doi:10.1088/1742-6596/1377/1/012015

Figure 7. Axial Velocity for varying ωt values

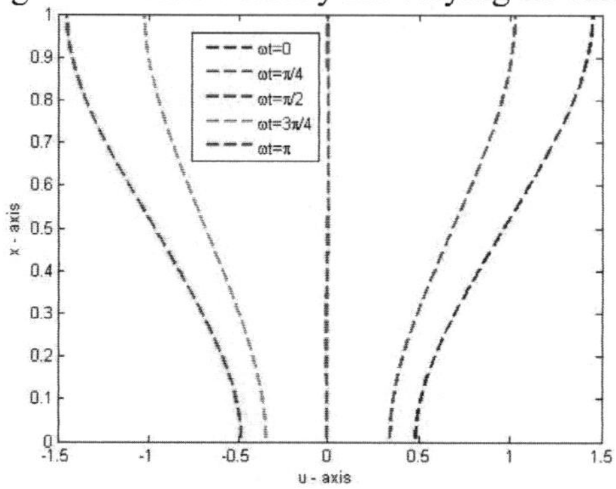

Figure 8. Axial Velocity for varying y values

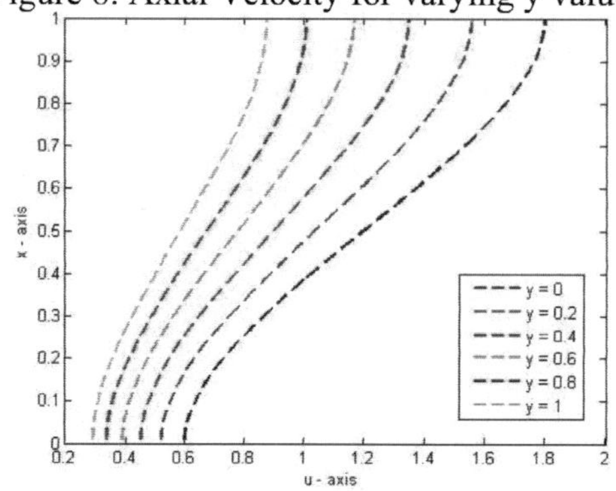

Figure 9. Transverse Velocity for different N values

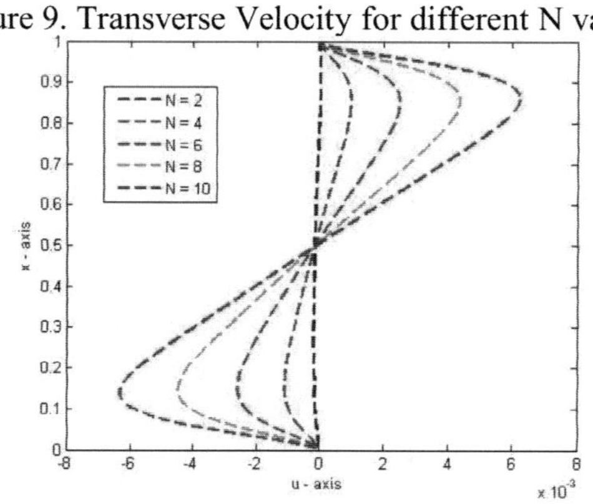

Figure 10. Density Distribution for varying ωt values

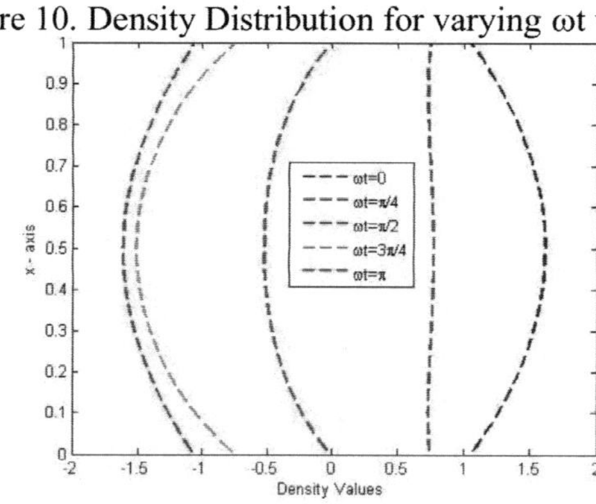

Figure 11. Density Distribution for varying y values

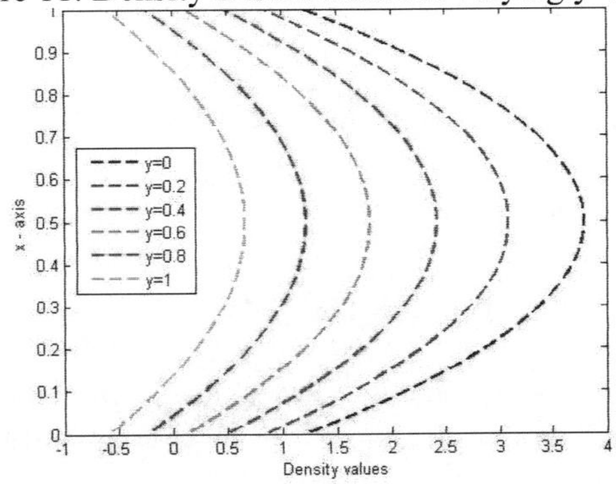

Figure 12. Transverse Velocity for different σ_e values

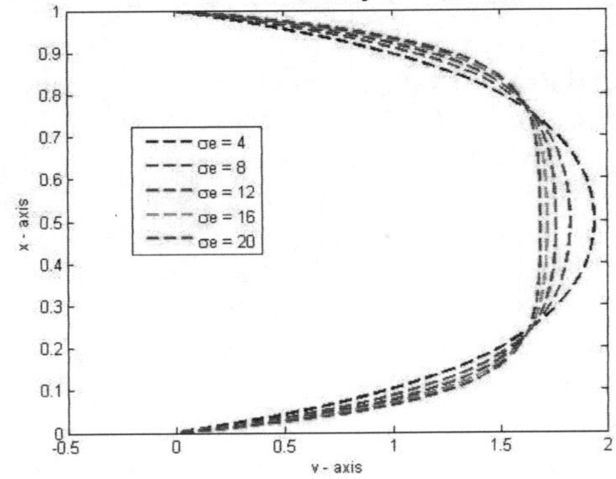

Geometric Brownian Motion in Stock Prices

K Suganthi* **and G Jayalalitha****

**Research Scholar, Department of Mathematics, VELS Institute of Science Technology &Advanced Studies(VISTAS) – Chennai, Tamil Nadu, India. E mail: suganthi19122015@gmail.com*

***Professor, Department of Mathematics, VELS Institute of Science Technology & Advanced studies(VISTAS) – Chennai, Tamil Nadu, India. E mail: g.jayalalithamaths.sbs@velsuniv.ac.in*

Abstract. Financial instability estimates the changes of the cost of a monetary instrument. It is a proportion of properties of the Stock prices stability. Fractal investigations are used to assess the money related instability. Forecasting of stock prices acts as an important challenge based on the Random Walk theory. This paper deals with comparison of two years 2013 -2014 and 2017(Jun to Nov) of stock prices. Explain the instability by the method of Box-Counting technique to find the Fractal dimensions of the Geometric Brownian Motion based on the Random Walk defective value. This creates the possibility that Fractal measurement is related with the monetary unpredictability. Its an essential instrument for both money related investigators and Financial specialists.

Keywords: Stock prices, Financial Volatility, Fractals, Fractal Dimension, Geometric Brownian Motion,
Random Walk.
AMS Classification 2000: 28A80, 60J60, 97M40, 97M99.

I. INTRODUCTION

1.1 STOCK PRICE

The stock price is the highest amount someone pays to sell the stock, or the minimum amount to buy the stocks [10]. Each shares are allocated a monetary value or price, when traded in public, companies issue shares of stock to the investors. Stock prices fluctuates depending on different factors [2].

1.2 FINANCIAL VOLATILITY

Unexpected events resulting in huge price movements with varying variable, consequently financial market creates an unexpected behaviour that may confuse the investors. This leads to financial volatility and an unexpected price may return [3]

** Corresponding Author

1.3 FRACTALS

Fractals was coined by Benoit Mandelbrot in 1975 and as derived from the Latin word fractus meaning *broken or fractured*. He was a father of Fractals [13]. It is endless models. It is immensely complex models that are self – alike during various levels. Examples are trees, rivers, coastlines, sea shells, Cantor set, Sierpinski Triangle, Von Koch curve, mountains etc.

1.4 RANDOM WALK

Random walk is a mathematical process and are referred as a stochastic otherwise random method to explain a route, which has a series of random steps on few mathematical spaces as like the integers [1].

Azme Khamis et.al explained Monetary Data using Geometric Brownian Motion based on Wiener process. But in this paper it explains the costs of the stock with Geometric Brownian Motion based Random Walk and Box Counting Dimension [3].

II.METHODS

2.1 FRACTAL DIMENSION

A Fractal dimension is a ratio which provides a statistical indicator of complexity comparing the model (exactly speaking, a Fractal model) gets changed through the scale on which it is calculated. The Hausdorff dimension that are much more broadly appropriate [12].

For example, Hausdorff dimension and the box counting dimension is described for any set and in these four example may be shown to equal the similarity dimension and Hausdorff and Box counting along with the methods used for their computation very roughly, a dimension provide a explanation of how much space a set occupies [12].It is a quantity of the importance of the abnormality of a set when looked upon at very small scales. A dimension has more details about the physical parameters of a set.

The Hausdorff measurement which characterizes the fractal measurement D as a partial quantity. This method does not finds an efficient solution and hence box counting method is preferred.

$$D \approx \frac{\log(N(s))}{\log(\frac{1}{s})} \tag{1}$$

The Hausdorff method expresses that the component of an article is connected with the quantity N(s) of circles of size s expected to over to question in a D_E-dimensional Euclidean space.

$$N(s) \sim \frac{1}{s} D_E \tag{2}$$

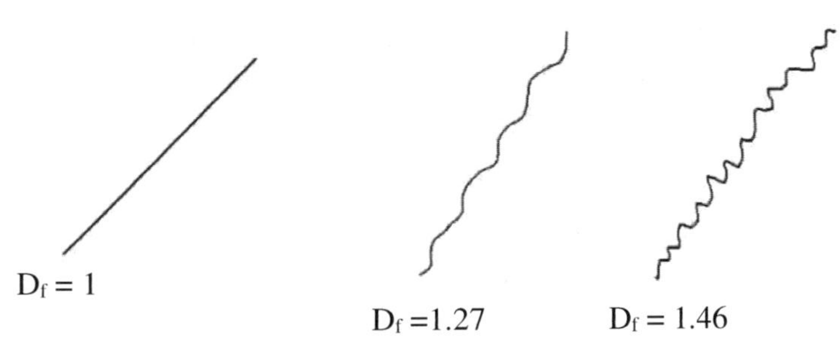

Figure 1. Fractal dimension of the different Random Walk

2.2 FRACTAL DIMENSION AND FINANCIAL VOLATILITY

Nowadays, there are different endeavours of utilizing fractal investigation in monetary market [3, 4]. Fractal measurement turned out to be an essential method for two reasons: As per observation of Mandelbrot, stocks exchange costs are fractal objects, second on the grounds that the fractal measurement is a proportion of the level of the pictures contour complication [11]. The random walk method recommends the alterations in stock prices, which has the similar distribution and not dependent on any other factor. Hence the previous movement or stock price or market will be used to forecast its future movement [8]. Profit values are plotted in fig 2. Mean values for the 2013 – 2014 years are calculated and the defective values present in Table1.

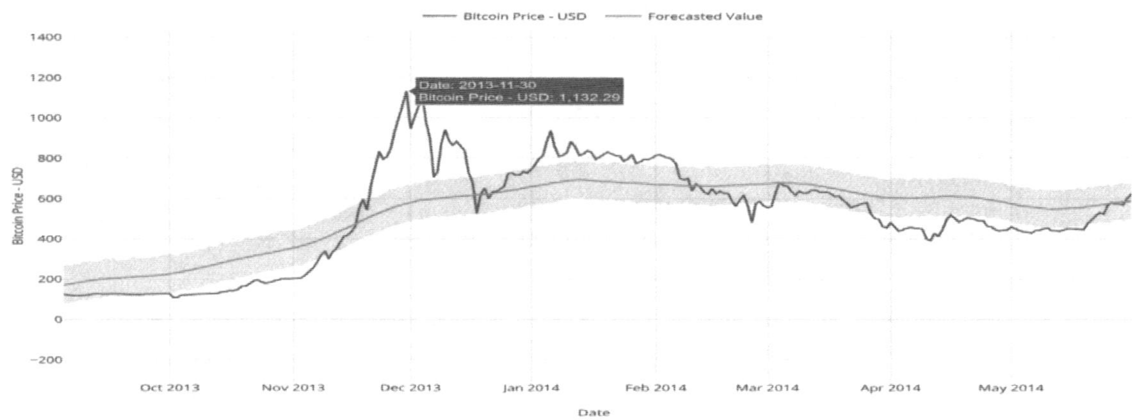

Figure 2. Forecast of Amazon Stock Prices in 2013-2014

Table 1. Actual and Forecast of the defective value in the Year 2013 – 2014.

Defective Values									
Month	Oct	Nov	Dec	Jan	Feb	Mar	Apr	May	Mean
Actual	150	200	950	720	800	560	500	450	541.25
Forecast	210	380	590	650	700	710	600	590	553.75

The forecasting method is applied and then compared with the stock buying and selling for years. The model forecasts increase and decrease of the stocks on each day. If it forecasts as decrease, stocks should not be bought and if it forecasts as increase, buying stocks would give more profit to buyers [6].The forecast and actual model for the year 2017 is shown in fig 3. The actual and forecast defective values present in Table 2.

The random walk theory is a financial theory which describes that the cost of the stock market evolves as per random walk and hence prediction is not possible [7] it is steady with the efficient market hypothesis. The random walk plot, the lower values are referred as defective Based on the mean value Poisson distribution is obtained.

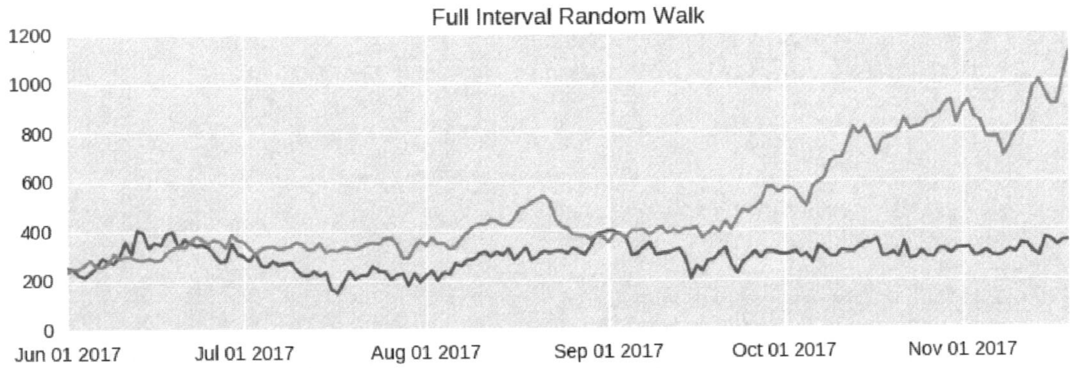

Figure 3. Company Actual Stock Price in 2017

Table 2 Actual and Forecast of the defective value in the Year 2017.

Defective Values							
Month	Jun	Jul	Aug	Sep	Oct	Nov	Mean
Actual	220	300	210	400	320	310	293.3
Forecast	210	380	390	395	592	850	469.5

2.3 GEOMETRIC BROWNIAN MOTION

Geometric Brownian motion are referred as (exponential Brownian motion) is a continuous-time stochastic process in which the logarithm of the randomly changing quantity results a Brownian motion with drift [5].

Geometric Brownian Motion X(t)is the solution of a stochastic differential equation

$$dX(t) = \mu X(t)dt + \sigma X(t)dw(t) \tag{3}$$

With initial value (0) = x_0,

The relative change is a grouping of a deterministic proportional development period alike to inflation or interest rate growth and a commonly distributed random transforms [10].

$$\frac{dX}{X} = rdt + \sigma dw \tag{4}$$

Random variable X is said to include the lognormal distribution (with Parameters r and σ)
The variance of Geometric Brownian Motion cannot be calculated by applying the normal distribution, along with the general formula

$$Var[X] \ = \ E[(X - E[X])^2] \ = \ E[X^2] - (E[X])^2 \tag{5}$$

This description is especially near toward the above equation, hence towards simulate the stock prices in this model. SDE or Stochastic Differential Equation of S_t (a stochastic process) is preferred.

The Stochastic Differential Equation implies that

$$S_t \ = \ \mu S_t t + \sigma S_t W_t \tag{6}$$

Where, S_t - stochastic process
μ - Percentage drift
σ -Percentage of volatility

W_t- Weiner's process or Brownian Motion
Let simulate the stock records from the equation by increasing interpretation.

$$S_t = S_{t-1} * ((\mu - (\frac{\sigma}{2})) * t + \sigma w_t \quad (7)$$

Where, S_t -Stock price at time t
S_{t-1} - Stock price at time t-1
μ-Mean
σ - Mean daily volatility
t - Time interval of the step
w_t-Random normal noise.

2.4. RANDOM WALK

The random walk is executed in the stock market. A random walker on a straight road (which is located with a point on the real line R) takes a step of length δ at time intervals of τ, moving forwards or backwards with similar probability $\frac{1}{2}$with the direction of the steps all independent [1].

Let $X_\tau(t)$ indicates the location of the walker at time t. Then, given the position $X_\tau(k\tau)$ at time kτ, the position $X_\tau((k+1)\tau)$ at time (k+1) τ is equal to be $X_\tau(k\tau)+ \delta$ or $X_\tau(k\tau)- \delta$. Considering that the walker begins at the origin at time 0,then for t>0,the position at time t is expressed by the random variable[1].

$$X_\tau (t) = \delta(Y_1+\ldots\ldots\ldots Y_{[t/\tau]}) \quad (8)$$

Where Y_1, Y_2.... are independent random variables, each taking the value of +1 with probability $\frac{1}{2}$ and -1 with probability$\frac{1}{2}$. Here [t/τ] denote the major integer less than or equal to t/ τ. We normalise the step length by the parameter $\delta = \sqrt{\tau}$so that

$$X_\tau (t) = \sqrt{\tau} (Y_1+\ldots\ldots\ldots Y_{[t/\tau]}) \quad (9)$$

Then the mean and variance of the walker's positions at time t are

$$E(X_\tau (t)) = \sqrt{\tau} (E (Y_1+\ldots\ldots\ldots (Y_{[t/\tau]})) = 0 \; and \quad (10)$$

$$var(X_\tau(t)) = \tau(var(Y_1) + \ldots\ldots. +var(Y_{[t/\tau]})) = \tau [t/ \tau] \sim t \quad (11)$$

If t is large compared with τ. Thus over a large number of random walks the average position at time t is 0 and the typical distance of the walker from the origin will be around$\sqrt{t}$, the standard deviations of X_τ (t). Random walk finds the defective value of the stock prices for the annual period of 2013-2014 and 2017.

Journal of Physics: Conference Series **1377** (2019) 012016 doi:10.1088/1742-6596/1377/1/012016

2.5 POISSON DISTRIBUTION

Poisson distribution is a discrete probability distribution and are most commonly in statistical work [9]. Poisson distribution are used in cases, in which the chance of any single incident even though a success is small. Poisson distribution is applied to calculate the drift of the defective value.

$$P(x) = \frac{e^{-\lambda}\lambda^{x}}{x!} \quad (12)$$

Where X=0, 1, 2, 3, 4...
e = 2.7183 (the base of natural Logarithms)
λ= Mean.

III. RESULTS

Prediction of stock prices plays an important role in share market, the stock price prediction based on Random walk based theory is analysed. This theory effectively analysis the forecasting of stock prices. Two years of stock prices was compared all together to find the instability. Geometric Brownian Motion method is also used to analyze the costs of the stocks. The defectives values are found for various month and shown in table 1 and table 2.The Actual and forecast defective value in the year 2013 - 2014 are tabulated in table 1. Similarly, the actual and forecast defective value in the year 2017(Jan to Nov) are tabulated in table 2. From the Poisson distribution minimum and maximum defective value can be found.

CASE 1: THE ACTUAL DEFECTIVE FOR THE YEAR 2013-2014

Random walk for defective value is ranging from 150 to 200. Its dimension can be calculated by Box counting method and its D_f is 1.7784 and is shown in the table 3 and fig 3a.

Random walk for the defective value ranges from 200 to 950. Its Box Counting dimension D_f is 1.9077 and is shown in the table 3 and fig 3b.

In the table 3, fig 3cshows the Random walk of the defective value is ranging from 950 to720 and its dimension D_f is 1.9603.

Random walk for defective value is ranging from 720 to 800. Its dimension can be calculated its D_f is 1.9101 and is shown in the table 3 and fig 3d.

In the table 3, fig 3eshowsthe random walk of the defective value is ranging from 800 to 560 and its dimension D_f is 1. 8376.

Random walk for defective value is ranging from 560 to 500. Its dimension can be calculated by Box counting method and its D_f is 1.9043 and is shown in the table 3 and fig 3f.

In the table 3, fig 3g shows the Random walk of the defective value ranges from 500 to 450 and its dimension D_f is 1.962.

Mean value is obtained and Poisson distribution is calculated by the equation (12).The maximum defective value is$4.6933 \times 10^{-233}$and minimum defective value is2.3414×10^{-220}.

CASE 2: THE FORECAST DEFECTIVE FOR THE YEAR 2013-2014

Random walk for defective value is ranging from210 to 380. Its dimension can be calculated by Box counting method and its D_f is 1.868 and is shown in the table 4 and fig 4a.

In the table 4, fig 4b shown the Random Walk for the defective value ranges from 380 to 590 and its dimension D_f is1.8419.

Random walk for defective value is ranging from 590 to 650.its dimension can be calculated by Box counting method and its D_f is 1.8794 and is shown in the table 4 and fig 4c.

In the table 4, fig 4d shows the Random walk of the defective value is ranging from 650 to 700 and its dimension D_f is1.9294.

Random walk for defective value is ranging from 700 to 710. Its dimension can be calculated by Box counting method and its D_f is 1. 953 and is shown in the table 4 and fig 4e.
In the table 4, fig 4f shows the Random walk of the defective value is ranging from 710 to 600 and its dimension D_f is1.9069.

Random walk for the defective value ranges from 600 to 590. Its dimension can be calculated by Box counting method and its D_f is 1.8537 and is shown in the table 4 and fig 4g.

Mean value is obtained and Poisson distribution is calculated by the equation (12).The maximum defective value isis1.7895×10^{-238} *and* minimum defective value is 1.2942×10^{-227}.

CASE 1: THE ACTUAL DEFECTIVE FOR THE YEAR 2017
Random walk for defective value is ranging from220 to 300. Its dimension can be calculated by Box counting method and its D_f is 1.8111 and is shown in the table5 and fig 5a.

In the table 5, fig 5b shows the Random walk of the defective value ranges from 300 to 210 and its dimension D_f is 1.8006.

Random walk for defective value is ranging from 210 to 400. Its dimension can be calculated by Box counting method and its D_f is 1.8797 and is shown in the table 5and fig 5c.
In the table 5, fig 5d shows the Random walk of the defective value is ranging from 400 to 320 and its dimension D_f is 1.7932.

Random walk for defective value is ranging from 320 to 310. Its dimension can be calculated by Box counting method and its D_f is 1.7215and is shown in the table 5 and fig 5e.

Mean value is obtained and Poisson distribution is calculated by the equation (12).The maximum defective value is 1.2267×10^{-125} *and* minimum defective value is 75649×10^{-118} .

CASE 2: THE FORECAST DEFECTIVE FOR THE YEAR 2017
Random walk for defective value is ranging from210 to 380. Its dimension can be calculated by Box counting method and its D_f is 1.868 and is shown in the table 6 and fig 6a.

In the table 6, fig 6b shows the Random walk of the defective value ranges from 380 to 390 and its dimension D_f is 1.8399.

Random walk for defective value is ranging from 390 to 395. Its dimension can be calculated by Box counting method and its D_f is1.8517 and is shown in the 6 table and 6c.

In the table 6, fig 6d shows the Random walk of the defective value is ranging from 395 to 592 and its dimension D_f is 1.8537.
Random walk for defective value is ranging from 592 to 850. Its dimension can be calculated by Box counting method and its D_f is 1. 8732 and is shown in the table 6 and fig 6e.

Mean value is obtained and Poisson distribution is calculated by the equation (12). The maximum defective value is5.8936×10^{-202}.*and*Minimum defective value2.3864×10^{-193}

IV. CONCLUSIONS

In this paper, Stock prices can be analysed by Geometric Brownian motion based on Random Walk and its dimension was calculated by Box Counting Method. The dimension of the Random Walk gives defective value of the Stock prices is non-integer. Hence it is fractal set. This value cannot be predicted. This creates the possibility that the fractal measurement is related with the monetary unpredictability.

REFERENCES

[1]. Fama, E. *Random walk in stock market prices*. Financial Analysts Journal, 51(1), 1995.
[2]. Reddy, Krishna and Vaughan, Clinton, *Simulating stock Price using Geometric Brownian Motion Evidence from Australian Companies*, Australian Accounting, Business and Finance Journal 10(3) .23-47(2016).
[3]. Azme, Khamis, MuhammedAkramAbdSukor, Maria Elena Nor, Siti Noor AsyikinMohdRazali, RohayuMohdSalleh, *Modelling and Forecasting Volatility of Financial Data using Geometric Brownian Motion*, Vol-**4**,ISSN:2350-0328,(2017).
[4]. Brewer, K.D Feng .Y Kwan ,C.C.Y, Geometric *Brownian Motion, Option pricing and Simulation Some Spread Sheet - Based Exercise in Financial Modelling, Spread Sheets in Education*, eJsiE,vol-**5**,no-5,(2012).
[5]. Laddle G. S L.Wu, Development and Modified GBM Models by Using *Stock Price Data and Basic Statistics Nonlinear Analysis*, Theory Methods and Application, 71(12): 1203-1208(2009).
[6]. Abidin S.and Jaffar, M. *Forecasting Share Prices of Small Size Companies in Bursa Malaysia* Using Geometric Brownian Motion. Appl. mat .inf and sci, 8(1), 107-112, (2014).
[7]. Kobeissi Y. H, Multifractal *Finance Markets: An Alternative Approach to Asset and Risk Management, springer*, New York. (2012).
[8]. Farida Agustini W, IkaRestuAffianti, Endah RM Putri, *Stock price prediction using geometric Brownian motion*, International Conference on Mathematics: Pure, IOP -974 012047(2018).
[9]. Di Asih I Maruddani, Trimono, *Modelling stock prices in a portfolio using multidimensional geometric Brownian motion*, 1025 -012122(2018).
[10]. NurAimiBadriah, N, SitiNazifah, Z. A. and Maheran, M. J., *Forecasting Share Prices Accurately For One Month Using Geometric Brownian Motion*, 26 (4): 1619 - 1635 (2018).
[11]. OseiAntwi, *Stochastic Modelling of Stock Price Behaviour on Ghana Stock Exchange*, 2(6): 116-125(2017).
[12]. Mandelbrot B.B., *Fractals: Form, chance and Dimension*, Freeman, San Francisco, CA, 1977.
[13]. Mandelbrot B.B. *The Fractals Geometry of Nature*, Freeman, San Francisco, CA, 1982.

Journal of Physics: Conference Series **1377** (2019) 012016 doi:10.1088/1742-6596/1377/1/012016

Table 3. Box Counting Dimension of the Actual defective value in the Year 2013-2014.

S.No	Range/Month	D_f	Figure
1	150-200/ Oct-Nov	1.7784	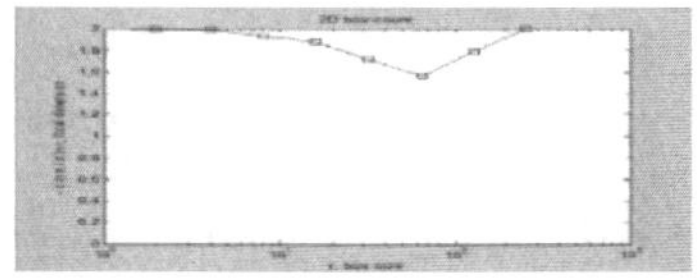 Figure 3a
2	200-950/ Nov-Dec	1.9077	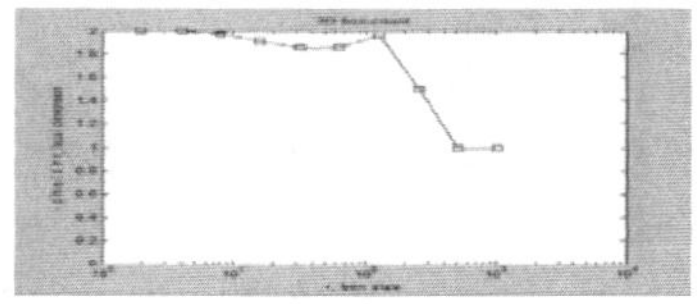 Figure 3b
3	950-720/ Dec-Jan	1.9603	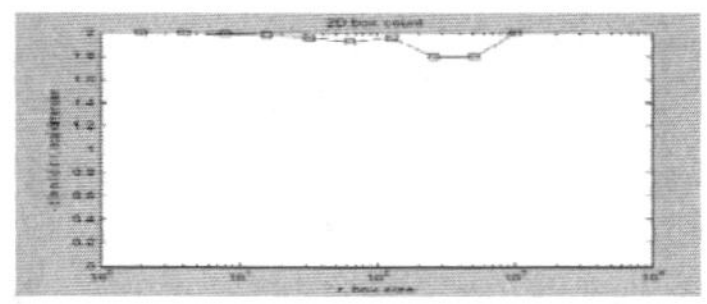 Figure 3c
4	720-800/ Jan-Feb	1.9101	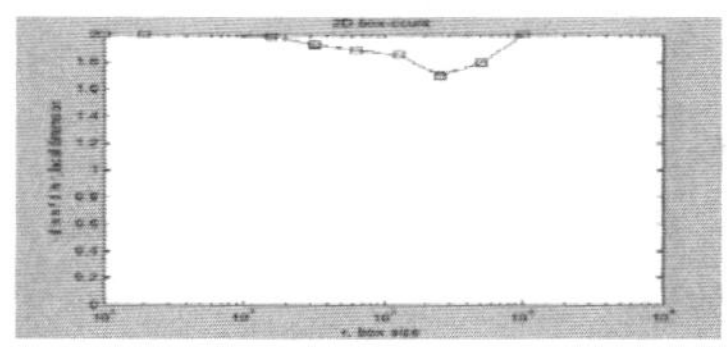 Figure 3d
5	800-560/ Feb-Mar	1.8376	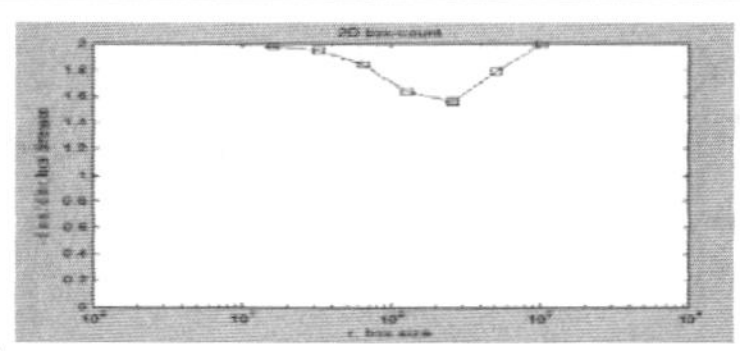 Figure 3e
6	560-500/ Mar-Apr	1.9043	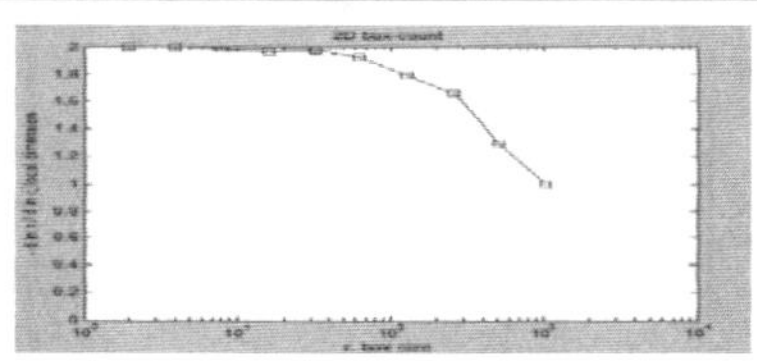 Figure 3f
7	500-450/ Apr-May	1.962	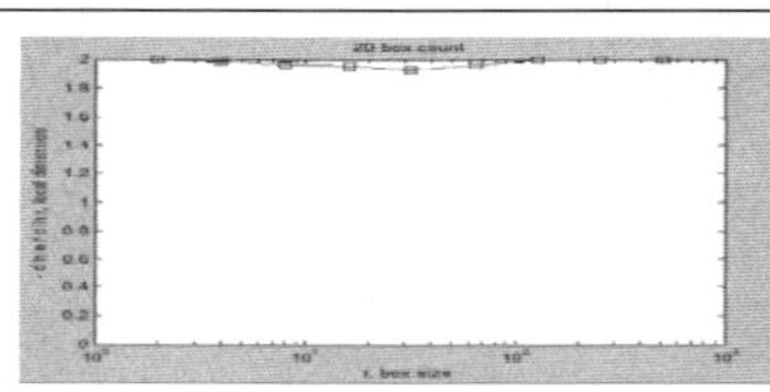 Figure 3g

Table 4. Box Counting Dimension of the Forecast defective value in the Year 2013-2014.

S.No	Range/month	D_f	Figures
1	210-380/ Oct-Nov	1.868	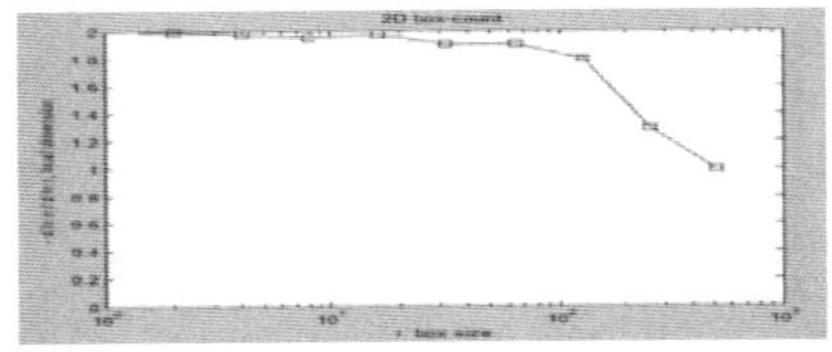 Figure 4a
2	380-590/ Nov-Dec	1.8419	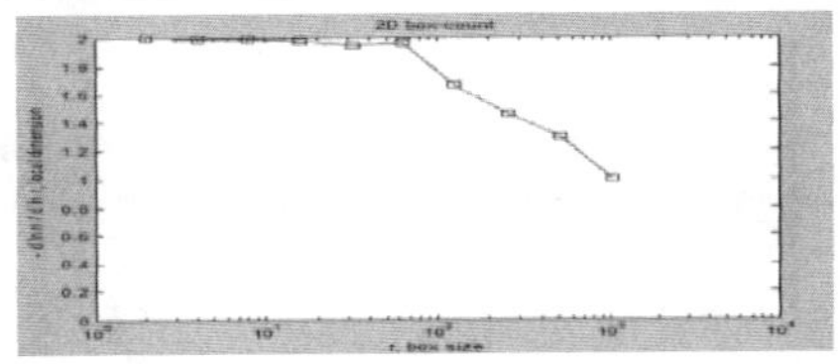 Figure 4b
3	590-650/ Dec-Jan	1.8794	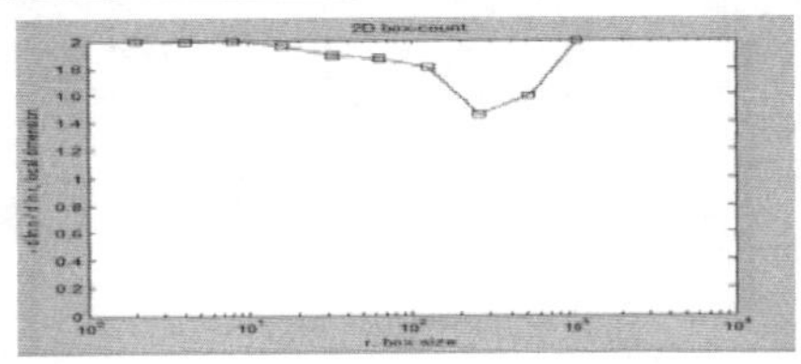 Figure 4c
4	650-700/ Jan-Feb	1.9294	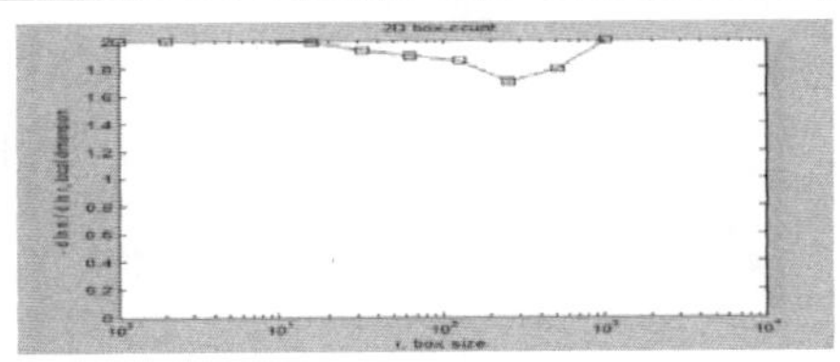 Figure 4d
5	700-710/ Feb-Mar	1.953	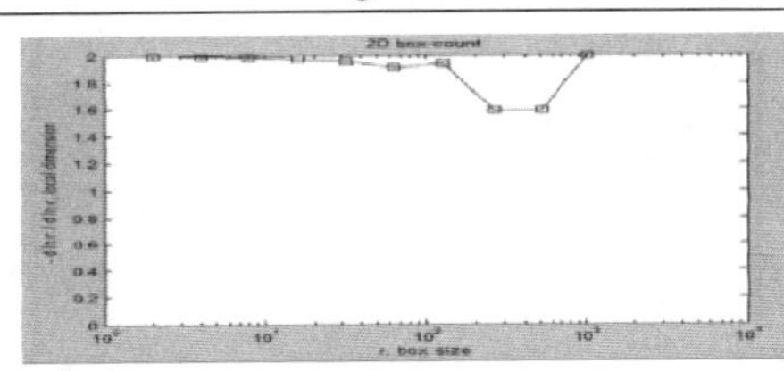 Figure 4e
6	710-600/ Mar-Apr	1.9069	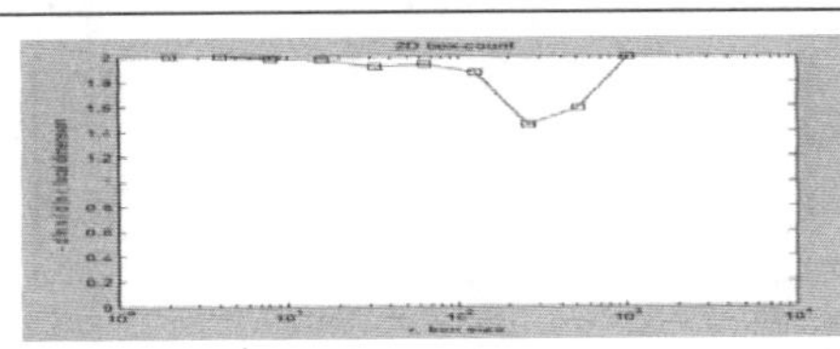 Figure 4f
7	600-590/ Apr-May	1.8537	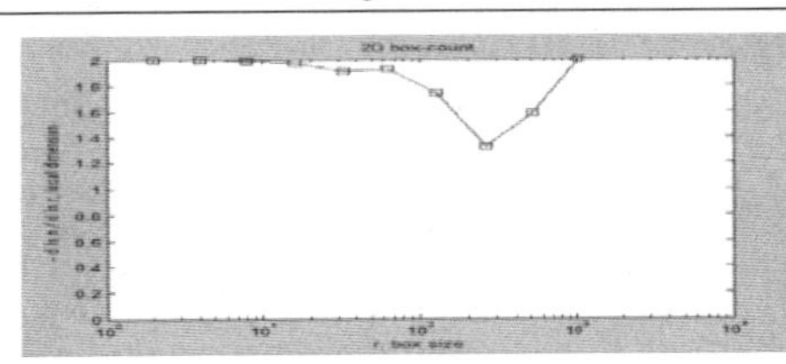 Figure 4g

Table 5. Box Counting Dimension of the Actual defective value in the 2017.

S.No	Range/Month	D_f	Figure
1	220-300/ Jun-July	1.8111	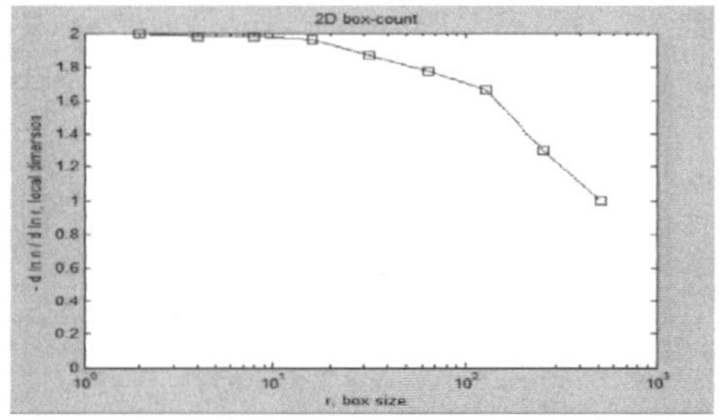 Figure 5a
2	300-210/ July-Aug	1.8006	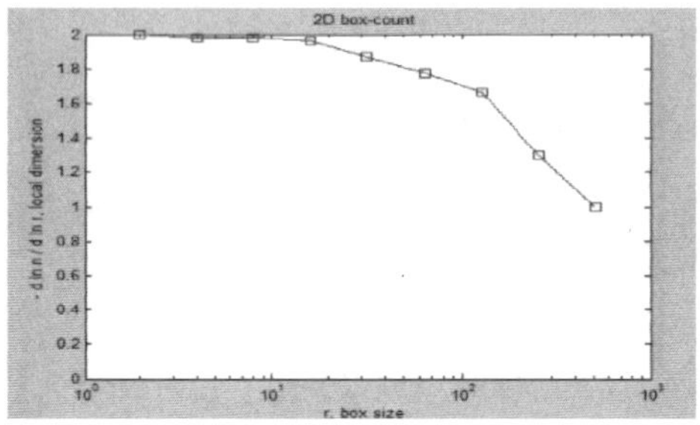 Figure 5b
3	210-400/ Aug-sep	1.8797	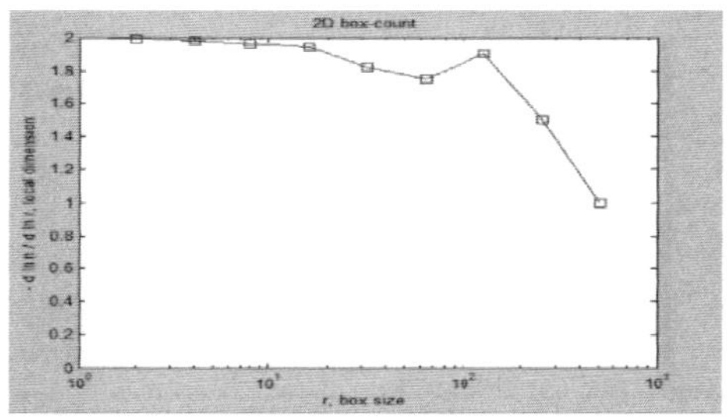 Figure 5c
4	400-320/ Sep-Oct	1.7932	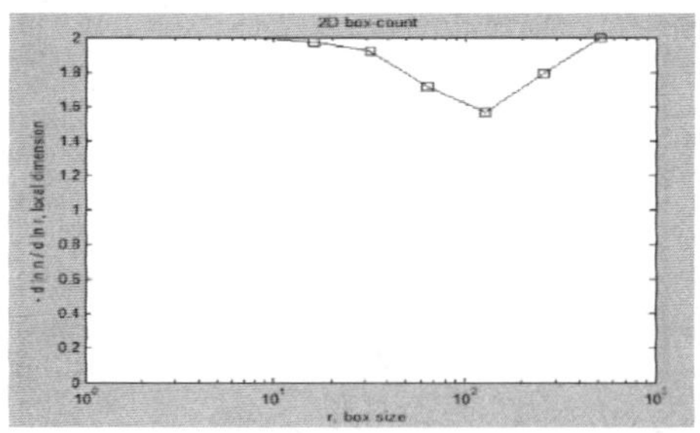 Figure 5d
5	320-310/ Oct-Nov	1.7215	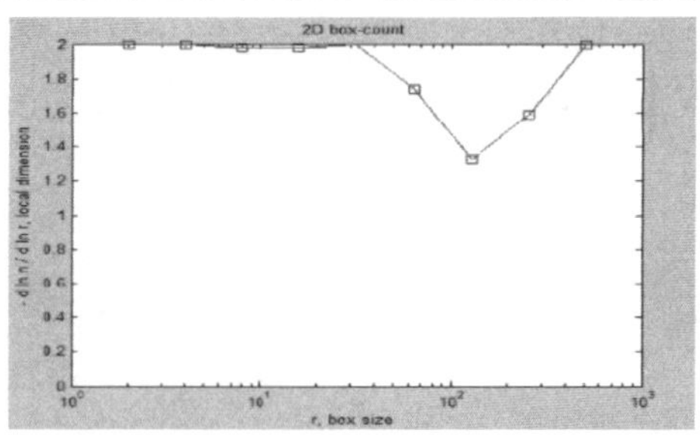 Figure 5f

Table 6. Box Counting Dimension of the Forecast defective value in the Year 2017.

S.No	Range/Month	D_f	Figure
1	210-380/ Jun-July	1.868	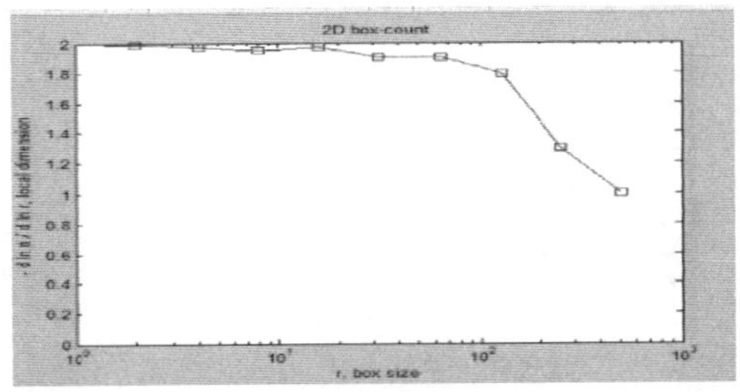 Figure 6a
2	380-390/ July-Aug	1.8399	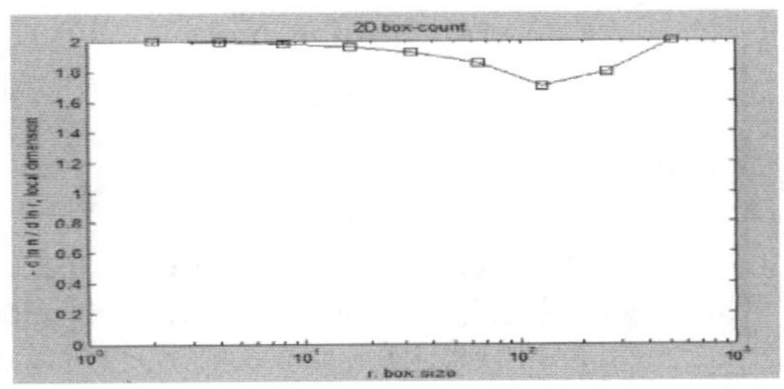 Figure 6b
3	390-395/ Aug-sep	1.8517	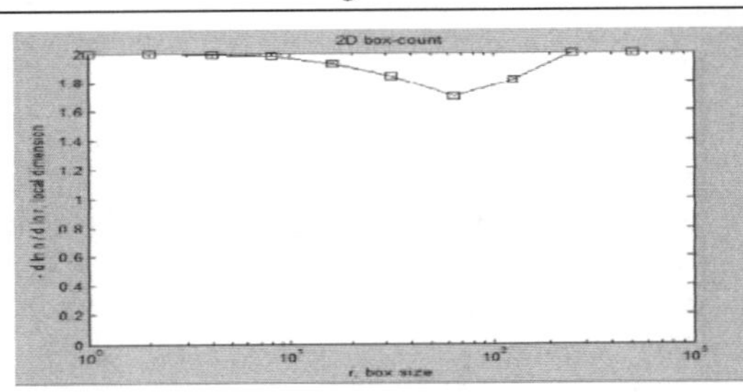 Figure 6c
4	395-592/ Sep-Oct	1.8537	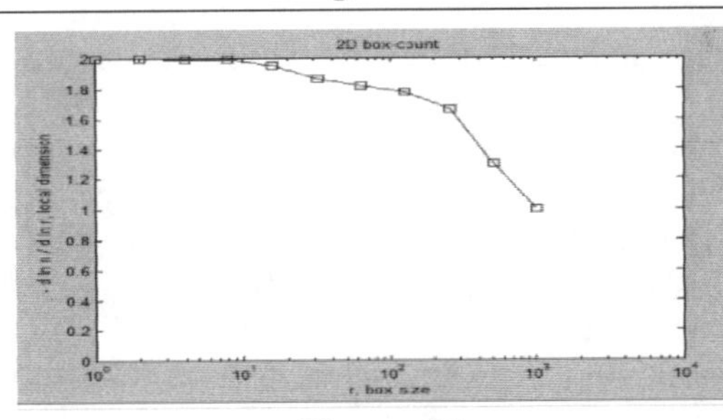 Figure 6d
5	592-850/ Oct-Nov	1.8732	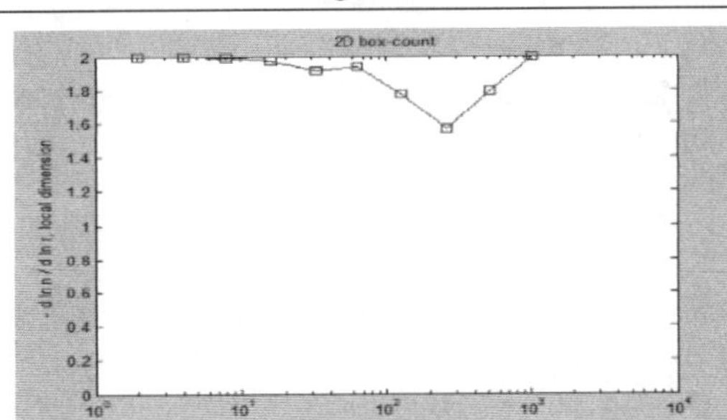 Figure 6e

Journal of Physics: Conference Series **1377** (2019) 012017 doi:10.1088/1742-6596/1377/1/012017

Approaching ϵ -closed sets via ideals

M John Peter[1,2] and R Manoharan[1]

[1]Sathyabama Institute of Science and Technology, Chennai-600119, India
[2]Panimalar Institute of Technology, Chennai-600123, India

E- mail: johnpmath@gmail.com

Abstract. Our intention of this paper is to initiate the new notions on ϵ_{I} -closed set and ϵs-closed sets. Some properties and relationships to other known closed sets are analyzed. Further we define and discuss ϵ_{I}-continuous and ϵs -continuous functions and their basic properties.

1. Introduction and Preliminaries

The space $X_{\mathcal{T}}$ is a TS (Topological space) with a topology $\mathcal{T}$ on X and no axioms are considered unless it is precisely specified. Let a point $x \in X_{\mathcal{T}}$, the collection of open neighborhoods of x is represented by $\mathcal{N}(x) = \{U \in \mathcal{T} : x \in U\}$. For any subset $A \subseteq X_{\mathcal{T}}$, we may designate closure of A as cl(A) and interior of A as int(A), for the given TS $X_{\mathcal{T}}$.

A nonempty system of subsets of a set X is called an ideal [2] on X, if the subsequent statements hold good: (i) If $L \in I$ and $M \subseteq L$, then $M \in I$; (ii) If $L \in I$ and $M \in I$, then $L \cup M \in I$. An ITS (Ideal topological space) or ideal space $(X_{\mathcal{T}}, I)$ means a TS $X_{\mathcal{T}}$ with an ideal I on X.

Let I be an ideal and $\mathcal{T}$ be a topology defined on a space X. Then for any subset A of X, $A^*(I, \mathcal{T}) = \{x \in X / A \cap U_x \notin I, \ \forall U \in \mathcal{N}(x)\}$ is said to be local function of A for the given ideal I and the topology $\mathcal{T}$ [2]. If there is no obscurity, we may denote $A^*(I)$ or simply A^* for $A^*(I, \mathcal{T})$. Also, $cl^*(A) = A^* \cup A$ denotes as a Kuratowski operator for the topology $\mathcal{T}^*$, which is finer than $\mathcal{T}$.

We initiate a definition for ϵ_{I} -closed sets that is associated to the collection of closed sets in the ITS. We discuss some properties on ϵ_{I} –closed sets and ϵs-closed sets using the operations of closure and interior with respect to ITS. Further we define and study ϵ_{I}-continuous and ϵs_{I}-continuous functions and relationships with same kind of continuous functions.

2. Some results on ϵ -Closed Sets

Talal Al-Hawary introduced and studied ϵ-closed sets [1]. In this subdivision we discuss some of the results about ϵ-closed sets and we observe that the equivalent conditions given for ϵ-closed set in [1] are incorrect.

2.1 Definition

Let $X_{\mathcal{T}}$ be topological space and $A \subseteq X_{\mathcal{T}}$. The ϵ-interior of A is the collection of all open sets of X whose closures are contained in cl(A), and it is represented by $int_{\epsilon}(A)$. The set A is said to be ϵ-open if $A = int_{\epsilon}(A)$ and the complement of a ϵ-open set is said to be ϵ -closed. Equivalently, a subset A of X is ϵ -closed if $A = cl_{\epsilon}(A)$, here $cl_{\epsilon}(A) = \{x \in X : cl(U) \cap cl(A) \neq \Phi, x \in U, U \in \mathcal{T}\}$.[1]

Clearly int(A) ⊆ int_ϵ(A) ⊆ cl(A) & A ⊆ cl(A) ⊆ cl_ϵ(A) and hence every ϵ-closed set is closed, but converse is not valid for both the statements. [1]

2.2 Remark
We see that the next example proves the definition given for ϵ--closed sets [1] are not equivalent.

2.3 Example
Let $Z=\{\theta_1, \theta_2, \theta_3, \theta_4\}$ and $\tau = \{\Phi, \{\theta_2\}, \{\theta_3\}, \{\theta_2, \theta_3\}, \{\theta_1, \theta_2\}, \{\theta_1, \theta_2, \theta_3\}, Z\}$.

Table 1:

A	int (A)	cl (A)	$int_\epsilon(A)$	$cl_\epsilon(A)$
Φ	Φ	Φ	Φ	Φ
$\{\theta_1\}$	Φ	$\{\theta_1, \theta_4\}$	Φ	Z
$\{\theta_2\}$	$\{\theta_2\}$	$\{\theta_1, \theta_2, \theta_4\}$	$\{\theta_1, \theta_2\}$	Z
$\{\theta_3\}$	$\{\theta_3\}$	$\{\theta_3, \theta_4\}$	$\{\theta_3\}$	Z
$\{\theta_4\}$	Φ	$\{\theta_4\}$	Φ	Z
$\{\theta_1, \theta_2\}$	$\{\theta_1, \theta_2\}$	$\{\theta_1, \theta_2, \theta_4\}$	$\{\theta_1, \theta_2\}$	Z
$\{\theta_1, \theta_3\}$	$\{\theta_3\}$	$\{\theta_1, \theta_3, \theta_4\}$	$\{\theta_3\}$	Z
$\{\theta_1, \theta_4\}$	Φ	$\{\theta_1, \theta_4\}$	Φ	Z
$\{\theta_2, \theta_3\}$	$\{\theta_2, \theta_3\}$	Z	Z	Z
$\{\theta_2, \theta_4\}$	$\{\theta_2\}$	$\{\theta_1, \theta_2, \theta_4\}$	$\{\theta_1, \theta_2\}$	Z
$\{\theta_3, \theta_4\}$	$\{\theta_3\}$	$\{\theta_3, \theta_4\}$	$\{\theta_3\}$	Z
$\{\theta_1, \theta_2, \theta_3\}$	$\{\theta_1, \theta_2, \theta_3\}$	Z	Z	Z
$\{\theta_1, \theta_2, \theta_4\}$	$\{\theta_1, \theta_2\}$	$\{\theta_1, \theta_2, \theta_4\}$	$\{\theta_1, \theta_2\}$	Z
$\{\theta_1, \theta_3, \theta_4\}$	$\{\theta_3\}$	$\{\theta_1, \theta_3, \theta_4\}$	$\{\theta_3\}$	Z
$\{\theta_2, \theta_3, \theta_4\}$	$\{\theta_2, \theta_3\}$	Z	Z	Z
Z	Z	Z	Z	Z

From the table 1, $\{\theta_3\}$ is ϵ- open set but it is given in [1] as not ϵ-open set. Also, the set $\{\theta_1, \theta_3\}$ is not ϵ-open which is wrongly mentioned as ϵ-open set.

In the above example, we notice that ϵ-open sets are $\{\Phi, \{\theta_3\}, \{\theta_1, \theta_2\}, Z\}$ and their complements are $\{\Phi, \{\theta_1, \theta_2, \theta_4\}, \{\theta_3, \theta_4\}, Z\}$. We observe that only Z, Φ are satisfying the condition A = $cl_\epsilon(A)$. The sets $\{\theta_1, \theta_2, \theta_4\}$and $\{\theta_3, \theta_4\}$ are ϵ -closed sets in one sense and not ϵ -closed sets in other sense.

2.4 Remark

In [1], author proved that arbitrary union of ϵ-open set is ϵ-open set. From example 2.3, we observe that the sets A={θ_3} and B= {θ_1 , θ_2} are ϵ-open sets but A∪B= {θ_1 , θ_2, θ_3} is not ϵ-open set.

3. ϵ -closed sets via ideal

In this chapter, we are going to initiate some open and closed sets in ITS.

3.1 Definition

Let $X_Ţ$ be a TS with topology Ţ and A ⊆ $X_Ţ$. The subset A is called ϵs-closed set of X, if A = $cl_{ϵs}$(A), where $cl_{ϵs}$(A) = {x ∈ X : cl(U) ∩cl(A)≠Φ ,x ∈U,U ∈ Ţ}. The complement of ϵs -closed set is ϵs -open set.

3.2 Definition

Let ($X_Ţ$,Ɨ) be an ITS with ideal Ɨ and the A⊆ ($X_Ţ$,Ɨ), the ϵĮ-interior of A is defined by $int_{ϵĮ}$(A)= ∪{ U ∈Ţ / cl(U) ⊆ cl*(A)} . The set A is said to be an ϵĮ-open set if A = $int_{ϵĮ}$(A) and the complement of ϵĮ – open set is ϵĮ-closed set.

3.3 Definition

Let A be a subset of an ITS ($X_Ţ$,Ɨ), The ϵsĮ -closed set is defined by A = $cl_{ϵsĮ}$*(A) and $cl_{ϵsĮ}$*(A)={x∈X: cl(U)∩cl*(A)≠Φ , x∈ U, U∈Ţ }. And the complement of ϵsĮ -closed set is ϵsĮ -open set.

3.4 Remark

Let $ϵ_c$($X_Ţ$) represent the system of all ϵ-closed sets in $X_Ţ$, $ϵs_c$($X_Ţ$) represent the system of all ϵs-closed sets in $X_Ţ$, $ϵĮ_c$($X_Ţ$, Ɨ) represent the system of all ϵĮ-closed sets in ($X_Ţ$,Ɨ), $ϵsĮ_c$($X_Ţ$, Ɨ) represent the system of all ϵsĮ-closed sets in ($X_Ţ$, Ɨ.).

3.5 Proposition

The subsequent statements are true for any ITS

i) $ϵĮ_c$($X_Ţ$, Ɨ) ⊆ $ϵ_c$($X_Ţ$) ii) $ϵsĮ_c$($X_Ţ$, Ɨ) ⊆ $ϵs_c$($X_Ţ$) but converse need not to be true.

3.6 Proposition

In any ITS, the following statements hold good
i) $ϵĮ_c$($X_Ţ$, Ɨ) = $ϵ_c$($X_Ţ$) ii) $ϵsĮ_c$($X_Ţ$, Ɨ) = $ϵs_c$($X_Ţ$) if and only if Ɨ = {Φ}

3.7 Remark

In the next example we are going to show that in any TS, ϵs-closed sets and ϵ-closed sets are independent.

3.8 Example

Let us consider the space Z = { θ_1 , θ_2 ,θ_3} with topology Ţ = { Φ, {θ_2 }, { θ_1 },{θ_1 ,θ_2 },Z}

and their complement is $Ţ^c$={ Φ,{ θ_2 ,θ_3},{θ_1 ,θ_3},{θ_3}, Z} , ideal with respect to topology is

Ɨ = {Φ, { θ_3}, {θ_1 ,θ_3}}

Tab le 2:

A	int A	cl A	A*	cl*A	$int_{\epsilon}(A)$	$cl_{\epsilon s}(A)$	$int_{\epsilon Ĩ}(A)$	$cl_{\epsilon sĨ}*(A)$
Φ	Φ	Φ	Φ	Φ	Φ	Φ	Φ	Φ
$\{\theta_1\}$	$\{\theta_1\}$	$\{\theta_1, \theta_3\}$	$\{\theta_1, \theta_3\}$	$\{\theta_1, \theta_3\}$	$\{\theta_1\}$	$\{\theta_2, \theta_3\}$	$\{\theta_1\}$	$\{\theta_2, \theta_3\}$
$\{\theta_2\}$	$\{\theta_2\}$	$\{\theta_2, \theta_3\}$	$\{\theta_2, \theta_3\}$	$\{\theta_2, \theta_3\}$	$\{\theta_2\}$	$\{\theta_1, \theta_3\}$	$\{\theta_2\}$	$\{\theta_2, \theta_3\}$
$\{\theta_3\}$	Φ	$\{\theta_3\}$	Φ	$\{\theta_3\}$	Φ	$\{\theta_3\}$	Φ	Z
$\{\theta_1, \theta_2\}$	$\{\theta_1, \theta_2\}$	Z	Z	Z	Z	Z	Z	Z
$\{\theta_1, \theta_3\}$	$\{\theta_1\}$	$\{\theta_1, \theta_3\}$	Φ	$\{\theta_1, \theta_3\}$	$\{\theta_1\}$	Z	$\{\theta_1\}$	Z
$\{\theta_2, \theta_3\}$	$\{\theta_2\}$	$\{\theta_2, \theta_3\}$	$\{\theta_2, \theta_3\}$	$\{\theta_2, \theta_3\}$	$\{\theta_2\}$	Z	$\{\theta_2\}$	Z
Z	Z	Z	Z	Z	Z	Z	Z	Z

From the table 2, we notice that the ϵ -open sets are$\{\{\theta_1\},\{\theta_2\}, Z, \Phi\}$,ϵ-closed set are $\{\{\theta_2, \theta_3\},\{\theta_1, \theta_2\},Z, \Phi\}$, ϵs-closed sets are $\{\{\theta_3\},Z,\Phi\}$, ϵs-open sets are $\{\{\theta_1, \theta_2\},Z,\Phi\}$, ϵĩ – open sets are $\{\{\theta_1\},\{\theta_2\},Z,\Phi\}$, ϵĩ – Closed sets are $\{\{\theta_2, \theta_3\},\{\theta_1, \theta_2\},Z,\Phi\}$, ϵsĩ -closed sets are$\{Z, \Phi\}$, ϵsĩ -open sets are $\{Z, \Phi\}$.The set $\{\theta_2, \theta_3\}$ is ϵ-closed set but not ϵs-closed set and the set $\{\theta_3\}$ is ϵs-closed set but not ϵ-closed set. Similarly ϵĩ – Closed sets and ϵs-closed set are not dependent.

Example 3.8 proves that every ϵĩ-closed set is ϵ-closed sets but converse is not valid. Every ϵsĩ-closed set is ϵs-closed set but converse is not valid. The set $\{\theta_3\}$ is ϵs-closed set but not ϵsĩ-closed set.

4. ϵs-continuous function and ϵsĩ -continuous function

4.1 Definition

A function F: $(X_{\Gamma}, Ɨ) \rightarrow (Y_{\Gamma'}, J)$ is called as i) ϵs –continuous function if F^{-1} ($\mathcal{P}$) is ϵs-closed set in $(X_{\Gamma}, Ɨ)$ for every closed set $\mathcal{P}$ in $(Y_{\Gamma'}, J)$, ii) ϵsĩ –continuous function if F^{-1} ($\mathcal{P}$) is ϵsĩ –closed set in $(X_{\Gamma}, Ɨ)$ for every ϵsĩ –closed set $\mathcal{P}$ *in* $(Y_{\Gamma'}, J)$.

4.2 Theorem

Every ϵsɪ –continuous function is ϵs –continuous function

Proof: Using the proposition 3.5 every ϵsɪ- closed set is ϵs – closed set but converse need not to be true.

4.3 Example

Let us consider the space Z={ θ_1 , θ_2 , θ_3 } with Topology Ʈ = {Φ, { θ_2 }, { θ_1 },{θ_1 , θ_2 }, Z } and their complement is $Ʈ^c$ = {Φ,{ θ_2 , θ_3 },{θ_1 , θ_3 },{θ_3 }, Z} , Ideal with respect to topology is

ɪ = {Φ, { θ_3 }, {θ_1 ,θ_3 }}. Now we define F: Z → Y such that F (θ_1) = F (θ_2) = θ_1 , F (θ_3) = θ_3

This implies that F^{-1}{θ_1 } = F^{-1}{θ_1 , θ_2 }=θ_1 , F^{-1}{θ_2 } = Φ , F^{-1}{ θ_3 } = θ_3 .

Now consider the closed sets F^{-1}{Z} = Z , F^{-1}{Φ} = Φ , F^{-1}{θ_2 ,θ_3 } = {θ_3 },

F^{-1}{θ_1 , θ_3 }={θ_1 , θ_2 , θ_3 } = Z, F^{-1}{θ_3 } = {θ_3 }, since {Φ,{ θ_3 },Z} are ϵs-closed sets ⇒ F is ϵs- continuous function. But {θ_3 } is not ϵsɪ –closed sets ⇒ F is not ϵsɪ- continuous function.

ϵsɪ-continuous function ⇒ ϵs- continuous function, but converse is not valid.

Acknowledgement

I would like to thank the referee for their comments and suggestion to improve this article.

References

[1] Talal Al-Hawary 2018 ϵ-closed sets *Thai J. of Math.* **16** Number 3: 675-681

[2] Manoharan R and Thangavelu P 2013 Some New Sets and Topologies in Ideal Topological Spaces *Chinese J. of Math.* Volume **2013** Article ID 973608 Hindawi publishing corporation

[3] Levine N 1963 Semi-open sets and semi-continuity in topological spaces *Amer. Math. Monthly* **70** 36-41

[4] Levine N 1970 Generalized closed sets in topology *Rend. Circ. Mat. Palermo* **19** (2) 89-96

[5] Talal Al-Hawary and Al-Omari A 2008 Generalized b-closed sets *Mutah Lil-Buhuth Wad-Dirasat* **5(1)** 27-39

[6] Tong J 1989 On Decomposition of continuity in topological spaces *Acta Math. Hung.* **54** (1-2) 51- 55

[7] Balachandran K, Sundaram P and Maki H 1991 On generalized continuous maps in topological spaces *Mem. Fac. Sci. Kôchi Univ. Ser. A Math* **12** 5-13.

[8] Julian Dontchev and Haruo Maki 1999 On θ-genralized closed sets *Internat. J. Math. & Math. Sci.* **22** No 2 (239-249)

[9] Talal Al-Hawary 2017 Fuzzy W-closed sets *Cogent Mathematics* **4**:1343518.

[10] Kuyucu F, Noiri T and Ozkurt A A 2008 A note on W-I-Continuous functions *Acta Math. Hungar.* **119(4)** (393-400)

A Study on Soft s-closedsets,soft s-dense sets via Soft Ideals

J.SebastianLawrence&R.Manoharan

Research scholar
SathyabamaInstitute of Science and Technology,Chennai-600119
Saveetha School of Engineering
Department of Science and Humanities
,
Department of Mathematics,
Sathyabama Institute of Science and Technology, Chennai-600119.

Email: lawrence.seba@gmail.com

Abstract
Thenotionofsoftsetsisintroducedasageneralmathematicaltool for dealing with uncertainty. In this work, we consider the concepts of soft closed set,soft dense setwere introducedinthesoftideal topologicalspace($SS(X)_E$, τ ,$\mathcal{J}$).Also,westudiedthe operations on this new soft s-closed and soft s-dense sets.

1 Introduction

Soft set was introduced by Molodtsov [11] as a mathematical approach to vagueness. Shabir et al[1,8] studied the new operations on soft set theory. D.N.Georgiou et al[2,3,5]developed the soft topological spaces and S.Hussain et al[7] studied some properties on soft topological spaces.Hamlett et al[6]construct the ideals in soft topological spaces and it isdenoted by SITS and SS for soft set ." A nonempty collection of subsets of a set X is said to bean ideal on X, if it satisfies the following two conditions: (i) IfA∈ I and B⊆A, then B∈ I; (ii) If A∈ I and B∈ I, then A∪B∈I.An ideal topological space (or ideal space) (X, τ,I)means a topological space (X, τ) with an ideal I defined on$X$. Let (X, τ)be a topological space with an ideal I definedon X. Then for any subset A of X, $A*(I, \tau) = \{x \in X/A \cap U \notin I \text{ for every } U \in N(x)\}$ is called the local function of$A$ with respect to I andτ. If there is no ambiguity, we willwrite $A*(I)$ or simply $A*$ for $A*(I, \tau)$. Also, $cl*(A) = A \cup A*$defines a Kuratowski closure operator for the topology $\tau*(I)$(or simply $\tau*$)which is finer thanτ.Manoharanet al. [10,12] introduced R*-topology which is finer than $\tau*$. Let (X, τ,I) be an ideal topological space. A subset A of X is said to R*-perfect if A*\ A∈I . The collection of all perfect sets satisfies the conditions of being a basis for some topology and it will be called as R*c(τ, I). we define R*(τ,I)={A⊆X / X \ A ∈R*c(τ,I)} on anonempty set .Clearly, R* (τ, I) is a topology if the set X is finite. The members of the collection R*(,I)will becalled R*-open sets. If there is no confusion about the topology τ and the ideal I , then R* (τ,I) can be called as R*-topology when X is finite". In this work we introduce soft s-closed set, soft s-dense set and soft F_C*-perfect set, study their properties and obtain some relation with other sets in soft ideal topological spaces.

2 Preliminaries

In this section, we consider the basic well-known definitions and results of the soft set theory that are useful for subsequent discussions which is foresaid in [1,2,8,9,10,12]. Throughout this work, X refers to an initial universe set, P(X) is the power set of X (that is the set of all subsets of X) and E is a set of parameters.

2.1*Definition*:A non-empty collection$\mathcal{J} \subseteq SS(\mathcal{X}, \mathcal{E})$ of SS over$SS(\mathcal{X}, \mathcal{E})$iscalledasoftideal,ifthefollowingconditionshold:
i)If $F_E \in \mathcal{J}$and $H_E \subseteq F_E$, then $F_E \in \mathcal{J}$.
ii)IfF_E and $H_E \in \mathcal{J}$, then $F_E \cup H_E \in \mathcal{J}$

3 "Softs-closed set", "soft s-dense set" and"F_C* perfect set"

In this section, we define three collections of soft subsets soft s-closed set,soft s-dense setand F_{C*}perfect in an SITS and study some of their properties.
3.1*Remark*:We denote soft s-closed set and soft s-dense set by ***S-cls*and*S-ds*** respectively.

3.2*Definition*:Let ($SS(\mathcal{X})_\mathcal{E}, \tau , \mathcal{J}$) be an SITS. A subset F_A of $SS(\mathcal{X})_\mathcal{E}$ is said to be***S-cls***-perfect if $F_{A*} \setminus F_A \in \mathcal{J}$.

3.3*Definition*:Let $(SS(\mathcal{X})\mathcal{E}, \tau ,\mathcal{J})$ be an SITS. A subset F_A of $SS(\mathcal{X})_\mathcal{E}$ is said to beS-ds -perfect if $F_A \setminus F_{A*} \in \mathcal{J}$.

3.4*Definition*:Let $(SS(\mathcal{X})_\mathcal{E}, \tau ,\mathcal{J})$ be an SITS. A subset F_A of $SS(\mathcal{X})_\mathcal{E}$ is said to beF_C*-perfect if $F_A$is both S-cls and S-ds.

3.5*Examples*:1.If $\mathcal{J} = \{ F_\phi \}$ then(F,E)*($\mathcal{J}$,τ) = cl(F,E) $\forall$(F,E) $\in SS(\mathcal{X})_\mathcal{E}$. Hence Cl*(F,E) = cl(F,E) and***S-cls*=*S-ds***.=τ*= τ.
2. If $\mathcal{J}=SS(\mathcal{X})_\mathcal{E}$, then (F,E)*($\mathcal{J}$, τ)={ F_ϕ }$\forall$(F,E) $\in SS(X)_E$.Hence Cl*(F,E) = (F,E) and ***S-cls*=*S-ds***.=τ *= τ=$SS(\mathcal{X})_\mathcal{E}$(the softdiscrete topology).
3. If$\mathcal{J} \subseteq \mathfrak{J}$, then (F,E)*($\mathfrak{J}$, τ)$\subseteq$(F,E)*($\mathcal{J}$, τ). Hence the *-soft topological space$(SS(\mathcal{X})_\mathcal{E},\tau*(\mathfrak{J}))$ is finer

than the ∗-soft topological space ($SS(\mathcal{X})_\varepsilon$, τ*($\mathcal{J}$)).

3.6*Remark*:The following example show that every τ * soft topology is S-dssoft topology and S-cls soft topology

3.7*Example*:Let X = {h₁,h₂,h₃}, E = {e} and τ ={$SS(\mathcal{X})_\varepsilon$, F_ϕ , (F,E)}where (F,E) is a SS over X defined by F(e) ={ h₂}. Then τ defines a soft topology on X. Let $\mathcal{J}$ ={ F_ϕ , (G,E)} be a soft ideal over X where (G,E) is aSS over X defined by G(e) = {h₂}. Then τ* ={$SS(\mathcal{X})_\varepsilon$, F_ϕ , (F_1E), (F_2,E)} where (F_1,E), (F_2,E) are SS over X where F_1(e) = {h₂} and F_2(e) = {h₁,h₃}.

3.8*Remark*:The following example show that every τ * soft topology is coincide with S-ds soft topology and S-cls soft topology.

3.9*Example*:Let X = { h₁, h₂, h₃}, E = {e} andτ = {$SS(\mathcal{X})_\varepsilon$, F_ϕ , (F_1,E), (F_2,E)} where (F_1,E), (F_2,E) are SS over X defined as follows:F_1(e) = { h₂}, F_2(e) = { h₁, h₂}. Then τ defines a soft topology on $SS(\mathcal{X})_\varepsilon$. Let $\mathcal{J}$ = { F_ϕ , (G_1,E), (G_2,E), (G_3,E)}be a soft ideal over X where (G_1,E), (G_2,E), (G_3,E) are SS over X defined by G_1(e) = { h₁},G_2(e) = {h₂} and G_3(e) = { h₁, h₂}. Then ***S-cls***= ***S-ds***= τ *= $SS(\mathcal{X})_\varepsilon$.

3.10*Proposition*:If a subset *A*of an SITS($SS(\mathcal{X})_\varepsilon$, τ ,$\mathcal{J}$)) is *C*∗-perfect, then$F_A \Delta F_A{}^* \in \mathcal{J}$.
*Proof.*Since $F_A$is both ***S-ds***-set and ***S-cls***-set,$F_A - F_A{}^* \in \mathcal{J}$ and $F_A{}^* - F_A \in \mathcal{J}$. By the finite additive property of ideals,($F_A - F_A{}^*$) ∪ ($F_A{}^* - F_A$) ∈$\mathcal{J}$. Hence $F_A \Delta F_A{}^* \in \mathcal{J}$.

3.11 *Proposition*:In an SITS($SS(\mathcal{X})_\varepsilon$, τ ,$\mathcal{J}$), every τ*-closed setis ***S-cls***set.
Proof. Let τ be a τ*-closed set. Therefore, $F_{A*} \subseteq F_{A*}$. Hence $F_{A*} - F_A$ = F_ϕ ∈$\mathcal{J}$.Therefore,F_A is an ***S-cls***set.

3.12*Corollary*:InSITS($SS(\mathcal{X})_\varepsilon$, τ ,$\mathcal{J}$),
(i) $SS(\mathcal{X})_\varepsilon$and F_ϕ are ***S-cls*** sets,
(ii) everyτ *-closed set is ***S-cls*** set,
(iii) for any subset F_A of an ideal topological space
($SS(\mathcal{X})_\varepsilon$, τ ,$\mathcal{J}$), *cl*($F_A$),$F_A$ ∗, *cl*∗(F_A) are ***S-cls***sets,
(iv) every regular-I-closed set is***S-cls*** set.
Proof. The proof follows from Proposition 3.10.

3.13*Proposition*: Let ($SS(\mathcal{X})_\varepsilon$, τ ,$\mathcal{J}$) be an SITS. Let F_A and F_B be two subsets of$SS(\mathcal{X})_\varepsilon$such that $F_A \subseteq F_B$ and***S-cls***(A)= ***S-cls***(B),then
(i) F_B is ***S-cls*** setif F_A is ***S-cls*** set.
(ii) F_A is ***S-ds***set if F_B is ***S-ds*** set.
Proof. (i) Let F_B be an F_R∗-perfect set. Then$F_A{}^* - F_A \in \mathcal{J}$.Now, $F_B{}^* - F_B = F_A{}^* - F_B \subseteq F_A{}^* - F_A$. By heredity property of ideals,$F_A{}^* - F_A \in \mathcal{J}$.Hence F_B is ***S-cls*** set.
(ii) Let F_A be an***S-ds***set. Then$F_B - F_B{}^* \in \mathcal{J}$. Now,$F_A - F_A{}^* = F_A - F_B{}^* \subseteq F_B - F_B{}^*$. By heredity property of ideals,$F_A - F_A{}^* \in \mathcal{J}$. Hence F_A is ***S-ds***set.

3.14*Corollary* : Let $(SS(\mathcal{X})_\varepsilon, \tau, \mathcal{J})$ be an SITS. Let F_A and F_B be two subsets of $SS(\mathcal{X})_\varepsilon$ such that $F_A \subseteq F_B \subseteq cl{*}F_A$; then
(i) F_B is ***S-cls*** set if F_A is ***S-cls set***;
(ii) F_A is ***S-ds*** set if F_B is ***S-ds*** set.
Proof. Since $F_A \subseteq F_B \subseteq cl{*}F_A$, $F_A{*} \subseteq F_B{*} \subseteq (cl{*}F_A){*} = F_A{*}$. Hence $F_A{*} = F_B{*}$. Therefore, the result follows from Proposition 3.11.

3.15*Proposition*:If a subset F_A of an SITS$(SS(\mathcal{X})_\varepsilon, \tau, \mathcal{J})$ is ***S-cls*** set and $F_A{*}$ is ***S-ds*** set, then $F_A \cap F_A{*}$ is ***S-ds*** set.
Proof. Since F_A is S-cls set, $F_A{*} - F_A \in \mathcal{J}$, for every $\mathcal{J} \in \mathcal{J}$, $(F_A \cup \mathcal{J}){*} = F_A{*} = (F_A - \mathcal{J}){*}$. Therefore, $(F_A{*} \cup (F_A - F_A{*})){*} = F_A{*} = (F_A{*} - (F_A - F_A{*})){*}$.
This implies $F_A{**} = (F_A \cap F_A{*})$. Therefore, we have $F_A \cap F_A{*} \subseteq F_A{*}$ with $(F_A \cap F_A{*}){*} = F_A{**}$. By Proposition 3.11, $F_A \cap F_A{*}$ is ***S-ds*** set if $F_A{*}$ is ***S-ds*** set.
Hence $F_A \cap F_A{*}$ is ***S-ds*** set.

3.16*Proposition*:If F_A and F_B are ***S-cls*** sets, then $F_A \cup F_B$ is an ***S-cls*** sets.
Proof. Let F_A and F_B be ***S-cls*** sets. Then $F_A{*} - F_A \in \mathcal{J}$ and $F_B{*} - F_B \in \mathcal{J}$. By finite additive property of ideals, $(F_A{*} - F_A) \cup (F_B{*} - F_B) \in \mathcal{J}$. Since $(F_A{*} \cup F_B{*}) - (F_A \cup F_B) \subseteq (F_A{*} - F_A) \cup (F_B{*} - F_B)$, by heredity property $(F_A{*} \cup F_B{*}) - (F_A \cup F_B) \in \mathcal{J}$. Hence $(F_A \cup F_B){*} - (F_A \cup F_B) \in \mathcal{J}$. This proves the result.

3.17*Corollary*: Finite union of ***S-cls*** sets is an ***S-cls*** set.
Proof. The proof follows from Proposition 3.16.

3.18*Proposition* : If F_A and F_B are ***S-ds*** sets, then $F_A \cup F_B$ is an ***S-ds*** set.
Proof. Let F_A and F_A be ***S-ds*** sets. Then $F_A - F_A{*} \in \mathcal{J}$ and $F_B - F_B{*} \in \mathcal{J}$. By finite additive property of ideals, $(F_A - F_A{*}) \cup (F_B - F_B{*}) \in \mathcal{J}$. Since $(F_A \cup F_B) - (F_A{*} \cup F_B{*}) \subseteq (F_A - F_A{*}) \cup (F_B - F_B{*})$, by heredity property $(F_A \cup F_B) - (F_A{*} \cup F_B{*}) \in \mathcal{J}$. Hence $(F_A \cup F_B) - ((F_A \cup F_B){*}) \in \mathcal{J}$. This proves the result.

3.19*Corollary*:Finite union of ***S-ds*** sets is an ***S-ds*** set.
Proof. The proof follows from Proposition 3.18.

3.20*Proposition*: If F_A and F_B are ***S-cls*** sets, then $F_A \cap F_B$ is an ***S-cls*** set.
Proof. Let F_A and F_A be ***S-cls*** sets. Then $F_A{*} - F_A \in \mathcal{J}$ and $F_B{*} - F_B \in \mathcal{J}$. By finite additive property of ideals, $(F_A{*} - F_A) \cap (F_B{*} - F_B) \in \mathcal{J}$. Since $(F_A{*} \cap F_B{*}) - (F_A \cap F_B) \subseteq (F_A{*} - F_A) \cup (F_B{*} - F_B)$, by heredity property $(F_A{*} \cap F_B{*}) - (F_A \cap F_B) \in \mathcal{J}$. Hence $(F_A \cap F_B){*} - (F_A \cap F_B) \in \mathcal{J}$. This proves the result.

3.21*Corollary*: Finite intersection of ***S-cls*** sets is an ***S-cls*** set.
Proof. The proof follows from Proposition 3.20.

3.22*Proposition*: If $(SS(\mathcal{X})_\varepsilon, \tau, \mathcal{J})$ is an SITS with τ being finite, then the collection R is a topology which is finer than the topology of $\tau{*}$-closed sets.
Proof. By Corollary 3.12, $SS(\mathcal{X})_\varepsilon$ and F_ϕ are ***S-cls*** sets. By Corollary 3.17, finite union of ***S-cls*** sets is an ***S-cls*** set, and by Corollary 3.21, finite intersection of ***S-cls*** sets is ***S-cls*** set. Hence the collection R is a topology if $SS(\mathcal{X})_\varepsilon$ is finite. Also, by Proposition 3.11 every $\tau{*}$-closed set is an ***S-cls*** set. Hence the topology R is finer than the topology of $\tau{*}$-closed sets if $SS(\mathcal{X})_\varepsilon$ is finite.

3.23*Theorem*:Let($SS(\mathcal{X})_{\mathcal{E}}$, τ ,$\mathcal{J}$) be a SITS.
Then the following are equivalent.
i) τ~$\mathcal{J}$.
ii)IfF_A has a cover of open sets each of whose intersection with F_A is I, then $F_A$is in$\mathcal{J}$.
iii)For every $F_A \subseteq SS(\mathcal{X})_{\mathcal{E}}$, $F_A \cap F_A* = F_\phi \Rightarrow F_A \in \mathcal{J}$.
iv)For ever$F_A \subseteq SS(\mathcal{X})_{\mathcal{E}}$, $F_A - F_A^* \in \mathcal{J}$.
v)For every $\tau*$-closed subset F_A,$F_A - F_A^* \in \mathcal{J}$.
vi)For every $F_A \subseteq SS(\mathcal{X})_{\mathcal{E}}$, if F_A contains no nonempty subset $F_B$with $F_B \subseteq F_B^*$, then $F_A \in \mathcal{J}$.
Proof. To prove this theorem, it is enough to prove (iv) ⇒(v)⇒(vi). Others follow fromTheorem 3.10 (iv)⇒(v) followsFromProposition 3.10 Suppose that$F_A \Delta F_A^* \in \mathcal{J}$. Since$F_A - F_A^* \subseteq F_A \Delta F_A^*$by heredity property$F_A - F_A^* \in \mathcal{J}$. Hence (v) ⇒ (vi).

References

[1] M. I. Ali, F. Feng, X. Liu, W. K. Min and M. Shabir, "*On some new operations in soft set theory", Comput. Math.Appl., 57, (2009), 1547- 1553.*

[2] A.AygunogluandH.Aygun,"*Somenotesonsofttopologicalspaces",Neu- ral. Comput. Appl., (2011),1-7.*

[3] N.Cagman;S.KaratasandS.Enginoglu,"*Softtopology",Comput.Math. Appl., 62, (2011), 351-358.*

[4] G.Choquet, *"Surlesnotionsdefiltreetgrille",ComptesRendusAcad.Sci. Paris, 224, (1974), 171-173.*

[5] D.N.GeorgiouandA.C.Megaritis, *"Onsoftsettopological",Appl.Math. Inf. Sci., 7 (5), (2013),1889-1901.*

[6] T. R. Hamlett and D. Jankovic, *"Ideals in topological spaces and the set operatory," Bollettinodell'UnioneMatematicaltaliana,vol. 7, pp. 863–874, 1990.*

[7] S.HussainandB.Ahmad, *" Somepropertiesofsofttopologicalspaces", Comput.Math. Appl., 62, (2011), 4058-4067.*

[8] P.K.Maji;R.BiswasandA.R.Roy *"Softsettheory",Compu.andMath. Appl., 45 (2003),555-562.*

[9] D.MandalandM.N.Mukherjee, *"Onatypeofgeneralizedclosedsets", Bol. Soc. Paran. Mat.,(3s.), 30(1), (2012), 67-76.*

[10] R.ManoharanandP.Thangavelu,"*SomeNewSetsandTopologiesinIdeal*

TopologicalSpaces",ChineseJournalofMathematics,vol.2013,Article ID 973608, 6 pages, 2013.doi:10.1155/2013/973608

[11] D.Molodtsov,*"Softsettheory-firstresults",Compu.andMath.Appl.,**37**, (1999),19- 31.*

[12] D. Jankovic and T. R. Hamlett, *"New topologies from old via ideals," The American Mathematical Monthly, vol. 97, pp. 295–310, 1990*

New finer topology from old via ideals

A.JosephineChristilda[1]&R. Manoharan[2]

1.Research Scholar,Department of Mathematics,Sathyabama Institute of Science and Technology, Chennai. jchristilda@yahoo.com

2.Department of Mathematics,SathyabamaInstitute of Science and Technology, Chennai. mano_rl@yahoo.co.in

Abstract. Using ideals defined, on a set many new sets and topologies were defined by many authors. We define a new set (τ_I- mini open set) in ideal space with the use of an ideal and later we proved the collection of that new sets has the properties of generating a new topology which we named as introduce a new topology called τ_I- topology which is containing the actual topology τ .

1. Introduction and Preliminaries

A sub collection of the power set of a nonempty set is called as an ideal I if any subset of an member of I is again an member of I and union of two members of I is again a member of I. A triplet consist of a non empty set S, topology τ and an ideal I is called an Ideal topological space (we use IDS) which we will denote simply as $\tau_I(S)$. In an IDS, for any subset A of the set S, $A^*(I, \tau)$ or simply it is usually referred as A^* contains the points for which every open set which contains the points whose intersection with A is not a member of I [1]. Also, $cl^*(A) = A \cup A^*$ which operator actually induces the topology τ^*_I (or simply τ^*) which contains the actual topology τ.

Many different topologies were introduced [3,6] in an IDS which are finer than the actual topology. In this work we introduced a new set called τ_I – mini open sets and later we will prove collection these sets have the properties of being a basis for a topology.

2. τ_I – mini open sets

In this section we introduce and study about τ_I – mini open sets. We develop some of the basic properties and compare this set with open set of an actual topology

Journal of Physics: Conference Series **1377** (2019) 012019 doi:10.1088/1742-6596/1377/1/012019

2.1 Definition: Let $\tau_I(S)$ be an IDS . Let A $\subseteq$ X then A is said to be a τ_I – mini open set if A \ int (A) $\in \tau$

For an example , take S = R, the set of all real numbers and is the usual topology on S with I is the ideal of finite sets. Take A = [c1,c2] where c1 and c2 are real numbers . Then A \ int (A)= {c1,c2} which is finite. Hence every closed interval as well as open intervals is τ_I – mini open sets. But closed intervals are not open in actual topology. Lawerence et al defined a set A is called L_I-perfect if A' \ A $\in$ I. With respect to the same ideal and topology on the set of real numbers, the set A=(c1,c2) is both τ_I – mini open and L_I-perfect. But if I= { φ }, then A is τ_I – mini open but not L_I-perfect.

*2.2Proposition:*Every τ –open set is τ_I – mini open..

2.3Corolarry:Let $\tau_I(S)$ be an IDS .Then (i) S and φ are τ_I –mini open . (ii) int (A) is τ_I – mini open for all subsets A of S.

2.4Example: let $\tau_I(S)$ be an IDS with S= {c1,c2,c3}, = { S , φ ,{c1}, {c2}, { c1,c2 } } and I = { φ , {c3} }. Then the set {c3},{ c1,c3} , {c2,c3} are τ_I – mini open sets but not τ - open. If we choose I= { φ , {c2} } then {c3} , {c1,c3 } , { c2, c3} are semi open[2] but not τ_I – mini open.

In the same way we can also construct examples to show that pre open [4], α -open sets[5] and τ_I - mini open sets are independent.

2.5Theorem :Union of two τ_I – mini open sets are again a τ_I – mini open set.

2.6Theorem:Intersection of two τ_I – mini open sets are again τ_I – mini open.

2.7corolarry:If A1,A2,A3…. An are τ_I – mini open sest, then (i) $\bigcup_{k=1}^{n} A_k$ is a τ_I – mini open set

(ii). $\bigcap_{k=1}^{n} A_k$ is a τ_I – mini open set.

2.8Theorem:Let $\tau_I(S)$ be an IDS with S is finite . Then the collection of all τ_I – mini open sets forms a topology and it contains the actual topology on τ.

Proof: Follows from Corollary 2.3, 2.7 and Proposition 2.2.

2.9Remark : In an IDS $\tau_I(S)$, if I = { φ }, then the collection of all τ_I – mini open sets coincides with the topology τ .

3. τ_I – open sets

By Theorem 2.5 and Theorem 2.6, the collection of all τ_I – mini open sets forms a basis for a topology. We will call it as τ_I- topology. The members of τ_I- topology are called as τ_I- open sets. A set B of an ideal space (X, τ , I) is τ_I- closed if its complement is τ_I- open.

*3.1Definition :*Let A$\subseteq$ $\tau_I(S)$. We define τ_I -interior of any subset of A which is we denote for further use as AI as A_I = {U/ U is an τ_I- open &Ui $\subseteq$ U for all τ_I- open sets of A}.

*3.2Definition :*Let A $\subseteq$ $\tau_I(S)$ A point x$\in$A is said to be an τ_I -interior point of the set A if there exists an τ_I -open set U of x such that x$\in$U $\subseteq$ A.

*3.3Definition :*Consider an IDS $\tau_I(S)$ and x$\in$S . We define τ_I- neighborhood of x as a τ_I -open set containing x. We denote the set of all τ_I- neighborhoods of x by τ_I -N(x).

3.4Remark :(i) Since every open set is an τ_I -open set, every neighborhood U of a point x $\in$ X is an τ_I - neighborhood of x.
(ii) If x $\in$ X is an interior point of a subset A of X, then x is an τ_I - interior point of A.
(iii) From (ii),we have int (A)$\subseteq$ τ_I -int (A) and int (A) = τ_I -int (A) when I= { φ }.

3.5Theorem: Let A and B be subsets of an IDS . Then the following properties hold:
(i).A_I = $\bigcup$ {U: U$\subseteq$ A and U is an τ_I -open set}.
(ii) τ_I -int(A) is the largest τ_I -open set of X contained in A.
(iii)A is τ_I-open if and only if then A = A_I.
(iv)$(A_I)_I$ = A_I .
(v)If A $\subseteq$ B, then A_I $\subseteq$ B_I

3.6Definition: Let A be a subset of an ideal topological space (X, τ , I) We define τ_I -closure of the set A as the smallest τ_I-closed set containing A. We will denote τ_I -closure of a set A by A_{cl}

3.7Remark : For any subset A of an ideal topological space (X, τ ,I) , τ_I -cl (A) $\subseteq$ cl(A). If I= { φ }, then A_{cl} =.cl(A) .

3.8Theorem: Let A ,B $\in$ $\tau_I(S)$ with S is finite. Then the following properties hold:
(i)A_{cl} = $\bigcap$ {F : A $\subseteq$ F and F is τ_I -closed set}.
(ii)A is τ_I -closed if and only if A = A_{cl} .
(iii)$(A_{cl})_{cl}$ = A_{cl}
(iv)If A $\subseteq$ B, then A_{cl} $\subseteq$ B_{cl}.

3.9Theorem: Let A be a subset of an ideal space (X, τ ,I) .Then the following properties hold:
(i) τ_I -int(X-A) = X \ (τ_I -cl(A)).
(ii) τ_I -cl (X-A) = X \ (τ_I -int (A)).

References

[1]D.Jankovic and T.R.Hamlett,"*New topologies from old via ideals", Amer. Math.Monthly,*.97 (1990), 295-310.

[2] N.Levine, *Semi-open sets and semi- continuity in topological spaces*, *Amer. Math.Monthly,* 70 (1961), 36-41. R. Manoharan and P. Thangavelu, *Some New sets and Topologies in Ideal Topological spaces, Chinese Journal of Mathematics.* Article ID 973608, 6 pages, 2013

[3] A.S.Mashhour, M.E. Abd El-Monsefand S.N.El-Deeb, *On pre- continuous and weak pre-continuous mappings*, Proc. Math. Phys. Soc.Egypt, 53 (1982), 47-53.

[4] R. Manoharan and P. Thangavelu, *Some New sets and Topologies in Ideal Topologicalspaces, Chinese Journal of Mathematics*. Article ID 973608, 6 pages, 2013

[5] O.Njastad, *On some classes of nearly open sets, Pacific J.Math.*,15 (1965), 961- 970..

[6] J. Sebastian Lawrence and R. Manoharan, *New sets and topologies in ideal topological spaces, International Journal of Pure and Applied Mathematics*, 118(20), 9398, 2018.

Flow Shop Scheduling Model for 5machine without Job Block Criteria Using NEH Technique

E.Janaki[1]and A. Mohamed Ismail[2]
[1]Research Scholar Sathyabama Institute of Science and Technology, Chennai, India
[2]Sathyabama Institute of Science and Technology, Chennai, India

Abstract: The present paper describes no job block criteria using NEH technique in a Flow Shop Scheduling Model for five machines. In flow shop, constraints demand that the jobs passing the machine are same. The probabilities are associated with their respective processing times, in which the optimization criteria is the utilization of 5 machines. For a group of jobs in 5 machines, with no job block and No-wait time is considered in the paper. Our agenda is to reduce the total completion time and the shortest job sequence path using NEH algorithm. Sample numerical illustrations are provided to understand the NEH algorithm easily.

Keywords: Flow shop scheduling, Flow Time, Make span, NEH Algorithm.

Introduction

Flow-shop scheduling problem (FSP) is one of the most important problems in industries to develop the manufacturing level with certain period of time and improve the customer's satisfactory level. The problem has a set of n jobs and a set of m machines. Each job should proceed in every m machines in the same order. In this paper, our objective is to find the optimal makespan in a flowshop scheduling problem. [1]Baskar, A., & Xavior, M. (2013) discussed Ruben Ruiz benchmark concepts for further development of obtain the makespan quickly in the usage NEH family of algorithms for scheduling problems. It has been usually general that, the NEH heuristic algorithm performs reasonably higher among the current easy heuristics. [2]Fernandez-Viagas, V., & Framinan, J. M. (2014) concluded that tie breaking mechanism to select the job sequential order while they are used NEH and iterated Greedy method. [3]Baskar, A., (2016) analyzed his paper, the job insertion technique employed by NEH algorithmic rule and proposes several new variants, the complexness level remains same. One hundred twenty numbers of downside instances planned by Taillard are used for the aim of validatory the algorithms. We referred Scheduling Theory, Algorithms, and Systems by [4]Pinedo, M. L. (2012). He explained different concepts about flow shop problems in that book.[5,6]Kalczynski, P. J., & Kamburowski, J. (2008) used concept of Johnson's algorithm for developing the solution for scheduling problem and finally he concluded that new priority sequential order for problem for all number of problems.
[7]Gajpal and Rajendran (2006) have found the best solution when compared the solution with ant colony and existing heuristic for the permutation PFSP. [8]Johnson's (1954) provided best algorithm for researchers to solve the various types of machine problem. [9]Campbell, Dudek, and Smith introduced heuristic algorithm to extend the Johnson's problem which pick the minimum completion time out of the (m-1) enumerations. The effective method was introduced by [10]Nawaz, M., Enscore, E. E., & Ham, I. (1983) which uses the nice job insertion technique after arranging the jobs in the non -increasing order of their total processing instances. In that method he Choose first two jobs as the initial partial sequence and other jobs are inserted one by one from the third job to

obtain a final optimal makespan and its corresponding sequence. [11]Taillard (1990) explained in detail about the bench mark problem by analyzing simple heuristics and it was finalized that the NEH algorithm is a better one for different sizes of problems varying from 9 to 50 jobs. Improvements had been finished on the NEH set of rules through the years to reduce the makespan in addition to the computing time.

Practical Situation

1 Assumptions:

1. Initially to start a process all the jobs are available.
2. Processing time associated with probabilities are included with set-up time for the machines
3. Job on a particular machine will be allowed only after completing the same job on the previous machine
4. Machines may be idle.
5. Processing time of the jobs on the schedule is maintained independently
6. At a time, in a machine multiple jobs cannot be processed.

2 Notations:
We are given n jobs to be processed on two stage flowshop scheduling problem and we have used the following notations

Jij - Processing time for ith job on machine Mj
Pij -Probability associated to the processing time Jij

3 Mathematical Development:
Consider n jobs say i=1, 2, 3 … n are processed on five machines M1, M2, M3, M4 & M5. A job J(i=1,2,3…n) has processing time Ji1,Ji2, Ji3 , Ji4 &Ji5 on each machine respectively, assuming their respective probabilities Pi1 , Pi2 , Pi3 , Pi4&Pi5such that $0 \leq Pij \leq 1$, $\Sigma Pij = 1$

Table 1

job	Machine M1		Machine M2		Machine M3		Machine M4		Machine M5	
i	Ji1	Pi1	Ji2	Pi2	Ji3	Pi3	Ji4	Pi4	Ji5	Pi5
1	J11	P11	J12	P12	J13	P13	J14	P14	J15	P15
2	J21	P21	J22	P22	J23	P23	J24	P24	J25	P25
3	J31	P31	J32	P32	J33	P33	J34	P34	J35	P35
4	J41	P41	J42	P42	J43	P43	J44	P44	J45	P45
5	J51	P51	J52	P52	J53	P53	J54	P54	J55	P55
	-	-	-	-	-	-	-	-	-	-
n	-	-	-	-	-	-	-	-	-	-

Our objective is to find the shortest makespan path by using NEH Algorithm

NEH Algorithm:

Step1: Calculate expected value of processing time on five machines Aij

i.e., the Total Expected value is $\sum_{i=1}^{4} A_{ij}$, j =1, 2, 3, 4 & 5.

Step2: Arrange the jobs by non increasing order
Step3: Select first 2 subsequence job from result of step 2
Step4: Choose the best sequence having minimum makespan.
Step5: Continue the process until all the jobs are available in the best sequence

Numerical Illustration:

Consider 4 jobs 5 machine flow shop problem whose processing time of the jobs on each machine is given below.

Table 2

job	Machine M1		Machine M2		Machine M3		Machine M4		Machine M5	
i	Ji1	Pi1	Ji2	Pi2	Ji3	Pi3	Ji4	Pi4	Ji5	Pi5
1	15	0.4	50	0.2	30	0.3	55	0.2	50	0.4
2	50	0.2	40	0.1	55	0.2	20	0.1	45	0.2
3	50	0.2	10	0.5	20	0.3	30	0.3	35	0.2
4	25	0.2	45	0.2	45	0.2	20	0.4	15	0.2

Solution: As per step1 the expected processing times for machines M1, M2, M3, M4& M5 are as follow:

Table3

job	Machine M1	Machine M2	Machine M3	Machine M4	Machine M5	Sum
i	Ai1	Ai2	Ai3	Ai4	Ai5	$\sum_{i=1}^{4} A_{ij}$
1	6	10	9	11	2	38
2	10	4	11	2	9	36
3	10	5	6	9	7	37
4	5	9	9	8	3	34

As per Step2, Arrange the jobs in non increasing order 1-3-2-4
As per Step3, select first 2 jobs from the above sequence
The possible sequence are 1-3 and 3 -1

Table 4

job	Machine M1	Machine M2	Machine M3	Machine M4	Machine M5
1	0-6	6-16	16-25	25-36	36-38
3	6-16	16-21	25-31	36-45	45-52

Table 5

job	Machine M1	Machine M2	Machine M3	Machine M4	Machine M5
3	0-10	10-15	15-21	21-30	30-37
1	10-16	16-26	26-35	35-46	46-48

The minimum makespan is 48, Hence the optimal sequence is 3-1

As per Step 4, Insert the job 2 in the above optimal sequence
The possible sequence are 2- 3 -1, 3-2-1 and 3-1-2

Table 6

job	Machine M1	Machine M2	Machine M3	Machine M4	Machine M5
2	0-10	10-14	14-25	25-27	27-36
3	10-20	20-25	25-31	31-40	40-47
1	20-26	26-36	36-45	45-56	56-58

Table 7

job	Machine M1	Machine M2	Machine M3	Machine M4	Machine M5
3	0-10	10-15	15-21	21-30	30-37
2	10-20	20-24	24-35	35-37	37-46
1	20-26	26-36	36-45	45-56	56-58

Table 8

job	Machine M1	Machine M2	Machine M3	Machine M4	Machine M5
3	0-10	10-15	15-21	21-30	30-37
1	10-16	16-26	26-35	35-46	46-48
2	16-26	26-30	35-46	46-48	48-57

The minimum makespan is 57, Hence the optimal sequence is 3-1-2

Insert Job 4 in the above optimal sequence and the possible sequences are

4-3-1-2, 3-4-1-2, 3-1-4-2 and 3-1-2-4

The minimum makespan is 62, Hence the optimal sequence is 4-3-1-2.

Pascal's Triangle Algorithm

Step1: Find the product of processing time of the machine with Pascal's triangle number.

Step2: Find the sum of the product.

Step3: 'n' job, 'm' machine problem is converted into' n' job, '2' machine problem.

Step4: Apply Johnson's algorithm to find the optimal solution.

New Imaginary machines Gi and Hi are arrived from Table 3 as per Pascals algorithm

Table 9

job	Gi	Hi
1	74	72
2	57	52
3	52	57
4	67	63

By Johnson's rule, the total completion time is 67 and the corresponding optimal sequence is 3-2-1-4.

CDS algorithm

Step1: Convert five machine into two fictitious machines

Step2: Apply Johnson's algorithm to find the optimal solution

New machines MH1 and MH2 are arrived from **Table 3** as per CDS algorithm

Table 10

job	MH1	MH2
i	M1+M2+M3+M4	M2+M3+M4+M5
1	36	32
2	27	26
3	30	27
4	25	29

Applying Johnson's algorithm, The optimal sequence is 1-4-3-2 and the Completion time is 69.

The below comparison shows NEH algorithm provides the best optimal sequence with minimal completion time

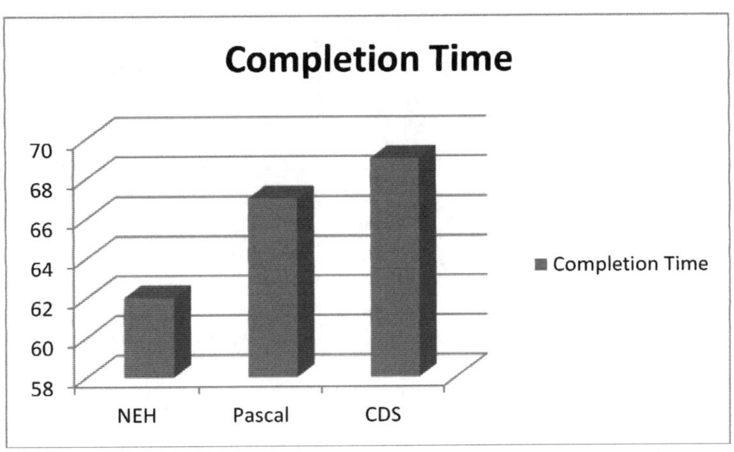

Conclusion

The objective was to minimize the total completion time and also satisfies the customer's satisfactory. This paper concludes that after using five machines without job block by using NEH method which minimizes the total elapsed time. The above illustration does not have any tie breaking scenarios. Tie breaking mechanism are applied to break the Tie.

References

1. Baskar, A., &Xavior, M. (2013). Optimization of makespan in flow shop scheduling problems using combinational NEH family of heuristics-An analysis. International Journal of Applied Engineering Research, 8(10), 1205-1217.
2. Fernandez-Viagas, V., &Framinan, J. M. (2014). On insertion tie-breaking rules in heuristics for the permutation flowshop scheduling problem. Computers & Operations Research, 45, 60-67.
3. Baskar, A., (2016) Revisiting the NEH algorithm- the power of job insertion technique for optimizing the makespan in permutation flow shop scheduling., International Journal of Industrial Engineering Computations 7(2016) 353–366
4.Pinedo, M. L. (2012). Scheduling Theory, Algorithms, and Systems. Springer Science& Business Media.
5. Kalczynski, P. J., &Kamburowski, J. (2008). An improved NEH heuristic to minimize makespan in permutation flow shops. Computers &Operations Research, 35(9), 3001-3008.
6. Kalczynski, P. J., &Kamburowski, J. (2007). On the NEH heuristic for minimizing the makespan in permutation flow shops. Omega, 35(1), 53-60.
7. Yuvraj Gajpal., & Chandrasekharan Rajendran (2006) an ant-colony optimization algorithm for minimizing the completion-time variance of jobs in flowshops.International Journal of Production Economics 101(2):259-272 · June 2006.
8. Johnson, S. M. (1954). Optimal two-and three-stage production schedules with setup times included. Naval Research Logistics Quarterly, 1(1), 61-68.
9. Campbell, H. G., Dudek, R. A., & Smith, M. L. (1970). A heuristic algorithm for the n job, m machine sequencing problem. Management Science, 16(10), B-630.
10. Nawaz, M., Enscore, E. E., & Ham, I. (1983). A heuristic algorithm for the m-machine, n-job flow-shop sequencing problem. Omega, 11(1), 91-95.
11. Taillard, E. (1993). Benchmarks for basic scheduling problems. European Journal of Operational Research, 64(2), 278-285.

A Method For Solving Balanced And Unbalanced Trapezoidal Intuitionistic Fuzzy Assignment Quandary

M Joseph Robinson[1] C Veeramani [2] and A Sudha Rani[3]

[1]**Assistant Professor, Department of Mathematics,**
Gojan School Of Business And Technology Redhills, Chennai-52.
[2]**Assistant Professor, Department of Applied Science,**
PSG College of Technology, Peelamedu, Coimbatore-4.
[3]**Assistant Professor, Department of Mathematics,**
SRMIST, Ramapuram, Chennai-89.

Abstract. In this paper, we discuss Trapezoidal intuitionistic fuzzy assignment quandary(TrIFAP). In classical assignment quandary, Cost always remains same. But, here we develop an method to solve Trapezoidal intuitionistic fuzzy assignment quandary where cost is not deterministic numbers but infelicitous ones. Here, the costs(profits) matrix have all the elements they are TrIFN and others have been immensely colossal demonstration against the classical method. Then its trapezoidal membership and non-membership functions are defined. In general compare the findings from different methods attested the same picture. Numerical examples show that an intuitionistic fuzzy ranking method is very efficacious and utilizable method for handling a TrIFAP.

Keywords: Trapezoidal Intuitionistic fuzzy, fuzzy assignment problem.

1. Introduction

The assignment quandary is one of the rudimental combinatorial analytics in the branch of operation research in mathematics. It consists of finding a maximum weight equal in numbers (or minimum weight perfect equal in numbers) The quandary cases has a number of person and a number of works. Any person can be assigned to perform any work, incurring some cost that may vary depending on the person-work assignment.

It is required to perform all works by assigning precisely one work to one person in such a way that the total cost of the assignment is minimized. If the numbers of person and works are identically tantamount and the total cost of the assignment for all works is identically tantamount to the sum of the costs for each person (or the sum of the costs for each work, which is identically tantamount thing in this case), then the quandary is called the linear assignment quandary

Conventionally, when verbalizing about of the assignment quandary without any adscititious qualification, then the linear assignment quandary is betokened. However, in authentic life situations, the inhibition of assignment quandary are inopportune numbers in lieu of fine-tuned authentic numbers because time/cost for doing a work by a facility (machine/person) modal vary due to different reasons.

[1] In 1965 Zadeh introduced fuzzy set it has been achieved prosperous applications in sundry fields. [6] In 1970, Belmann and Zadeh introduce the concepts of fuzzy set theory into the decision making quandaries involving dubiousness and imprecision. Amir Kumar et al investigated Assignment and Travelling Salesman Quandaries with cost coefficients as LR fuzzy parameters [7] and Fuzzy

linear programming approach for solving fuzzy conveyance quandaries with transhipment [8] R.ZahirHussain, P.Senthil Kumar Method for solving plenarily fuzzy assignment quandaries utilizing trapezoidal fuzzy numbers[9].[10] In Sathi Mukherjee et al presented an Application of fuzzy ranking method for solving assignment quandaries with fuzzy costs. Lin and Wen [11] proposed an efficient algorithm predicated an labeling method for solving the linear fractional programming case.

Y.L.P.Thorani and N.RaviSankar did Fuzzy assignment quandary with generalized fuzzy numbers [4]. Different kinds of fuzzy assignment quandaries are worked out in the many papers. The concept of Intuitionistic Fuzzy Sets (IFSs) proposed by Atanassow[2] in 1986 is found to be highly utilizable to deal with vagueness. Jahir Hussain et al presented An Optimal More-for-Less Solution of Commixed Constraints Intuitionistic Fuzzy Conveyance Quandaries.

[12], This paper is framed as follows: Section 2 contain with some rudimentary definitions and ranking of trapezoidal intuitionistic fuzzy numbers, In section 3, yield not only the definition of intuitionistic fuzzy assignment quandary but withal Fundamental Theorems of an Intuitionistic Fuzzy Assignment Quandary. Section 4 describes the solution procedure of an intuitionistic fuzzy assignment quandary, In section 5, to illustrate the proposed method a numerical example with results is discussed and followed by the conclusions which are given in Section 6.

2 Preliminaries. **Definition 2.1 [1, 2] FUZZY SET:** Let X be a universal set. The fuzzy set $\check{A}$ contain in X is defined by the set of tuples as $\widetilde{A^S} = \{(x, \mu_A(x)); \mu_A(x): X \to [0,1]\}$ .The membership function $\mu_A(x)$ of a fuzzy set $\check{A}$ is a function with mapping $\mu_A(x): X \to [0,1]$.So every element x in X has membership function $\mu_A(x)$ in $[0, 1]$ which is real number.

Definition 2.2[1, 2] INTUITIONISTIC FUZZY SET: Let X be a universe of discourse, then an intuitionistic fuzzy set A in X is given by a set of ordered triples.

$$\tilde{A}^{IFs} = \{< x,\ \mu_A(x), \vartheta_A(x) >; x \in X\}$$

Where $\mu_A,\ \vartheta_A: X \to [0,1]$, are functions such that $\leq\ \mu_A(x) + \vartheta_A(x) \leq 1, \forall x \in X$. For each x the membership $\mu_A(x)$ and $\vartheta_A(x)$represent the degree of membership and the degree of non-membership of the element $x \in X$ to $A \subset X$ respectively.

Definition 2.3 [1, 2] An intuitionistic fuzzy number $\tilde{A}^{TrIFN} = (a_2, a_3, \alpha_1, \beta_1)(a_2, a_3, \alpha_1{}', \beta_1{}')$ is said to be trapezoidal intuitionistic fuzzy number (TrIFN) if its membership and non-membership functions are respectively given by,

$$\mu_A(x) = \begin{cases} \dfrac{x-\alpha_1}{a_2-\alpha} \alpha_1 \leq x \leq a_2 \\ 1 \quad a_2 \leq x \leq a_3 \\ \dfrac{\beta_1 - x}{a_4 - a_3} a_3 \leq x \leq \beta_1 \\ 0 \quad otherwise \end{cases}$$

$$\vartheta_A(x) = \begin{cases} \dfrac{a_2 - x}{a_2 - \alpha_1{}'}, \alpha_1{}' \leq x \leq a_2 \\ 0 \quad a_2 \leq x \leq a_3 \\ \dfrac{a_3 - x}{a_3 - \beta_1{}'}, a_3 \leq x \leq \beta_1{}' \\ 1 \quad otherwise \end{cases}$$

Definition 2.4 Let $\tilde{A}^{TrIFN}$ *and* $\widetilde{\boldsymbol{B}}^{TrIFN}$ be two TrIFNs. The ranking of $\tilde{A}^{TrIFN}$ and $\widetilde{\boldsymbol{B}}^{TrIFN}$ by the $R(.)$on U, the set of TrIFNs is defined as follows:

i. $R(\tilde{A}^{TrIFN}) > R(\widetilde{\boldsymbol{B}}^{TrIFN})$ iff $\tilde{A}^{TrIFN} > \widetilde{\boldsymbol{B}}^{TrIFN}$
ii. $R(\tilde{A}^{TrIFN}) < R(\widetilde{\boldsymbol{B}}^{TrIFN})$ iff $\tilde{A}^{TrIFN} < \widetilde{\boldsymbol{B}}^{TrIFN}$

iii. $R(\tilde{A}^{TrIFN}) = R(\tilde{\boldsymbol{B}}^{TrIFN})$ iff $\tilde{A}^{TrIFN} = \tilde{\boldsymbol{B}}^{TrIFN}$

iv. $R(\tilde{A}^{TrIFN} + \tilde{\boldsymbol{B}}^{TrIFN}) = R(\tilde{A}^{TrIFN}) + R(\tilde{\boldsymbol{B}}^{TrIFN})$

v. $R(\tilde{A}^{TrIFN} - \tilde{\boldsymbol{B}}^{TrIFN}) = R(\tilde{A}^{TrIFN}) - R(\tilde{\boldsymbol{B}}^{TrIFN})$

Arithmetic Operations

Let $\tilde{A}^{TrIFN} = (a_2, a_3, \alpha_1, \beta_1)(a_2, a_3, \alpha_1', \beta_1')$ and $\tilde{\boldsymbol{B}}^{TrIFN} = (b_2, b_3, \gamma_1, \delta_1)(b_2, b_3, \gamma_1', \delta_1')$ be any two TrIFNs then the following arithmetic operations as follows:

Addition:

$$\tilde{A}^{TrIFN} + \tilde{\boldsymbol{B}}^{TrIFN} = (\alpha_1 + \gamma_1, a_2 + b_2, a_3 + b_3, \beta_1 + \delta_1)(\alpha_1' + \gamma_1', a_2 + b_2, a_3 + b_3, \beta_1' + b\delta_1')$$

Subtraction

$$\tilde{A}^{TrIFN} - \tilde{\boldsymbol{B}}^{TrIFN} = (\alpha_1 - \delta_1, a_2 - b_2, a_3 - b_3, \beta_1 - \gamma_1)(\alpha_1' - \delta_1', a_2 - b_2, a_3 - b_3, \beta_1' - \gamma_1')$$

Ranking of Trapezoidal Intuitionistic Fuzzy Numbers

The Ranking of a Trapezoidal intuitionistic fuzzy number $\tilde{A}^{TrIFN} = (a_2, a_3, \alpha_1, \beta_1)(a_2, a_3, \alpha_1', \beta_1')$ is defined

$$R(\tilde{A}^{NTrIFN}) = \frac{1}{4}\left[\frac{(\beta_1' - \alpha_1')(a_2 + a_3 - 3\alpha_1' - 3\beta_1') + (\beta_1 - \alpha_1)(\alpha_1 + a_2 + a_3 + \beta_1) + 4(\beta_1^2 - \alpha_1^2)}{(\beta_1 - \alpha_1' + \beta_1 - \alpha_1)}\right]$$

The ranking technique is, If $R(\tilde{A}^{NTrIFN}) \le R(\tilde{B}^{TrIFN})$ iff $\tilde{A}^{TrIFN} \le \tilde{B}^{TrIFN}$

Example: Let $\tilde{A}^{TrIFN} = (9, 10, 10, 11)(9, 10, 7, 12)$ and $\tilde{\boldsymbol{B}}^{TrIFN} = (18, 19, 17, 20)(18, 19, 16, 21)$ be any two TrIFN, then its rank is defined by $R(\tilde{A}^{TrIFN}) = 9.485$ and $R(\tilde{\boldsymbol{B}}^{TrIFN}) = 18.489$ this implies $\tilde{A}^{TrIFN} < \tilde{\boldsymbol{B}}^{TrIFN}$.

3 Trapezoidal Intuitionistic Fuzzy Assignment Quandary. Consider the solution of assigning n person to n works and each person is capable of doing any works at different costs. Let $\tilde{c}_{ij}^{TrIFN}$ TrIFN be an intuitionistic fuzzy cost of assigning intuitionistic fuzzy cost of assigning the jth work to the ith person. Let x_{rc} be the decision variable denoting the assignment of the person i to the work j. The objective is to minimize the total cost. This situation is kenned as balanced intuitionistic fuzzy assignment quandary.

Minimize $\tilde{Z}^{TrIFN} = \sum_{i=1}^{n}\sum_{j=1}^{n} \tilde{c}_{rc}^{TrIFN} x_{rc}$

Subject to

$$\sum_{c=1}^{n} x_{rc} = 1, \quad i = 1, 2, \dots, n$$
$$\sum_{r=1}^{n} x_{rc} = 1, \quad j = 1, 2, \dots, n$$
$$\tilde{c}_{rc}^{TrIFN} = \left(\tilde{c}_{rc}^{a2}, \tilde{c}_{rc}^{a3}, \tilde{c}_{rc}^{\alpha_1} \tilde{c}_{rc}^{\beta_1}\right)(\tilde{c}_{rc}^{a2}, \tilde{c}_{rc}^{a3}, \tilde{c}_{rc}^{\alpha_1'}, \tilde{c}_{rc}^{\beta_1'})$$
$$x_{ij} \in [0,1]$$

Where,

$$x_{rc} = \begin{cases} 1, & if\ r^{th}\ person\ is\ assigned\ to\ c^{th}\ work \\ 0, & if\ r^{th}\ person\ is\ not\ assigned\ to\ c^{th}\ work \end{cases}$$

3.1 Fundamental Theorems Of An Intuitionistic Fuzzy Assignment Quandary. The Solution of an intuitionistic fuzzy assignment quandary is primary source predicated on the following theorems.

Theorem 1

In an intuitionistic fuzzy assignment quandary, if we sum or difference an intuitionistic fuzzy number to every element of any row (or column) of the intuitionistic fuzzy cost matrix $[\tilde{c}_{rc}^{TrIFN}]$, then an assignment that minimizes the total intuitionistic fuzzy cost on one matrix withal minimizes the total intuitionistic fuzzy cost on the other matrix. In other words if $x_{rc} = x_{rc}^*$

minimizes, $\tilde{Z}^{TrIFN} = \sum_{r=1}^{n}\sum_{c=1}^{n}\tilde{c}_{rc}^{TrIFN}x_{rc}$ with $\sum_{c=1}^{n}x_{rc} = 1$, $\sum_{c=1}^{n}x_{rc} = 1$, $x_{rc} = 0\ or\ 1$ then $x_{rc}{}^{*}$ also minimizes $\tilde{Z}^{TrIFN^*} = \sum_{r=1}^{n}\sum_{c=1}^{n}\tilde{c}_{rc}^{TrIFN^*}x_{rc}$, where $\tilde{c}_{rc}^{TrIFn^*} = \tilde{c}_{rc}^{TrIFN} - \tilde{u}_r^{TrIFN} - \tilde{v}_c^{TrIFN}$ for all $r, c = 1,2, \dots, n$ and $\tilde{u}_r^{TrIFN}$, $\tilde{v}_c^{TrIFN}$ are some real trapezoidal intuitionistic fuzzy numbers.

Proof: $\tilde{Z}^{TrIFN^*} = \sum_{r=1}^{n}\sum_{c=1}^{n}\tilde{c}_{rc}^{TrIFN^*}x_{rc}$

$$= \sum_{r=1}^{n}\sum_{c=1}^{n}(\tilde{c}_{rc}^{TrIFN} - \tilde{u}_r^{TrIFN} - \tilde{v}_c^{TrIFN})x_{rc}$$

$$= \sum_{r=1}^{n}\sum_{c=1}^{n}\tilde{c}_{rc}^{TrIFN}x_{rc} - \sum_{r=1}^{n}\tilde{u}_r^{TrIFN}\sum_{c=1}^{n}x_{rc} - \sum_{c=1}^{n}x_{rc}\sum_{c=1}^{n}\tilde{v}_c^{TrIFN}$$

$$= \tilde{Z}^{TrIFN} - \sum_{r=1}^{n}\tilde{u}_r^{I} - \sum_{r=1}^{n}\tilde{v}_r^{I}$$

Since $\sum_{c=1}^{n}x_{rc} = 1$, $\sum_{c=1}^{n}x_{rc} = 1$,

This display that the minimization of the new objective function $\tilde{Z}^{TrIFN^*}$ submit the same solution as the minimization of a primary form objective function $\tilde{Z}^{TrIFN}$ because $\sum_{r=1}^{n}\tilde{u}_r^{TrIFN}$ and $\sum_{c=1}^{n}\tilde{v}_c^{TrIFN}$ are independent of x_{rc}.

Theorem 2

In an intuitionistic fuzzy assignment problem with cost $[\tilde{c}_{rc}^{\mathrm{TrIFN}}]$, if all $[\tilde{c}_{rc}^{\mathrm{TrIFN}}] \geq \tilde{0}^{\mathrm{TrIFN}}$ then a suitable solution x_{rc} which satisfies $\sum_{r=1}^{n}\sum_{c=1}^{n}\tilde{c}_{rc}^{\mathrm{TrIFN}}x_{rc} = \tilde{0}^{\mathrm{TrIFN}}$, x_{rc} is best for the problem.

Proof:

Since all $[\tilde{c}_{rc}^{\mathrm{TrIFN}}] \geq \tilde{0}^{\mathrm{TrIFN}}$ and $[x_{rc}] \geq 0$,

The objective function $\tilde{Z}^{\mathrm{TrIFN}} = \sum_{r=1}^{n}\sum_{c=1}^{n}\tilde{c}_{rc}^{\mathrm{TrIFN}}x_{rc}$ cannot be negative.

The minimum possible value that TrIFN can attain $\tilde{0}^{\mathrm{TrIFN}}$.

Thus, any suitable solution $[x_{rc}]$ that satisfies $\sum_{r=1}^{n}\sum_{c=1}^{n}\tilde{c}_{rc}^{\mathrm{TrIFN}}x_{rc} = \tilde{0}^{\mathrm{TrIFN}}$, x_{rc} will be an best.

4 The Computational Procedure For trapezoidal Intuitionistic Fuzzy Assignment Problem.

To find optimal solution to TrIFN Assignment problem there two stages

STAGE 1:

Step:1 Obtain new matrix by doing the following

ROW REDUCTION:

From each row, find the row minimum TrIFN, subtract it from all entries on that row

Step 2:

COLUMN REDUCTION:

Next from each column, find the column minimum TrIFN; subtract it from all entries on that Column.

STAGE 2:

Step 3:

Draw minimum number of lines covered all zeros of theTrIFN matrix. The procedure to do it is follows

(a) **ROWSCANNING**

Starting from first Row, check for precisely on Zero in the row, if so square it and draw a vertical line passing through that zero otherwise skip the row . After consummating all row, check all zero's are strike, if so go to step-4, otherwise do column scanning

(b) **COLUMN SCANNING**

Starting from first column check for precisely one zero's in the column(only not strike zero)

Journal of Physics: Conference Series **1377** (2019) 012021 doi:10.1088/1742-6596/1377/1/012021

If so square it and draw a horizontal line passing through it otherwise skip the column.After scanning all column, check whether all zero's are strike,If yes go to step-4, otherwise again do row scanning

Step 4:

Check whether the number of square marked is equipollent to the number of rows of the matrix. If yes optimality is reached treat the marked zero's as assignment, Otherwise go to Step 5.

Step 5:

Identify the minimum value of the undeleted cell values .obtain the next matrix by doing the following

1. Copy the ingressions on the line except the intersection points of the present matrix, as such to the incipient matrix
2. Add the minimum undelete cell value to the intersection point and facsimile to incipient matrix
3. Subtract the minimum undeleted cell value from all undeleted cell and copy then to the incipient matrix.

Step 6:

Go to Step 3.

5 Numerical Examples

Example 5.1 Let us consider an intuitionistic fuzzy assignment quandary with rows representing 3 persons namely A,B,C and columns representing 3 $Jobs$,namely 1,2,3. The cost matrix $[\tilde{c}_{rc}^{TrIFN}]$ is given whose elements are TrplFN. The problem is to find the Correct assignment so that the total cost of	1	2	3

job assignment becomes minimum.			
A	(10,11,9,12)(10,11,8,13)	(8,9,7,10)(8,9,6,11)	(12,13,11,14)(12,13,10,15)
B	(18,19,17,20)(16,18,19,16,21)	(6,7,5,8) (6,7,4,9)	(14,15,,13,16)(13,14,12,17)
C	(6,7,5,8) (6,7,4,9)	(4,5,3,6)(4,5,2,7)	(2,3,1,4)(2,3,0,5)

Solution:

The above intuitionistic fuzzy assignment quandary can be formulated in the following mathematical programming form

Min $[(10,11,9,12)(10,11,8,13)x_{A1} + (8,9,7,10)(8,9,6,11)x_{A2} + (12,13,11,14)(12,13,10,15)x_{A3} + (18,19,17,20)(18,19,16,21)x_{B1} + (6,7,5,8)(6,7,4,9)x_{B2} + (14,15,13,16)(13,14,12,17)x_{B3} + (6,7,5,8)(6,7,4,9)x_{C1} + (4,5,3,6)(4,5,2,7)x_{C2} + (2,3,1,4)(2,3,0,5)x_{C3}]$

Subject to $x_{A1} + x_{A2} + x_{A3} = 1, x_{B1} + x_{B2} + x_{B3} = 1, x_{C1} + x_{C2} + x_{C3} = 1$

$x_{A1} + x_{B1} + C_{31} = 1, x_{A2} + x_{B2} + x_{C2} = 1, x_{A3} + x_{B3} + x_{C3} = 1$

$x_{rc} \in [0,1]$

Now, using the Step 1 of the intuitionistic fuzzy Hungarian assignment method, we have the following reduced intuitionistic fuzzy assignment table.

	1	2	3
A	(2,2,-1,5)(2,2,-3,7)	(0,0,-3,3)(-5,0,0,5)	(4,4,1,7)(4,4,-1,7)
B	(12,12,9,15)(12,12,717)	(0,0,-3,3)(0,0,-5,5)	(8,8,5,11)(8,8,3,13)
C	(4,4,1,7)(-1,4,47)	(2,2,-1,5)(-3,2,2,7)	(0,0,-3,3)(0,0-5,5)

Now, using the Step 2 of the intuitionistic fuzzy Hungarian assignment method, we have the following reduced intuitionistic fuzzy assignment table.

	1	2	3
A	(0,0,-6,6)(0,0-10,10)	(0,0,-3,3)(0,0,-5,5)	(4,4,1,7)(4,4,-1,7)
B	(10,10,4,16)(0,10,10,20)	(0,0,-3,3)(0,0,-5,5)	(8,8,5,11)(8,8,3,13)
C	(2,2,-4,8)(2,2,-8,12)	(2,2,-1,5)(2,2,-3,7)	(0,0,-3,3)(0,0,-5,5)

The optimal solution is,

$$x_{A1}{}^* = x_{B2}{}^* = x_{C3}{}^* = 1, \quad x_{A2}{}^* = x_{A3}{}^* = x_{B1}{}^* = x_{B3}{}^* = x_{C1}{}^* = x_{C2}{}^* = 0$$

With the optimal objective value $R(\tilde{Z}^{TrIFN^*}) = 18.125 \cong 18$ which represents the optimal total cost. In other words the optimal assignment is $A \rightarrow 1\ B \rightarrow 2, C \rightarrow 3$

The intuitionistic fuzzy minimum total cost is calculated as,

$\tilde{c}_{A1}^{TrIFN} + \tilde{c}_{B2}^{TrIFN} + \tilde{c}_{C3}^{TrIFN} = (9,10,11,12)(8,10,11,13) + (5,6,7,8)(4,6,7,9) + (1,2,3,4)(0,2,3,5)$

$= (15, 18, 21,24)(12, 18, 21, 27)$

Also we find that $R(\tilde{Z}^{TrIFN^*}) = R[(15,18,21,24)(12,18,21,27)] = 19.512$

Example 5.2 Let us consider an intuitionistic fuzzy assignment problem with rows representing 4 persons namely1,2,3,4 and columns representing 4 Works namely A,B,C,D . The cost matrix $[\tilde{c}_{rc}^{TrIFN}]$

is given whose elements are TrplFN. The problem is to find the optimal assignment so that the total cost of job assignment becomes minimum

	1	2	3	4
A	(18,19,17,20)18,19,16 ,21)	(24,25,23,26)(24,2522 ,,27)	(28,29,27,30)(28,29,2 6,31)	(32,33,31,34)(32,33,3 0,35)
B	(8,9,7,10)(8,9,6,11)	(13,14,12,15)(13,14,1 1,16)	(17,18,16,19)(17,18,1 5,20)	(19,20,18,21)(19,20,1 7,22)
C	(10,11,9,12)(10,11,8, 13)	(15,16,1417)(15,16,13 ,18)	(19,20,18,21)(19,20,1 7,22)	(22,23,21,24)(22,23,2 0,25)
D	(0,0,-1,1)(0,0,-3,3)	(0,0,-1,1)(0,0,-3,3)	(0,0,-1,1)(0,0,-3,3)	(0,0,-1,1)(0,0,-3,3)

Solution:

The above intuitionistic fuzzy assignment quandary can be formulated in the following linear programming problem form

$$\text{Min}[(17,18,19,20)(16,18,19,21)x_{A1} + (23,24,25,26)(22,24,25,27)x_{A2} + (27,28,29,30)(26,28,29,31) x_{A3} + (31,32,33,34)(30,32,33,35)x_{A4} + (7,8,9,10)(6,8,9,11)x_{B1} + (12,13,14,15)(11,13,14,16)x_{B2} + (16,17,18,19)(15,17,18,20)x_{B3} + (18,19,20,21)(17,19,20,22)x_{B4} + (9,10,11,12)(8,10,11,13)x_{C1} + (14,15,16,17)(13,15,16,18)x_{C2} + (18,19,20,21)(17,19,20,22)x_{C3} + (21,22,23,24)(20,22,23,25)x_{C4} + (-1,0,0,1)(-3,0,0,3)x_{D1} + (-1,0,0,1)(-3,0,0,3)x_{D2} + (-1,0,0,1)(-3,0,0,3)x_{D3} + (-1,0,0,1)(-3,0,0,3)x_{D4}$$

Subject to $x_{A1} + x_{A2} + x_{A3} + x_{A4} = 1, x_{A1} + x_{B1} + x_{C1} + x_{D1} = 1$

$x_{B1} + x_{B2} + B + x_{B4} = 1, x_{A2} + x_{B2} + x_{C2} + x_{D2} = 1$

$x_{C1} + x_{C2} + x_{C3} + x_{C4} = 1, x_{A3} + x_{B3} + x_{C3} + x_{D3} = 1$

$x_{D1} + x_{D2} + x_{D3} + x_{D4} = 1, x_{A4} + x_{B4} + x_{C4} + x_{D4} = 1, x_{rc} \in [0,1]$

Now, using the Step 1 of the intuitionistic fuzzy Hungarian assignment method, we have the following bring down intuitionistic fuzzy assignment table.

	1	2	3	4
A	(0,0,-3,3)(0,0,-5,5)	(6,6,3,9)(6,6,1,11)	(10,10,7,13)(10,10,5,15)	(14,14,111,17)(14,14,9,19)
B	(0,0,-3,3)(0,0,-5,5)	(5,5,2,8)(5,5,0,10)	(9,9,6,12)(9,9,4,14)	(11,11,8,14)(11,11,6,16)
C	(0,0,-3,3)(0,0,-5,5)	(5,5,2,8)(5,5,0,10)	(9,9,6,12)(9,9,4,14)	(12,12,9,15)(12,12,7,17)
D	(0,0,-1,1)(0,0,-3,3)	(0,0,-1,1)(0,0,-3,3)	(0,0,-1,1)(0,0,-3,3)	(0,0,-1,1)(0,0,-3,3)

Now, using the Step 3 to the Step 6 of the intuitionistic fuzzy Hungarian assignment method, we have the following bring down intuitionistic fuzzy assignment table.

	1	2	3	4
A	(0,0,-3,3)(0,0,-5,5)	(1,1,-5,7)(1,1,-9,11)	(5,5,-1,11)(5,5,-515)	(9,9,7,15)(9,9,-9,19)
B	(0,0,-3,3)(0,0,-5,5)	(0,0,-6,6)(0,0,-10,10)	(4,4,-2,10)(4,4,-6,14)	(6,6,0,12)(6,6,-4,16)
C	(0,0,-3,3)(0,0,-5,5)	(0,0,-6,6)(0,0,-10,10)	(4,4,-2,10)(4,4,-6,14)	(7,7,1,13)(7,7,-3,17)
D	(5,5,2,8)(5,5,0,10)	(0,0,-1,1)(0,0,-3,3)	(0,0,-1,1)(0,0,-3,3)	(0,0,-1,1)(0,0,-3,3)

The solutions is not optimal. Now, again using the Step 3 to the Step 6 of the intuitionistic fuzzy Hungarian assignment method, we have the following bring down intuitionistic fuzzy assignment table.

	1	2	3	4

A	(0,0,-3,3)(0,0,-5,5)	(-1,1,-5,7)(1,1,-9,11)	(1,1,-11,13)(1,1,-19,21)	(5,5,.-3,17)(5,5,-15,25)
B	(0,0,-3,3)(0,0,-5,5)	(0,0,-6,6)(0,0,-10,10)	(0,0,-12,12)(0,0,-20,20)	(2,2,-10,14)(2,2,-18,22)
C	(0,0,-3,3)(0,0,-5,5)	(0,0,-6,6)(0,0,-10,10)	(0,0,-12,12)(0,0,-20,20)	(3,3,-11,15)(3,3,-17,23)
D	(9,9,0,18)(9,9,-6,24)	(4,4,-3,11)(4,4,-917)	(0,0,-1,1)(0,0,-3,3)	(0,0,-1,1)(0,0,-3,3)

The optimal solution is,

$$x_{A1}{}^* = x_{B2}{}^* = x_{C3}{}^* = x_{D4}{}^* = 1,$$

$$x_{A2}{}^* = x_{A3}{}^* = x_{A4}{}^* = x_{B1}{}^* = x_{B3}{}^* = x_{B4}{}^* = x_{C1}{}^* = x_{C2}{}^* = x_{C4}{}^* = x_{D1}{}^* = x_{D2}{}^* = x_{D3}{}^* = 0$$

With the optimal objective value $R(\tilde{Z}^{TrIFN^*}) = 50.123 \cong 50$ which represents the optimal total cost. In other words the optimal assignment is $A \to 1, B \to 2, C \to 3, D \to 4$

The intuitionistic fuzzy minimum total cost is calculated as,

$$\tilde{c}_{A1}^{TrIFN} + \tilde{c}_{B2}^{TrIFN} + \tilde{c}_{C3}^{TrIFN} + \tilde{c}_{D4}^{TrIFN}$$

$$= (17,18,19,20)(16,18,19,21) + (12,13,14,15)(11,13,14,16) + (18,19,20,21)(17,19,20,22) + (-1,0,0,1)(-3,0,0,3)$$

$$= (42, 46, 49, 53)(37, 46, 49, 58)$$

Also we find that $R(\tilde{Z}^{TrIFN^*}) = R[(42, 46, 49, 53)(37, 46, 49, 58)] = 50.113 \cong 50.$

6. Conclusion

In this paper, we discussed to bring out a solution of an assignment quandary in which cost coefficients are trapezoidal intuitionistic fuzzy numbers. The total cost is optimal by our method stay with same as that obtained by transform the total intuitionistic fuzzy cost by assigning the ranking technique. This technique can additionally be applied in solving variants of quandaries like, engenderment orchestrating , conveyance quandary and network flow ,Dihydrogen monoxide line quandaries.

References

[1] Zadeh L. A, 1965 *Information and computation* **8** 338-53

[2] Atanassov K. T, 1965 *Intuitionistic fuzzy sets and systems* **20(1)** 87-96

[3] ZahirHussain R. , Senthil Kumar P., 2014 *Int.J.Contemp.Math.Sciences* **4(12)** 897-903

[4] Thorani Y.L.P. , RaviSankar N., 2013 *App.Mat.Sci* **7(71)** 3511-37

[5] Amit Kumar, Amarpreet Kaur, Anila Gupta 2011 *J Math Model Algor* **10** 163-180

[6] Bellman, R., Zadeh L.A., 1970 *management sci* **17(B)** 141-164

[7] Amit Kumar, Anila Gupta., 2012 *International Journal of Applied Science and Engineering* **10** 155-170

[8] Amit Kumar, Amarpreet Kaur, Anila Gupta., 2011 *J Math Model Algor* **10** 163-180

[9] ZahirHussain, R., Senthil Kumar P., 2014 *Int.J.Contemp.Math.Sciences* **12** 897-903

[10] Sathi Mukherjee, KajlaBasu., 2010 *Int.Jour.Comp and Appl. Mathematics* **5** 359-368

[11] Lin Chi-Jen, Wen Ue-Pyng., 2004 *Fuzzy Sets and Systems* **14** 373-391

[12] JahirHussain, Senthil Kumar P., 2013 *Int.J.Contemp.Math.Sciences* **8(12)** 565-576

MULTIPLE ATTRIBUTE GROUP DECISION MAKING METHODS USING NUMERICAL METHODS OF INTUITIONISTIC TRIANGULAR FUZZY SETS

Akila .S[1] , John Robinson .P[2]
[1]Assistant Professor, Theivanai Ammal College for women, Villupuram, India.
akila2829@gmail.com
[2]Assistant Professor, Bishop Heber College, Tiruchirappalli, India.
robijohnsharon@gmail.com

Abstract: Solving Multiple Attribute Group Decision Making (MAGDM) problems has become one of the most important research in the recent trends. The information or data is in the form of Intuitionistic Triangular Fuzzy Number (ITrFN). The weights are derived from the numerical solutions of partial differential equations of Laplace's equation and poisson`s equation. The weights obtained from the methods are applied in decision making problems. The Intuitionistic Triangular Fuzzy Ordered Weighted Averaging (ITrFOWA) operator and the Intuitionistic Triangular Fuzzy Hybrid Aggregation (ITrFHA) operator are used to combine the decision matrix. Distance function is used as a tool for ranking the best alternatives. Numerical illustration is proposed to show the effectiveness of the method.

Key words: MAGDM, Intuitionistic Triangular Fuzzy Number, Partial Differential Equation, ITrOWA, ITrHA

INTRODUCTION:

In Intuitionistic Fuzzy sets, Triangular Intuitionistic Fuzzy sets is a special case with two characterizations namely the Triangular Fuzzy Characterization and the Intuitionistic Fuzzy Characterization. In Fuzzy set theory Atanassov.K (1986, 1989, 1994) has introduced the concept of Intuitionistic Fuzzy sets and Operators over interval-valued intuitionistic fuzzy sets. Chen, Y., & Li, B. (2011) provided some new techniques for Multiple Attribute Group Decision Making Problems with Triangular Intuitionistic Fuzzy information, attribute values in the form of Triangular Intuitionistic Fuzzy number. Here the information about the attribute weights are derived from the Numerical solution of Partial Differential Equation. Balagurusamy.E, (2009) & Arumugam.S & Thangapandi.A. (2012) has solved various Numerical Methods with Application of computer. The Numerical Illustration is utilized from Shouzhen & Zeng et.al, (2016) and it is used in Multiple Attribute Decision Making Problems. Li, D.F. (2005) has developed the multiattribute decision making models and methods using intuitionistic fuzzy sets. Chen, Z., et, al. (2016) proposed in fuzzy linguistic term set for multiple criteria group decision making models. Li, D. F. (2010) introduced to find the new method to ranking the triangular Intuitionistic Fuzzy numbers to MAGDM Problems.

Aggregation operators to be recognized as a new process of combining many Numerical values in to a single value. These aggregating operators are used in the field of Mathematics, Physics, Economics, Natural Sciences, Computer Engineering and so on. To aggregate the Intuitionistic Fuzzy number, the Intuitionistic Fuzzy Ordered Weighted Aggregation (IFOWA) operator, Intuitionistic Fuzzy Hybrid Aggregation (IFHA) operator, and Intuitionistic Fuzzy Ordered Weighted Geometric (IFWG) operator are commonly used. Wei, G. (2008, 2010) has developed some arithmetic aggregation operators with intuitionistic fuzzy sets and trapezoidal fuzzy numbers to Multiple Attribute Group Decision Making. Yager, R. R. (1988) has also introduced the ordered weighted averaging aggregation operators in multicriteria decision making. Xu, Z. S., & Yager, R. R. (2006) has introduced some geometric aggregation operators based on Intuitionistic Fuzzy sets.

Xu, Z.S., (2007): proposed for aggregating interval valued intuitionistic fuzzy sets and its applications. Robinson, J.P., & Amirtharaj, E.C.H, (2011b, 2012b) have developed the Intuitionistic Triangular Fuzzy Ordered Weighted Averaging (TrIFOWA) operator and the Triangular Intuitionistic Fuzzy Hybrid Aggregation (TrIFHA) operator. He also proposed Trapezoidal Intuitionistic Fuzzy sets for Multiple Attribute Group Decision Making. Robinson, J.P, & Jeeva, S. (2018): has introduced Double Sumudu Transform in MAGDM problems with Intuitionistic Triangular Fuzzy Sets and its applications. Robinson, J.P. (2018) defined the new distance function in Multiple Attribute Group Decision Making Analysis.

In this paper the methods are combined with Laplace's and Poisson's equations proposed for determining the weights and it is used for decision making problems under

Triangular Fuzzy sets. The feasibility and effectiveness of the Numerical methods are illustrated with example.

Definition 1:

Let $\tilde{a_j} = ([a_j, b_j, c_j]; \mu_{a_j}, \gamma_{a_j})$ for all j = 1, 2, 3, n be a collection of triangular intuitionistic fuzzy values. The intuitionistic Triangular Fuzzy Weighted Arithmetic Averaging (ITrFWAA) operator, ITrFWAA: $Q^n \rightarrow Q$ is defined as:

ITrFWAA $(\tilde{a_1}, \tilde{a_2}, \ldots, \tilde{a_{n1}}) =$

$$\sum_{j=1}^{n} \tilde{a}_j w_j = \left(\left[\sum_{j=1}^{n} a_j w_j, \sum_{j=1}^{n} b_j w_j, \sum_{j=1}^{n} c_j w_j \right]; 1 - \prod_{j=1}^{n} (1 - \mu_{a_j})^{w_j}, \prod_{j=1}^{n} (\gamma_{a_j})^{w_j} \right)$$ where

$w = (w_1, w_2, \ldots, w_n)^r$ be the weighted vector of decision makers with $\sum_{j=1}^{n} w_j = 1$.

Definition 2:

Let $\tilde{a_j} = ([a_j, b_j, c_j]; \mu_{a_j}, \gamma_{a_j})$ for all j = 1, 2, 3, n be a collection of intuitionistic triangular fuzzy values. The intuitionistic Triangular Fuzzy Hybrid Aggregation (ITrFHA) operator, ITrFHA: $Q^n \rightarrow Q$ is defined as:

$$\text{ITrFHA}_{w,w}(\tilde{a_1}, \tilde{a_2}, \ldots, \tilde{a_n}) = \sum_{j=1}^{n} \dot{\tilde{a}}_{\sigma(j)} w_j =$$

$$\left(\left[\sum_{j=1}^{n} \dot{a}_{\sigma(j)} w_j, \sum_{j=1}^{n} \dot{b}_{\sigma(j)} w_j, \sum_{j=1}^{n} \dot{c}_{\sigma(j)} w_j \right]; 1 - \prod_{j=1}^{n} \left(1 - \dot{\mu}_{\tilde{a}_{\sigma(j)}} \right)^{w_j}, \prod_{j=1}^{n} \left(\dot{\gamma}_{\tilde{a}_{\sigma(j)}} \right)^{w_j} \right)$$

where $w = (w_1, w_2, \ldots, w_n)^T$ is the associated vector such that $w_j > 0$ and $\sum_{j=1}^{n} w_j = 1$.

$w = (w_1, w_2, \ldots, w_n)^T$ be the weight vector of $\tilde{a_j}$ for all $j = 1, 2, \ldots, n$ such that $w_j > 0$ and $\sum_{j=1}^{n} w_j = 1$.

Theorem 1:

The ITrFWAA operator is special case of the ITrFHA operator.

Proof: Let $w = \left(\frac{1}{n}, \frac{1}{n}, \frac{1}{n}, \ldots, \frac{1}{n}\right)$. *Then*

$$\text{ITrFHA}_{w,w}(\tilde{a_1}, \tilde{a_2}, \ldots, \tilde{a_n}) = \sum_{j=1}^{n} \dot{\tilde{a}}_{\sigma(j)} w_j = \frac{1}{n}$$

$\sum_{j=1}^{n} \dot{\tilde{a}}_{\sigma(j)} = \sum_{j=1}^{n} \tilde{a}_{\sigma(j)} w_j = ITrFWAA_w(\tilde{a}_1, \tilde{a}_2, \ldots, \tilde{a}_n)$ which competes the proof of the theorem.

Theorem 2:

The ITrFOWA operator is special case of the ITrFHA operator.

Proof: Let $w = \left(\frac{1}{n}, \frac{1}{n}, \frac{1}{n}, \ldots, \frac{1}{n}\right)$. *Then* $\text{ITrFHA}_{w,w}(\tilde{a_1}, \tilde{a_2}, \ldots, \tilde{a_n}) = \sum_{j=1}^{n} \dot{\tilde{a}}_{\sigma(j)} w_j = \frac{1}{n}$

$\sum_{j=1}^{n} \dot{\tilde{a}}_{\sigma(j)} = \sum_{j=1}^{n} \tilde{a}_{\sigma(j)} w_j = ITrFOWA_w(\tilde{a}_1, \tilde{a}_2, \ldots, \tilde{a}_n)$ which competes the proof of the theorem.

ALGORITHM FOR MAGDM USING ITrFHA OPERATOR AND CORRELATION COEFFICIENT OF ITrFS

MAGDM problems based on ITrFWAA and ITrFHA operators where both attribute weights and the expert weights take the form of real numbers, attribute values take the form of triangular intuitionistic fuzzy numbers, are investigated as follows:

Let $A = \{A_1, A_2, ..., A_m\}$ be a discrete set of alternatives, and $G = \{G_1, G_2, ..., G_n\}$ be the set of attributes, $W = \{W_1, W_2, ..., W_n\}$ is the weighted vector of the attribute G_j $(j = 1, 2, ..., n)$, Where $w_j \in [0,1], \sum_{j=1}^{n} w_j = 1$. Let $D = \{D_1, D_2, ..., D_t\}$ be the set of decision makers $w = \{w_1, w_2, ..., w_t\}$ be the weighting vector of decision makers, with $w_k \in [0,1], \sum_{k=1}^{t} w_k = 1$.

Suppose that, $\widetilde{R_K} = \left(\tilde{r}_{ij}^{(k)}\right)$ m $\times n = \left(\left[a_{ij}^{(k)}, b_{ij}^{(k)}, c_{ij}^{(k)}\right]; \mu_{ij}^{(k)}, \gamma_{ij}^{(k)}\right)$ m × n $0 \le a_{ij}^{(k)} \le b_{ij}^{(k)} \le c_{ij}^{(k)} \le 1, 0 \le \mu_{ij}^{(k)} + \gamma_{ij}^{(k)} \le 1$, is the normalized intuitionistic triangular fuzzy decision matrix, where $i = 1, ..., m; j = 1, 2 ..., n; k = 1, 2 ...., t$. Then the intuitionistic triangular fuzzy positive ideal value and intuitionistic triangular fuzzy negative ideal value are defined as follows:

$$r^{\sim +} = \left(\left[a^+, b^+, c^+\right]; \mu^+, \gamma^+\right) = \left([1,1,1]; 1, 0\right),$$

$$r^{\sim -} = \left(\left[a^-, b^-, c^-\right]; \mu^-, \gamma^-\right) = \left([0,0,0]; 0, 1\right).$$

In the following , The ITrFWAA and ITrFHA operator are applied to MAGDM problem based on intuitionistic triangular fuzzy information .

Step 1:

Utilize the decision information given in the intuitionistic triangular fuzzy decision matrix

$r_i^{\sim(k)} = \left([a^{(k)}, b^{(k)}, c^{(k)}]; \mu^{(k)}, \gamma^{(k)}\right) = ITrFWAA_w\left(r_{i1}^{\sim(k)}, r_{i2}^{\sim(k)}, \ldots, r_{in}^{\sim(k)}\right)$ where $i = 1, 2, \ldots, m$;

k = 1,2,…,t, to derive the individual overall preference intuitionistic triangular fuzzy values $r_i^{\sim(k)}$ of the alternative A_i.

Step 2:

Utilize the ITrFHA operator $r_i^{\sim} = \left([a_i, b_i, c_i]; \mu_i, \gamma_i\right) = ITrFHA_w\left(r_i^{\sim(1)}, r_i^{\sim(2)}, \ldots, r_i^{\sim(t)}\right)$ to derive the collective overall preference intuitionistic triangular fuzzy values $r_i^{\sim}$, I = 1, 2,….,m of the alternative A_i., where $w = \{w_1, w_2, \ldots\ldots, w_n\}$ be the weighted vector of decision makers with $\sum_{j=1}^{n} w_j = 1.$; and $v = \{v_1, v_2, \ldots\ldots, v_n\}$ is the associated weighting vector of the ITrFHA operator with $v_j > 0$, and $\sum_{j=1}^{n} v_j = 1.$

Step 3:

Calculate the distance between the collective overall values

$r_i^{\sim} = \left([a_i, b_i, c_i]; \mu_i, \gamma_i\right)$, and the intuitionistic triangular fuzzy positive ideal value $r^{\sim+} = \left([a^+, b^+, c^+]; \mu^+, \gamma^+\right) = ([1,1,1]; 1, 0)$,

$$d\left(r_i^{\sim}, r^{\sim+}\right) = \frac{1}{8}\Big(\left|(1+\mu_i-\gamma_i+\pi_i)a_1 - (1+\mu^+-\gamma^++\pi^+)a_2\right| + \left|(1+\mu_i-\gamma_i+\pi_i)b_1 - (1+\mu^+-\gamma^++\pi^+)b_2\right| +$$
$$\left|(1+\mu_i-\gamma_i+\pi_i)c_1 - (1+\mu^+-\gamma^++\pi^+)c_2\right|\Big)$$

Step 4:

Journal of Physics: Conference Series **1377** (2019) 012022 doi:10.1088/1742-6596/1377/1/012022

Rank the alternatives and choose the best one based on the shortest distance obtained in *Step 3*

Numerical Illustration

Many physical phenomena in applied science and engineering formulated into mathematical models drop into a category of systems is Partial Differential Equations. A Partial Differential equation is a Differential Equation involving more than one independent variable. These variable determine the behavior of the dependent variables. Here the three elliptic equations namely Laplace's equation, Liebmann's iterative method and Poisson's equation are used.

Laplace's Equation and Poisson's equation

A second order equation involving two independent variables are

$a\frac{\partial^2 f}{\partial x^2} + b\frac{\partial^2 f}{\partial x \partial y} + c\frac{\partial^2 f}{\partial y^2} = F(x,y,f,\frac{\partial f}{\partial x},\frac{\partial f}{\partial y})$ when a= 1, b= 0, c= 1 and

$F(x, y, f, f_x, f_y) = 0.$ $\frac{\partial^2 f}{\partial x^2} + \frac{\partial^2 f}{\partial y^2} = \nabla^2 f = 0.$ is the Laplacian operator and

$\frac{\partial^2 f}{\partial x^2} + \frac{\partial^2 f}{\partial y^2} = \nabla^2 f$ is the poisson's equation.

(i) Consider a steel plate of size 15cm x 15cm. If two of the sided are held at (i)100°C and the other two sided are held at 0°C, what are the steady state temperature at interior points assuming a grid size of 5cm X 5cm.

A Problem with the values known on each boundary is Dirichlet boundary conditions.

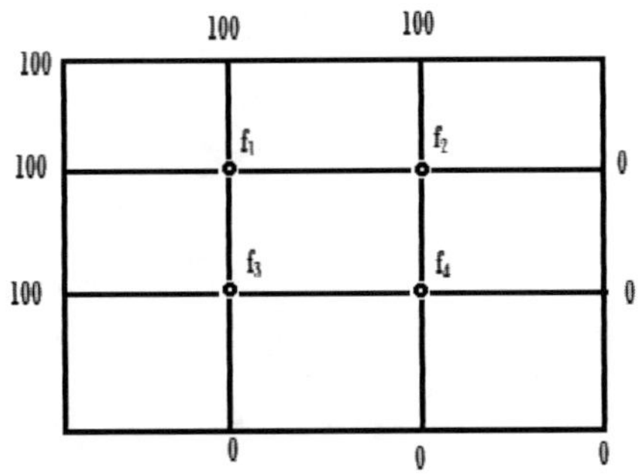

The system of equation is

At point 1: $-4f_1 + f_2 + f_3 + 100 + 100 = 0$

At point 2: $f_1 - 4f_2 + f_4 + 100 = 0$

At point 3: $f_1 - 4f_3 + f_4 + 100 = 0$

At point 4: $f_2 + f_3 - 4f_4 = 0$

f_i	Exact solution	Weighting Vector
f_1	75	0.375
f_2	50	0.25
f_3	50	0.25
f_4	25	0.125

(ii) *Liebmann's Iterative Method*

Consider a steel plate of size 15cm x 15cm. If two of the sided are held at 200°C and the other two sided are held at 0°C, what are the steady state temperature at interior points assuming a grid size of 5cm X 5cm.

Journal of Physics: Conference Series **1377** (2019) 012022 doi:10.1088/1742-6596/1377/1/012022

A Problem with the values known on each boundary is Dirichlet boundary conditions

At point 1: $-4f_1 + f_2 + f_3 = -400$

At point 2: $f_1 - 4f_2 + f_4 = -200$

At point 3: $f_1 - 4f_3 + f_4 = -200$

At point 4: $f_2 + f_3 - 4f_4$

By using the iterative method we get

f_i	Exact solution	Weighting Vector
f_1	149.804688	0.376349362
f_2	99.902344	0.249263984
f_3	99.902344	0.249263984
f_4	49.609375	0.125122669

(iii) Solve the Poisson equation $\nabla^2 f = 2x^2y^2$ and Gauss Seidel iteration method over the square domain $0 \leq x \leq 3$ and $0 \leq y \leq 3$ with f = 0 on the boundary and h = 1.

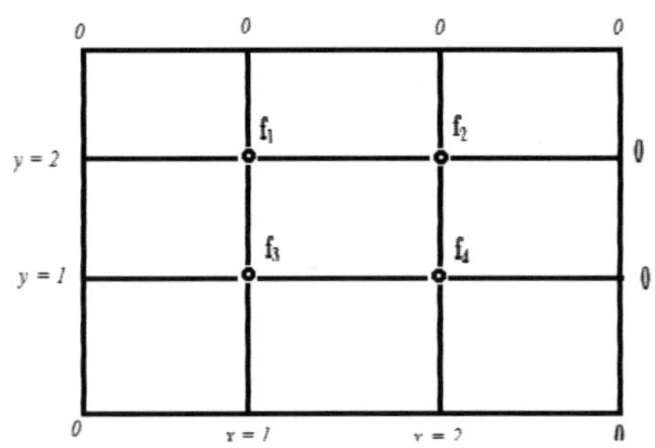

At point 1: $-4f_1 + f_2 + f_3 = 8$

At point 2: $f_1 - 4f_2 + f_4 = 32$

At point 3: $f_1 - 4f_3 + f_4 = 2$

At point 4: $f_2 + f_3 - 4f_4 = 8$

Solving the equations by elimination method, we get the required exact solution.

f_i	Poisson equation Exact solution	Weighting Vector
f_1	- 5.5	0.22
f_2	-10.75	0.43
f_3	-3.25	0.13
f_4	-5.5	0.22

Numerical Example:

Assume that a company wants to invest some money in another company after analyzing the information the board of directors considers five possible investments

A_1: Chemical Company

A_2: Computer Company

A_3: Car Company

A_4: Pharmaceutical Company

In order to evaluate these investments, the group of experts considers the key factor is the economic situation of the next year The five possible alternatives are to be evaluated using intuitionistic triangular fuzzy numbers by the three decision makers whose weighting vectors are $\gamma = (0.375, 0.25, 0.25, 0.125)^T$, $\omega = (0.3763, 0.2492, 0.2492, 0.1251)^T$ under the four attributes numerical partial differential equation weighting vector $\omega =$

$(0.22, 0.43, 0.13, 0.22\ )^T$, the decision matrices as listed in the following matrices R = $(r_{2ij}^{(k)})_{4x4}$ (k= 1,2,3,4)

T_1: Negative Growth rate

T_2: Low Growth rate

T_3: Medium Growth rate

T_4: High Growth rate

$$R_1 = \begin{pmatrix} ([0.4,0.5,0.7];0.6,0.2) & ([0.5,0.6,0.7];0.1,0.3) & ([0.4,0.5,0.7]);0.3,0.6) & ([0.1,0.5,0.7];0.4,0.3 \\ ([0.1,0.3,0.5];0.8,0.1) & ([0.2,0.3,0.4];0.4,0.5) & ([0.3,0.4,0.5];0.7,0.1) & ([0.4,0.7,0.9];0.6,0.2 \\ ([0.3,0.4,0.5];0.4,0.3) & ([0.4,0.6,0.8];0.5,0.2) & ([0.6,0.7,0.8);0.5,0.3) & ([0.5,0.6,0.7];0.5,0.4 \\ ([0.3,0.5,0.6];0.5,0.3) & ([0.1,0.2,0.3];0.7,0.2) & ([0.2,0.4,0.6];0.6,0.2) & ([0.4,0.7,0.8];0.6,0.3 \end{pmatrix}$$

$$R_2 = \begin{pmatrix} ([0.3,0.4,0.7];0.4,0.2) & ([0.3,0.6,0.8];0.4,0.2) & ([0.3,0.6,0.8]);0.5,0.4) & ([0.3,0.5,0.6];0.4,0.3 \\ [0.2,0.4,0.5];0.7,0.2) & ([0.5,0.7,0.9];0.6,0.3) & ([0.4,0.6,0.8];0.6,0.3) & [0.2,0.4,0.7];0.5,0.2 \\ ([0.5,0.6,0.7];0.5,0.2) & ([0.2,0.6,0.8];0.4,0.4) & ([0.5,0.6,0.7);0.4,0.4) & ([0.3,0.4,0.5];0.2,0.4 \\ ([0.2,0.4,0.6];0.4,0.4) & ([0.1,0.3,0.5];0.5,0.2) & ([0.2,0.4,0.6];0.5,0.3) & ([0.5,0.6,0.7];0.2,0.7 \end{pmatrix}$$

$$R_3 = \begin{pmatrix} ([0.2,0.4,0.6];0.3,0.7) & ([0.5,0.6,0.7];0.5,0.3) & ([0.2,0.6,0.7]);0.5,0.4) & ([0.3,0.4,0.5];0.5,0.2 \\ [0.3,0.5,0.7];0.9.0.1) & ([0.1,0.3,0.5];0.7,0.2) & ([0.5,0.6,0.7];0.6,0.3) & [0.7,0.8,0.9];0.6,0.1 \\ ([0.5,0.6,0.9];0.6,0.3) & ([0.4,0.6,0.8];0.5,0.3) & ([0.3,0.4,0.5);0.3,0.2) & ([0.5,0.7,0.9];0.4,0.2 \\ ([0.6,0.7,0.8];0.4,0.4) & ([0.7,0.8,0.9];0.4,0.3) & ([0.1,0.5,0.6];0.3,0.6) & ([0.3,0.6,0.7];0.3,0.6 \end{pmatrix}$$

$$R_4 = \begin{pmatrix} ([0.3,0.6,0.7];0.5,0.3) & ([0.4,0.6,0.8];0.4,0.4) & ([0.1,0.3,0.5]);0.7,0.2) & ([0.5,0.7,0.9];0.3,0.4) \\ ([0.5,0.6,0.8];0.3.0.2) & ([0.3,0.5,0.7];0.5,0.3) & ([0.4,0.6,0.7];0.3,0.3) & ([0.4,0.6,0.8];0.5,0.2) \\ ([0.4,0.7,0.9];0.5,0.4) & ([0.2,0.3,0.4];0.6,0.3) & ([0.4,0.5,0.6);0.4,0.3) & ([0.4,0.5,0.6];0.5,0.3) \\ ([0.1,0.3,0.5];0.3,0.4) & ([0.6,0.7,0.8];0.3,0.3) & ([0.2,0.3,0.4];0.3,0.2) & ([0.5,0.6,0.7];0.4,0.3) \end{pmatrix}$$

Step 1:

Journal of Physics: Conference Series **1377** (2019) 012022 doi:10.1088/1742-6596/1377/1/012022

Utilize the decision information given in the intuitionistic Triangular Fuzzy Decision Matrix $\widetilde{R_k}$ and the ITrWAA operator to derive the individual overall preference intuitionistic triangular fuzzy values $\widetilde{R_i^k}$ of the alternative A_i

$$ITrFWAA(a_1, a_2, ..., a_n) = \sum_{j=1}^{4} a_j w_j = ([\sum_{j=1}^{4} a_j w_j, \sum_{j=1}^{4} b_j w_j, \sum_{j=1}^{4} c_j w_j], 1 - \prod_{j=1}^{4} (1-\mu_{\alpha_j})^{w_j}, \prod_{j=1}^{4} (\gamma_{\alpha_j})^{w_j})$$

$r_1^{(1)}$= [(0.3875, 0.525, 0.7), 0.4072, 0.3064]

$r_2^{(1)}$= [(0.2125, 0.375, 0.525), 0.6823, 01631]

$r_3^{(1)}$= [(0.425, 0.55, 0.6750), 0.4646, 0.2810]

$r_4^{(1)}$= [(0.2375, 0.425, 0.55), 0.5953, 0.2449

$r_1^{(2)}$= [(0.3, 0.5125, 0.7375), 0.4267, 0.2502]

$r_2^{(2)}$= [(0.325, 0.525, 0.7), 0.6307, 0.2449]

$r_3^{(2)}$= [(0.4, 0.575, 0.7), 0.4191, 0.3084]

$r_4^{(2)}$= [(0.2125, 0.4, 0.5875), 0.4322, 0.3357]

$r_1^{(3)}$= [(0.2875, 0.5, 0.6375), 0.4328, 0.4211]

$r_2^{(3)}$= [(0.35, 0.5125, 0.675), 0.7787, 0.1565]

$r_3^{(3)}$= [(0.425, 0.5625, 0.7755), 0.4882, 0.2577]

$r_4^{(3)}$= [(0.4625, 0.6625, 0.7630), 0.3764, 0.4334]

$r_1^{(4)}$= [(0.3, 0.5375, 0.5), 0.5197, 0.3020]

$r_2^{(4)}$= [(0.4125, 0.575, 0.75), 0.3830, 0.2449]

$r_3^{(4)}$= [(0.35, 0.525, 0.6630), 0.5051, 0.3342]

$r_4^{(4)}$= [(0.3, 0.4375, 0.5753), 0.3134, 0.3020]

Step 2:

Utilize the ITrFHA operator to derive the collective overall preference intuitionistic Triangular fuzzy values $\breve{R}_i$ of the alternative A_i.

$$ITrFWAA(a_1, a_2, ..., a_n) = \sum_{j=1}^{4} \overset{\circ}{a}_{\sigma(j)} w_j = ([\sum_{j=1}^{4} \overset{\circ}{a}_{\sigma(j)} w_j, \sum_{j=1}^{4} \overset{\circ}{b}_{\sigma(j)} w_j, \sum_{j=1}^{4} \overset{\circ}{c}_{\sigma(j)} w_j], 1 - \prod_{j=1}^{4} (1 - \overset{\circ}{\mu}_{a\sigma(j)})^{w_j}, \prod_{j=1}^{4} (\gamma_{a\sigma(j)})^{w_j})$$

$r_1^{(1)}$= [(0.3814, 0.5335. 0.6592), 0.6550, 0.2981]

$r_2^{(1)}$= [0.3794, 0.5618, 0.6979), 0.5693, 0.3422]

$r_3^{(1)}$= [(0.4833, 0.6327, 0.7365) 0.6063, 0.3488]

$r_4^{(1)}$= [(0.4249, 0.5778, 0.6597), 0.4910, 0.3532]

Step 3:

Distance between the collective overall values $r_i = ([a_i, b_i, c_i]; \mu_i, \gamma_i)$ and the intuitionistic triangular Fuzzy positive ideal Values

The new distance functions are

$\rho(r_1, \widetilde{r_1)} = 0.473784803$

$\rho(r_2, \widetilde{r_{2)}} = 0.480450005$

$\rho(r_3, \widetilde{r_{3)}} = 0.448413$

$\rho(r_4, \widetilde{r_{4)}} = 0.48118992$

Step 4:

Rank all the alternatives A_i (i = 1, 2, 3, 4) based on the distance measure obtained from step3

$A_4 > A_2 > A_1 > A_3.$

Hence, the best alternative is A_5.

Conclusion:

In this paper Laplace's and Poisson's equations are proposed to obtain the Numerical solution of Partial Differential Equations. The MAGDM Problems in which the attribute weights are in the form of real numbers, attribute values are taken in the form of Triangular Fuzzy Number. The Arithmetic operators ITrFOWA and ITrFHA are used to aggregate the weighted arguments into a collective data. TrIFNs distance function is proposed for ranking the best alternatives. Finally an Illustrative example was presented to show the effectiveness of the proposed method.

References:

1. K.Atanassov, (1986). Intuitionistic fuzzy sets. *Fuzzy Sets and Systems*, *20*(1), 87–96. doi:10.1016/S0165- 0114(86)80034-3
2. K Atanassov, (1989). More on intuitionistic fuzzy sets. *Fuzzy Sets and Systems*, *33*(1), 37–45. doi:10.1016/0165-0114(89)90215-7
3. K. Atanassov, (1994). Operators over interval-valued intuitionistic fuzzy sets. *Fuzzy Sets and Systems*, *64*(2), 159–174. doi:10.1016/0165-0114(94)90331-X
4. Y.Chen, & B.Li, (2011). Dynamic multiattribute decision making model based on triangular intuitionistic fuzzy numbers, Scientia Iranica, 18(2), 268-274

5. Z.Chen, K.Chin, & Y.Li, Y.Yang, (2016). Proportional hesistant fuzzy linguistic term set for multiple criteria group decision making, Information Sciences, 357, 61-87.
6. M.Dong, S. Li, & H. Zhang, (2015). Approaches to group decision making with incomplete information based on power geometric operators and triangular fuzzy AHP, Expert Systems with Applications, 42(21), 7846-7857
7. D. F. Li, (2005). Multi attribute decision making models and methods using intuitionistic fuzzy sets. *Journal of Computer and System Sciences*, *70*(1), 73–85. doi:10.1016/j.jcss.2004.06.002
8. D. F. Li, (2010c). A ratio ranking method of triangular intuitionistic fuzzy numbers and its application to MADM problems. *Computers & Mathematics with Applications (Oxford, England)*, *58*(6), 1557–1570. doi:10.1016/j.camwa.2010.06.039
9. P.D.Liu, & T. J. Wang, (2007). A method for multiple attribute decision making with triangular fuzzy Number and partial attribute weight information. *Journal of Information and Computational Science*, *4*(3), 1017–1022.
10. Shouzhen & Zeng et.al, (2016) Intuitionistic Fuzzy Generalized Probabilistic ordered weighted Averaging operator and its Application to Group Decision Making. Technological and Economic Development of Economy 22(2) doi:10.3846/20294913,2014.984253
11. E.Balagurusamy, (2009) Numerical Methods Tata McGraw-Hill Education Private Limited New Delhi.
12. S.Arumugam, & A.Thangapandi, (2012) Numerical Methods. SciTech Publications India pvt limited, Chennai.
13. J. P. Robinson, & E. C. H. Amirtharaj, (2011b). Extended TOPSIS with correlation coefficient of Triangular Intuitionistic fuzzy sets for Multiple Attribute Group Decision Making. *International Journal of Decision Support System Technology*, *3*(3), 15–40. doi:10.4018/jdsst.2011070102
14. J.P. Robinson, Amirtharaj, E.C.H, (2012b). A Search for the Correlation coefficient of Triangular and Trapezoidal intuitionistic Fuzzy sets for Multiple Attribute Group Decision Making. In *Mathematical Modeling and Scientific Computation* (pp. 333-342).

15. J. P.Robinson, & Amirtharaj, E. C. H. (2015). MADM problems with correlation coefficient of Triangular fuzzy IFS. *International Journal of Fuzzy System Applications*, *4*(1), 1–32. doi:10.4018/IJFSA.2015010101
16. J.P.Robinson, & Jeeva, S. (2018): Application of Double Sumudu Transform in MAGDM problems with Intuitionistic Triangular Fuzzy Sets. International Journal of Research in Advent Technology, 6(7), pp. 1620-1628.
17. P.John Robinson, (2016) *Multiple Attribute Group Decision Analysis for Intuitionistic Triangular and Trapezoidal Fuzzy Numbers.* International Journal of Fuzzy System Applications, 5(3), 45-76. DOI: 10.4018/IJFSA.2016070104
18. G.Wei, (2008) *Induced intuitionistic fuzzy ordered weighted averaging operator and its application to MAGDM, LNAI-5009*. Springer-Verlag Berlin Heidelberg.3
19. G.Wei, (2010). Some arithmetic aggregation operators with intuitionistic trapezoidal fuzzy numbers and their application to group decision making. *Journal of Computers*, *5*(3), 345–351. doi:10.4304/jcp.5.3.345-351
20. Z.S. Xu, (2007): Methods for aggregating intervalued intuitionistic fuzzy information and their application to decision making. Control and Decision, 22, pp. 215-219.
21. Z.S.Xu, & R. R.Yager, (2006). Some geometric aggregation operators based on Intuitionistic Fuzzy sets. *International Journal of General Systems*, *35*(4), 417–433. doi:10.1080/03081070600574353
22. R. R.Yager, (1988). On ordered weighted averaging aggregation operators in multicriteria decision making. *IEEE Transactions on Systems, Man, and Cybernetics*, *18*(1), 183–190. doi:10.1109/21.87068

Triangular fuzzy graceful labeling on star graph

N Sujatha[1a], C Dharuman[2], K Thirusangu[3]

[1,2] Department of Mathematics, SRM Institute of Science and Technology, Ramapuram, Chennai-89, India.
[3] Department of Mathematics, S.I.V.E.T College, Gowrivakkam, Chennai-73, India.

Abstract. In this paper, we present an algorithm to find triangular fuzzy star graph $K_{1,n}$ atmost 1241 edges. We show that it admits triangular fuzzy labeling. Also we prove that it satisfies triangular fuzzy graceful labeling.

Keywords. Triangular Fuzzy Labeling, Triangular Fuzzy Graceful Labeling.

AMS Mathematics Subject Classification: 05C72, 05C78.

1. Introduction

In 1965, fuzzy sets were orginated by Lofti.A. Zadeh[1] which had an outstanding improvement in mathematical design in case of ambiguity. A. Rahmani, F. Hosseinzadeh Lotfi, M. Rostamy-Malkhalifeh, and T. Allahviranloo[2] discussed a new method for defuzzification and ranking of fuzzy numbers based on the statistical beta distribution.

Narsingh deo[3] developed the notion of graph theory which are associated between vertices and edges . A. Solairaju and S. Ambika [4] concluded some results on gracefulness of a new class of stars merged with trees. Fuzzy relations on fuzzy sets had been leading in an admirable way to compose a fuzzy graph model when there is an vagueness in vertices and edges.

Fuzzy graph model has been refined by K.R.Bhutani, J.N.Moderson and A.Rosenfield[5]. S.Mathew and M.Sunitha[6] extended the concepts of fuzzy graphs in to various graphs.

A. Nagoorgani, Muhammed Akram and D.Rajalakshmi(a)Subhasini [7,8] brought up the notion of labeling in fuzzy graph and proved some properties in fuzzy graph labeling . Existence of fuzzy edge vertex graceful labeling in some special graphs has been explained by S.Vimala and R.Jebesty Shajila [9]. Fuzzy graceful labeling technique has been demonstrated in an algorithmic approach for extended duplicate graph by S.Bala, M.L.MorslinLifin Lee, K.Thirusangu [10]. K.Ameenal Bibi, M.Devi[11] have been proceeded the results in fuzzy vertex graceful labeling on some special graphs.

In this paper, we present an algorithm to find triangular fuzzy star graph $K_{1,n}$ atmost 1241 edges. we show that it admits triangular fuzzy labeling .Also we prove that it satisfies triangular fuzzy graceful labelling.

2. Preliminaries

[a] Corresponding author:2004july19@gmail.com

Definition [1] 2.1. Let X be the space of points(objects),with a generic element of X denoted by x. Thus X={x}. A fuzzy set (class) A in X is characterized by a membership (characteristic) function $f_A(x)$ which associates with each point in X a real number in the interval [0,1], with the value of $f_A(x)$ at x , representing the "grade of membership" of x in A.

Definition [2] 2.2. A fuzzy number is a fuzzy set in the form of $\tilde{a} : R \to [0, 1]$ that satisfies the following conditions: (1) $\tilde{a}$ is upper semi continuous, (2) $\tilde{a}(x)$ is zero outside the interval $[l, u]$, (3) there exist real numbers m_1, m_2 such that $l \le m_1 \le m_2 \le u$ and (3.1) $\tilde{a}(x)$ is increasing on $[l, m_1]$, (3.2) $\tilde{a(x)}$ is decreasing on $[m_2, u]$, (3.3) $\tilde{a}(x) = 1$, $m_1 \le x \le m_2$.If $m_1 = m_2 = m$ then the fuzzy number $\tilde{a} = (l, m, u)$ is called the triangular fuzzy number and is defined as follows:

$$\tilde{a}(x) = \begin{cases} \frac{x-l}{m-l}, & l \le x \le m \\ \frac{u-x}{u-m}, & m \le x \le u \\ 0, & Other\ wise \end{cases}$$

Definition [3] 2.3. A Linear graph (or simply a graph) G = (V, E) consists of set of objects $V=\{v_1,v_2,... v_n\}$ called vertices, and another set $E =\{e_1,e_2,.. e_m\}$ whose elements are called edges such that each edge e_k is identified with an unordered pair of vertices (v_i,v_j).

Definition [4] 2.4. The star graph S_n of order 'n' is a tree on 'n' nodes with one node having vertex degree (n-1) and the other (n-1) node having vertex degree 1. Star is a complete bi graph $K_{1,n}$.

Definition [5] 2.5. Let U and V be two sets. Then ρ is said to be a fuzzy relation from U into V if ρ is a fuzzy set of $U \times V$.A fuzzy graph $G = (\sigma, \mu)$ is a pair of functions $\sigma : V \to [0,1]$ and $\mu : V \times V \to [0, 1]$, where for all u, v $\in V$, we have $\mu(u, v) \le \sigma(u) \wedge \sigma(V)$.

Definition [7] 2.6. A graph $G = (\sigma, \mu)$ is said to be a fuzzy labeling graph if $\sigma : V \to [0,1]$ and $\mu : V \times V \to [0,1]$ is bijective such that the membership value of edges and vertices are distinct and $\mu(u, v) < \sigma(u) \wedge \sigma(v)$ for all u, v $\in V$.

Definition 2.7. A fuzzy labeling graph $G = (\sigma, \mu)$ is said to be a fuzzy graceful graph if $\sigma : V \to [0,1]$ $\mu : V \times V \to [0,1]$ such that $\mu(u, v) = |\sigma(u) - \sigma(v)|$ and $\mu(u, v)$ are distinct for all u, v $\in V$.

Definition 2.8. A fuzzy labeling of a graph is said to be a triangular fuzzy labeling if it admits the triangular fuzzy number.

3. Main results

In this section , we present an algorithm to find triangular fuzzy star graph $K_{1,n}$ atmost 1241 edges . we have shown that it admits triangular fuzzy labeling .Also we prove that it satisfies triangular fuzzy graceful labeling .

3.1 Triangular fuzzy star graph $K_{1,n}$

In this section ,we develop an algorithm for triangular fuzzy star graph $K_{1,n}$, at most 1241 edges such as membership functions of the apex vertex $\sigma(v_0) = (a_0, b_0, c_0)$, pendent vertices $\sigma(v_i) = (a_i, b_{ij}, c_{jk})$ and pendent edges $\mu(v_0, v_i) = (x_i, y_{ij}, z_{jk})$.

Algorithm 3.1.1

Input : Star Graph

Procedure {Triangular fuzzy star graph ($K_{1,n}$) ,n < 1241}

$v_0 \leftarrow$ Apex vertex of $K_{1,n}$
$\sigma(v_0) = (a_0, b_0, c_0) = [8,9,10]$
for i = 1 to 15
{ $a_i = \frac{10a_0-12+i}{10}$
{ for j = 1 to (16-i)
{ $b_{i,j} = \frac{10a_i+j}{10}$
{ for k = 1 to 32-(2i-1)-2j
{ $c_{j,k} = \frac{10b_{i,j}+k}{10}$
$(v_{i,j,k}) \leftarrow$ Pendent vertices of $K_{1,n}$
$\sigma(v_{i,j,k}) = (a_i, b_{i,j}, c_{j,k})$
$(v_0, v_{i,j,k}) \leftarrow$ Internal edges of $K_{1,n}$
$\mu(v_0, v_{i,j,k}) = (u_i, v_{i,j}, w_{j,k})$
$u_i = |a_0 - c_{j,k}|$
$v_{i,j} = |b_{i,j} - b_0|$
$w_{j,k} = |c_0 - a_i|$
}
}
}
}
}
end procedure.

3.1.2 ILLUSTRATIONS

S.No.	Membership function of apex vertex of $K_{1,n}$ $\sigma(v_0) = (a_0, b_0, c_0)$			Membership function of pendent vertices of $K_{1,n}$ $\sigma(v_{i,j,k}) = (a_i, b_{ij}, c_{jk})$			No of .Edges with $\mu(v_0, v_{i,j,k}) = (x_i, y_{ij}, z_{jk})$, $x_i = \|a_0 - c_{j,k}\|$ $y_{ij} = \|b_{ij} - b_0\|$ $z_{jk} = \|c_0 - a_i\|$	Total No. of edges
	a_0	b_0	c_0	a_i	b_{ij}	c_{jk}		
1	8	9	10	8.3	8.4	8.5	1	1
2	8	9	10	8.2	8.3	8.4 to 8.6	3	4
	8	9	10	8.2	8.4	8.5	1	
3	8	9	10	8.1	8.2	8.3 to 8.7	5	9
	8	9	10	8.1	8.3	8.4 to 8.6	3	
	8	9	10	8.1	8.4	8.5	1	
4	8	9	10	8	8.1	8.2 to 8.8	7	16

	8	9	10	8	8.2	8.3 to 8.7	5	
	8	9	10	8	8.3	8.4 to 8.6	3	
	8	9	10	8	8.4	8.5	1	
	8	9	10	7.9	8	8.1 to 8.9	9	
	8	9	10	7.9	8.1	8.2 to 8.8	7	
5	8	9	10	7.9	8.2	8.3 to 8.7	5	25
	8	9	10	7.9	8.3	8.4 to 8.6	3	
	8	9	10	7.9	8.4	8.5	1	
	8	9	10	7.8	7.9	8.0 to 9.0	11	
	8	9	10	7.8	8	8.1 to 8.9	9	
	8	9	10	7.8	8.1	8.2 to 8.8	7	
6	8	9	10	7.8	8.2	8.3 to 8.7	5	36
	8	9	10	7.8	8.3	8.4 to 8.6	3	
	8	9	10	7.8	8.4	8.5	1	
	8	9	10	7.7	7.8	7.9 to 9.1	13	
	8	9	10	7.7	7.9	8.0 to 9.0	11	
	8	9	10	7.7	8	8.1 to 8.9	9	
7	8	9	10	7.7	8.1	8.2 to 8.8	7	49
	8	9	10	7.7	8.2	8.3 to 8.7	5	
	8	9	10	7.7	8.3	8.4 to 8.6	3	
	8	9	10	7.7	8.4	8.5	1	
	8	9	10	7.6	7.7	7.8 to 9.2	15	
	8	9	10	7.6	7.8	7.9 to 9.1	13	
	8	9	10	7.6	8	8.0 to 9.0	11	
8	8	9	10	7.6	8	8.1 to 8.9	9	64
	8	9	10	7.6	8.1	8.2 to 8.8	7	
	8	9	10	7.6	8.2	8.3 to 8.7	5	
	8	9	10	7.6	8.3	8.4 to 8.6	3	
	8	9	10	7.6	8.4	8.5	1	
9	8	9	10	7.5	7.6	7.7 to 9.3	17	81

	8	9	10	7.5	7.7	7.8 to 9.2	15	
	8	9	10	7.5	7.8	7.9 to 9.1	13	
	8	9	10	7.5	8	8.0 to 9.0	11	
	8	9	10	7.5	8	8.1 to 8.9	9	
	8	9	10	7.5	8.1	8.2 to 8.8	7	
	8	9	10	7.5	8.2	8.3 to 8.7	5	
	8	9	10	7.5	8.3	8.4 to 8.6	3	
	8	9	10	7.5	8.4	8.5	1	
10	8	9	10	7.4	7.5	7.6 to 9.4	19	100
	8	9	10	7.4	7.6	7.7 to 9.3	17	
	8	9	10	7.4	7.7	7.8 to 9.2	15	
	8	9	10	7.4	7.8	7.9 to 9.1	13	
	8	9	10	7.4	7.9	8.0 to 9.0	11	
	8	9	10	7.4	8	8.1 to 8.9	9	
	8	9	10	7.4	8.1	8.2 to 8.8	7	
	8	9	10	7.4	8.2	8.3 to 8.7	5	
	8	9	10	7.4	8.3	8.4 to 8.6	3	
	8	9	10	7.4	8.4	8.5	1	
11	8	9	10	7.3	7.4	7.5 to 9.5	21	121
	8	9	10	7.3	7.5	7.6 to 9.4	19	
	8	9	10	7.3	7.6	7.7 to 9.3	17	
	8	9	10	7.3	7.7	7.8 to 9.2	15	
	8	9	10	7.3	7.8	7.9 to 9.1	13	
	8	9	10	7.3	8	8.0 to 9.0	11	
	8	9	10	7.3	8	8.1 to 8.9	9	
	8	9	10	7.3	8.1	8.2 to 8.8	7	
	8	9	10	7.3	8.2	8.3 to 8.7	5	
	8	9	10	7.3	8.3	8.4 to 8.6	3	
	8	9	10	7.3	8.4	8.5	1	
12	8	9	10	7.2	7.3	7.4 to 9.6	23	144

	8	9	10	7.2	7.4	7.5 to 9.5	21	
	8	9	10	7.2	7.5	7.6 to 9.4	19	
	8	9	10	7.2	7.6	7.7 to 9.3	17	
	8	9	10	7.2	7.7	7.8 to 9.2	15	
	8	9	10	7.2	7.8	7.9 to 9.1	13	
	8	9	10	7.2	7.9	8.0 to 9.0	11	
	8	9	10	7.2	8	8.1 to 8.9	9	
	8	9	10	7.2	8.1	8.2 to 8.8	7	
	8	9	10	7.2	8.2	8.3 to 8.7	5	
	8	9	10	7.2	8.3	8.4 to 8.6	3	
	8	9	10	7.2	8.4	8.5	1	
	8	9	10	7.1	7.2	7.3 to 9.7	25	
	8	9	10	7.1	7.3	7.4 to 9.6	23	
	8	9	10	7.1	7.4	7.5 to 9.5	21	
	8	9	10	7.1	7.5	7.6 to 9.4	19	
	8	9	10	7.1	7.6	7.7 to 9.3	17	
	8	9	10	7.1	7.7	7.8 to 9.2	15	
13	8	9	10	7.1	7.8	7.9 to 9.1	13	169
	8	9	10	7.1	8	8.0 to 9.0	11	
	8	9	10	7.1	8	8.1 to 8.9	9	
	8	9	10	7.1	8.1	8.2 to 8.8	7	
	8	9	10	7.1	8.2	8.3 to 8.7	5	
	8	9	10	7.1	8.3	8.4 to 8.6	3	
	8	9	10	7.1	8.4	8.5	1	
	8	9	10	7	7.1	7.2 to 9.8	27	
	8	9	10	7	7.2	7.3 to 9.7	25	
	8	9	10	7	7.3	7.4 to 9.6	23	
14	8	9	10	7	7.4	7.5 to 9.5	21	196
	8	9	10	7	7.5	7.6 to 9.4	19	
	8	9	10	7	7.6	7.7 to 9.3	17	
	8	9	10	7	7.7	7.8 to 9.2	15	
	8	9	10	7	7.8	7.9 to 9.1	13	

	8	9	10	7	7.9	8.0 to 9.0	11	
	8	9	10	7	8	8.1 to 8.9	9	
	8	9	10	7	8.1	8.2 to 8.8	7	
	8	9	10	7	8.2	8.3 to 8.7	5	
	8	9	10	7	8.3	8.4 to 8.6	3	
	8	9	10	7	8.4	8.5	1	
	8	9	10	6.9	7	7.1 to 9.9	29	
	8	9	10	6.9	7.1	7.2 to 9.8	27	
	8	9	10	6.9	7.2	7.3 to 9.7	25	
	8	9	10	6.9	7.3	7.4 to 9.6	23	
	8	9	10	6.9	7.4	7.5 to 9.5	21	
	8	9	10	6.9	7.5	7.6 to 9.4	19	
	8	9	10	6.9	7.6	7.7 to 9.3	17	
15	8	9	10	6.9	7.7	7.8 to 9.2	15	225
	8	9	10	6.9	7.8	7.9 to 9.1	13	
	8	9	10	6.9	7.9	8.0 to 9.0	11	
	8	9	10	6.9	8	8.1 to 8.9	9	
	8	9	10	6.9	8.1	8.2 to 8.8	7	
	8	9	10	6.9	8.2	8.3 to 8.7	5	
	8	9	10	6.9	8.3	8.4 to 8.6	3	
	8	9	10	6.9	8.4	8.5	1	

Example 3.1.3:

The triangular fuzzy star graph $K_{1,n}$ at most 1241 edges is shown in the following Figure (1),(2), and (3) .

if $\sigma(v_{i,j,k}) = (a_i, b_{i,j}, c_{j,k})$ with $a_i = 6.9$, $b_{i,j} = (7.0 \text{ to } 8.4)$

$c_{j,k} = (7.1 \text{ to } 9.9 \ldots\ldots 8.4 \text{ to } 8.5 \text{ },8.5)$

then $K_{1,n}$ has atmost 225 edges which is shown in Figure (1)

Journal of Physics: Conference Series **1377** (2019) 012023 doi:10.1088/1742-6596/1377/1/012023

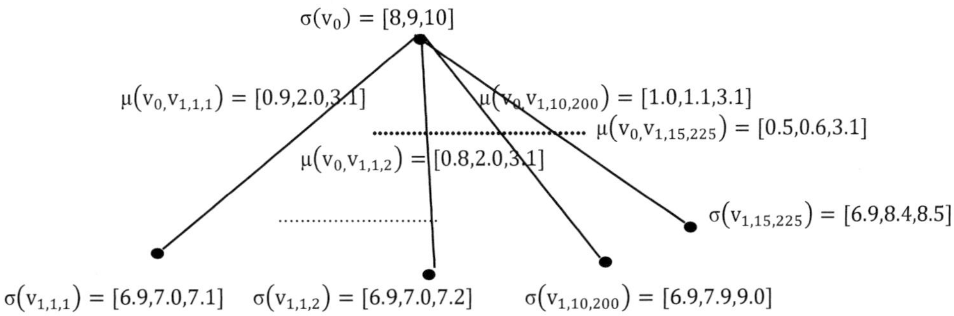

if $\sigma(v_{i,j,k}) = (a_i, b_{i,j}, c_{j,k})$ with $a_i = 7.0$, $b_{i,j} = (7.1 \text{ to } 8.4)$
$c_{j,k} = (7.2 \text{ to } 9.8 \ldots 8.4 \text{ to } 8.5, 8.5)$
then $K_{1,n}$ has at most 196 edges which is shown in Figure (2)

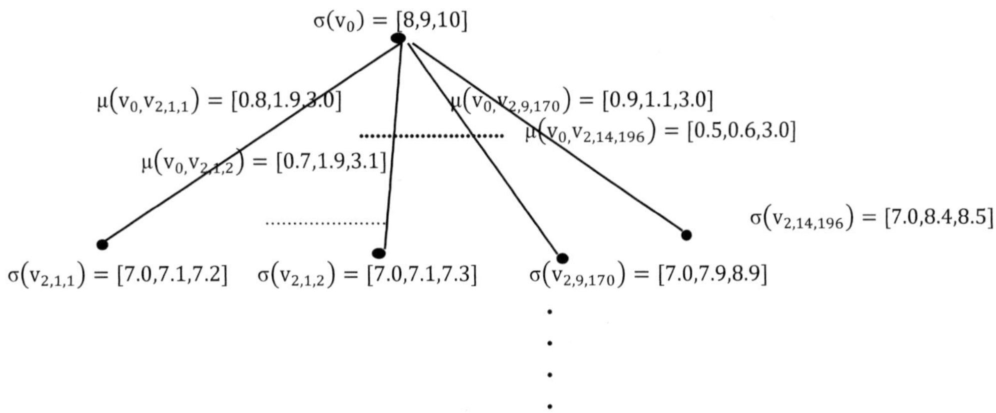

if $\sigma(v_{i,j,k}) = (a_i, b_{i,j}, c_{j,k})$ with $a_i = (8.2)$, $b_{i,j} = (8.3, 8.4)$ $c_{j,k} = (8.4 \text{ to } 8.6, 8.5)$
then $K_{1,n}$ has atmost 4 edges which is shown in Figure (3).

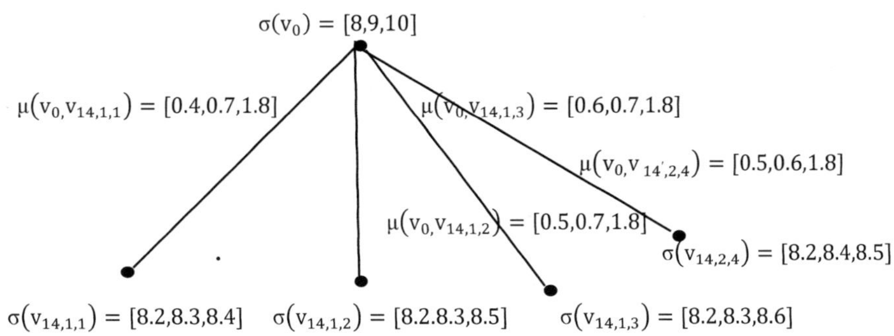

Figure(1) ,(2) and (3) Triangular Fuzzy star graph $K_{1,n}$.

Theorem 3.1.4:

The triangular fuzzy star graph $K_{1,n}$ with (n < 1241) admits triangular fuzzy labeling.

Proof:

Let $K_{1,n}$ with (n < 1241) be the star graph with 'n+1' vertices and 'n' edges. To prove $K_{1,n}$ admits triangular fuzzy labeling. That is to prove that triangular fuzzy star graph satisfies $\mu(v_0, v_{i,j,k}) < \sigma(v_0)$ *and* $\mu(v_0, v_{i,j,k}) < \sigma(v_{i,j,k})$,

i = 1 to 15 , j = 1 to 16-i, k = 1 to 32-(2-i)-2j..................................(1)

From Algorithm 3.1.1, membership function of the apex vertex 'v_0', the pendent vertices 'v_i' and the pendent edges $\mu(v_0, v_i)$ defined as follows.

$$\sigma : v_0 \to [0,1] \ni \sigma(v_0) = [a_0, b_0, c_0] = [8, 9, 10]$$

$$\sigma : v_{i,j,k} \to [0,1] \ni \sigma(v_{i,j,k}) = \lfloor a_i, b_{i,j}, c_{j,k}, \rfloor, \quad i = 1 \ to \ 15, \quad j = 1 \ to \ 16 - i, \quad k = 1 \ to \ 32 - (2i-1) - 2j$$

where

$$a_i = \frac{10\, a_0 - 12 + i}{10} \text{.........................(2)}$$

$$b_{i,j} = \frac{10 a_i + j}{10} \text{..............................(3)}$$

$$c_{j,k} = \frac{10\, b_{ij} + k}{10} \text{............................(4)}$$

$$\mu : v_0 \times v_{i,j,k} \to [0,1] \ni \mu(v_0, v_{i,j,k}) = \lfloor u_i, v_{i,j}, w_{j,k}, \rfloor, \quad i = 1 \ to \ 15, \quad j = 1 \ to \ 16 - i, \quad k = 1 \ to \ 32 - (2i-1) - 2j$$

where

$$u_i = | a_o - c_{j,k} | \text{...........................(5)}$$

$$v_{i,j} = | b_o - b_{i,j} | \text{...............................(6)}$$

$$w_{j,k} = | c_o - a_i | \text{.............................(7)}$$

Now,

$$u_i = | a_o - c_{j,k} |$$

$$= \left| a_o - \left[\frac{10 b_{ij} + k}{10} \right] \right|$$

$$= \left| a_o - \left[\frac{10\left(\frac{10 a_i + j}{10} \right) + k}{10} \right] \right|$$

$$= \left| a_o - \left[\frac{(10a_i + j) + k}{10} \right] \right|$$

$$= \left| a_o - \left[\frac{10\left(\frac{10a_0 - 12 + i}{10}\right) + j + k}{10} \right] \right|$$

$$= \left| a_o - \left[\frac{10a_0 - 12 + i + j + k}{10} \right] \right|$$

$$u_i = \left| \left[\frac{12 - i - j - k}{10} \right] \right| \quad \text{.......(8)}$$

$$v_{i,j} = | b_o - b_{i,j} |$$

$$= \left| b_o - \left[\frac{(10a_i + j)}{10} \right] \right|$$

$$= \left| b_o - \left[\frac{10\left(\frac{10a_0 - 12 + i}{10}\right) + j)}{10} \right] \right|$$

$$= \left| b_o - \left[\frac{10a_0 - 12 + i + j)}{10} \right] \right|$$

$$v_{i,j} = \left| \left[\frac{10b_0 - 10a_0 + 12 - i - j)}{10} \right] \right| \quad \text{.......(9)}$$

$$w_{j,k} = | c_o - a_i |$$

$$= \left| c_o - \left(\frac{10a_0 - 12 + i}{10} \right) \right|$$

$$w_{j,k} = \left| \left(\frac{10c_0 - 10a_0 + 12 - i}{10} \right) \right| \quad \text{.......(10)}$$

Clearly , from (8),(9) and (10) ,
we have $\mu(v_0, v_{i,j,k}) < \sigma(v_0)$ *and* $\mu(v_0, v_{i,j,k}) < \sigma(v_{i,j,k})$, for all i, j and k.
Hence the triangular fuzzy star graph $K_{1,n}$ with (n < 1241) admits triangular fuzzy labeling .

Theorem 3.1.5:
The triangular fuzzy star labeled graph $K_{1,n}$ with (n < 1241) satisfies triangular fuzzy graceful labeling.

Proof:
Let $K_{1,n}$ with (n < 1241) be the star graph with 'n+1' vertices and 'n' edges.
To prove $K_{1,n}$ admits triangular fuzzy graceful labeling. That is to prove that edges of triangular fuzzy labeled star graph has the distinct values.
Using theorem 3.1.4, we have

$$\mu : v_0 \times v_{i,j,k} \to [0,1] \ni \mu(v_0, v_{i,j,k}) = \lfloor u_i,\ v_{i,j},\ w_{j,k}, \rfloor \ , \ i=1 \ to \ 15, \quad j=1 \ to\ 16-i\ , \quad k=1 \ to \ 32-(2i-1)-2j$$

$$where \quad u_i = \left| \left\lceil \frac{12-i-j-k}{10} \right\rceil \right|.$$

$$v_{i,j} = \left| \left\lceil \frac{10b_0 - 10a_0 + 12 - i - j)}{10} \right\rceil \right|$$

$$w_{j,k} = \left| \left(\frac{10c_0 - 10a_0 + 12 - i}{10} \right) \right|$$

Therefore , for any l ,m, n, p, q and r ϵ i ,j and k with (l,m,n)≠(p,q,r)

$$\mu(v_0, v_{l,m,n}) \neq \mu(v_0, v_{p,q,r}) \qquad \text{.......................................(11)}$$

Hence the triangular fuzzy star labeled graph $K_{1,n}$ with (n < 1241) satisfies triangular fuzzy graceful labelling.

4. Conclusion

We discussed triangular fuzzy star graph $K_{1,n}$ at most 1241 edges. It has been shown that triangular fuzzy star graph admits triangular fuzzy labeling . Also proved that $K_{1,n}$ satisfies triangular fuzzy graceful labeling. Further research work to be extended on some special graphs.

5. References

[1] Zadeh A 1965 Fuzzy sets *Information and control* (8)

[2] Rahmani A, Hosseinzadeh Lotfi F, Rostamy-Malkhalifeh M, and Allahviranloo T 2016 A New Method for Defuzzification and Ranking of Fuzzy Numbers Based on the Statistical Beta Distribution *Advances in Fuzzy Systems* 2016

[3] Narsingh Deo 1974 *Graph Theory with applications to Engineering and Computer science* Prentice-Hall, Inc. Englewood Cliffs, N.J

[4] Solairaju A and Ambika S 2012 Gracefulness of a New Class of Stars Merged with Trees, *International Journal of Fuzzy Mathematics and Systems* 2(1) 83–94

[5] Bhutani KR, Moderson J and Rosenfield A 2004 On degrees of end nodes and cut nodes in fuzzy graphs *Iranian Journal of Fuzzy Systems* (1) 57–64

[6] Mathew S and Sunitha M S 2009 Types of arcs in fuzzy graph *Information Sciences* 179 1760–1768

[7] Nagoorgani A and Rajalakshmi(a)Subhasini D 2014 A note on fuzzy labeling *International Journal of Fuzzy Mathematical Archive*

[8] Nagoorgani A, Muhammed Akram and Rajalakshmi(a)Subhasini D 2014 Novel properties of fuzzy labeling graphs *Research article* Hindawi

[9] Vimala S and Jebestry Shajila R 2016 A note on edge vertex graceful labeling of star and Helmgraph, *Advances in Theoriticle and Applied Mathematics*

[10] Bala S, MorslinLifin Lee M L and Thirusangu K 2018 Fuzzy graceful labeling for the extended duplicate graphs *International journal of Mathematics trends and Technology*

[11] Ameenal Bibi K and Devi M 2017 A note on fuzzy vertex graceful labeling on special graphs, *International Journal of Advanced Research in Computer Science* 8(6)

Modelling Electric Circuit Problem with Fuzzy Differential Equations

S. Sindu Devi[1] and K. Ganesan[2]

[1]Department of Mathematics, Faculty of Engineering and Technology, SRM IST, Ramapuram, Chennai - 600 089, India.
[2]Department of Mathematics, Faculty of Engineering and Technology, SRM IST, Kattankulathur, Chennai - 603 203, India.

sindudes@srmist.edu.in; ganesank@srmist.edu.in

Abstract

In this paper we propose a new idea for the readers on the use of fuzzy in a real-life application in the triangular rule for solving the analytical solution of the second-order linear fuzzy differential equation (LFDE) under Hukuhara differentiability, especially increasing length of support. The case study is provided to show the advantages in the future.

Introduction

Fuzzy Differential Equations are applied in various modeling problems including engineering and science like Aerospace, Business, Defence, Electronics, Finance, Manufacturing, Industrial Sector, Medical, Transportation & Psychology. They are used to analyze the behavior of phenomena that are subject to vagueness or uncertain factor. Chang and Zadeh [5] initialized the fuzzy derivative. At that point, in light of the extension principle standard Dubois and Prade [8] displayed the fuzzy derivative idea. In 1987, Kandel and Byatt [9] presented the fuzzy differential equations. And afterward, Puri and Ralesu [10] made Hukuhara derivative. Following a summed up H - differentiability of fuzzy functions was presented by Bede and Gal [2,3]. The solution of Linear fuzzy Laplace Transform and some of their properties are introduced by Salahshour and Allanviranloo [11] (2013). They have proposed the fuzzy Laplace transforms for solving first order fuzzy differential equations under generalized H-differentiability and pointed out that under what conditions the fuzzy- valued functions can possess the fuzzy Laplace Transform. Allanviranioo and ahmadi [1] applied the fuzzy Laplace transform in an analytic solution method under a strong generalized differentiability concept for some fuzzy differential equations. Zuohua Ding & etal [7] proposed a numerical solution of Fuzzy Differential Equations in the performance analysis of service using composition approximate reasoning rules and fuzzy simulation principle. S.Chakraverty and Smita Tapaswini [6] have presented a fuzzy continuum reaction-diffusion model to analyze the fire propagation, the complex fuzzy arithmetic and computation are used to solve hyperbolic reaction-diffusion equation using Euler and improved Euler type methods and Homotopy perturbation method to solve Fuzzy Differential Equations. The rest of the paper is composed as pursues: In section 2, we will survey the essential idea of fuzzy set, Fuzzy derivative and initial value problem required for our concern are displayed. In section 3, we research the existence of an analytic solution utilizing the Fuzzy Laplace Transform technique. We give some genuine use of the issue are talked about and ongoing development identified with its answer in section4. In the last, we present our decision.

Preliminaries

In this section, some basic definition and concepts which will be highly useful throughout this paper are given.

Definition 1. *A fuzzy set $\widetilde{a}$ defined on the set of real numbers R is said to be a fuzzy number if its membership function $\widetilde{a} : R \to [0,1]$ has the following:*

(i) $\widetilde{a}$ is convex, i.e, $\widetilde{a}\{\lambda x_1 + (1-\lambda)x_2\} \geq \min\{\widetilde{a}(x_1), \widetilde{a}(x_2)\}$, for all $x_1, x_2 \in R$ and $\lambda \in [0,1]$

(ii) $\widetilde{a}$ is normal i.e. there exists an $x \in R$ such that $\widetilde{a}(x) = 1$

(iii) $\widetilde{a}$ is piecewise continuous. studied.

Definition 2. *A triangular fuzzy number is denoted as $\widetilde{a} = (a_1, a_2, a_3)$ and is defined by the membership function*

$$\widetilde{a}(x) = \begin{cases} 0, & x \leq a, \\ \frac{x-a_1}{a_2-a_1}, & a_1 \leq x \leq a_2, \\ \frac{a_3-x}{a_3-a_2}, & a_2 \leq x \leq a_3, \\ 0, & x \geq a_3. \end{cases}$$

Parametric form of triangular fuzzy numbers

A triangular fuzzy number $\widetilde{a} \in F(R)$ can also be represented as a pair $\widetilde{a} = (\underline{a}, \overline{a})$ of functions $\underline{a}(r)$ and $\overline{a}(r)$ for $0 \leq r \leq 1$ which satisfies the following requirements:

(i) $\underline{a}(r)$ is a bounded monotonic increasing right continuous function.

(ii) $\overline{a}(r)$ is a bounded monotonic decreasing left continuous function.

(iii) $\underline{a}(r) \leq \overline{a}(r)$, $0 \leq r \leq 1$.

Definition 3. *The Fuzzy Laplace transform of fuzzy-valued function $\widetilde{f}(t)$ is defined as following:*
$\widetilde{F}(s) = L(\widetilde{f}(t)) = \int_0^\infty e^{-st}\widetilde{f}(t) = \lim_{h\to\infty}\int_0^h e^{-st}\widetilde{f}(t)dt$ *whenever the limits exist. Where $L(\widetilde{f}(t))$ denote the fuzzy Laplace transform of fuzzy-valued function $\widetilde{f}(t)$, and the integral is the fuzzy Riemann improper integral and generates a new fuzzy-valued function $\widetilde{F}(s) = L(\widetilde{f}(t))$. Consider fuzzy-valued function $\tilde{f}$, then the lower and upper fuzzy Laplace transform of this function and denoted, base on the lower and upper of fuzzy-valued function $\tilde{f}$ as following:*
$\widetilde{F}(s,\alpha) = L(\widetilde{f}(t,\alpha)) = [L(\underline{f}(t,\alpha)), L(\overline{f}(t,\alpha))]$ *where*

$$L(\underline{f}(t,\alpha)) = \int_0^\infty e^{-st}\underline{f}(t,\alpha)dt = \lim_{h\to\infty}\int_0^h e^{-st}\underline{f}(t,\alpha)dt, 0 \le \alpha \le 1$$
$$L(\overline{f}(t,\alpha)) = \int_0^\infty e^{-st}\overline{f}(t,\alpha)dt = \lim_{h\to\infty}\int_0^h e^{-st}\overline{f}(t,\alpha)dt, 0 \le \alpha \le 1$$

Theorem 1. *[1] Let $\widetilde{f}'(t)$ be an integrable fuzzy valued function, and $\widetilde{f}(t)$ is the primitive of $\widetilde{f}'(t)$ on $[0,\infty)$.*
Then (a) $L[\widetilde{f}'(t)] = sL[\widetilde{f}(t)]\Theta\widetilde{f}(0)]$ where $\widetilde{f}$ is (i) differentiable
(b) $L[\widetilde{f}'(t)] = (-\widetilde{f}(0))\Theta(-sL[\widetilde{f}(t)])$ where $\widetilde{f}$ is (ii) differentiable

Theorem 2. *[11] Suppose that $\widetilde{f}$ and $\widetilde{f}'$ are continuous fuzzy valued functions on $[0,\infty)$ and of exponential order and that $\widetilde{f}''$ is piecewise continuous fuzzy valued function on $[0,\infty)$, then:*

1. $L[\widetilde{f}''(t) = s^2L[\widetilde{f}(t)]\Theta s\widetilde{f}(0)\Theta\widetilde{f}'(0)$, *if $\widetilde{f}$ and $\widetilde{f}'$ are (i) differentiable*
2. $L[\widetilde{f}''(t) = -\widetilde{f}'(0)\Theta(-s^2)L[\widetilde{f}(t)] - s\widetilde{f}(0)$ *if $\widetilde{f}$ is (i) differentiable and $\widetilde{f}'$ is (ii) differentiable,*
3. $L[\widetilde{f}''(t) = -s\widetilde{f}(0)\Theta(-s^2)L[\widetilde{f}(t)]\Theta\widetilde{f}'(0)$, *if $\widetilde{f}$ is (ii) differentiable and $\widetilde{f}'$ is (i) differentiable,*
4. $L[\widetilde{f}''(t) = s^2L[\widetilde{f}(t)]\Theta s\widetilde{f}(0) - \widetilde{f}'(0)$, *if $\widetilde{f}$ and $\widetilde{f}'$ are (ii) differentiable.*

Numerical Examples

Example 1. *[4] Consider the circuit model shown in figure, where $L = 1h, R = 2\Omega, C = 0.25f$ and $E(t) = 50\cos(t)$ if $Q(t)$ is the charge on the capacitor at time $t > 0$, then*

$Q''(t) + 2Q'(t) + 4Q(t) = 50\cos t$
$Q(0,\alpha) = [4+\alpha, 6-\alpha]$
$Q'(0,\alpha) = [\alpha, 2-\alpha]$

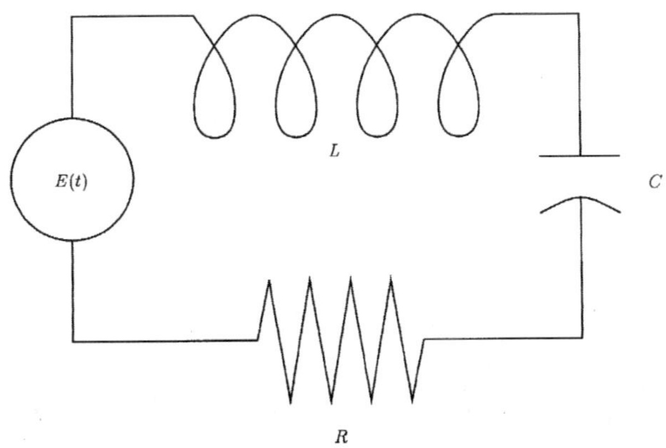

Figure 1: Electrical circuit model problem

By using fuzzy Laplace Transform, we have
$l(Q''(t)) + 2l(Q'(t)) + 4l(Q(t)) = 50l(\cos t)$
Now the classical Fuzzy Laplace Transform of the above equation is
$l(\underline{Q}''(t,\alpha)) = s^2 l(\underline{Q}(t,\alpha)) - s\underline{Q}(0) - \underline{Q}'(0)$
$l(\overline{Q}''(t,\alpha)) = s^2 l(\overline{Q}(t,\alpha)) - s\overline{Q}(0) - \overline{Q}'(0)$
and $l(\underline{Q}'(t,\alpha)) = sl(\underline{Q}(t,\alpha)) - \underline{Q}(0)$
$l(\overline{Q}'(t,\alpha)) = sl(\overline{Q}(t,\alpha)) - \overline{Q}(0)$
Applying the above and the initial conditions,
$$l(\underline{Q}(t,\alpha)) = \frac{s(4+\alpha)+(8+3\alpha)}{s^2+2s+4} + \frac{50s}{(s^2+1)(s^2+2s+4)}$$
$$l(\overline{Q}(t,\alpha)) = \frac{s(6-\alpha)+(14-3\alpha)}{s^2+2s+4} + \frac{50s}{(s^2+1)(s^2+2s+4)}$$
After performing partial fractions,

$$\begin{aligned} l(\underline{Q}(t,\alpha)) =& (4+\alpha)\left(\frac{s}{s^2+2s+4}\right) + (8+3\alpha)\left(\frac{1}{s^2+2s+4}\right) - \frac{150}{13}\left(\frac{s}{s^2+2s+4}\right) \\ & - \frac{400}{13}\left(\frac{1}{s^2+2s+4}\right) + \frac{150}{13}\left(\frac{s}{s^2+1}\right) + \frac{100}{13}\left(\frac{1}{s^2+1}\right) \\ l(\overline{Q}(t,\alpha)) =& (6-\alpha)\left(\frac{s}{s^2+2s+4}\right) + (14-3\alpha)\left(\frac{1}{s^2+2s+4}\right) - \frac{150}{13}\left(\frac{s}{s^2+2s+4}\right) \\ & - \frac{400}{13}\left(\frac{1}{s^2+2s+4}\right) + \frac{150}{13}\left(\frac{s}{s^2+1}\right) + \frac{100}{13}\left(\frac{1}{s^2+1}\right) \end{aligned}$$

By using inverse Laplace Transform on both sides,

$$\begin{aligned}\underline{Q}(t,\alpha) =&(4+\alpha)e^{-t}\cos\sqrt{3}t + \frac{2(2+\alpha)}{\sqrt{3}}e^{-t}\sin\sqrt{3}t - \frac{150}{13}e^{-t}\cos\sqrt{3}t \\ &- \frac{250}{13\sqrt{3}}e^{-t}\sin\sqrt{3}t + \frac{150}{13}\cos t + \frac{100}{13}\sin t \\ \overline{Q}(t,\alpha) =&(6-\alpha)e^{-t}\cos\sqrt{3}t + \frac{2(4-\alpha)}{\sqrt{3}}e^{-t}\sin\sqrt{3}t - \frac{150}{13}e^{-t}\cos\sqrt{3}t \\ &- \frac{250}{13\sqrt{3}}e^{-t}\sin\sqrt{3}t + \frac{150}{13}\cos t + \frac{100}{13}\sin t\end{aligned}$$

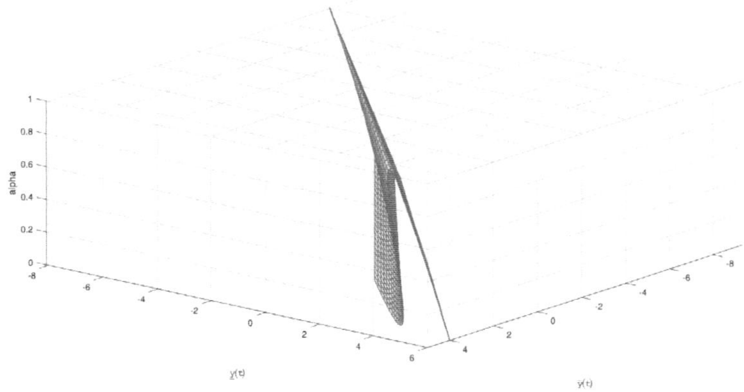

Figure 2: Solution of electrical circuit model problem by using generalized differentiability.

Example 2. *[4] Consider the vibrating mass (m=1 slug) in figure, the spring constant is* $\frac{4lb}{ft}$*, there is no damping force and the forcing function is* $100\cos\xi t$ *for* $\xi > 0$.

Solution:
$y''(t) + 4y(t) = 100\cos(\xi t)$
$y(0) = [-1+\alpha, 1-\alpha]$
$y'(0) = [-1+\alpha, 1-\alpha]$
By using fuzzy Laplace Transform,
$l(y''(t)) + 4l(y(t)) = 100l(\cos(\xi t))$
Applying formulas & initial conditions we get

$$L(\underline{y}) = \frac{s(-1+\alpha)+(-1+\alpha)}{s^2+4} + \frac{100s}{(s^2+\xi^2)(s^2+4)}$$

$$L(\overline{y}) = \frac{s(1-\alpha)+(1-\alpha)}{s^2+4} + \frac{100s}{(s^2+\xi^2)(s^2+4)}$$

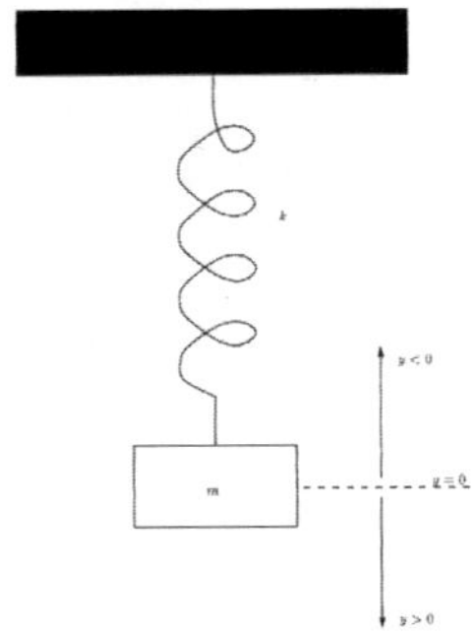

Figure 3: Vibrating mass on a string

After performing partial fraction,

$$L(\underline{y}) = (-1+\alpha)\left(\frac{s}{s^2+4}\right) + (-1+\alpha)\left(\frac{1}{s^2+4}\right) + \frac{100}{4-\xi^2}\left(\frac{s}{s^2+\xi^2}\right) - \frac{100}{4-\xi^2}\left(\frac{s}{s^2+4}\right)$$

$$L(\overline{y}) = (1-\alpha)\left(\frac{s}{s^2+4}\right) + (1-\alpha)\left(\frac{1}{s^2+4}\right) + \frac{100}{4-\xi^2}\left(\frac{s}{s^2+\xi^2}\right) - \frac{100}{4-\xi^2}\left(\frac{s}{s^2+4}\right)$$

$$\underline{y} = (-1+\alpha)\cos 2t + \frac{(-1+\alpha)}{2}\sin 2t + \frac{100}{4-\xi^2}[\cos\xi t - \cos 2t]$$

$$\overline{y} = (1-\alpha)\cos 2t + \frac{(1-\alpha)}{2}\sin 2t + \frac{100}{4-\xi^2}[\cos\xi t - \cos 2t]$$

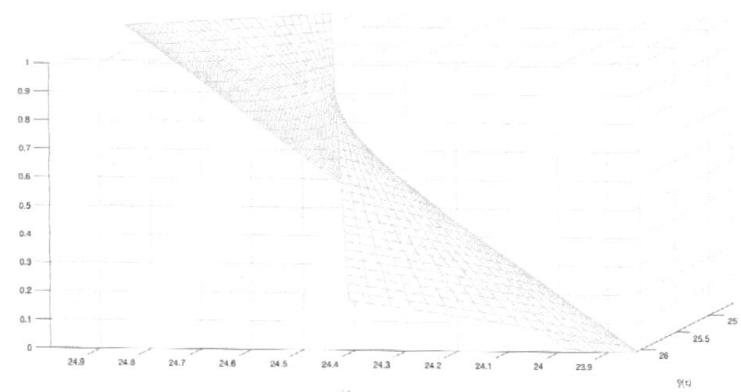

Figure 4: Exact solution of the system

Conclusion

Modeling circuit problem has been shown to find the analytical solution successfully and easily. Further, this can be used to solve many mathematical equations which will reflect better result in our future.

References

[1] Allahviranloo.T, T.Ahmadi MB, Fuzzy Laplace Transforms, Soft computing, 14, (2010), pp 235-243.

[2] Bede. B. and Gal, S.G 2005 Fuzzy Sets and Systems. 151, (2005), pp.581-599.

[3] Bede. B., Rudas, I.J. and Bencsik, 2007 Inform.Sci. 177, (2007), pp.1648-1662.

[4] Buckley J.J, Feuring .T, Fuzzy initial value problem for nth-order linear differential equations, Fuzzy Sets and Systems 121 (2001) 247255. http://dx.doi.org/10.1016/S0165-0114(00)00028-2.

[5] Chang. S. L, Zadeh. L. A IEEE Trans Systems Man cybernet. 2, (1972), pp.30-34.

[6] Chakraverty.S and Smita Tapaswini Numerical Solution of Fuzzy Differential Equations and its applications, (2014) Pg 23 DOI: 10.4018/978-1-4666-4991-D-Ch007.

[7] Ding Z, Shen H, Applying Fuzzy Differential Equations to the performance Analysis of service composition. Advanced intelligent computing Theories and Applications. ICIC 6215 (2010) pp 118-125

[8] D. Dubois, H. Prade, 1982 Fuzzy Sets Systems. 8, (1982), pp.225-233.

[9] Kandel. A Byatt W. J, 1980 Fuzzy Sets and Systems, 110, (1980), pp. 117-152.

[10] Puri. M. L, Ralesu. D. A, 1983 J.Math.Anal.Appl. 91, (1983), pp.552-558.

[11] Salahshour. Allahviranloo.T: Applications of fuzzy Laplace Transforms, soft Computing,17(1), (2013), pp 145-158.

Impact of Heat Transfer and Viscous Dissipation on Unsteady MHD oscillatory Dusty Fluid Flow Through a Vertical Porous Channel

R. Vijayalakshmi[1] and A. Govindarajan[2]
[1]SRM Institute of Science and Technology, Ramapuram.
[2]SRM Institute of Science and Technology, Kattankulathur.
E-mail: [1]`vijayal1@srmist.edu.in`, [2]`hod.maths.ktr@srmist.edu.in`

Abstract. This work discusses the effect of heat transfer and viscous dissipation on unsteady MHD (Magneto Hydro Dynamic or Hydromagnetic) oscillatory dusty fluid flow through a vertical porous channel. The governing equations for both fluid and particle phase are solved separately to get the exact solution for velocity and temperature profiles. The effect of various dimensionless parameters are displayed graphically and discussed qualitatively.

1. Introduction

For a viscous fluid flow, the fluid takes energy from the motion of the fluid and convert it into internal energy. This phenomenon is called viscous dissipation.

The existence of viscous dissipation stimulates the rate of heat transfer in many industrial problems. Megahed et al. [1] analysed the influence of Joule and viscous dissipation on MHD oscillatory dusty fluid. Cookey et al. [11] studied the impact of viscous dissipation and thermal radiation on unsteady hydromagnetic flow with suction based on time.

Later, Chen [2] discussed the heat and mass transfer effects with viscous dissipation in MHD oscillatory vertical surface. Abo et al. [3] made a research on the influence of viscous dissipation and Joule heating effects in the presence of Hall current. Mansour et al. [4] studied the chemical reaction effects and viscous dissipation on hydromagnetic flows through porous medium. Prasad et al. [5] investigated the influence of thermal radiation and mass transfer on an unsteady MHD semi-infinite vertical porous plate. Sekhar et al. [6] worked out the impact of chemical reaction and viscous dissipation on hydromagnetic oscillatory flow.

Uwanta et al. [9] investigated the problem of chemical reaction effects and viscous dissipation on semi-infinite porous plate. Vyas et al. [12] discussed the impact of thermal radiation and viscous dissipation on MHD transient flow along a vertical porous channel. Vyas et al. [13] also studied the influence of thermal radiative viscous dissipation through porous medium.

Neild [14] made a resolution of a paradox with viscous dissipation and non-linear drag. Salawu [10] discussed the effects of thermal radiation and viscous dissipation in an non-Darcy medium.

Saffman[7] discussed the stable nature of the laminar dusty fluid flow, in which the dust particles are distributed evenly. Cogley [8] assessed the differential approximation for radiative transfer in a non-grey gas near equilibrium.

Journal of Physics: Conference Series **1377** (2019) 012025 doi:10.1088/1742-6596/1377/1/012025

To be more spefic, Kiran Kumari et al. [15] discussed the effects of Viscous Dissipation and Mass Transfer on hydromagnetic oscillatory fluid flow through a vertical porous channel.

Upto the knowledge of the researcher, the topic Impact of Heat Transfer and Viscous Dissipation on Unsteady MHD oscillatory Dusty Fluid Flow Through a Vertical Porous Channel is not yet carried out by any researcher so far. This paper deals with this topic.

2. Mathematical Formulation

Consider the unsteady flow of a viscous, incompressible, hydromagnetic, oscillatory, optically thin dusty fluid moving along two vertical porous parallel channel.

Figure 1 explains the flow geometry in Cartesian coordinates (x, y) where x component is taken along the centre of the channel and y component is the distance calculated normal to it.

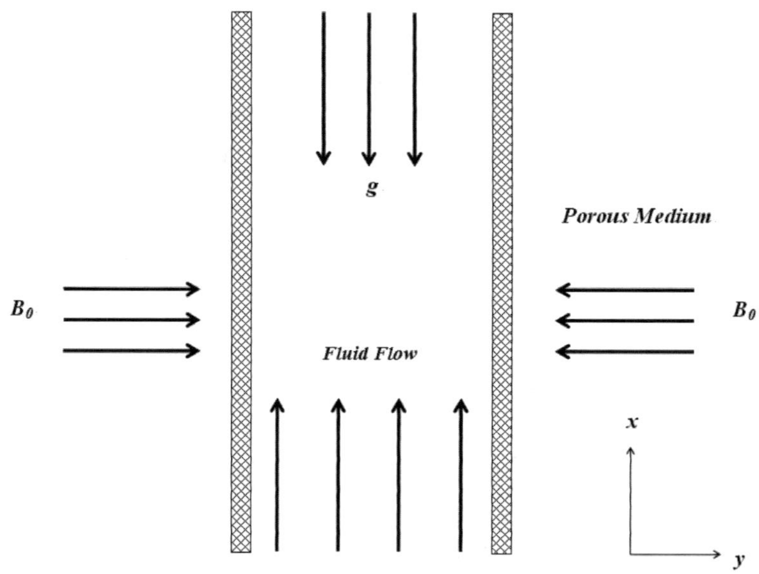

Figure 1. Geometry of Flow

The following presumptions are made here.

(i) A magnetic field of uniform strength is put perpendicular to the plane of the plates
(ii) Boussinesq approximation is considered while framing the governing equations.
(iii) The impact of Joule's dissipation is ignored in the energy equation.
(iv) The dust particles are considered to be spherical in shape and are spread uniformly through out the fluid.
(v) Chemical reactions and heat radiations are not present in the dust particles.
(vi) The magnetic Reynolds number is meagre, so that Hall effect is ignored.
(vii) u and u_d denote the velocity components of the fluid and dust particles in the x-direction.
(viii) The walls of the channel are maintained at temperatures T_0 and T_w.
(ix) Viscosity ν and Darcy's resistance terms Da are taken into account.
(x) In addition, the heat transfer from the dust particle to the fluid is taken for solving the problem.

GOVERNING EQUATIONS

Fluid Phase

Momentum equation

$$\frac{\partial u^*}{\partial t^*} = -\frac{1}{\rho}\frac{\partial P^*}{\partial x^*} + \nu\,\frac{\partial^2 u^*}{\partial y^{*2}} - \frac{\nu}{K^*}u^* - \frac{\sigma_e\,B_0^2\,u^*}{\rho} + g\,\beta_T\,(T^* - T_0{}^*) + \frac{K_0\,N_0\,(u_d{}^* - u^*)}{\rho} \tag{1}$$

Energy equation

$$\frac{\partial T^*}{\partial t^*} = \frac{k}{\rho\,C_p}\frac{\partial^2 T^*}{\partial y^{*2}} + \frac{1}{\rho\,C_p}\frac{\partial q^*}{\partial y^*} + \frac{\mu}{\rho\,C_p}\left(\frac{\partial u^*}{\partial y^*}\right)^2 + \frac{\rho_d\,C_d\,(T_d{}^* - T^*)}{\rho\,C_p\,\gamma_T} \tag{2}$$

Particle Phase

Momentum equation

$$m_d\,\frac{\partial u_d{}^*}{\partial t^*} = K_0\,(u^* - u_d{}^*) \tag{3}$$

Energy equation

$$\frac{\partial T_d{}^*}{\partial t^*} = -\frac{1}{\gamma_T}\,(T_d{}^* - T^*) \tag{4}$$

Boundary Conditions

$$u^* = 0, u_d{}^* = 0, T^* = T_0{}^*, T_d{}^* = T_0{}^*, \text{ at } y^* = 0 \tag{5}$$

$$u^* = 0, u_d{}^* = 0, T^* = T_w{}^*, T_d{}^* = T_w{}^*, \text{ at } y^* = h \tag{6}$$

According to Cogley [8],

$$\frac{\partial q^*}{\partial y^*} = 4\,\alpha^2\,(T^* - T_0{}^*) \tag{7}$$

The governing equations are changed into non-dimensional form by using the dimensionless parameters.

Dimensionless parameters

$$x = \frac{x^*}{h},\ y = \frac{y^*}{h},\ t = \frac{t^*\,U}{h},\ \omega = \frac{\omega^*\,h}{U},\ u = \frac{u^*}{U},\ u_d = \frac{u_d{}^*}{U},\ Re = \frac{U\,h}{\nu},\ \theta = \frac{T^* - T_0{}^*}{T_w{}^* - T_0{}^*},$$
$$\theta_d = \frac{T_d{}^* - T_0{}^*}{T_w{}^* - T_0{}^*},\ \phi = \frac{C^* - C_0{}^*}{C_w{}^* - C_0{}^*},\ Da = \frac{K^*}{h^2},\ s^2 = \frac{1}{Da},\ Sc = \frac{U\,h}{D_m},\ M^2 = \frac{\sigma_e\,B_0^2\,h^2}{\rho\,\nu},$$
$$P = \frac{P^*\,h}{\rho\,\nu\,U},\ Pr = \frac{\gamma\,\rho\,C_p}{k},\ N^2 = \frac{4\,\alpha^2\,h^2}{k},\ K_c = \frac{K_c{}^*\,h}{U},\ Gr = \frac{g\,\beta_T\,(T_w{}^* - T_0{}^*)\,h^2}{\nu\,U},$$
$$Gc = \frac{g\,\beta_C\,(C_w{}^* - C_0{}^*)\,h^2}{\nu\,U},\ R = \frac{K_0\,N_0\,h^2}{\rho\nu},\ G = \frac{m_d\,\nu}{K_0\,h^2},\ L_0 = \frac{h}{U\,\gamma_T},\ \gamma_T = \frac{3\,Pr\,\gamma_p\,C_d}{2\,C_p},$$
$$\gamma_p = \frac{2\,\rho_s\,D^2}{9\,\mu},\ \rho_s = \frac{3\,\rho_d}{4\,\pi\,D^3\,N_0},\ Ec = \frac{U^2}{C_p\,(T_w^* - T_o^*)} \tag{8}$$

where ρ density of the fluid, ρ_d density of dust particles, γ_T temperature relaxation time parameter, C_d specific heat capacity of dust particles, C_p specific heat capacity of fluid at constant pressure, m_d average mass of dust particles, $K_0 = 6\pi\rho\nu D$ Stoke's constant, γ_p velocity relaxation time parameter, ρ_s material density of dust particles, Re Reynolds number, Gr Grashoff number for heat transfer, Ec Eckert number respectively.

3. Solution of the problem

After applying the dimensionless parameters, equations (1 - 6) become

Fluid Phase

$$Re\,\frac{\partial u}{\partial t} = -\frac{\partial P}{\partial x} + \frac{\partial^2 u}{\partial y^2} - \left(s^2 + M^2 + R\right)u + R\,u_d + Gr\,\theta \tag{9}$$

$$Re\,Pr\,\frac{\partial \theta}{\partial t} = \frac{\partial^2 \theta}{\partial y^2} + N^2\,\theta + Ec\,Pr\left(\frac{\partial u}{\partial y}\right)^2 + \frac{2}{3}R\,(\theta_d - \theta) \tag{10}$$

Particle Phase

$$Re\,G\,\frac{\partial u_d}{\partial t} = u - u_d \tag{11}$$

$$\frac{\partial \theta_d}{\partial t} = -L_0\,(\theta_d - \theta) \tag{12}$$

Boundary Conditions

$$u = 0, u_d = 0, \theta = 0, \theta_d = 0, \text{ at } y = 0 \tag{13}$$

$$u = 0, u_d = 0, \theta = 1, \theta_d = 1, \text{ at } y = 1 \tag{14}$$

Equations (9 - 14) are coupled, non-linear partial differential equations and these cannot be solved in closed form. However, these equations can be reduced to a set of ordinary differential equations, which can be solved analytically.
Since the flow is purely oscillatory and Eckert number ($Ec << 1$), assume

$$-\frac{\partial P}{\partial x} = A_0 + A_1\,Ec\,e^{i\omega t};\; u(y,t) = u_0(y) + u_1(y)Ec\,e^{i\omega t};\; u_d(y,t) = u_{d_0}(y) + u_{d_1}(y)\,Ec\,e^{i\omega t};$$
$$\theta(y,t) = \theta_0(y) + \theta_1(y)\,Ec\,e^{i\omega t};\; \theta_d(y,t) = \theta_{d_0}(y) + \theta_{d_1}(y)\,Ec\,e^{i\omega t}; \tag{15}$$

Apply equation (15) in equations (9 - 14).

$$\theta_{d_0} = \theta_0 \tag{16}$$

$$u_{d_0} = u_0 \tag{17}$$

$$\theta_{d_1} = \left(\frac{L_0}{L_0 + i\,\omega}\right)\theta_1 \tag{18}$$

$$u_{d_1} = \left(\frac{1}{1 + i\,\omega\,Re\,G}\right)u_1 \tag{19}$$

$$\theta_0'' + N^2\,\theta_0 = -Ec\,Pr\,u_0'^2 \tag{20}$$

$$u_0'' - m_2^2\,u_0 = -A_0 - Gr\,\theta_0 \tag{21}$$

$$\theta_1'' + m_1^2\,\theta_1 = -2\,Ec\,Pr\,u_0'\,u_1' \tag{22}$$

$$u_1'' - m_3^2\,u_1 = -A_1 - Gr\,\theta_1 \tag{23}$$

Boundary Conditions

$$u_0 = 0, u_1 = 0, u_{d_0} = 0, u_{d_1} = 0, \theta_0 = 0, \theta_1 = 0, \theta_{d_0} = 0, \theta_{d_1} = 0 \text{ at } y = 0 \tag{24}$$

$$u_0 = 0, u_1 = 0, u_{d_0} = 0, u_{d_1} = 0, \theta_0 = 1, \theta_1 = 0, \theta_{d_0} = 1, \theta_{d_1} = 0 \text{ at } y = 1 \tag{25}$$

where $m_1^2 = N^2 + \dfrac{2\,R\,L_0}{3\,(L_0 + i\,\omega)} - \dfrac{2}{3}\,R - i\,\omega\,Re\,Pr;\quad m_2^2 = s^2 + M^2$

$$m_3^2 = s^2 + M^2 + R + i\,\omega\,Re - \frac{R}{1 + i\,\omega\,Re\,G}$$

Solve equations (16 - 23) using equations (24 - 25).

TEMPERATURE

Fluid Phase

$$\theta(y,t) = \theta_0(y) + \theta_1(y)\,Ec\,e^{i\,\omega\,t} \tag{26}$$

Particle Phase

$$\theta_d(y,t) = \theta_{d_0}(y) + \theta_{d_1}(y)\,Ec\,e^{i\,\omega\,t} \tag{27}$$

VELOCITY

Fluid Phase

$$u(y,t) = u_0(y) + u_1(y) Ec\, e^{i\,\omega\,t} \tag{28}$$

Particle Phase

$$u_d(y,t) = u_{d_0}(y) + u_{d_1}(y)\,Ec\,e^{i\,\omega\,t} \tag{29}$$

where

$$u_0(y) = C_1\,e^{m_2\,y} + C_2\,e^{-m_2\,y} + t_1 + t_2\,\theta_{d_0} \tag{30}$$

$$\theta_0(y) = C_3\,cos\,(Ny) + C_4\,sin\,(Ny) - C_1^2\,t_3\,t_4\,e^{2\,m_2\,y} - C_2^2\,t_3\,t_4\,e^{-2\,m_2\,y} + t_3\,t_5\,t_6 \tag{31}$$

$$u_1(y) = C_5\,e^{m_3\,y} - C_6\,e^{-m_3\,y} + t_7 + t_8\,t_9\,\theta_{d_1} \tag{32}$$

$$\theta_1(y) = C_7\,cos\,(m_1\,y) + C_8\,sin\,(m_1\,y) - t_{10}\,e^{(m_2+m_3)\,y} - t_{11}\,e^{(m_2-m_3)\,y} + t_{12}\,e^{-(m_2-m_3)\,y} + t_{13}\,e^{-(m_2+m_3)\,y} \tag{33}$$

and $t_1 = \dfrac{A_0}{m_2^2};\ t_2 = \dfrac{Gr}{m_2^2};\ t_3 = Ec\,Pr\,m_2^2;\ t_4 = \dfrac{1}{4\,m_2^2 + N^2};\ t_5 = \dfrac{1}{N^2};\ t_6 = 2\,C_1\,C_2;$

$$t_7 = \frac{A_1}{m_3^2};\ t_8 = \frac{Gr}{m_3^2};\ t_9 = \frac{L_0 + i\,\omega}{L_0};\ t_{10} = \frac{2Ec\,Pr\,C_1\,C_5\,m_2\,m_3\,e^{(m_2+m_3)y}}{(m_2+m_3)^2 + m_1^2};$$

$$t_{11} = \frac{2Ec\,Pr\,C_1\,C_6\,m_2\,m_3\,e^{(m_2-m_3)y}}{(m_2-m_3)^2 + m_1^2};\ t_{12} = \frac{2\,Ec\,Pr\,C_2\,C_5\,m_2\,m_3\,e^{-(m_2-m_3)y}}{(m_2-m_3)^2 + m_1^2};$$

$$t_{13} = \frac{2\,Ec\,Pr\,C_2\,C_6\,m_2\,m_3\,e^{-(m_2+m_3)y}}{(m_2+m_3)^2 + m_1^2};$$

$$C_1 = \frac{-t_1\,e^{-m_2} + t_1 + t_2}{e^{-m_2} - e^{m_2}};\ C_2 = \frac{-t_1 - t_2 + t_1\,e^{m_2}}{e^{-m_2} - e^{m_2}};\ C_3 = C_1^2\,t_3\,t_4 + C_2^2\,t_3\,t_4 - t_3\,t_5\,t_6;$$

$$C_4 = \frac{1}{sin\,(N)}\,[1 - C_3\,cos\,(N) + C_1^2\,t_3\,t_4\,e^{2\,m_2} + C_2^2\,t_3\,t_4\,e^{-2\,m_2} - t_3\,t_5\,t_6];$$

$$C_5 = \frac{-t_7\,e^{-m_3} + t_7}{e^{-m_3} - e^{m_3}};\ C_6 = \frac{-t_7 + t_7\,e^{m_3}}{e^{-m_3} - e^{m_3}};\ C_7 = t_{10} + t_{11} - t_{12} - t_{13};$$

$$C_8 = \frac{1}{sin\,(m_1)}\,[t_{10}\,e^{(m_2+m_3)} - t_{11}\,e^{(m_2-m_3)} + t_{12}\,e^{-(m_2+m_3)} + t_{13}\,e^{-(m_2+m_3)} - C_7\,cos\,(m_1)]$$

SKIN FRICTION
Fluid Phase

$$\tau = \left[\mu \, \frac{\partial u}{\partial y}\right] \text{ at } y = 0, 1. \tag{34}$$

$$\tau = \left[\mu \, \frac{\partial}{\partial y} \left(u_0 + Ec \, u_1 \, e^{\, i \, \omega \, t}\right)\right] \text{ at } y = 0, 1. \tag{35}$$

Particle Phase

$$\tau_d = \left[\mu \frac{\partial u_d}{\partial y}\right] \text{ at } y = 0, 1. \tag{36}$$

$$\tau_d = \left[\mu \, \frac{\partial}{\partial y} \left(u_{d_0} + Ec \, u_{d_1} \, e^{\, i \, \omega \, t}\right)\right] \text{ at } y = 0, 1. \tag{37}$$

NUSSELT NUMBER (Rate of Heat Transfer)
Fluid Phase

$$Nu = -\left[\frac{\partial \theta}{\partial y}\right] \text{ at } y = 0, 1. \tag{38}$$

$$Nu = -\left[\frac{\partial}{\partial y} \left(\theta_0 + Ec \, \theta_1 \, e^{\, i \, \omega \, t}\right)\right] \text{ at } y = 0, 1. \tag{39}$$

Particle Phase

$$Nu = -\left[\frac{\partial \theta_d}{\partial y}\right] \text{ at } y = 0, 1. \tag{40}$$

$$Nu = -\left[\frac{\partial}{\partial y} \left(\theta_{d_0} + Ec \, \theta_{d_1} \, e^{\, i \, \omega \, t}\right)\right] \text{ at } y = 0, 1. \tag{41}$$

4. Graphical Results and Discussions

The above mathematical expressions are resolved analytically and the computed results are graphically plotted using MATLAB 8.3. For all numerical computations, take Prandtl number as $\mathrm{Pr} = 0.71$, that corresponds to air at $20°$ C.

Figures 2 - 11 illustrate the impact of various significant parameters on velocity of fluid and dust particles. The parabolic nature of velocity profiles is noted in all the graphs. Generally, the velocity profiles of fluid and dust particles are maximum at the middle of the channel and minimum near the walls of the channel. (*i.e.*) All the velocity profiles accelerate steadily from the left end of the plate ($y = 0$), attain maximum at its centre and converges near the other end of the channel ($y = 1$), satisfying the prescribed boundary conditions.

Figures 12 - 21 show the impact of various physical parameters on temperature of fluid and dust particles. It is also observed that the nature of temperature profiles is parabolic. Moreover, the temperature profiles of fluid and dust particles reaches its maximum in the centre of the channel ($y = 0.5$).

The skin friction coefficient and the rate of heat transfer for fluid and dust particles by varying various parameters are tabulated in **Tables 1 - 4**.

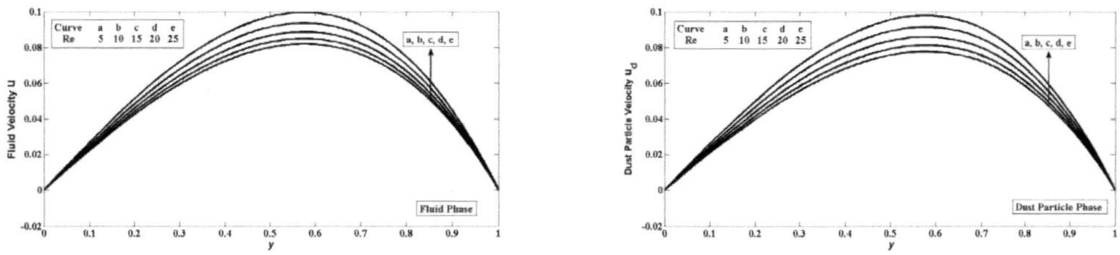

(a) (b)

Figure 2. Impact of Re on **(a)** Velocity of Fluid (u), **(b)** Velocity of Dust Particles (u_d), with $Pr = 0.71$; $Gr = 2$; $M = 3$; $N = 3$; $G = 3$; $R = 3$; $L_0 = 1$; $s = 2$; $A = 0.1$; $w = 1$; $t = 0.5$; $Ec = 0.5$.

From **Figure 2**, it is noted that rise in the value of Reynolds number (Re) increases the velocity of fluid and dust particles. Re has the same influence on velocity of fluid and dust particles.

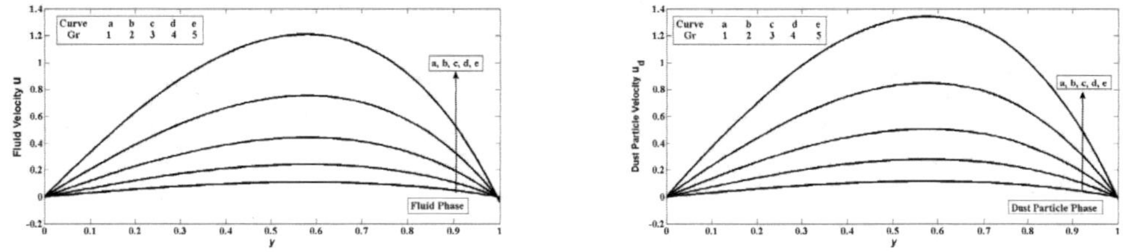

(a) (b)

Figure 3. Impact of Gr on **(a)** u, **(b)** u_d, with $Pr = 0.71$; $Re = 2$; $M = 3$; $N = 3$; $G = 3$; $R = 3$; $L_0 = 1$; $s = 2$; $A = 0.1$; $w = 1$; $t = 0.5$; $Ec = 0.5$.

From **Figure 3**, it is clear that while augmenting Grashof number for heat transfer (Gr), the velocity of fluid and dust particles" gradually increase due to buoyancy force. Gr has minor impact on velocity of fluid and dust particles. For the same flow geometry, this result coincides with Kiran Kumari et al. [15] for the MHD oscillatory fluid flow without dust parameters.

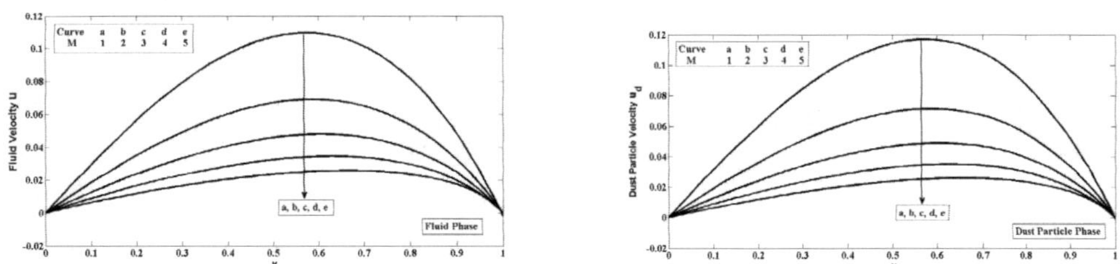

(a) (b)

Figure 4. Impact of M on **(a)** u, **(b)** u_d, with $Pr = 0.71$; $Re = 2$; $Gr = 2$; $N = 3$; $G = 3$; $R = 3$; $L_0 = 1$; $s = 2$; $A = 0.1$; $w = 1$; $t = 0.5$; $Ec = 0.5$.

From **Figure 4**, it is seen that the velocity of fluid and dust particles" decelerate for a gradual rise in the value of Hartmann number (M). It is true that the transverse magnetic field produces a resistive force called Lorentz force, which retards the fluid motion. M has significant effect on velocity of fluid and dust particles. For the same flow geometry, this result coincides with Kiran Kumari et al. [15] for the MHD oscillatory fluid flow without dust parameters.

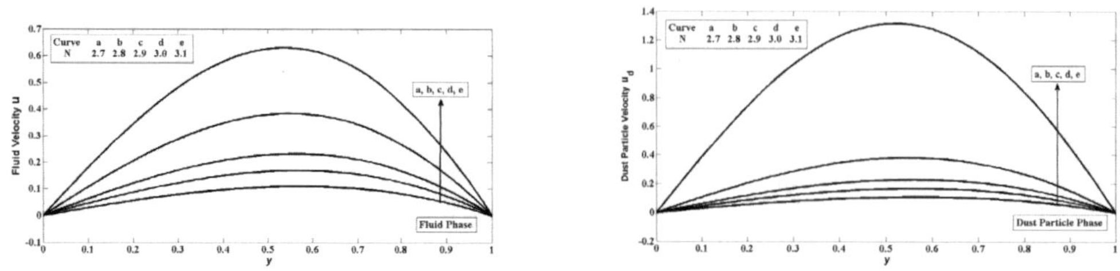

(a) **(b)**

Figure 5. Impact of N **(a)** u, **(b)** u_d, with $Pr = 0.71$; $Re = 2$; $Gr = 2$; $M = 3$; $G = 3$; $R = 3$; $L_0 = 1$; $s = 2$; $A = 0.1$; $w = 1$; $t = 0.5$; $Ec = 0.5$.

From **Figure 5**, it is apparent that the velocity of fluid and dust particles accelerate gradually, while enhancing the value of radiation parameter (N). N shows a major variation on velocity of fluid and dust particles.

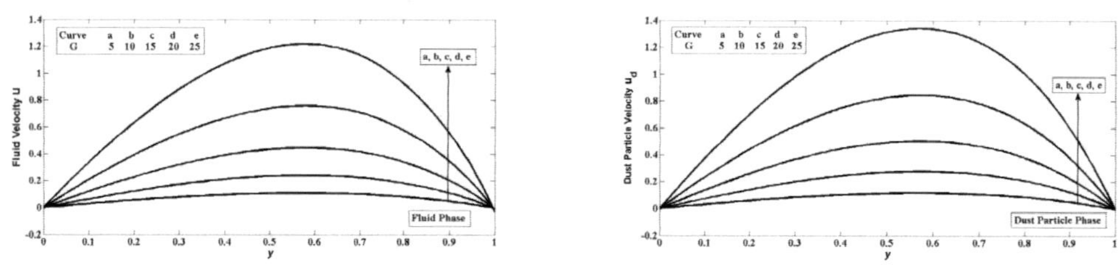

(a) **(b)**

Figure 6. Impact of G on **(a)** u, **(b)** u_d, with $Pr = 0.71$; $Re = 1$; $Gr = 1$; $M = 3$; $N = 1.1$; $R = 3$; $L_0 = 1$; $s = 2$; $A = 0.1$; $w = 1$; $t = 0.5$; $Ec = 0.5$.

From **Figure 6**, it is seen that a gradual increase in particle mass parameter (G) shows that the velocity of fluid and dust particles increases steadily. G has significant effect on velocity of fluid and dust particles.

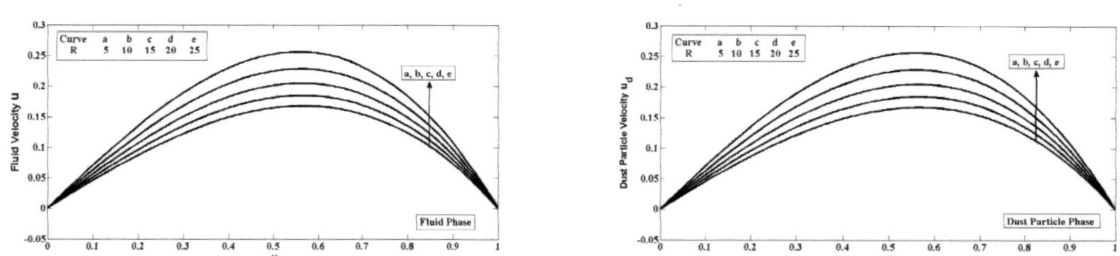

(a) **(b)**

Figure 7. Impact of R on **(a)** u, **(b)** u_d, with $Pr = 0.71$; $Re = 2$; $Gr = 2$; $M = 3$; $N = 3$; $G = 3$; $L_0 = 1$; $s = 2$; $A = 0.1$; $w = 1$; $t = 0.5$; $Ec = 0.5$.

From **Figure 7**, it is apparent that while augmenting the value of particle concentration parameter (R), the velocity of fluid and dust particles increase. R has a minor impact on velocity of fluid and dust particles.

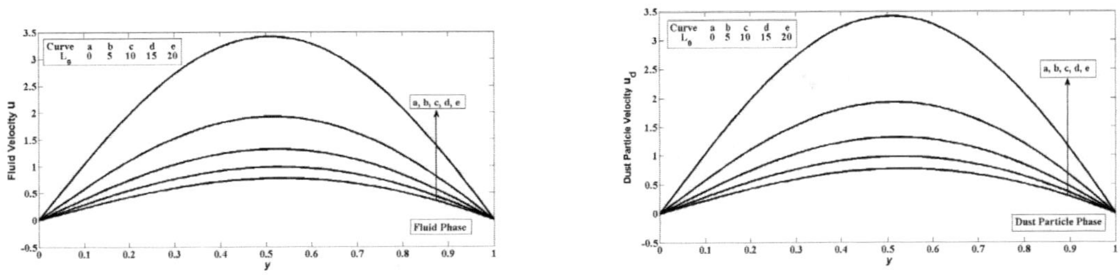

(a) (b)

Figure 8. Impact of L_0 on **(a)** u, **(b)** u_d, with $Pr = 0.71$; $Re = 2$; $Gr = 2$; $M = 3$; $N = 3$; $G = 3$; $R = 3$; $s = 2$; $A = 0.1$; $w = 1$; $t = 0.5$; $Ec = 0.5$.

From **Figure 8**, it is noteworthy to mention that an increase in temperature relaxation time parameter (L_0) rises the velocity of fluid and dust particles. L_0 has the same influence on velocity of fluid and dust particles.

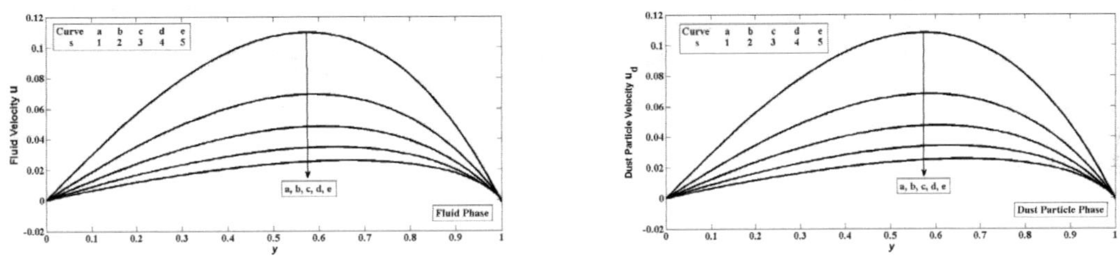

(a) (b)

Figure 9. Impact of s on **(a)** u, **(b)** u_d, with $Pr = 0.71$; $Re = 2$; $Gr = 2$; $M = 3$; $N = 3$; $G = 3$; $R = 3$; $L_0 = 1$; $A = 0.1$; $w = 1$; $t = 0.5$; $Ec = 0.5$.

From **Figure 9**, it is clear that a rise in the value of porous medium shape factor (s) reduces the velocity of fluid and dust particles steadily. s shows a minor variation on velocity of fluid and dust particles.

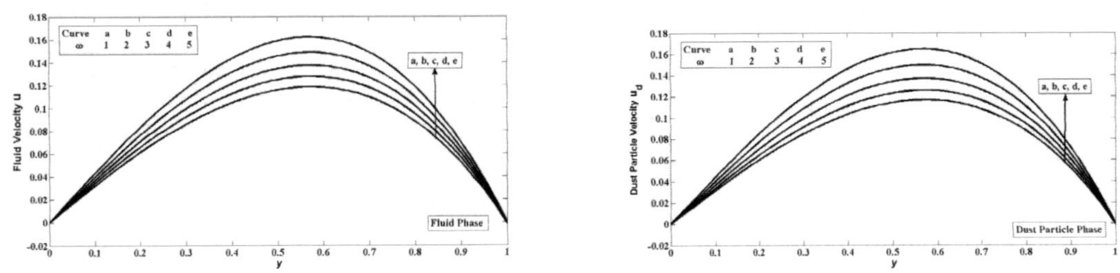

(a) (b)

Figure 10. Impact of ω on **(a)** u, **(b)** u_d, with $Pr = 0.71$; $Re = 2$; $Gr = 2$; $M = 3$; $N = 3$; $G = 3$; $R = 3$; $L_0 = 1$; $s = 2$; $A = 0.1$; $t = 0.5$; $Ec = 0.5$.

From **Figure 10**, it is clear that augmenting the value of frequency of oscillation parameter (ω), the velocity of fluid and dust particles rise steadily. ω shows a slight change on velocity of fluid and dust particles.

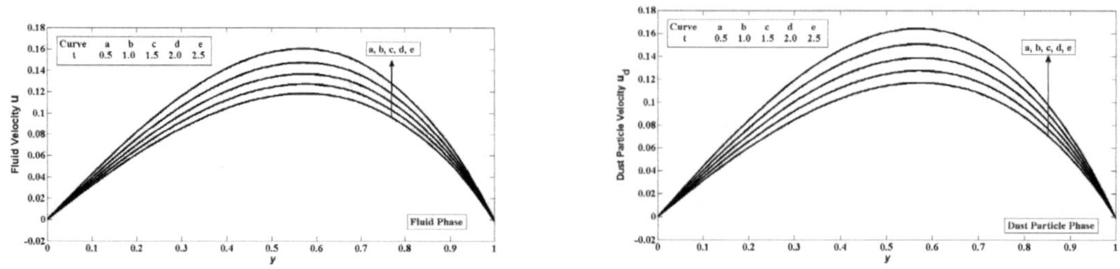

(a) **(b)**

Figure 11. Impact of t on **(a)** u, **(b)** u_d, with $Pr = 0.71$; $Re = 2$; $Gr = 2$; $M = 3$; $N = 3$; $G = 3$; $R = 3$; $L_0 = 1$; $s = 2$; $A = 0.1$; $w = 1$; $Ec = 0.5$.

From **Figure 11**, it is apparent for an unsteady flow, as time (t) increases, the velocity of fluid and dust particles increase. Time 't' has the same impact on velocity of fluid and dust particles.

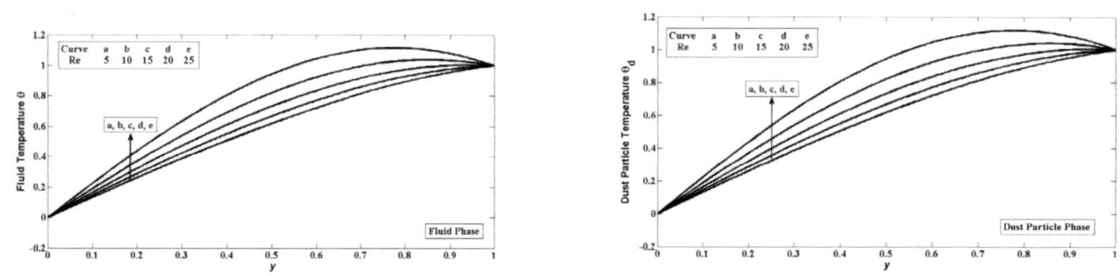

(a) **(b)**

Figure 12. Impact of Re on **(a)** Temperature of Fluid (θ), **(b)** Temperature of Dust Particles (θ_d), with $Pr = 0.71$; $Gr = 2$; $M = 3$; $N = 3$; $G = 3$; $R = 3$; $L_0 = 1$; $s = 2$; $A = 0.1$; $w = 1$; $t = 0.5$; $Ec = 0.5$.

From **Figure 12**, it is evident that the temperature of fluid and dust particles increase with an increase in value of Reynolds number (Re). Re has slight effect on the temperature of fluid and dust particles.

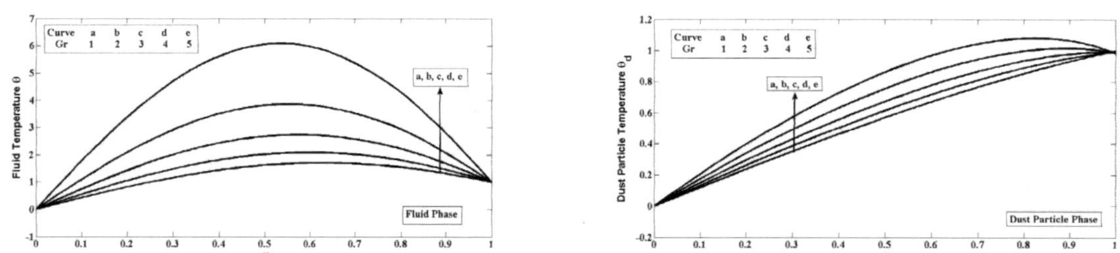

(a) **(b)**

Figure 13. Impact of Gr on **(a)** θ, **(b)** θ_d, with $Pr = 0.71$; $Re = 2$; $M = 3$; $N = 3$; $G = 3$; $R = 3$; $L_0 = 1$; $s = 2$; $A = 0.1$; $w = 1$; $t = 0.5$; $Ec = 0.5$.

From **Figure 13**, it is visible that the temperature of fluid and dust particles increase with an increase in the value of Grashof number for heat transfer (Gr). Gr has major influence on the temperature of fluid and dust particles.

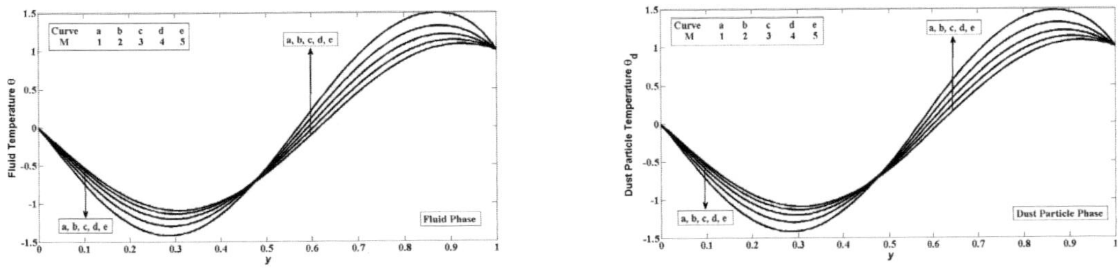

(a) **(b)**

Figure 14. Impact of M on **(a)** θ, **(b)** θ_d, with $Pr = 0.71$; $Re = 2$; $Gr = 2$; $N = 3$; $G = 3$; $R = 3$; $L_0 = 1$; $s = 2$; $A = 0.1$; $w = 1$; $t = 0.5$; $Ec = 0.5$.

From **Figure 14**, it is seen that for a gradual increase in Hartmann number (M) or magnetic parameter, the temperature of fluid and dust particles decrease in the left part of the channel and increase in the right side. M has the same influence on the temperature of fluid and dust particles.

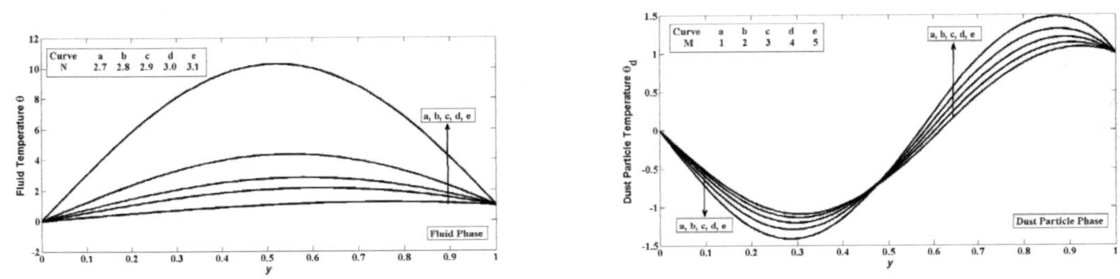

(a) **(b)**

Figure 15. Impact of N on **(a)** θ, **(b)** θ_d, with $Pr = 0.71$; $Re = 2$; $Gr = 2$; $M = 3$; $G = 3$; $R = 3$; $L_0 = 1$; $s = 2$; $A = 0.1$; $w = 1$; $t = 0.5$; $Ec = 0.5$.

From **Figure 15**, it is apparent that the temperature of fluid and dust particles increase with an increase in the value of radiation parameter (N). N has major impact on the temperature of fluid when compared to dust particles.

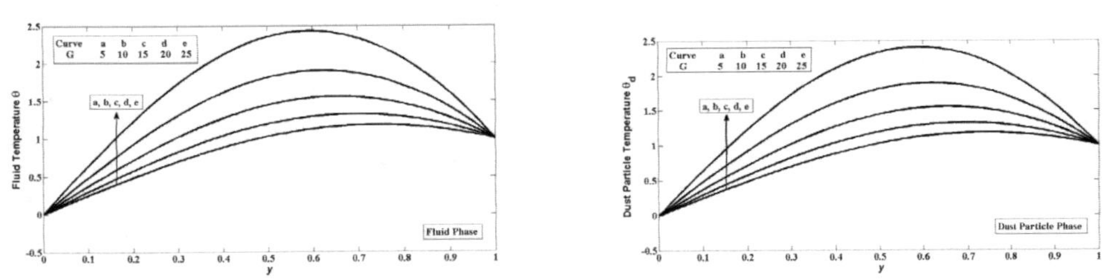

(a) **(b)**

Figure 16. Impact of G on **(a)** θ, **(b)** θ_d, with $Pr = 0.71$; $Re = 2$; $Gr = 2$; $M = 3$; $N = 3$; $R = 3$; $L_0 = 1$; $s = 2$; $A = 0.1$; $w = 1$; $t = 0.5$; $Ec = 0.5$.

From **Figure 16**, it is clear that the temperature of fluid and dust particles increase for an increase in the value of particle mass parameter (G). G has the same influence on the temperature of fluid and dust particles.

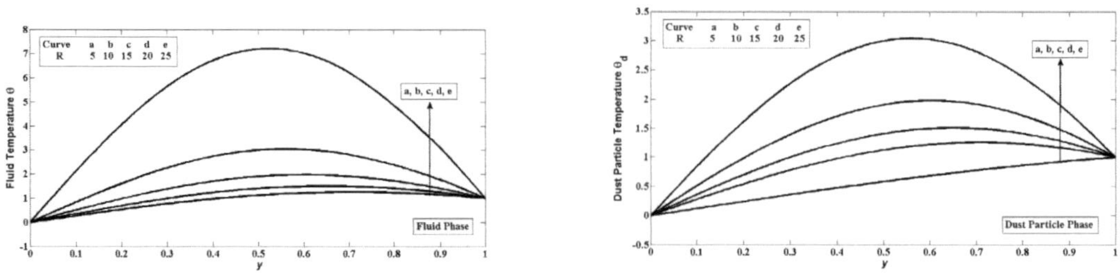

(a) (b)

Figure 17. Impact of R on **(a)** θ, **(b)** θ_d, with $Pr = 0.71$; $Re = 2$; $Gr = 2$; $M = 3$; $N = 3$; $G = 3$; $L_0 = 1$; $s = 2$; $A = 0.1$; $w = 1$; $t = 0.5$; $Ec = 0.5$.

From **Figure 17**, it is evident that the temperature of fluid and dust particles increase for a rise in the value of particle concentration parameter (R). R has significant effect on temperature of fluid compared to that of dust particles.

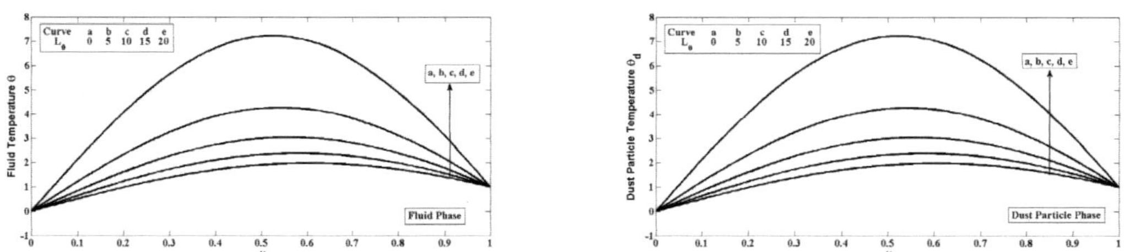

(a) (b)

Figure 18. Impact of L_0 on **(a)** θ, **(b)** θ_d, with $Pr = 0.71$; $Re = 2$; $Gr = 2$; $M = 3$; $N = 3$; $G = 2$; $R = 2$; $s = 1$; $A = 0.1$; $w = 1$; $t = 0.5$; $Ec = 0.5$.

From **Figure 18**, it is noteworthy to mention that an increase in temperature relaxation time parameter (L_0) rises the temperature of fluid and dust particles gradually. L_0 has the same influence on temperature of fluid and dust particles.

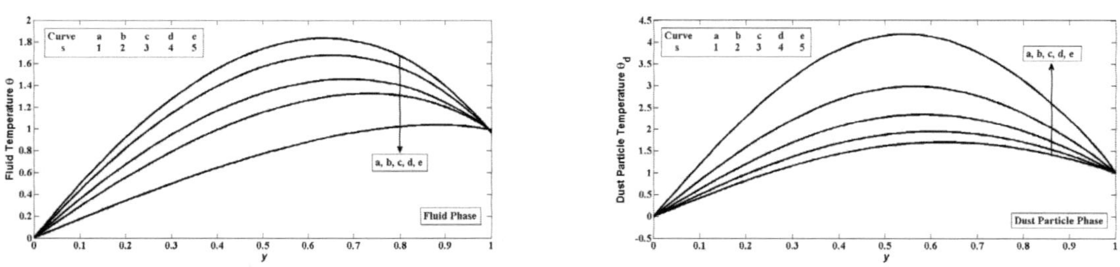

(a) (b)

Figure 19. Impact of s on **(a)** θ, **(b)** θ_d, with $Pr = 0.71$; $Re = 2$; $Gr = 2$; $M = 3$; $N = 3$; $G = 3$; $R = 3$; $L_0 = 1$; $A = 0.1$; $w = 1$; $t = 0.5$; $Ec = 0.5$.

From **Figure 19**, it is notable that an increase in the value of porous medium shape factor (s) decreases the temperature of fluid, whereas it shows an opposite effect for the dust particles. s has significant effect on temperature of fluid and dust particles.

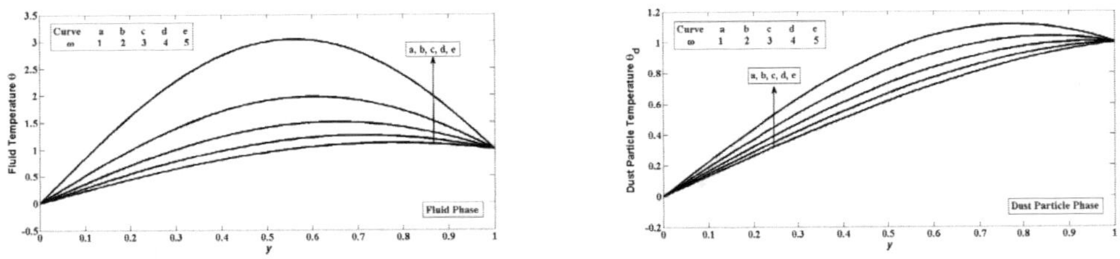

(a) (b)

Figure 20. Impact of ω on **(a)** θ, **(b)** θ_d, with $Pr = 0.71$; $Re = 2$; $Gr = 2$; $M = 3$; $N = 3$; $G = 3$; $R = 3$; $L_0 = 1$; $s = 1$; $A = 0.1$; $t = 0.5$; $Ec = 0.5$.

From **Figure 20**, it is noteworthy to mention that while augmenting the value of frequency of oscillation parameter (ω), the temperature of fluid and dust particles increase gradually. ω shows a major variation on temperature of fluid compared to dust particles.

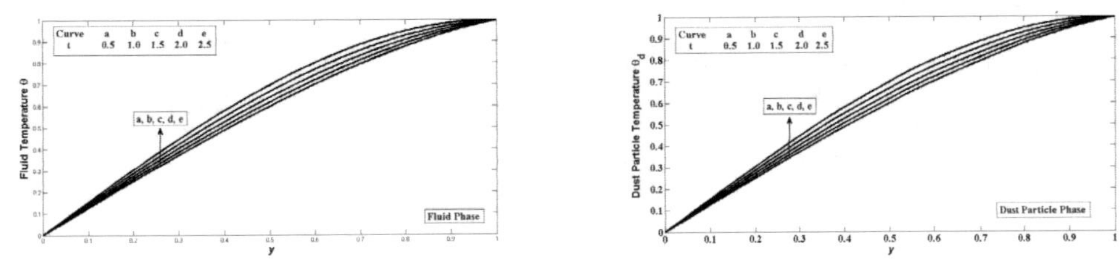

(a) (b)

Figure 21. Impact of t on **(a)** θ, **(b)** θ_d, with $Pr = 0.71$; $Re = 2$; $Gr = 2$; $M = 3$; $N = 3$; $G = 3$; $R = 3$; $L_0 = 1$; $s = 1$; $A = 0.1$; $w = 1$; $Ec = 0.5$.

From **Figure 21**, it is noted that as time (t) increases, the temperature of fluid and dust particles enhance steadily, in the case of unsteady flow. Time 't' has the same influence on temperature of fluid and dust particles.

5. Tables and Discussions

Table 1. Skin friction coefficient for Fluid Phase (τ) and Dust Particle Phase (τ_d) by varying t with $Pr = 0.71$; $Re = 2$; $Gr = 2$; $M = 3$; $N = 3$; $G = 3$; $R = 3$; $L_0 = 1$; $s = 2$; $A = 0.1$; $w = 1$; $Ec = 0.5$.

t	Fluid Phase			Dust Particle Phase		
	τ at $y = 0$	τ at $y = 0.5$	τ at $y = 1$	τ_d at $y = 0$	τ_d at $y = 0.5$	τ_d at $y = 1$
0.1	0.2832	0.0474	−0.6470	0.3006	0.0619	−0.6136
0.2	0.2830	0.0467	−0.6481	0.2990	0.0605	−0.6169
0.3	0.2828	0.0464	−0.6487	0.2974	0.0592	−0.6199
0.4	0.2827	0.0462	−0.6487	0.2960	0.0579	−0.6226
0.5	0.2817	0.0461	−0.6488	0.2948	0.0568	−0.6251

It is evident from **Table 1** that as time (t) increases, the skin friction coefficient of fluid and dust particles decrease at all positions across the channel, maximum at the left end of the channel ($y = 0$) and minimum at the right end ($y = 1$) and decrease along the ends and the middle of the channel.

Table 2. Skin friction coefficient for Fluid Phase (τ) and Dust Particle Phase (τ_d) by R and G with $Pr = 0.71$; $Re = 2$; $Gr = 2$; $M = 3$; $N = 3$; $L_0 = 1$; $s = 1$; $A = 0.1$; $w = 1$; $t = 0.6$; $Ec = 0.5$.

R	G	Fluid Phase			Dust Particle Phase		
		τ at $y = 0$	τ at $y = 0.5$	τ at $y = 1$	τ_d at $y = 0$	τ_d at $y = 0.5$	τ_d at $y = 1$
1	1	0.2846	0.0465	−0.6470	0.2938	0.0558	−0.6272
2	1	0.2876	0.0502	−0.6398	0.2961	0.0585	−0.6226
3	1	0.2899	0.0529	−0.6347	0.2978	0.0605	−0.6193
3	2	0.2922	0.0555	−0.6315	0.3058	0.0654	−0.6066
3	3	0.2930	0.0563	−0.6305	0.3086	0.0669	−0.6019

It is evident from **Table 2** that by varying particle concentration parameter (R) and particle mass parameter (G), the skin friction coefficient of fluid and dust particles decrease at all positions across the channel, maximum at the left end of the channel ($y = 0$) and minimum at the right end ($y = 1$) and increase along the ends and the middle of the channel.

Table 3. Nusselt number (Nu) for Fluid Phase and Dust Particle Phase by varying t with $Pr = 0.71$; $Re = 2$; $Gr = 2$; $M = 3$; $N = 3$; $G = 3$; $R = 3$; $L_0 = 1$; $s = 1$; $A = 0.1$; $w = 1$; $Ec = 0.5$.

t	Fluid Phase			Dust Particle Phase		
	Nu at $y = 0$	Nu at $y = 0.5$	Nu at $y = 1$	Nu at $y = 0$	Nu at $y = 0.5$	Nu at $y = 1$
0.1	−2.4950	−1.2195	1.4142	−2.4937	1.2195	1.3744
0.2	−2.4949	−1.2194	1.4125	−2.4936	1.2194	1.3741
0.3	−2.4948	−1.2193	1.4099	−2.4935	1.2193	1.3613
0.4	−2.4947	−1.2192	1.4065	−2.4934	1.2192	1.3539
0.5	−2.4946	−1.2191	1.4022	−2.4933	1.2191	1.3458

It is noted from **Table 3** that as time (t) increases, the rate of heat transfer for fluid and dust particles increase at all positions across the channel, minimum at the left end of the channel ($y = 0$) and maximum at the right end ($y = 1$) and increase along ($y = 0$), decrease along ($y = 1$).

Table 4. Nusselt Number (Nu) for Fluid Phase and Dust Particle Phase by varying L_0 and Pr with $Re = 2$; $Gr = 2$; $M = 2$; $N = 2$; $G = 2$; $R = 2$; $s = 2$; $A = 0.1$; $w = 1$; $t = 0.5$; $Ec = 0.5$.

L_0	Pr	Fluid Phase			Dust Particle Phase		
		Nu at $y = 0$	Nu at $y = 0.5$	Nu at $y = 1$	Nu at $y = 0$	Nu at $y = 0.5$	Nu at $y = 1$
0.1	0.71	−2.4947	−1.2188	1.3933	−2.4930	1.2188	1.3286
0.2	0.71	−2.4947	−1.2188	1.3937	−2.4931	1.2189	1.3358
0.3	0.71	−2.4948	−1.2188	1.3942	−2.4933	1.2190	1.3434
0.3	1	−2.5202	−1.2223	1.4633	−2.5179	1.2225	1.3897
0.3	6.7	−3.0190	−1.3012	2.7049	−2.9981	1.2899	2.2171

It is evident from **Table 4** that by varying temperature relaxation time parameter (L_0) and Prandtl number (Pr), the rate of heat transfer for fluid and dust particles increase at all positions across the channel, very low at one end of the channel ($y = 0$) and gradually rise at the other end ($y = 1$) and decrease along ($y = 0$), increase along ($y = 1$).

6. Conclusion

In view of the results discussed above, the following conclusion is obtained:

- Increase in Grashof number (Gr), radiation parameter (N), particle mass parameter (G), particle concentration parameter (R), temperature relaxation parameter (L_0), frequency of oscillation parameter (ω) and time (t) increase the velocity and temperature of fluid and dust particles.
- Increase in porous medium shape factor (s) decreases the velocity and temperature of fluid and dust particles.
- Increase in Reynolds number (Re) increases the velocity of fluid and dust particles.
- As time (t) increases, the skin friction coefficient of fluid and dust particles decrease at all positions across the channel, maximum at the left end of the channel ($y = 0$) and minimum at the right end ($y = 1$).
- By varying particle concentration parameter (R) and particle mass parameter (G), the skin friction coefficient of fluid and dust particles decrease at all positions across the channel, maximum at the left end of the channel ($y = 0$) and minimum at the right end ($y = 1$).
- As time (t) increases, the rate of heat transfer for fluid and dust particles increase at all positions across the channel, minimum at the left end of the channel ($y = 0$) and maximum at the right end ($y = 1$).
- By varying temperature relaxation time parameter (L_0) and Prandtl number (Pr), the rate of heat transfer for fluid and dust particles increase at all positions across the channel, very low at one end of the channel ($y = 0$) and gradually rise at the other end ($y = 1$).

Limiting Case: In the absence of dust parameters, the results obtained in this chapter coincide with those of Kiran Kumari [15] for MHD oscillatory flow through a vertical porous channel.

References

[1] A.A. Megahed, A.L. Aboul-Hassan, H. Sharaf El-Din, Effect of Joule and viscous dissipation on temperature distributions through electrically conducting dusty fluid, Fifth Miami International Symposium on Multi-Phase Transport and Particulate Phenomena, Miami, FL, USA, vol. 3, 1988, pp. 111.

[2] C.-H. Chen, Combined heat and mass transfer in MHD free convection from a vertical surface with Ohmic heating and viscous dissipation, International Journal of Engineering Science, vol. 42, no. 7, pp. 699713, 2004.

[3] E. M. Abo-Eldahab and M. A. El Aziz, Viscous dissipation and Joule heating effects on MHD-free convection from a vertical plate with power-law variation in surface temperature in the presence of Hall and ion-slip currents, Applied Mathematical Modelling, vol. 29, no. 6, pp. 579595, 2005.

[4] Mansour, M. A., ElAnssary, N. F., and Aly, A. M., 2008 Effects of chemical reaction and viscous dissipation on MHD natural convection flows saturated in porous media, Int. J. Appl. Math. Mech., 4, 6076.

[5] Prasad, V. R., and Reddy, N. B., 2008, Radiation and mass transfer effects on an unsteady MHD free convection flow past a semi-infinite vertical permeable moving plate with viscous dissipation, Indian J. of Pure Appl. Phys., 46, pp. 8192.

[6] Sekhar, D. V., and Reddy, G. V., 2012, Effects of chemical reaction on MHD free convective oscillatory flow past a porous plate with viscous dissipation and heat sink, Advances in Appl. Science Research, 3 (5), pp. 32063215.

[7] P. G. Saffman, On the stability of laminar flow of a dusty gas, *J. Fluid Mech.*, **Vol. 13**, (1962), pp. 120129.

[8] A. C. Cogley, S. E. Gilles and W. G. Vincent, Differential approximation for radiative transfer in a nongrey gas near equilibrium, *American Institute of Aeronautics and Astronautics*, **Vol. 6**, (1968), pp. 551 - 553.

[9] I. J. Uwanta, Effects of chemical reaction and radiation on heat and mass transfer past a semi-infinite vertical porous plate with constant mass flux and dissipation, European Journal of Scientific Research, vol. 87, no. 2, pp. 190200, 2012.

[10] Salawu, S.O., and Dada, M.S., 2016, Radiative heat transfer of variable viscosity and thermal conductivity effects on inclined magnetic field with dissipation in a non-Darcy medium, J. Nigerian Mathe. Society, 35, pp. 93106.

[11] C. Israel-Cookey, A. Ogulu, V. B. Omubo-Pepple, Influence of viscous dissipation and radiation on unsteady MHD free-convection flow past an infinite heated vertical plate in a porous medium with time-dependent suction, Int. J. of Heat and Mass Transfer, 46, (2003), 2305-2311.

[12] Paresh Vyas, Nupur Srivastava, Oscillatory Flow in a Vertical Channel Filled with Porous Medium with Radiation and Dissipation, Walailak J. Sci. & Tech., 2013, 10, 5, 2013.

[13] P. Vyas, N. Srtivasta, On dissipative radiative MHD boundary layer flow in a porous medium over a non-isothermal stretching sheet, J. Applied Fluid Mech., 2012, 5, 23-31.

[14] D. A. Neild, Resolution of a paradox involving viscous dissipation and non-linear drag in a porous medium, Transport Media, 2000, 41, 349-357.

[15] Kiran Kumaria, Mamta Goyal, Viscous Dissipation and Mass Transfer Effects on MHD Oscillatory Flow in a Vertical Channel with Porous Medium, Advances in Dynamical Systems and Applications, Volume 12, Number 2, (2017) pp. 205-216.

A Comparative Study of Hybrid Evolutionary Based Algorithms with Machine Learning Classifiers for the Prediction of Medical Database

S Poongothai[1,*], C Dharuman[2] and P Venkatesan[3]

[1]Asst. Professor, Department of Mathematics, SRM IST, Ramapuram Campus, Chennai, India.
[2] Professor, Department of Mathematics, SRM IST, Ramapuram Campus, Chennai, India.
[3]Faculty of Research, Sri Ramachandra University, Chennai, India.

*Corresponding author: poongothaikannan25@gmail.

Abstract. In the current world, day-by-day, huge amount of data being generated at a rate far higher than by which it can be examined by human comprehension alone, in this case, data mining became an important task for extracting useful information from large amount of data. Especially in these cases, the standard techniques of datamining plays in a better way such as Evolutionary approaches performs more efficient as well as more accurate. The objective of this paper is to compare Fuzzy based Evolutionary Algorithm with other machine learning classifiers to improve the classification of medical dataset of heart disease.

1. Introduction

In today medical world, Artificial Intelligence played a vital role in diagnosing the disease [1-3]. One of the major challenging diseases to the human society is heart problems. In this paper dataset of Heart disease is taken from UCI machine learning repository for classification. Genetic algorithm is used to find the relevant variables and combined with fuzzy to find the optimal solution [4] and also combined with ensemble methods like Bagging, Adaboost and Stacking. The performance of all the methods are compared and discussed.

1.1. Database

The dataset of heart disease is taken from UCI machine learning repository for the proposed model [4]. The dataset contains 294 instances with 14 variables including class variable. The variables are age, sex, chestpain, trestbps, chol, fbs, restecg, thalach, exang, oldpeak, slope, ca, thal and num.

2. Preliminaries

2.1. Fuzzy Logic

Despite of many forefathers in the mathematical field, Lofti A. Zadeh, a professor at University of California, fixed his place in the field of uncertainty or vagueness concepts. He proposed a theory of fuzzy sets and logic namely 'Fuzzy Logic' in the year 1965. He defined the fuzzy set as a set in which each element have membership values that lies between 0 and 1.

The first and foremost step of fuzzy logic is defining the linguistic variables. Next step is designing the membership function based on the problem. The other main components of Fuzzy logic controller are fuzzification system, inference system and defuzzification system [5,6]. The process of converting a crisp set into a fuzzy set i.e., changing crisp input values to linguistic variables is called Fuzzification. Support Fuzzification Method (SFM) and Grade Fuzzification Method (GFM) are the two methods attempted for fuzzification. In SFM method, using the following relation, the set can be fuzzified.

$$\widetilde{A} = \mu_1 S(y_1) + \mu_2 S(y_2) + \cdots + \mu_n S(y_n) \quad (1)$$

Out of many fuzzy classifier systems, most famous classifier systems are Fuzzy Inference System, Fuzzy C means, and Fuzzy Grid Partition System. The fuzzy inference system (FIS) follows the technique as "If antecedent then consequent". Fuzzy C means (FCM) is a data clustering technique in which a database is divided into n clusters such that every point in the database belongs to every cluster to some degree. In the simple fuzzy partition grid methods, each variable can be divided by various linguistic values [7]. The extraction of features from the set of attributes obtained by Genetic Algorithms acts as an input to the fuzzy system and simple fuzzy grid system yields the excellent output. Also Fuzzy Unordered Rule Induction Algorithm (FURIA) is used for classification. FURIA method is expertise in differentiate and separating every class from other ones, so no alternative rule is applied and the classes order are not relevant [8]. This algorithm is developed from RIPPER Algorithm which applies pruning technology while creating the replacement and the revision rule.

2.2. Ensembled systems

Ensembled systems is a combination of various machine learning algorithms to get good optimal solution i.e., ensemble learning is a machine learning technique in which many models are considered for training to solve one problem and combined to obtain a good output. The main idea behind the ensemble methods are when the weak models are correctly identified and then combined gives more better results. Many experimental studies proves that by combining the results of various classifiers gives accurate results and minimize the errors [9-12]. Most commonly used ensemble algorithms are Bagging, Adaboost and Stacking.

2.2.1.Bagging. The process of making many subsets of data from a sample (used for training) randomly chosen with replacement is termed as Bagging. It is also called **B**ootsrap **agg**regat**ing** which was developed by Leo Breiman in 1994 [13].

2.2.2.AdaBoost. It is the shortest form of Adaptive Boosting and was proposed by Freund and Schapire in the year 1996. The process of boosting the weaker classifiers to strong classifiers is termed as Adaboost[14].

2.2.3.Stacking. The concept of stacking was developed by Leo Breiman in the year 1996 [15]. The process of finding new algorithm by combining the prediction of several other learners is termed as stacking.

Journal of Physics: Conference Series **1377** (2019) 012026 doi:10.1088/1742-6596/1377/1/012026

2.3. Evolutionary Algorithms

I.Rechenberg introduced the concept of evolutionary computation in his paper "*Evolutionary strategies*" in the year 1960. Based on Darwinian model, EAs are developed and it is a class of stochastic optimization algorithm. The basic concept of EAs are survival of the fittest. Selection, Crossover and Mutation are the major operators of EAs. Genetic Algorithms one of the branch of Evolutionary Algorithms was introduced and developed by Holland in 1975 [16] inspired by Darwin's theory of evolution which states that the survival of an organism is affected by rule "the strongest species that survives". In Selection process, The individuals of the population should be selected at random in initial stage. Roulette Wheel Process is one of the common selection operator in which each member has ability to survive for the next generation. New offsprings are obtained by crossover method by crossing a couple of individuals from the current individuals. Mutation is a unary operator i.e., it will change individuals randomly in the current population. Then the original population are replaced by the newly created offsprings (variables). This replacement always gives the best set of variables deleting the worst ones. Every generation is better than its previous one so that the possibility of getting the optimal solution is better. The entire process should be repeated till the desired output or stopping criteria met.

3. Methods and Results

In this paper, by using GA, the variables in the dataset of heart disease are reduced to 6 namely Sex, Chestpain, FBS, Exang, Oldpeak and Slope. These variables are considered as relevant information to classify the database. Then the reduced variables are applied to fuzzy systems and ensembled methods such as Bagging, Adaboost and Stacking for classification. Suppose that fuzzy rules $r_1^j, \ldots r_n^j$ have been trained to learn for class λ_j. For a new case y, the support of this class is described by

$$s_j(y) = \sum_{i=1}^{n} \mu_{r_i^{(j)}}(y)\, cf\left(r_i^{(j)}\right) \quad (2)$$

where $cf\left(r_i^{(j)}\right)$ is the *certainty factor* of the rule $r_i^{(j)}$. It is defined as follows

$$cf\left(r_i^{(j)}\right) = \frac{\frac{\left|D_T^{(j)}\right|}{|D_T|} + \frac{\sum_{y\in D_T^{(j)}} \mu_{r_i^{(j)}}(y)}{2}}{1 + \frac{\sum_{y\in D_T} \mu_{r_i^{(j)}}(y)}{2}} \quad (3)$$

where $D_T^{(j)}$ denotes the subset of training instances with label λ_j. According to these equatioins the rules accepts to model more flexible boundaries which improves the classification rate.

In bagging, each collection of data is applied to get their respective decision trees. By this method, the resultant obtained is ensemble of various models. Finding the average of all the output from various decision trees gives the desired result of the problem taken. Consider the dataset ***D*** of size *r*, then create *n* new subsets of data $\boldsymbol{D_i}$ of size r_i'. Few observations may get repeated in each $\boldsymbol{D_i}$. If r_i'=*r*, then the expected value of the set is given by $1-\frac{1}{e}$ being distinct subsets of ***D*** and the remaining sets are duplicates [14]. These types of datasets are bootstrap sample. Afterwards, that *n* dataset models are fitted by the *n* samples of bootstrap. Then the results are combined by finding the mean of all the output.

Journal of Physics: Conference Series **1377** (2019) 012026 doi:10.1088/1742-6596/1377/1/012026

In Adaboost, consider the dataset $\{(x_1,y_1),\ldots,(x_r,y_r)\}$ where each x_j has an associated class $y_j\epsilon\{-1,1\}$. Consider the set of weak classifiers $\{w_1,w_2,\ldots,w_n\}$ which gives the output $w_i(x_j)\epsilon\{-1,1\}$ for each object. After the (m-1)th iteration, boosted classifier will be a linear combination of weak classifiers which is of the form

$$C_{(m-1)}(x_i) = \gamma_1 w_1(x_i) + \cdots + \gamma_{m-1} w_{m-1}(x_i) \quad (4)$$

At mth iteration, by extending this to a better classifier by adding another weak classifier w_m With weight γ_m,

$$C_m(x_i) = C_{(m-1)}(x_i) + \gamma_m w_m(x_i) \quad (5)$$

The process continues to find which weak classifier is the good choice for w_m with weight γ_m. Total error TE can be defined as summation of its exponential loss on each object.

$$TE = \sum_{i=1}^{N} e^{-y_i C_m(x_i)}$$

$$= \sum_{i=1}^{N} e^{-y_i C_{m-1}(x_i)}\, e^{-y_i\gamma_m w_m(x_i)} \quad (6)$$

By considering $K_i(m) = \left\{\begin{matrix} 1, if\ m=1 \\ e^{-y_i C_{m-1}(x_i)},\ if\ m>1 \end{matrix}\right\}$. So we have

$$TE = \sum_{i=1}^{N} K_i(m)\, e^{-y_i\gamma_m w_m(x_i)} \quad (7)$$

Splitting the summation between correctly classified and misclassified object considering by $y_i w_m(x_i) = 1\ and - 1$ respectively.

$$TE = \sum_{y_i=w_m(x_i)} K_i(m)\, e^{-\alpha_m} + \sum_{y_i\neq w_m(x_i)} K_i(m)\, e^{\alpha_m} \quad (8)$$

To find the desired weight α_m, which minimizes the total error by differentiating the total error w.r.t. α_m, we get

$$\frac{d(TE)}{d\alpha_m} = \frac{d\left(\sum_{y_i=w_m(x_i)} K_i(m)e^{-\alpha_m} + \sum_{y_i\neq w_m(x_i)} K_i(m)e^{\alpha_m}\right)}{d\alpha_m} \quad (9)$$

Equating the above equation to be zero and solving, we get

$$\alpha_m = \frac{1}{2}log\left(\frac{\sum_{y_i=w_m(x_i)} K_i(m)}{\sum_{y_i\neq w_m(x_i)} K_i(m)}\right) \quad (10)$$

In stacking, the step by step procedure is given by

Step 1: The training set is separated into two disjoint sets.
Step 2: Train the set (base level) present in the first set.
Step 3: Using the training, test the remaining set (base level) present in the second set.
Step 4: Using the results obtained from step 3 as inputs and the perfect responses as the outputs, training should be updated on higher level learner.

Table 1 shows the classification rate of heart disease database with and without reduction of variables. Also the results are compared. Figure 1 shows the performance of classification rate of dataset before and after reduction of variables. It is obviously seen that hybrid techniques performs better than the normal techniques. Also GA with Fuzzy is good in classification compared to other techniques.

Table 1. Comparison of hybrid classification of Fuzzy and Ensembled methods.

Classification Techniques	Attributes used	Rate of accuracy (%)
Adaboost	14	77.89
Bagging	14	78.57
Stacking	14	63.94
Fuzzy	14	78.93
GA-Adaboost	6	81.97
GA-Bagging	6	80.03
GA-Stacking	6	64.93
GA-Fuzzy	6	82.61

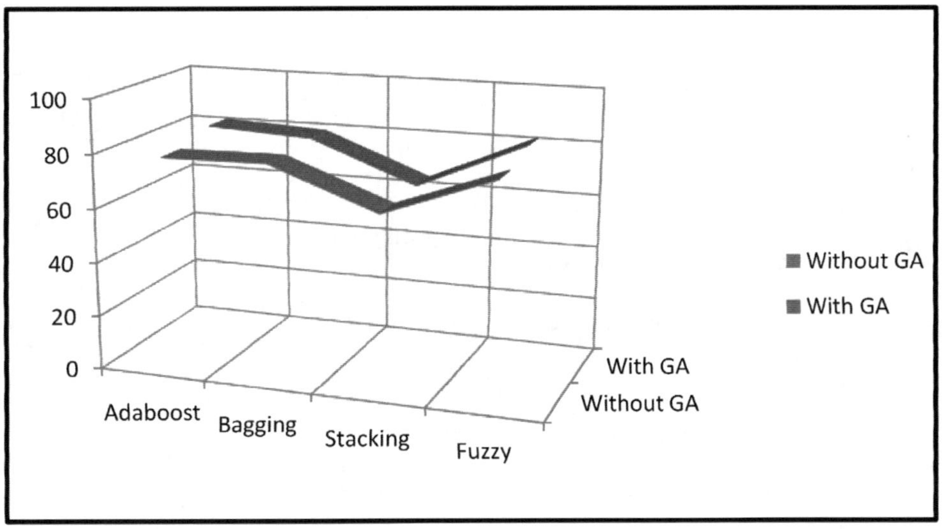

Figure 1. Performance of classification rate of dataset with and without GA.

4. Conclusion

By the proposed method the hybrid based models performs better in classifying the data even if it is noisy. GA used to reduce the variables then the classification is done by fuzzy and ensemble methods with and without GA. By reducing the variables it is cost effective and time effective to solve the particular case. The results are compared and discussed briefly which shows hybrid techniques are good in classifying the database. Also among all hybrid techniques GA with fuzzy is good. In future, some more social relevant problems are going to be discussed by applying fuzzy relation based algorithms with advanced version of evolutionary computations having hope that it would perform better in classification.

References

[1] Srimani P K and Koti M S 2014 Knowledge discovery in medical data by using rough set rule induction algorithms *Indian Journal of Science and Technology* vol 1 pp 905–15

[2] Overiu M and Simon D 2010 Biogeography-based optimization of neuro-fuzzy system parameters for diagnosis of Cardiac disease *Proceedings of the 12th annual conference on Genetic and evolutionary computation* pp 1235–42

[3] Rajeswari K, Vaithiyanathan V and Amirtharaj P 2011 Prediction of risk score for heart disease in India using Machine intelligence *International Conference on Information and Network Technology* pp19–22

[4] UCI : Machine Learning Repository, http://archive.ics.uci.edu/ml/.

[5] Bezdek J C, Keller J, Krisnapuram R and Pal N R 2005 Fuzzy Models and Algorithms for Pattern Recognition and Image Processing (The Handbooks of Fuzzy Sets). Secaucus, NJ, USA: Springer- Verlag NewYork, Inc.

[6] Verikas A, Guzaitis J, Gelzinis A and Bacauskiene M 2010 A general framework for designing a fuzzy rule-based classifier *Knowledge and Information Systems* pp 1-19

[7] Poongothai S, Dharuman C and Venkatesan P 2016 Fuzzy Evolutionary Computing in Biological Data Mining *Global Journal of Pure and Applied Mathematics* vol 12 pp 306-309

[8] del Jesus M, Hoffmann F, Navascues L and Sánchez L 2004 Induction of fuzzy rule-based classifiers with evolutionary boosting algorithms *IEEE Transactions on Fuzzy Systems*, vol 12 p 96

[9] Bauer E and Kohavi R 1999 An Empirical Comparison of Voting Classification Algorithms: Bagging, Boosting, and Variants *Machine Learning* vol 35 pp 1-38

[10] Domingos P 1996 Using Partitioning to Speed Up Specific-to-General Rule Induction. In Proceedings of the AAAI-96 Workshop on Integrating Multiple Learned Models pp 29-34

[11] Opitz D and Maclin R 1999. Popular Ensemble Methods: An Empirical Study *Journal of Artificial Research* vol 11 pp 169-198

[12] Quinlan J R 1996 Bagging , Boosting, and C4.5 *In Proceedings of the Thirteenth National Conference on Artificial Intelligence* pp 725-730

[13] Breiman, Leo 1994 Bagging Predictors (PDF) Department of Statistics University of California Berkeley. Technical Report No. 421. Retrieved 2019-07-28

[14] Aslam, Javed A, Popa, Raluca A and Rivest R L 2007 On Estimating the Size and Confidence of a Statistical Audit

[15] Breiman, Leo 1996 Stacked Regressions *Machine Learning* vol 24 pp 49–64

[16] Holland J 1975 Adaptation in Natural and Artificial Systems The University of Michigan Press Ann Arbor.

MEAN SQUARE CORDIAL LABELING ON STAR RELATED GRAPHS

S Dhanalakshmi[1], N. Parvathi[2]

[1]Department of Mathematics, Faculty of Engineering and Technology, SRM IST, Chennai -600089, India
[2]Department of Mathematics, Faculty of Engineering and Technology SRM IST, Chennai -603203, India
*[3]Corresponding author, Email: parvathn@srmist.edu.in

Abstract. A Mean Square Cordial Labeling (MSCL) of a Graph G(V,E) with p vertices and q edges is a bijection from V to {0, 1} such that each edge uv is assigned the label $(\lceil (f(u)^2+f(v)^2)/2 \rceil)$ where $\lceil x \rceil$ (ceil(x)) is the least integer greater than or equal to x with the condition that the number of vertices labeled with 0 and the number of vertices labeled with 1 differ by at most 1 and the number of edges labeled with 0 and the number of edges labeled 0 and the number of edges labeled with 1 differ by at most 1 . In this paper it is analyzed that some star related graphs like Bistar graph B(k,k), Subdivision of a star graph $S(S_k)$, Coconut tree CT(k,k), Merge graph C_k*S_k,, Graph $P_m\Theta\, K_{1,n}$ and Banana tree BT(k , k) admits mean square cordial labeling.

I. INTRODUCTION

Graph theory plays an important role for automatic graph generation in computer science technology applications such as database design, software engineering, circuit designing, networks and data mining. Due to the involvement of the researchers for the past 60 years, over 200 graphs labeling techniques [1] have been discussed in thousands of research papers. For basics terms and notations we follow Harary [2]. Cordial labeling was introduced by Cahit [3] and Ponraj et al[4] were initiated the mean cordial labeling of a graph .Mean square cordial labeling introduced by A.Nellai murugan et al and they have discussed it for some special graphs[5].Moreover they have discussed the mean square cordial labeling for some tree and cycle related graphs[6,7]. Dhanalakshmi et al have discussed mean square cordial labeling related to some cyclic and acyclic graphs and its rough approximations [8,9].In this paper it is analyzed that some star related graphs like Bistar graph B(k,k), Subdivision of a star graph $S(S_k)$, Coconut tree CT(k,k), Merge graph C_k*S_k,, Graph $P_m\Theta\, K_{1,n}$ and Banana tree BT(k, k) admits mean square cordial labeling.

II. PRELIMINARIES

Definition 1: Let G = (V,E) be a graph with p vertices and q edges. "A Mean Square Cordial labeling of a Graph G(V,E) with p vertices and q edges is a bijection from V to {0, 1} such that each edge uv is assigned the label $(\lceil (f(u)^2+f(v)^2)/2 \rceil)$ where $\lceil x \rceil$ (ceil(x)) is the least integer greater than or equal to x with the condition that the number of vertices labeled with 0 and the number of vertices labeled with 1 differ by at most 1 and the number of edges labeled with 0 and the number of edges labeled 0 and the number of edges labeled with 1 differ by at most 1 .

Definition 2: Bistar graph $B_{n,n}$ is obtained by joining the apex(centre) vertices of two $K_{1,n}$ by an edge.

Definition 3: Banana tree as defined is graph obtained by connecting one leaf of each of n copies of an K-star graph with single root vertex that is distinct from all the stars.

Definition 4: The graph $P_m\Theta\, K_{1,n}$ is obtained by replacing each vertex of a path graph by a central vertex of a star graph

Definition 5: Merge graph $C_n*K_{1,n}$ is obtained by attaching a apex vertex of a star graph $K_{1,n}$ with any i^{th} vertex of a cycle C_n

Definition 6: Let G be a graph. The subdivision graph S(G) is obtained from G by subdividing each edge of G with a vertex.

Definition 7: Coconut tree CT(m,n) is the graph obtained from the path P_m by adding a star graph S_n at the end vertex of a path P_m

III. MAIN RESULTS

Theorem 1: Bistar graph B(k, k) admits mean square cordial labeling , $\forall k \geq 2$
Proof: Let $V(B(k, k)) = \{(u_0,v_0,u_i, v_i)$: i varies from 1 to k$\}$ and
$E(B(k, k)) =\{[(u_0u_i)$: i varies from 1 to k$] \cup [(v_0v_i)$: i varies from 1 to k$] \cup [(u_0v_0)]\}$ where u_0 and v_0 are the apex vertices and u_i and v_i be the pendent vertices corresponding to the apex vertices u_0 and v_0 respectively.
Here $|V| = 2k+2 \, and \, |E| = 2k+1$
Define f maps from V(G) to { 0,1}
$f(u_0) = 0$
$f(u_i) = 0$, i varies from 1 to k
$f(v_0) = 1$, i varies from 1 to k
$f(v_i) = 1$, i varies from 1 to k
Hence the labeling of edges are
$f(u_0 u_i) = 0$, i varies from 1 to k
$f(v_0v_i) = 1$, i varies from 1 to k
$f(u_0 v_0) =1$, i varies from 1 to k
Table: 1 vertex and edge cardinality for MSCL of a Bisstar graph B(k,k)

T	0	1
$\lvert v_f(t)\rvert$	$k+1$	$k+1$
$\lvert e_f(t)\rvert$	k	$k+1$

It is very clear that above labeling pattern satisfies the condition of MSCL. Hence the theorem is proved.
Illustration 1*:* MSCL of a Bistar graph B(4,4) is shown in figure 4.2.2.

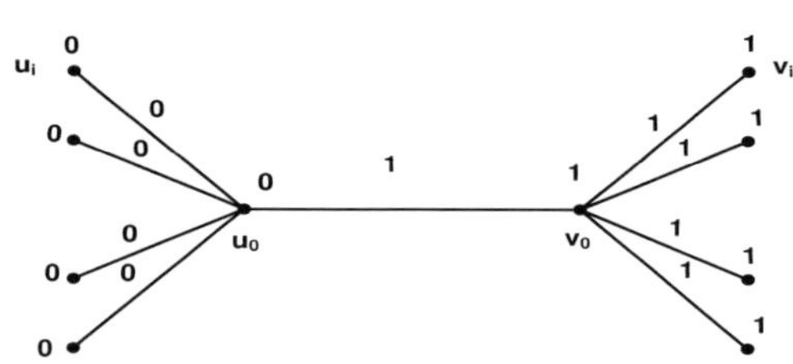

Figure(1)

Journal of Physics: Conference Series **1377** (2019) 012027 doi:10.1088/1742-6596/1377/1/012027

Theorem 2: Subdivision of a star graph $S(S_k)$ admits mean square cordial labeling , $\forall k \geq 2$

Proof: Let $V[S(S_k)]= \{(u_0,u_i,v_i)$: i varies from 1 to k$\}$ and
$E[S(S_k)] =\{[(u_0u_i)$: i varies from 1 to k$] \cup [(u_iv_i)$: i varies from 1 to k$] \}$ where u_0, u_i and v_i are the apex vertices, v_i be the pendent vertices and u_i be the subdivision vertex of u_0v_i respectively.
Here $|V| = 2k+1 \, and \, |E| = 2k$
Define f maps from V(G) to { 0,1}
$f(u_0) = 0$
$f(u_i) = 0$, i varies from 1 to k
$f(v_i) = 1$, i varies from 1 to k
Hence the labeling of edges are
$f(u_0\, u_i) = 0$, i varies from 1 to k
$f(v_0v_i) = 1$, i varies from 1 to k
Table: 2 vertex and edge cardinality for MSCL of a Subdivision of a star graph ,$S(S_k)$

T	0	1
$\|v_f(t)\|$	$k+1$	k
$\|e_f(t)\|$	k	k

It is very clear that above labeling pattern satisfies the condition of MSCL. Hence the theorem is proved.

Illustration 2: MSCL of a Subdivision of a star graph $S(S_5)$ is shown in figure 4.2.3

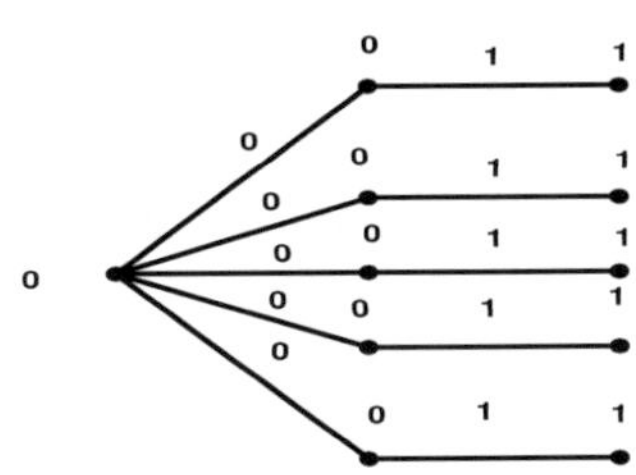

Figure(2)

Theorem 3: Coconut tree CT (k,k) admits mean square cordial labeling , $\forall k \geq 2$

Proof: Let $V[CT(k,k)]= \{(u_i,v_i)$: i varies from 1 to k$\}$ and
$E[CT(k,k)] =\{[(u_iu_{i+1})$: i varies from 1 to k$] \cup [(u_kv_i)$: i varies from 1 to k$] \}$ where u_i and v_i are the vertices on a path and the pendent vertices of a star graph respectively.
Here $|V| = 2k \, and \, |E| = 2k-1$
Define f maps from V(G) to { 0,1}
$f(u_i) = 0$, i varies from 1 to k
$f(v_i) = 1$, i varies from 1 to k
Hence the labeling of edges are
$f(u_i\, u_{i+1}) = 0$, i varies from 1 to k-1
$f(u_kv_i) = 1$, i varies from 1 to k

Table: 3 vertex and edge cardinality for MSCL of a coconut tree CT(k,k)

T	0	1
$\|v_f(t)\|$	k	k
$\|e_f(t)\|$	$k-1$	k

It is very clear that above labeling pattern satisfies the condition of MSCL. Hence the theorem is proved.

Illustration 3: MSCL of a Coconut tree CT(4,4) is shown in figure 4.2.4

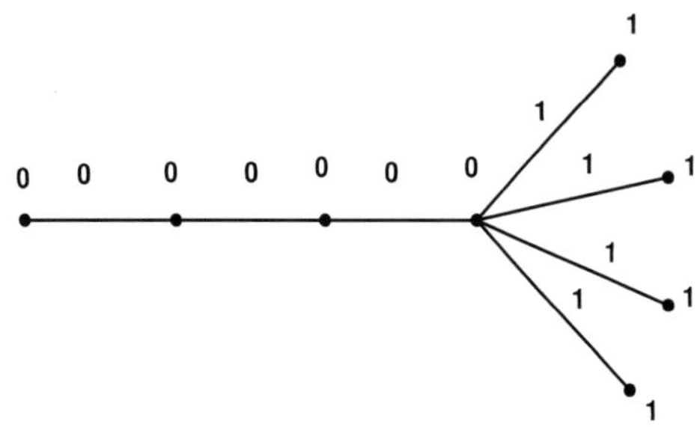

Figure(3)

Theorem 4: The merge graph C_k*S_k admits mean square cordial labeling , $\forall k \geq 3$

*Proof:*Let $V(C_k* S_k) = \{ u_i,v_i$:i varies from 1 to k } and E(G) ={ [$(u_iu_{i+1}$: i varies from 1 to k-1] ∪ [u_1u_k] ∪ [(u_kv_i): i varies from 1 to k] } where u_i and v_i are the vertices on a cycle and pendent vertices of a star graph respectively.

Here $|V| = 2k$ *and* $|E| = 2k$

Here f maps $V(C_k\Theta S_k)$ to {0,1} .
$f(u_i) = 0$,i varies from 1 to k
$f(v_i) = 1$, i varies from 1 to k
Hence the labeling of edges are

$f((u_iu_{i+1}) =0$,i varies from 1 to k -1
$f(u_1u_k) = 0$
$f(u_kv_i) = 1$,i varies from 1 to k

Table: 4 vertex and edge cardinality for MSCL of a merge graph C_k*S_k

t	0	1
$\|v_f(t)\|$	k	k
$\|e_f(t)\|$	k	k

Journal of Physics: Conference Series **1377** (2019) 012027 doi:10.1088/1742-6596/1377/1/012027

It is very clear that the above labeling pattern satisfies the condition of MSCL. Hence the theorem is proved.

Illustration 4*:* MSCL of a merge graph C_6*S_6 is shown in figure4.2.5

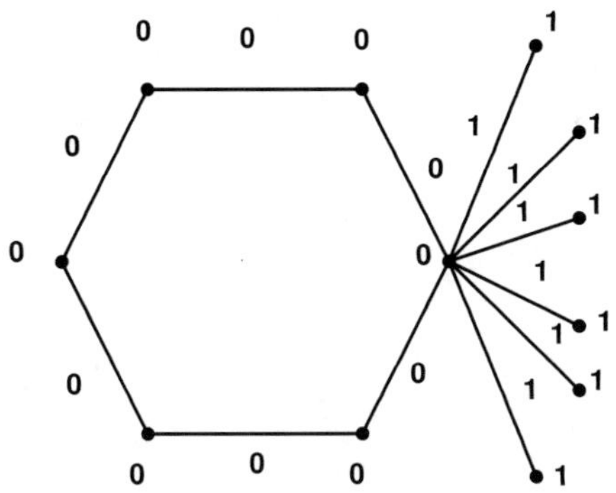

Figure(4)

Theorem 5: The graph $P_m \Theta K_{1,n}$ admits mean square cordial labeling , $\forall m \geq 1, n \geq 2$

*Proof:*Let $V(P_m \Theta K_{1,n})$ = { v_i u_j^i ,i varies from 1 to m; ,j varies from 1 to n } and E(G) ={ [($v_i v_{i+1}$: i varies from 1 to m-1] $\cup$ [($v_i u_j$): $v_i u_j^i$ i varies from 1 to m, j varies from 1 to n] } where v_i and u_j^i are the vertices of path and pendent vertices of a star graph attached to the vertices of the path graph respectively.

Here $|V| = m + nm$ *and* $|E| = m + nm - 1$

Here f maps $V(P_m \Theta K_{1,n})$ to {0,1} .

Case (i) : m and n are even

$$f(v_i) = 0, i \text{ varies from } 1 \text{ to } \frac{m}{2}$$

$$1, i \text{ varies from } \frac{m+2}{2} \text{ to } m$$

$$f(u_j^i) = 0, i \text{ varies from } 1 \text{ to } \frac{m}{2}, j \text{ varies from } 1 \text{ to } n$$

$$1, i \text{ varies from } \frac{m}{2}+1 \text{ to } m, j \text{ varies from } 1 \text{ to } n$$

Hence the labeling of edges are

$$f(v_i v_{i+1}) = 0, i \text{ varies from } 1 \text{ to } \frac{m}{2}-1$$

$$1, i \text{ varies from } \frac{m}{2} \text{ to } m-1$$

$f(v_i u_j^i) = 0, i \text{ varies from } 1 \text{ to } \frac{m}{2}, j \text{ varies from } 1 \text{ to } n$

$1, i \text{ varies from } \frac{m}{2}+1 \text{ to } m, j \text{ varies from } 1 \text{ to } n$

Table: 5.1 vertex and edge cardinality for MSCL of a $P_m \Theta K_{1,n}$,m and n are even

T	0	1
$\lvert v_f(t) \rvert$	$\frac{m+nm}{2}$	$\frac{m+nm}{2}$
$\lvert e_f(t) \rvert$	$\frac{m+nm}{2}-1$	$\frac{m+nm}{2}$

Case (ii) : m is odd and n is even

$f(v_i) = 0, i \text{ varies from } 1 \text{ to } \frac{m+1}{2}$

$1, i \text{ varies from } \frac{m+3}{2} \text{ to } m$

$Subcase\,(i): If\ i = \frac{m+1}{2}$

$f(u_j^i) = 0, j \text{ varies from } 1 \text{ to } n/2$

$1, j \text{ varies from } (n/2)+1 \text{ to } n$

$Subcase\,(ii): If\ i \neq \frac{m+1}{2}$

$f(u_j^i) = 0, i \text{ varies from } 1 \text{ to } \frac{m-1}{2}, j \text{ varies from } 1 \text{ to } n$

$1, i \text{ varies from } \frac{m+3}{2} \text{ to } m, j \text{ varies from } 1 \text{ to } n$

Hence the labeling of edges are

$f(v_i v_{i+1}) = 0, i \text{ varies from } 1 \text{ to } \frac{m-1}{2}$

$1, i \text{ varies from } \frac{m+1}{2} \text{ to } m-1$

$Subcase\,(i): If\ i = \frac{m+1}{2}$

$f(v_i u_j^i) = 0, j \text{ varies from } 1 \text{ to } n/2$

$\quad 1, j \text{ varies from } (n/2)+1 \text{ to } n$

$Subcase\,(ii): If\ i \neq \frac{m+1}{2}$

$f(v_i u_j^i) = 0, i \text{ varies from } 1 \text{ to } \frac{m-1}{2}, j \text{ varies from } 1 \text{ to } n$

$\quad 1, i \text{ varies from } \frac{m+3}{2} \text{ to } m, j \text{ varies from } 1 \text{ to } n$

$f(v_i u_j^i) = 0, i \text{ varies from } 1 \text{ to } \frac{m-1}{2}, j \text{ varies from } 1 \text{ to } n$

$\quad 1, i \text{ varies from } \frac{m+3}{2} \text{ to } m, j \text{ varies from } 1 \text{ to } n$

Table: 5.2 vertex and edge cardinality for MSCL of a $P_m \Theta K_{1,n}$,m is odd and n is even

T	0	1
$\lvert v_f(t)\rvert$	$\frac{m+nm+1}{2}$	$\frac{m+nm-1}{2}$
$\lvert e_f(t)\rvert$	$\frac{m+nm-1}{2}$	$\frac{m+nm-1}{2}$

Case (iii) : m is odd and n is odd

$f(v_i) = 0, i \text{ varies from } 1 \text{ to } \frac{m+1}{2}$

$\quad 1, i \text{ varies from } \frac{m+3}{2} \text{ to } m$

$Subcase\,(i): If\ i = \frac{m+1}{2}$

$f(u_j^i) = 0, j \text{ varies from } 1 \text{ to } (n-1)/2$

$\quad 1, j \text{ varies from } (n+1)/2 \text{ to } n$

$Subcase\,(ii): If\ i \neq \frac{m+1}{2}$

$f(u_j^i) = 0, i \text{ varies from } 1 \text{ to } \frac{m-1}{2}, j \text{ varies from } 1 \text{ to } n$

$\quad 1, i \text{ varies from } \frac{m+3}{2} \text{ to } m, j \text{ varies from } 1 \text{ to } n$

Hence the labeling of edges are

$$f(v_iv_{i+1}) = 0, i \text{ varies from } 1 \text{ to } \frac{m-1}{2}$$

$$1, i \text{ varies from } \frac{m+1}{2} \text{ to } m-1$$

$$Subcase\,(i): If\ i = \frac{m+1}{2}$$

$$f(v_iu_j^i) = 0, j \text{ varies from } 1 \text{ to } (n-1)/2$$

$$1, j \text{ varies from } (n+3)/2 \text{ to } n$$

$$Subcase\,(ii): If\ i \neq \frac{m+1}{2}$$

$$f(v_iu_j^i) = 0, i \text{ varies from } 1 \text{ to } \frac{m-1}{2}, j \text{ varies from } 1 \text{ to } n$$

$$1, i \text{ varies from } \frac{m+3}{2} \text{ to } m, j \text{ varies from } 1 \text{ to } n$$

Table: 5.3vertex and edge cardinality for MSCL of a $P_m\Theta\ K_{1,n}$,m is odd and n is odd

T	0	1
$\lvert v_f(t)\rvert$	$\frac{m+nm}{2}$	$\frac{m+nm}{2}$
$\lvert e_f(t)\rvert$	$\frac{m+nm}{2}-1$	$\frac{m+nm}{2}$

Case (iv) : m is even and n is odd

$$f(v_i) = 0, i \text{ varies from } 1 \text{ to } m$$

$$f(u_j^i) = 0, i \text{ varies } 1 \text{ to } m,\ j \equiv o \bmod 2$$

$$1, i \text{ varies } 1 \text{ to } m,\ j \equiv 1 \bmod 2$$

Hence the labeling of edges are

$$f(v_iv_{i+1}) = 0, i \text{ varies from } 1 \text{ to } m-1$$

$$f(v_i u_j^i) = 0, i \text{ varies } 1 \text{ to } m,\ j \equiv o \bmod 2$$

$$1, i \text{ varies } 1 \text{ to } m,\ j \equiv 1 \bmod 2$$

Table: 5.4. vertex and edge cardinality for MSCL of a $P_m\Theta\ K_{1,n}$,m is even and n is odd

Journal of Physics: Conference Series **1377** (2019) 012027 doi:10.1088/1742-6596/1377/1/012027

T	0	1
$\lvert v_f(t) \rvert$	$\frac{m+nm}{2}$	$\frac{m+nm}{2}$
$\lvert e_f(t) \rvert$	$\frac{m+nm}{2}-1$	$\frac{m+nm}{2}$

It is very clear that the above labeling pattern satisfies the condition of MSCL. Hence the theorem is proved.

Illustration 5.1*:* MSCL of a graph $P_4\Theta$ $K_{1,4}$ is shown in figure 4.2.6.1

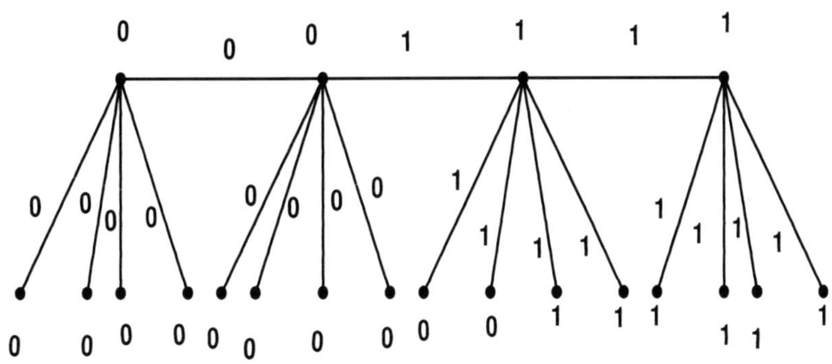

Figure(5.1)

Illustration 5.2*:* MSCL of a graph $P_3\Theta$ $K_{1,4}$ is shown in figure 4.2.6.2

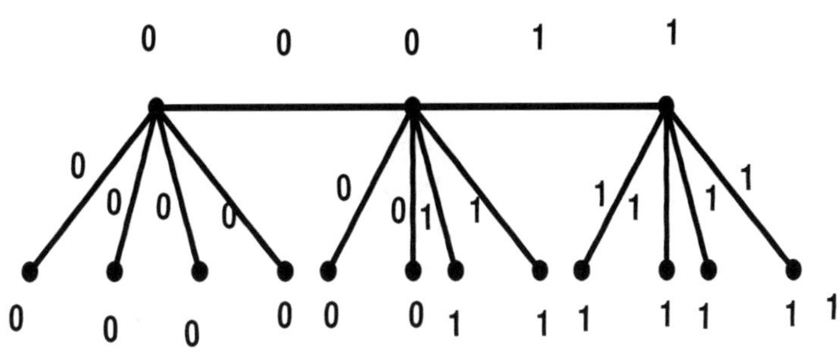

Figure(5.2)

Illustration 5.3: MSCL of a graph $P_5\Theta\ K_{1,3}$ is shown in figure *4.2.6.3*

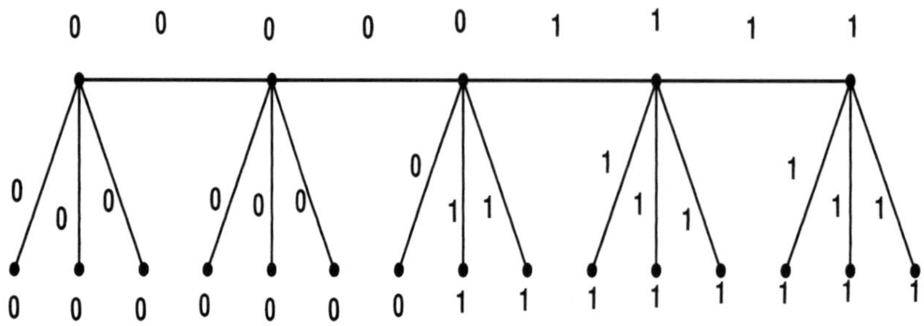

Figure(5.3)

Theorem 6: The Banana graph BT(m,n), where m=n admits mean square cordial labeling , $\forall m \geq 2, n \geq 2$

*Proof:*Let V(BT(m,n)) = { u_0, u_j, u_j^i, u_j^1 , j varies from 1 to m; i varies from 1 to n } and E(G) ={ [($u_0 u_j^1$: j varies from 1 to m] $\cup$ [(u_j u_j^i): i varies from 1 to m, j varies from 1 to n] } where u_0, u_j , u_j^1 and u_j^i are the root vertex, central vertex of a star graph, one leaf of a star graph attached to the single root vertex and the remaining pendent vertices of a star graph respectively.

Here $|V| = m + nm + 1 and \ |E| = m + nm$

Here f maps V(BT(m,n)) to {0,1} .

Case (i) : m and n are even

$f(u_0) = 0$

$f(u_j) = 0, j \text{ varies from } 1 \text{ to } \frac{m}{2}$

$\quad 1, i \text{ varies from } \frac{m+2}{2} \text{ to } m$

$f(u_j^i) = 0, i \text{ varies from } 1 \text{ to } n, j \text{ varies from } 1 \text{ to } \frac{m}{2}$

$\quad 1, i \text{ varies from } 1 \text{ to } n, j \text{ varies from } \frac{m}{2}+1 \text{ to } n$

Hence the labeling of edges are

$f(u_0 u_j^1) = 0, j \text{ varies from } 1 \text{ to } \frac{m}{2}$

$\quad 1, i = 1, j \text{ varies from } \frac{m}{2}+1 \text{ to } n$

$f(u_j u_j^i) = 0, i \text{ varies from } 1 \text{ to } n, j \text{ varies from } 1 \text{ to } \frac{m}{2}$

$1, i \text{ varies from } 1 \text{ to } n, j \text{ varies from } \frac{m}{2}+1 \text{ to } n$

Table: 6.1 vertex and edge cardinality for MSCL of a Banana tree BT(k,k) ,k is even

T	0	1
$\lvert v_f(t)\rvert$	$\frac{m+nm+2}{2}$	$\frac{m+nm}{2}$
$\lvert e_f(t)\rvert$	$\frac{m+nm}{2}$	$\frac{m+nm}{2}$

Case (ii) : m and n are odd

$f(u_0) = 0$

Subcase (i): j=m/2

$f(u_j^i) = 0, i \text{ varies from } 1 \text{ to } \frac{n-1}{2}$

$1, i \text{ varies from } \frac{n+1}{2} \text{ to } n$

$f(u_j) = 0$

Hence the labeling of edges are

$f(u_0 u_j^1) = 0$

$f(u_j u_j^i) = 0, i \text{ varies from } 1 \text{ to } \frac{n-1}{2}$

$1, i \text{ varies from } \frac{n+1}{2} \text{ to } n$

Subcase (ii): $j \neq m/2$

$f(u_0) = 0$

$f(u_j) = 0, j \text{ varies from } 1 \text{ to } \frac{m-1}{2}$

$1, j \text{ varies from } \frac{m+3}{2} \text{ to } m$

$$f(u_j^i) = 0,\ j \text{ varies from } 1 \text{ to } \frac{m-1}{2}, i \text{ varies from } 1 \text{ to } n$$

$$1,\ j \text{ varies from } \frac{m+3}{2} \text{ to } m, i \text{ varies from } 1 \text{ to } n$$

Hence the labeling of edges are

$$f(u_0 u_j^1) = 0,\ j \text{ varies from } 1 \text{ to} \frac{m-1}{2}$$

$$1,\ j \text{ varies from } \frac{m+3}{2} \text{ to } m$$

$$f(u_j u_j^i) = 0,\ j \text{ varies from } 1 \text{ to } \frac{m-1}{2}, i \text{ varies from } 1 \text{ ton}$$

$$1,\ j \text{ varies from } \frac{m+3}{2} \text{to } m, i \text{ varies from } 1 \text{ ton}$$

Table: 6.2 vertex and edge cardinality for MSCL of a Banana tree BT(k,k) ,k is odd

T	0	1
$\lvert v_f(t)\rvert$	$\frac{m+nm+2}{2}$	$\frac{m+nm}{2}$
$\lvert e_f(t)\rvert$	$\frac{m+nm}{2}$	$\frac{m+nm}{2}$

It is very clear that the above labeling pattern satisfies the condition of MSCL. Hence the theorem is proved.

Illustration 6.1: MSCL of a Banana tree BT(5,5) is shown in figure4.2.7.2

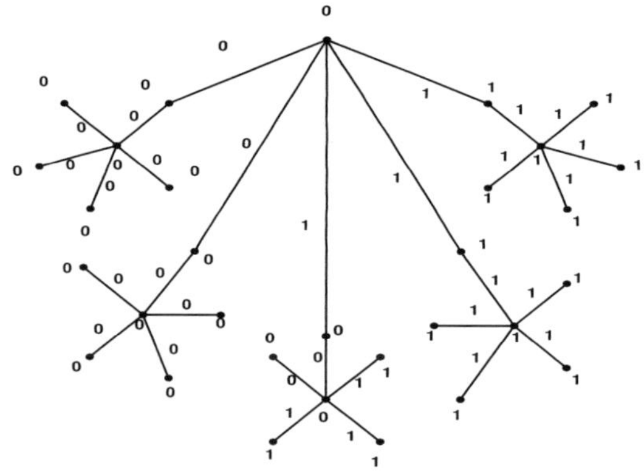

Figure(6.1)

Illustration 6.2*:* MSCL of a Banana tree BT(4,4) is shown in figure 6.2

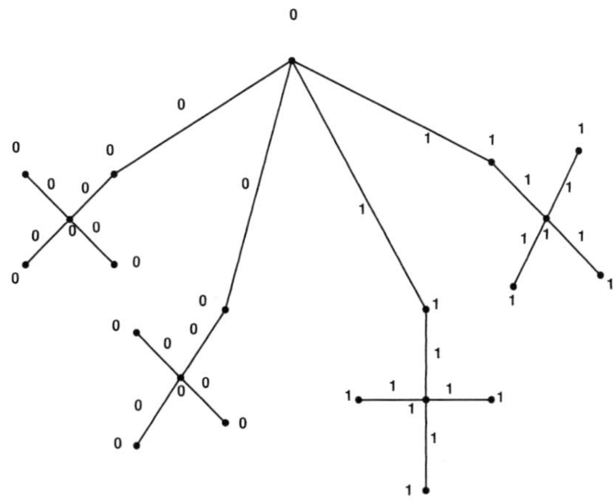

Figure(6.2)

IV. CONCLUSION

Mean square cordial labeling is widely used in social networks such as Facebook, Research gate, Orkut, Linked in etc., In this paper it is analyzed that some star related graphs like Bistar graph B(k,k), Subdivision of a star graph $S(S_k)$, Coconut tree CT(k,k), Merge graph C_k*S_k, Graph $P_m\Theta K_{1,n}$ and Banana tree BT(k, k) admits mean square cordial labeling.

V. FUTURE SCOPE

Graph operations like union, intersection, corona of two graphs etc., can also be discussed for mean square cordial labeling in future.

ACKNOWLEDGEMENTS

We offer our sincere thanks to the referee for the valuable suggestion for revision of the paper.

REFERENCES

[1]A. Gallian A Dynamic Survey of Graph Labeling Electronic Journal of Combinatorics, Vol. 18, 2011, pp. 1-219

[2] F. Harary Graph Theory Narosa Publishing House, New Delhi, 1988.

[3]I. Cahit Cordial Graphs: A Weaker Version of Graceful and Harmonious Graphs Ars Combinatoria, Vol. 23, No. 3, 1987, pp. 201-207.

[4]Ponraj, M.Sivakumar and M.Sundaram, Mean cordial labeling of graphs ,Open journal of Discrete Mathematics, **2**(4)(2012), 145-14

[5]A.Nellai Murugan, S.Heerajohn Special Class of Mean Square Cordial Graphs, International Journal of Applied Research 2015; 1(11): 128-131

[6]A.Nellai Murugan, S.Heerajohn Tree Related Mean Square Cordial Graphs ,outreach IX 2016 126-131 ,A multidisciplinary refereed journal

[7]A.Nellai Murugan, S.Heerajohn Cycle related of Mean Square Cordial Graphs, International Journal of research and development organization.

[8]S.Dhanalakshmi and N.Parvathi Mean square cordial labelling related to some cyclic graphs , International Journal of Engineering and Technology.

[9]S.Dhanalakshmi and N.Parvathi Mean square cordial labelling related to some acyclic graphs and its rough approximations , National Conference on Mathematical Techniques and its Applications (NCMTA 18) IOP

Publishing IOP Conf. Series: Journal of Physics: Conf. Series 1000 (2018) 012040 doi :10.1088/1742-6596/1000/1/01204

****

A Comparison of Radial Basis Function and Multilayer Perceptron Network as tool for classification of Medical Data

S. Leoni Sharmila[1], C. Dharuman[2], P.Venkatesan[3]

[1,2] Department of Mathematics, SRM IST, Ramapuram, Chennai.
[3] Department of Community Medicine, Sri Ramachandra University, Chennai.

sharmilamartin169@gmail.com,

Abstract. Artificial Neural Network has become a popular tool in developing systems that encircles human proficiency. The importance of exact detection is exceptionally important for proper treatment and preserve of disease. Clinical cytology has improved tremendously in disease diagnosis. In this paper, two Artificial Neural Network (ANN) methods, Multilayer Perceptron (MLP) and Radial Basis Function (RBF) network are compared. The RBF network predicts comparatively better accuracy in compared to MLP methods. Also it was detected that the RBF method requires a lesser amount of time for the development of the model, this is because there is no repetition to reach the favourable parameters in the model.

1. Introduction

Artificial Intelligence (AI) provides learning ability to the computer in real world problem [1]. Machine learning is a subset of AI that was developed to mimic human brain in computer to investigate the details from the collect data. In many situations of clinical test, ML system is considered as a development to diagnosis disease and analyze that imitates like an experts knowledge in medical field. Also ML method makes the computer learn from datasets, which in turn develops a model to identify common patterns and this can able to make decisions based on availability of knowledge, more over the imperfectness medical database does not affect the classification [2]. In medical diagnosis, classification is a famous ML technique among which the most usually applied technique is Artificial Neural Networks (ANNs). The practice of ML technique in disease classification has increased very frequent [3-5].

Applications of this type have used the Multilayer Perceptron (MLP) ANN method which is combined with the Back Propagation (BP) algorithm [6,7]. However , the nonlinearity of MLP with its parameters and steepest descent method used in BP algorithm do not converge to a globally optimum set of parameters always. Several attempts of trial and error method are necessary to choose the best parameter from a set of locally optimum parameters. Another alternate method to MLP is RBF network [8,9] that consists of linear parameters and has established many applications in areas like electrical and electronic engineering. [10] have proved theoretically that this type of ANNs are applicable for global approximations and also can learn without local minima, therefore strictly convergence to optimum parameters. For complicated situations, [11] proved that the learning rate of

Journal of Physics: Conference Series **1377** (2019) 012028 doi:10.1088/1742-6596/1377/1/012028

RBF is faster than MLP networks. This study is to apply the RBF and MLP network approach for classification of medical data.

2. RADIAL BASIS FUNCTION (RBF) TYPE NETWORK

2.1. Network Architecture and Composition

An RBF is a feed-forward two-layered neural network where the basis function transforms (i/p) (*i/p*) o the (h/l) layer(*h/l*). Responses of the (*h/l*) are added in linear combinations to form the (o/p) in the (*o/p*)layer. In the (h/l) layer, the basis functions are radially symmetric. For the performance of the network, selecting the basis function is not important but, commonly used function is the Gaussian function which is given by mean and standard deviation. A representation diagram of an RBF network with I, H and O respectively of (*i/p*), (*h/l*) and (*o/p*) layer nodes for the transformation of ID points of I $(I^1,...,I^k,...I^{ID})$ in the (*i/p*) layer to the points $O(O^1,...,O^k,...O^{ID})$ in the (*o/p*) layer is shown in Figure 1.

In RBF, the parameters are the centers (C_j) and range (φ_j) in the basis functions at the (*h/l*) nodes, and the synaptic weights (w_{kj}) of the (*o/p*) nodes. The RBF centers are also present in the (*i/p*) layer. The basis function responds to an (*i/p*) depends on the distance between the (i/p) (I^k) and the RBF center (C_j). If (*i/p*) falls in a small localized region of the basis functions' center then the (*h/l*) layer nodes of RBF will produce non-zero replies only [12]

2.1.1. Network functioning[12]: The connections between the (*i/p*) and (*h/l*) layers are not weighted in RBF networks. Therefore, the (*i/p*) reach the (*h/l*) layer without any change. For an (*i/p*) I^k the *j*th (h/l) node produces a response r_j given by,

$$r_j = \exp\left\{\frac{-\left\|I^k - C_j\right\|}{2\varphi_j^2}\right\} \tag{1}$$

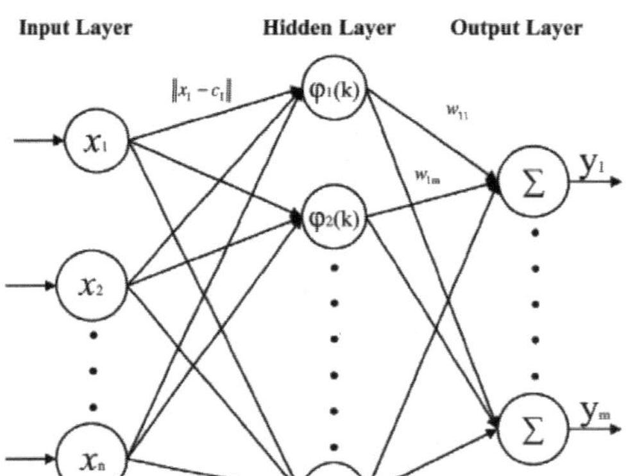

Fig-1 RBF- Schematic diagram

Journal of Physics: Conference Series **1377** (2019) 012028 doi:10.1088/1742-6596/1377/1/012028

where $\left\| I^k - C_j \right\|$ is the distance between the point representing the input I^k and the center of the *j*th (*h/l*) node as measured by some norm. In this study the Euclidean norm is used. The (o/p) y_{in} of the network at the (*o/p*) node is given by,

$$y_{in} = \sum_{j=1}^{H} r_j w_{nj} \tag{2}$$

In some cases the number of (*h/l*) layer nodes is equal to the number of data in the training set *(H =ID)* and the RBF centers coincide with the (i/p) s $(C_j = I^k, where, k = j = 1,2,...ID)$, the (*h/l*) layer response according to equation (1) becomes one, $j = k$. When the basis functions are accurately localized, the response of the other (*h/l*) layer nodes will be closer zero. It is also be seen that equation (2) gives the exact (*o/p*) when the (*o/p*) layer weight is equal to the (*o/p*). In the best case, RBF network can be made to map points in *N*-dimensional (*i/p*) space precisely on to points in *M*-dimensional (*o/p*) space [12]

3. RADIAL BASIS FUNCTION (RBF) TYPE NETWORK

The Type of neural network is commonly known and normally used type.when there are more than one layer in artificial neural network that contains forward connections of *(i/p)* s to (*o/p*) s is known as Multilayer Perceptron (MLP) or Multilayered Feed-forward Neural Networks. An MLP consists of a set of (*i/p*) terminal, an (*o/p*) terminal, and one or more terminals of (*h/l*) nodes between the (*i/p*) terminals and the (*o/p*) terminal.[13]

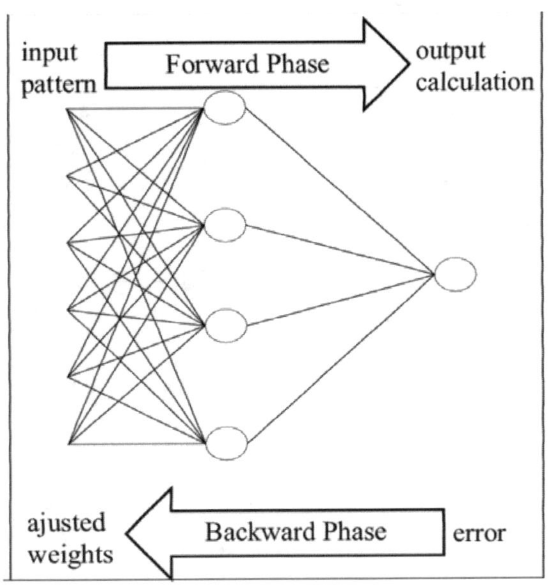

Figure2- Multilayer Perceptron model

Journal of Physics: Conference Series **1377** (2019) 012028 doi:10.1088/1742-6596/1377/1/012028

3.1. Back Propagation Algorithm[13]:

Learning in this network is usually done through supervised method. This environment consists of learning background that contains both the learning model and desired (*o/p*) model corresponding to the (*i/p*) . Minimizing the error between the network and desired (*o/p*) is the main aim of this learning. This causes a back propagation all over the network due to this reason this algorithm is called back-propagation learning algorithm. The algorithm for a MLP with one (*h/l*) layer is given by the following [13]:

Step 1: *Initializing* : Initially the network weights and thresholds are distributed evenly with random small range of values.

Step 2: *A new training set is create:* Here the weights are adjusted that is all the gradients of the weights and the current error are initially considered as 0 $(\Delta w_{ij} = 0 \,\&\, E_r = 0)$.

Step 3: *The forward pass of the signal* : The (*h/l*) layer neurons (o/p) are calculated here:

$$v_j(q) = f\left(\sum_{i=1}^{m} u_i(q).w_{ij} - \theta_j \right) \tag{3}$$

here f the sigmoid activation function and m is the (*i/p*) s number for the jth neuron from the (*h/l*) layer. The network (*o/p*) s are calculated as:

$$v_k(q) = f\left(\sum_{i=1}^{n} u_{jk}(q).w_{jk}(q) - \theta_k \right) \tag{4}$$

That is k belongs to the (*o/p*) layer and n is the (*i/p*) s number of the neuron. The error per epoch is updated as follows:

$$E_r = E_r + \frac{(e_k(q))^2}{2} \tag{5}$$

Step 4: The adjustments of the weights are done in backward propagation according to the errors: Here single-pole sigmoid is used (equation 3, is derived as:

$$f'(u) = \frac{e^{-u}}{\left(1 + e^{-u}\right)^2} = f(u).(1 - f(u)) \tag{6}$$

The weights present between the (*h/l*) layer and the (*o/p*) layer are updated as:

$$\Delta w_{jk}(q) = \Delta w_{jk}(q) + v_j(q).\,\phi_k(q) \tag{7}$$

The weights between the (*i/p*) layer and the (*h/l*) layer are updated:

$$\Delta w_{ij}(q) = \Delta w_{ij}(q) + u_i(q).\,\phi_j(q) \tag{8}$$

Step 5: *A new iteration*: At this stage when the test vectors are identified, immediately the training epoch will go to the 3rd step otherwise, according to the gradient weights, connection weight will be updated.

$$w_{ij} = w_{ij} + \eta \,.\, \Delta w_{ij} \tag{9}$$

where η represents the rate of learning . The criteria for termination is tested after an epoch is completed i.e. ($E_r < E_{\max}$ or when number of training epochs reaches maximum). If no, move to 2nd step else if yes end the algorithm [13].

4. EXPERIMENTS RESULTS AND DISCUSSIONS

RBF network requires more number of neurons for training. When there is a high number of training vectors, this network performs better. Unlike other network, the entire (*i/p*) space not react at the same time in RBF. Here, the center of the *(i/p)* layer is calculated first, next the (*i/p*) s which are close to this center are reacted. Because of this, these networks answer to local (*i/p*)s quickly. The two layers of RBF are the radial basis layer and the next is linear (*o/p*) layer. Competitive learning or clustering is done in the training process. Spread number and goal number are included in network parameters. By changing these parameters, performance of the network is improved. Distance between (*i/p*) vectors and weight vectors are calculated by applying (*i/p*) s to the network and multiplying the calculated values by bias values obtains the vector product. At that moment, these values generate several neurons as (*i/p*) s by the corresponding functions and as a final point, the (*o/p*) layer obtains the (*o/p*) values [14]. For classification, Hepatitis data base is chosen from UCI Repository. The data set consists of 155 cases with 17 variables among these 123variables are classified as live and 32 as die . When MLP method is used were the learning algorithm is Back propagation method which is used for training the neurons and is based on descent gradient technique. Classifies an accuracy of 80% . The same data is classified using RBF network with a classification rate of 85.80% accuracy .

	Accuracy Rate
MLP Network	80%
RBF Network	85.80%

Table1- Classification Rate

5. CONCLUSION

In Artificial Neural Network models, Radial Basis Function type was developed to predict hepatitis diseases. When classification accuracy is considered, RBF along with k-means clustering technique performs better when compared with MLP with back propagation method. Since the RBF network models are linear in the parameters, there is an assurance in convergence to their optimum values. Finally this sums up that RBF minimizes the error in minimum trial or in other words it needs minimum time and effort .

References

[1] N. Sandhya, K.R. Charanjeet, A review on Machine Learning Techniques, International Journal on Recent and Innovation Trends in Computing and Communication, 2016,

[2] G.D. Magoulas and A. Prentza, Machine learning in medical applications. In Machine Learning and its applications, Springer Berlin Heidelberg, 2001.

[3] Badnjevic A, Cifrek M, Koruga D, Osmankovic D,Neuro-fuzzy classification of asthma and chronic obstructive pulmonary disease" BMC Medical Informatics andDecision Making Journal, 2015.

[4] Badnjevic A, Koruga D, Cifrek M, Smith HJ, Bego T. „Interpretation of pulmonary function test results in relation to asthma classification using integrated software suite", IEEE MIPRO, 2013.

[5] Avdakovic S., Omerhodzic I., Badnjevic A., Boskovic D., „Diagnosis of Epilepsy from EEG Signals using Global Wavelet Power Spectrum", MBEC, 07-11. September 2014.,

[6] Isobe, I., Ohkado, T., Hanyuda, H., Oda, S. & Gotoh, Y. (1994) The development of forecasting system of the water levels of rivers by neural networks (in Japanese). /. *Japan Soc. Hydrol. & Wat. Res.* 7(2), 90-97.

[7] Jayawardena, A. W. & Fernando, D. A. K. Hydrological forecasting using artificial Neural networks. In Proc.Second Int. Study Conf. on GEWEX in Asia and GAME (Thailand), 1995.

[8] Bianchini, M., Frasconi, P. & Gori, M. Learning without local minima in radial basis function networks. *IEEETrans. Neural Networks* 6(3), 1995.

[9] Chen, S., Cowan, C. F. N. & Grant, P. M. (1991) Orthogonal least squares learning algorithm for radial basis function networks. *IEEE Trans. Neural Networks* 2(2), 1991.

[10] Park, J. & Sandberg, I. W. Universal approximations using Radial-Basis-Function networks. Neural Computation 3(2), 1991.

[11] Moody, J. & Darken, C. J. Fast learning in networks of locally-tunes processing units. Neural Computation 1(2), 1989.

[12] Jayawardena.A.W, Fernando.D.A.K & Zhou. M.C. Comparison of Multilayer Perceptron and Radial Basis Function networks as tools for flood forecasting, Proceedings of the Conference held at Anaheim, California,1997.

[13] Halagundegowda G. R. and Abhishek Singh, Multilayer Perceptron Method of Artificial Neural Network for Classification of Farmers Based on Adoption of Drought Coping Mechanisms, International Journal of Pure and applied Bioscience,2008.

[14] Fesghandis. G.S, Pooya.A, Kazemi.M, and Azimi . Z. N.Comparison of Multilayer Perceptron and Radial Basis Function Neural Networks in Predicting the Success of New Product Development , Engineering, Technology & Applied Science Research, 2017.

[15] www.ics.uci.edu/pub/ml-repos/machine-learningdatabases/hepatitisb-disease/

Analysis of Psychology of Students by Graphical Representation

P.Lalitha [1] , L. Tamilselvi[2]

[1]Assistant Professor, Department of Mathematics, SRM IST

[2]Associate Professor, Department of Mathematics, AVIT, Chennai

Corresponding Author: P.Lalitha, Assistant Professor, Department of Mathematics, SRM IST, Ramapuram, Chennai, Tamilnadu , India

Mail ID: arthimohan2010@gmail.com, ltamilselvi@avit.ac.in

Abstract. Graph theoretical analysis plays a vital role in various fields. In Graph theory, Graph labeling is the major area of application. Graph labeling is the assignment of labels to the vertices and edges of a graph pertaining to certain conditions. Graph theory finds its applications in Coding theory, Radar, Astronomy, Circuit designs, Communication networks, X-ray crystallography, Data base management and a major role in medical field like Brain network, Forensics and Cephalometric analysis. Here we would like to enhance the graph theory applications in the field of Psychology. This paper gives an overview of graph techniques applied in Sociometry using Sociogram. Sociometry is a quantitative method for measuring social relationships and Sociogram is a tool to find Sociometry of a social space. In this article, an analysis in a classroom climate with of a group of 50 college students was done after they were subjected to answer a series of questions probing for affiliations with other classmates. A graphical representation (sociogram) of the social links of each student was constructed by plotting the structure of interpersonal relations in the group. This graphical representation depicts the social relations, channels of influence, attraction towards others, hatred, isolation and lines of communication of thegroup.

Keywords: Graph theory, graph Labeling, psychology, sociometry, sociogram, clique, isolate, star, chain, island, ghost, triangles and circles

1. INTRODUCTION

Graph theory deals with graphs which are a combination of vertices and edges represented by the set G= (V, E) where V denotes the set of all vertices of the graph and E denotes the set of all edges of the graph. A graph which shows the direction of the edges which connects the respective vertices is known as Directed graph .The vertices which are connected to each other are known as Adjacent vertices. The degree of a vertex is determined by the number of adjacent vertices to that particular vertex.

The social relationship which exists between individuals of a group is termed as Sociometry.The attractions or attentions among these members of the group can be described by analysing whom they select or reject using Sociometric techniques. This technique can be implemented in educational institutions, companies or factories, offices etc. to learn the people's social adjustment, their discipline, group dynamics and social relations.

To learn the interpersonal relationship of a team in form of a graph, Sociogram is used. It is applied by class teachers, psychologists, councillors etc. It is helpful in identifying the person with leadership quality among the group members, the isolated individuals in the group and the common interests of thegroup.

Classroom environment plays a vital role in a student's ability to acquire knowledge. It is an important aspect that a student should feel safe and supported in order to be comfortable to try new and learn new skills .So a healthy classroom environment is necessary for learning to occur. Even though class teachers observe the students and also have practical knowledge about them, formal data collection of these students social relationship can help to create positive classroomculture.

Graph theoretical analysis using Sociograms are tools that can make things easier to gain clear knowledge on Social relationship of the class.

2. SOCIAL GRAPHANALYSIS

A Sociogram has two categories:

The first one is the Individual Phenomena which describes individual's attributes like Stars, Isolates and Ghosts.

The second one is the Group Phenomena which describes attributes of Social interaction in a group like Clique, Mutual Choices, Chains, Islands and Triangles and Circles.

3. CONSTRUCTION OF A SOCIOGRAM BY APPLYING GRAPHTHEORY

A class Sociogram is constructed by asking the students to fill up a questionnaire confidentially with questions based on their friendship, their foe, the person they like the most and hate the most, the students who are unapproachable and the respective reasons for their attitude towards their class members.

These data are filled in a chart with the name of each student and their top five priorities and the students whom they are unable to connect with or hate to connect.

We use a directed graph to construct the Sociogram. Each student is considered as a vertex of the graph and the students who have affinity towards other students are connected by directed edges pointing towards that student. This gives a visual impact of the social relationship among members of a group and facilitates to discern the various degrees popularity.

Journal of Physics: Conference Series **1377** (2019) 012029 doi:10.1088/1742-6596/1377/1/012029

GROUP COHESION

Group cohesion describes the attraction structure of the group. It exhibits not only individual friendship but also how the whole group is attracted towards individuals in the group. In such kind of cohesive environment, people value their friends, care about one another, help each other and feel proud of their membership in the group. This group cohesiveness is a driving factor for the progress of the group.

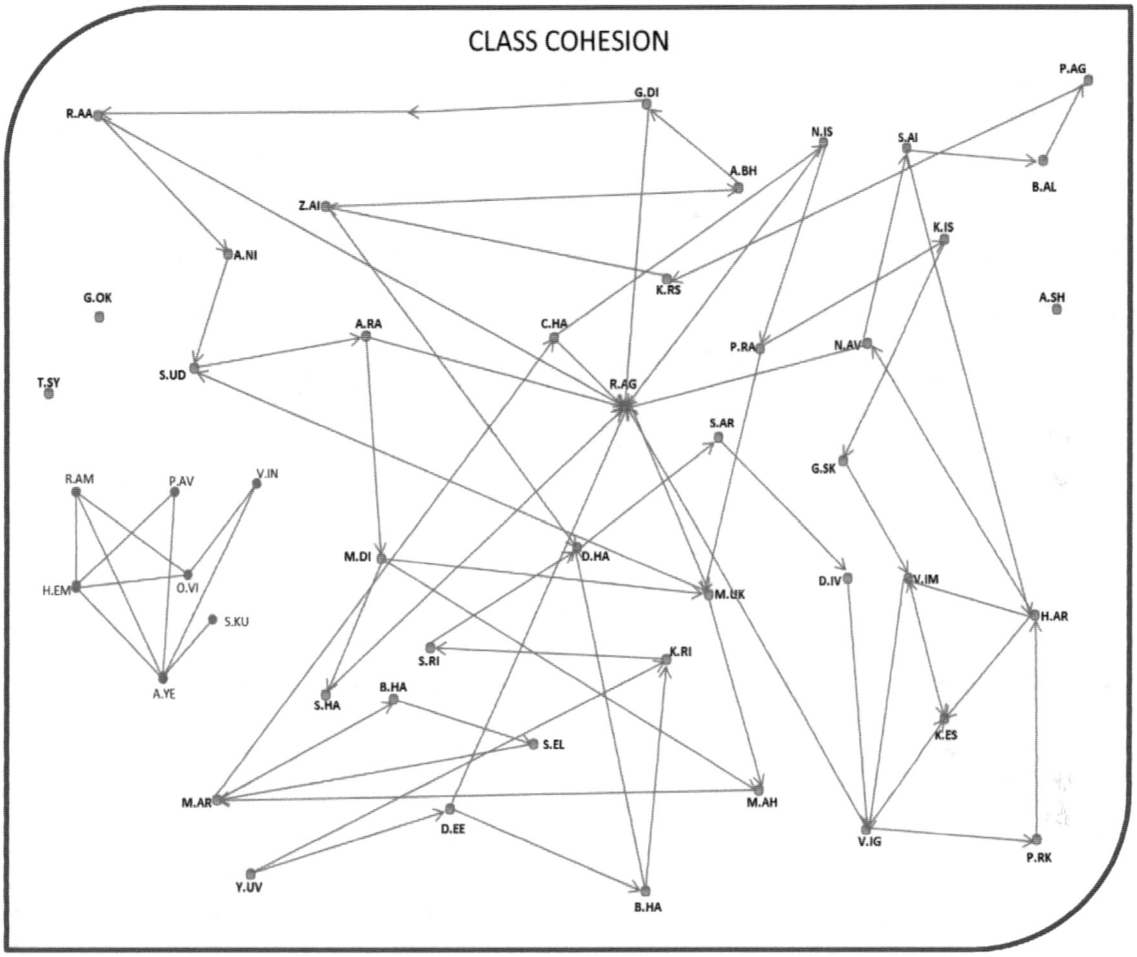

CLIQUE

Clique refers to a small team of a few members of the group who choose among themselves one another .Clique has highly connected nodes. A clique is a stronger version of community. A set of nodes forms a clique if all possible connections between these nodes exist. A two node clique is simply two connected nodes. Graph also contains maximal cliques which are complete subgraphs such that no other node can be added while maintaining completeness. Depending on structure, graphs may contain a large number of cliques.

Thus, clique helps to trace individuals who actively join groups and socialise, thus becoming a highly connected person that can build Cross functional relationships providing value to them.

Journal of Physics: Conference Series **1377** (2019) 012029 doi:10.1088/1742-6596/1377/1/012029

STAR

Star refers to an individual who is chosen as the most desirable person by almost all the people in the group. Thus a person who is a Star in group possess positive attitude towards others, helps in times of need, optimistic and treats everyoneequally.

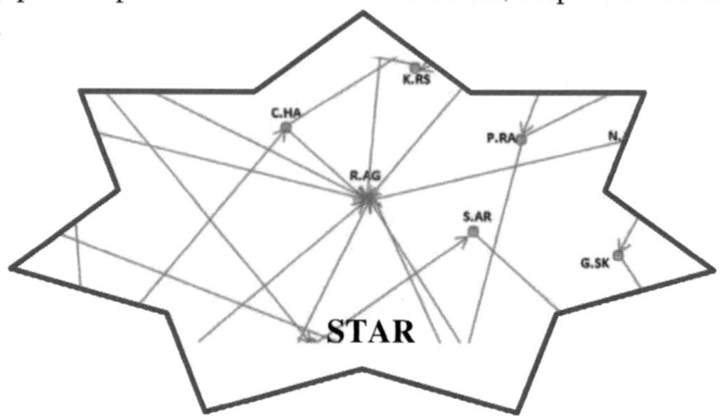

ISOLATES

On the contrary, Isolate refers to an individual who is seldom chosen by the people in the group. They are not positively nominated by anyone in the group. This shows that these individuals need major attention and care. They also need to develop an enrichedpersonality.

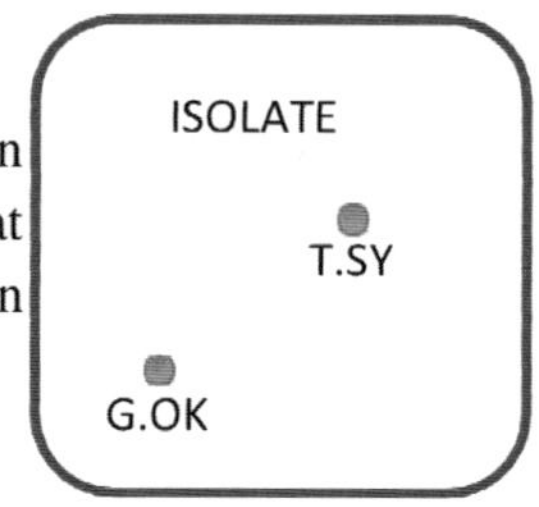

GHOST

Ghosts refer to the individual who is not even acknowledged as existing in the group. These kinds of people are neither positively nor negatively nominated. These people are most probably low in assertiveness and needs personal care and support to overcome their introvertpersonality.

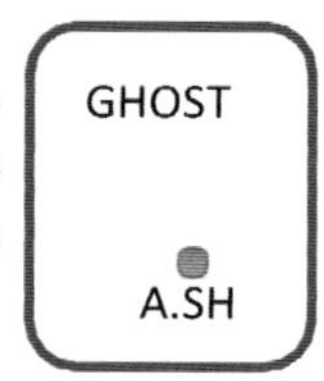

MUTUAL CHOICES

This consists of a pair of people who chose each other .The more there are mutual choices in a group, the more congenial the group is leading to higher positive social climate in the group.

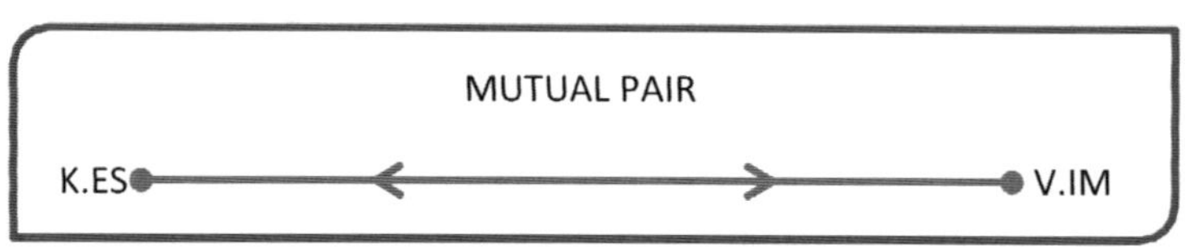

CHAIN

It refers to a person choosing another one in the group who in turn nominates another individual in the group and so on, chains often leads toward a “Star.” Chain represents the common interest these set of people have and are attracted by one or many characters of other individuals in this chain.

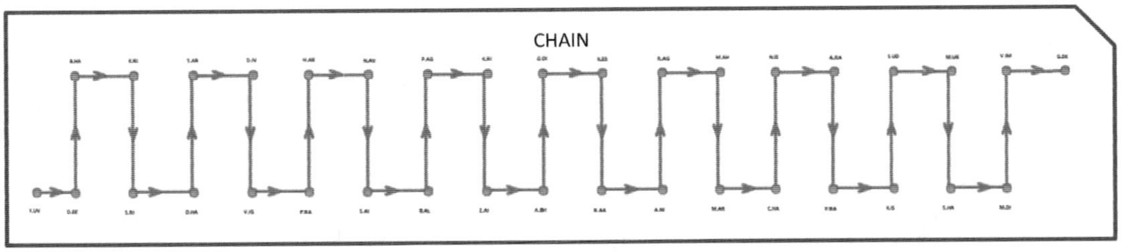

ISLAND

If pairs or small team are separated from the group and the members of this team are not selected by anyone in other teams, then these teams are described as Island. This team requires consideration and kindness from other cliques. In order to improvise their social network standards, this team has to pay more attention on interacting and communicating with every individual in the group creating a friendlyatmosphere.

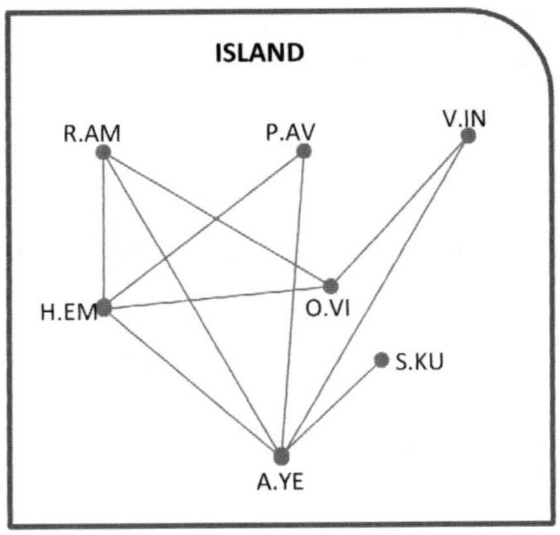

TRIANGLES AND CIRCLES

When a chain ends at the starting person, it is referred as circle. If it has only three members, then it is known as a triangle. A three node clique in a group is called as a triangle. The members of triangles and circles are highly interconnected and their group activity scores the best.

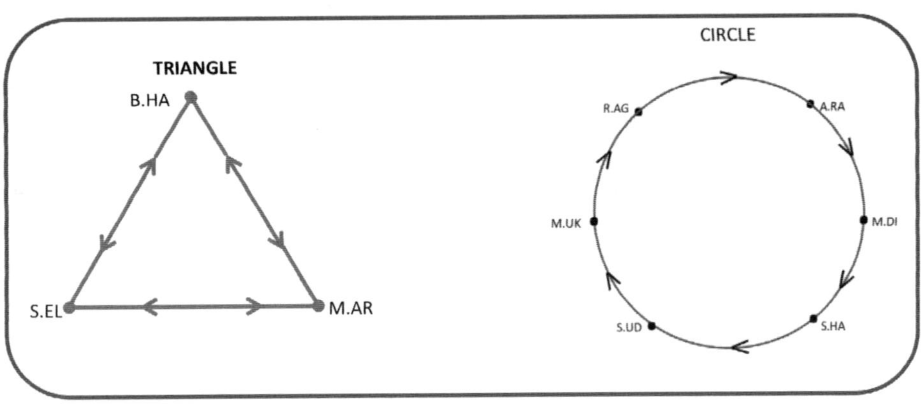

4. INFERENCE

The individual who has the highest vertex degree is considered as “Star “of the class and that with the lowest vertex degree as “Isolate” or “Ghost”. This identification paves the way for the class teacher to concentrate more on Isolates andGhosts.

Necessary action should be implemented to make sure that this kind of students feel supported and connected in the classroom. Their emotional oscillations and mood swings has to be taken under consideration. Slow and steady counselling given to such students enhances major transformation of their personality.

Thus, Sociogram constructed using Graph Theory leads to proper understanding and gives awareness about the psychology of a group. It also serves as guidance for the teacher to set up a supportive learning environment for every individual in the group.

5. CONCLUSION

Node-link graph can help in data visualisation of the various attributes and the level of interaction between these attributes. The various features of the individual nodes and the nature of their social networks, its depth and breadth, its composition its different connections, its path, its clustering, its evolution can be easily analysed using the Sociometric graph. The sub-cliques or the subnetworks maybe used for investigating this group’s character. The Islands and Ghosts are the focal points for research of their respective Social links which needs improvisation. Thus, this Sociogram enable multi-dimensional viewing of a classroom climate.

REFERENCES

1. Gallian .J.A A Dynamic survey of graph labeling (2007), Electronics journal ofCombinatorics.

2. S.W. Golomb, How to number a graph, in Graph Theory and Computing, R.C. Read, ed., Academic Press, New York (1972)23-37.

3. Loomis, Charles P., and Harold B. Pepinsky. "Sociometry, 1937-1947: Theory and Methods." Sociometry 11, no. 3 (1948):262-86.

4. K.M. Evans ; 1963 ; sociometry in school -1 , sociomtric techniques, Educational Research, 6:1, 50-58

Application of Fractals to Detect Breast Cancer

Debabrata Datta[1] and Shakeela Sathish[2]

Radiological Physics and Advisory division[1]
Bhabha Atomic Research Centre
Mumbai – 400085
Department of Mathematics,
SRM Institute of Science and Technology[2],
Ramapuram Chennai – 600089. INDIA
***Corresponding author:* shakeela_sathish@yahoo.co.in**

ABSTRACT
Breast cancer is one of the major causes of death among women. Small clusters of micro calcifications appearing as collection of white spots on mammograms show an early warning of breast cancer. Image segmentation is an important element of Digital Image processing that subdivides the image into discrete regions/objects, each identified by the property of homogeneity of pixels. X-ray mammography is used as diagnostic tool for diagnosis of breast cancer. Edge detection of micro calcification clusters in mammogram images is the main issue of early detection of breast cancer. Fractals are of rough or fragmented geometric shape that can be subdivided in parts, each of which is a reduced similar of the whole. Fractal dimension and Hurst exponent are used to locate the micro calcifications in the mammogram. The concept of fractal is associated with geometrical objects satisfying criteria such as self-similarity and fractal dimensionality. Present method of edge detection is superior compared to conventional Sobel method, in which detection of edges of the regions or objects in an image takes place by the Fudge factor. The paper presents the application of fractals for early detection of breast cancer.

Key words: Mammogram, edge detection, fractal dimension, Hurst exponent

1. Introduction

Breast cancers are commonly associated with a high incidence and a high mortality rate in the female population worldwide. Mammography associated with clinical and self examination of breast is the only viable and efficient method at present for mass screening to detect breast cancer. Breast cancer is the main cause of death among women [1]. However, at a microscopic and molecular level, breast cancer is not a homogeneous disease. Accordingly, an early detection of breast cancer is an important research component for improving its prognosis. Screening mammography is the best tool available as on today to detect cancerous before appearance of any sorts of clinical symptoms appear. In this context, radiologist can judge the likelihood of presence of malignancy on the basis of shape and arrangement of micro-calcifications. Malignant calcifications are typically of different varieties, such as (1) clustered, (2) small and (3) dot-like or elongated.All the varieties are variable in size, shape and density. Literature study reveals that benign group of calcifications are generally larger, more rounded, smaller in number, more diffusely distributed and more homogeneous in size and shape [1]. Tests and procedures used to diagnose breast cancer include (a) breast exam, (b) mammogram, (c) breast ultrasound, (d) removal of a sample of breast cells for biopsy test and (e) breast magnetic resonance imaging. Basically, a mammogram is an X-ray of the breast which helps doctor to detect and diagnose breast cancer up to two years before the occurrence of tumor. Probability of long term survival of a patient suffered from breast cancer can be improved by early detection of the disease and early detection is in turn enhanced by an accurate diagnosis. Even though risk factors like ageing, genetic disorder, family history, menstrual periods, not having children, obesity, alcohol, overweight, etc, are known by scientists, but appropriate reasoning of breast cancer is unknown in medical science and accordingly, prognostic methods for early detection is a challenging task. Symptoms of cancer include a lump in the breast or underarm that persists after menstrual cycle, swelling in the armpit, pain or tenderness in the breast, any change in the size, contour, texture, or temperature of the breast, a marble-like area under the skin. Some of the characteristics of malignant tumors are: clustered calcification, isolated ducts, poorly defined mass, etc. Several methods have been proposed for the characterization of ROI such as classical image filtering and local thresholding to contour region of interest. Researchers have attempted several computerized techniques such as method of fuzzy c–means, support vector regression, artificial neural network, naïve Bayes to detect the presence of microcalcification [2]. Nevertheless, the most work reported in the literature employs neural networks for cluster characterization. Each of such techniques has pros and cons. Therefore, an appropriate decision before diagnosis is an important issue. With the advent of artificial intelligence and machine learning technology, a large number of classification (benign or malignant) methods of breast cancer diagnosis are also envisaged. It can be easily said that characteristics of symptoms of breast cancer are of an irregular geometrical objects. Determination of dimension of these irregular objects by contouring or edge detection is the main motivation for developing an innovative method for early detection of breast cancer. In this context, we have developed a fractal technique as a tool for early detection of breast cancer. The paper presents the mathematical structure of fractals, physics of fractals and their application for early detection of breast cancer. Irregular geometry of breast cancer symptoms is due to high frequency component of image spectrum and the fractal dimension of this geometrical shape is evaluated by box counting method. Fractals being self similar object are allowed to grow to contour or detect the edge of the malignant tumor and fractal dimension quantifies the dimension of the target tumor. The remaining part of the paper is organized into four more sections. Section 2 describes mammography. In section 3, we have presented physics of fractals with their mathematical structure. Section 4 presents methodology of early detection of breast cancer using fractal technique and results of case study. Section 5 presents conclusion of the paper.

2. Background of Mammography

Mammography is known as the most efficient method to detect breast cancer. Experts separate the mammographic images in normal and abnormal images for detecting breast cancer. In order to seek to identify the characteristics of the affected tumor like: shape, definition of edges, roughness of contour, density and size for rating it as benign or malignant tumor, the corresponding images are studied more carefully. However, it is not appropriate to have conclusive results by this examination. In addition, limitations of mammography devices, viewing images can be impaired. Furthermore, there is not much variation between the contrast of the nodule and breast tissue in many cases. These aspects may interfere with the identification of the nodules, and in many cases, it can lead to misinterpretations of the expert examination. The benign tumors are well defined and usually rounded, have soft edges, of low density, while malignant tumors are often poorly defined edges and irregular contours. Malignant tumors often examined as self similar pattern with the

Journal of Physics: Conference Series **1377** (2019) 012030 doi:10.1088/1742-6596/1377/1/012030

presence of specula, thus allowing users to apply fractals for the detection [3]. Fractal dimension and Hurst exponent may be the quantification scheme of the detection.

3. Physics of Fractals

A fractal is a mathematical object that is both self-similar and chaotic. Self-similarity of fractal signifies that if we magnify, we can see the object over and over again in its parts. In other words, fractals are of rough or fragmented geometric shape subdivided in parts. They are creased objects and are most often characterized by their fractal dimension. A fractal dataset is known by its characteristic of being self-similar. The dataset has roughly the same properties for a wide variation in scale or size i.e., parts of any size of the fractal are almost similar to the whole fractal. Fractals are forms that have fractional dimension, not obeying the traditional Euclidean geometry where the objects have entire dimensions.

3.1 Edge detection

Edge detection is one of the main tasks for contouring of the affected tumor. Edge detection technique uses the abrupt intensity changes among the pixels of an image. Edge is a set of connected pixels that lie on the boundary between two regions. The magnitude of the first derivative is used to detect an edge. The sign of the second derivative can determine whether an edge pixel is on the dark or light side of an edge. The first order derivatives in an image are computed using the gradient operator. The second-order derivatives are obtained using the Laplacian operator.

3.2 Fractal dimension

The fractal dimension contains information about the geometrical shape of fractals and therefore it is an important characteristic of fractal. The dimension of the object never changes whatever the object undergoes any transformation. This type of dimension is known as topological dimension, symbolized by d, introduced by Mandelbrot. The fractal dimension is a statistical quantity which measures the degree of fractal boundary fragmentation or irregularity over multiple scales. It also differentiates fractal from Euclidean objects (point, line, plane, circle etc.). A fractal can be classified according to their degree of irregularity. Fractal dimension is an effective measure for complex objects. It is widely applied in the fields of image segmentation and shape recognition.

3.3 Mathematical Definition

Mathematically, fractal dimension is defined by equation (1).

$$FD = \frac{\log N}{\log\left(\frac{L}{U}\right)} \qquad (1)$$

where, FD represents the fractal dimension, N, the number of parts for each iteration, L signifies the initial length of the object , and U is the length of each segment. Now, in order to calculate FD, we have adopted Box-Counting technique in which the object is quantified on a plan with grid. Fractal dimension by Box-Counting technique is formulated mathematically by equation (2).

$$FD = \frac{\left(log(N_{n+1}(U)) - log(N_n(U))\right)}{log\left(\frac{1}{U_{n+1}}\right) - log\left(\frac{1}{U_n}\right)} \quad (2)$$

The fractal dimension can be any non-negative real number and can be estimated in several ways.

4. Methods

We have used the database of breast imaging-reporting and data system (BI-RADS). BI-RADS is a unitary system designed for helping medical professionals assess, interpret and classify mammography, ichnography and magnetic resonance imaging results in a concise and ambiguous and standardized way, by assigning numbers or numerical codes to different risk categories. For example, (a) BIRADS-I means that the breast is almost entirely fatty, (b) BIRADS-II signifies that there is some fibro-glandular tissue, (c) BIRADS-III means that the breast is heterogeneously dense and (d) BIRADS-IV means that the breast is extremely dense. All these four varieties are as shown in Fig. 1.

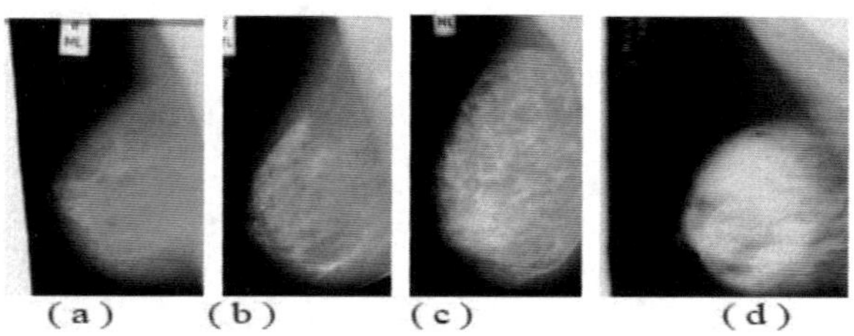

Fig. 1 BIRADS classification

The assessment categories are numbered from 0 to 5 describing the diagnostic status as follows: 0: Incomplete, 1: Negative, 2: Benign finding(s), 3: Probably benign, 4: Suspicious abnormality and 5: Highly suggestive of malignancy. Each image found in the database has a coordinate (x, y) and RADIUS, to facilitate the identification of the lesion area. An image has undergone preprocessing such as histogram equalization with balancing grayscale and contrast adjustment that facilitates the visualization of regions of interest. Equalization process is followed by the identification of the region of interest. With the fractal approach, instead of one quantity or measure, ω, describing the phenomenon in all scales (as in case of fractals), a set of measures (weight factors) depicting statistically the same phenomenon in different scales, has to be used for characterizing such structures. The signal value within the box or region, is embedded into the process of signal characterization.

At the first step, the quantity δ called roughness exponent is derived. The roughness exponent δ corresponds to the fractal dimension of the measure and it is close to the corresponding fractal dimension of the structure under observation. Finally, each image having different features is segmented by improved thresholding variable. In addition to identify the tumor, the segmentation of the image, also identified regions that are not part of the area of our interests. In this study we

Journal of Physics: Conference Series **1377** (2019) 012030 doi:10.1088/1742-6596/1377/1/012030

have used the box-counting technique to quantify the nodules to compute the fractal dimension of the region of lesion. Images were also analyzed for their edges and contours of the region of interest without concerning the mass density or texture.

4.1 Box counting

Box counting is a method of accumulating data or analyzing complex patterns by dividing adataset, object, image, etc. into smaller and smaller pieces, typically "box"-shaped, and analyzing the pieces at each smaller scale [4].

4.2 Results

An algorithm is developed for detection of microcalcification clusters from a mammogram image. Input of this algorithm is a 2-dimensional mammogram image, I and output is an outlined edges of microcalcification in I. Algorithm written is as follows:

Algorithm:

Step 1: Read a 2-Dimensional input mammogram image I

Step 2: [M, N] ← SIZE [I]

Step 3: If M > N then r ← M Else r ← N

Step 4: Compute fractal dimension D

Step 5: Dilate and erode I

Step 6: Compute Hurst coefficient H

Step 7: Detect the edges of I using Sobel operator with H

Step 8: Dilate the edge detected image

Step 9: Fill the interior gaps in the edge detected image

Step 10: Smoothen the image by erosion

Step 11: Outline the segmented image to detect the clusters

Step 12: Stop.

We have first found the fractal dimension D for the input mammogram image (I). We have used box counting method to estimate the fractal dimension, D. Thereafter, morphological operations such as dilation and erosion are applied on I and results are shown in Fig. 2.

Journal of Physics: Conference Series **1377** (2019) 012030 doi:10.1088/1742-6596/1377/1/012030

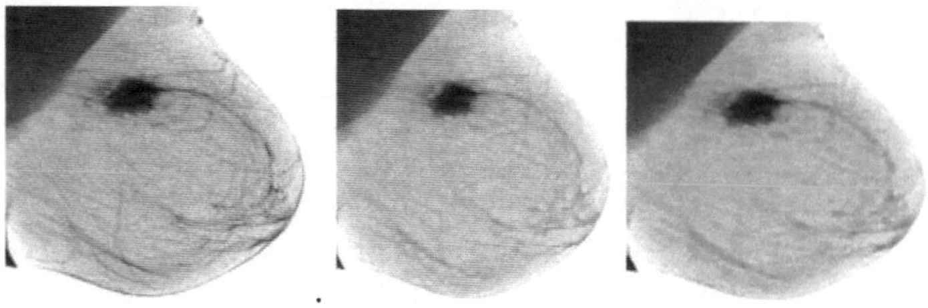

Fig. 2

(a) Original (b) Dilated (c) Eroded

As the fractal analysis is sensitive to noise, application of morphological operations tend to suppress noise if any, in addition to image enhancement on the input image. In addition to estimation of Hurst coefficient factor the edge detection is done using Sobel operator. Both the results are analyzed. The edges constructed using Hurst coefficient is far accurate and are found to be confined to the microcalcification clusters. Results are shown in Fig. 3.

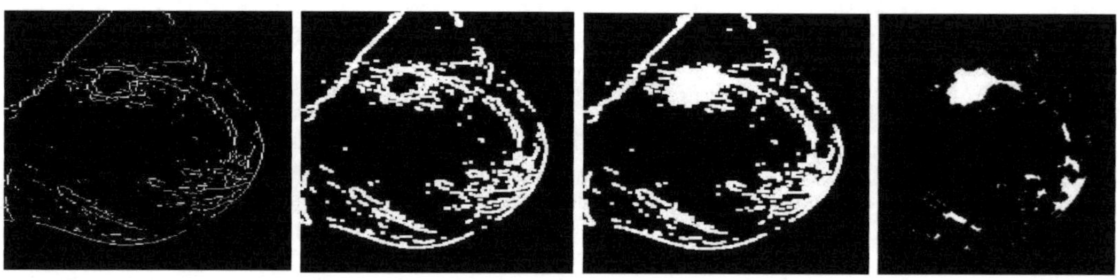

(a) (b) (c) (d)

Fig. 3

a) Edges detected using Sobel operator with Hurst co-efficient, b) Dilated image of (a)
c) Image (b) with filled holes, d) Segmented output Image.

The result of detection of microcalcification using fractal Hurst coefficient is shown in Fig. 4.

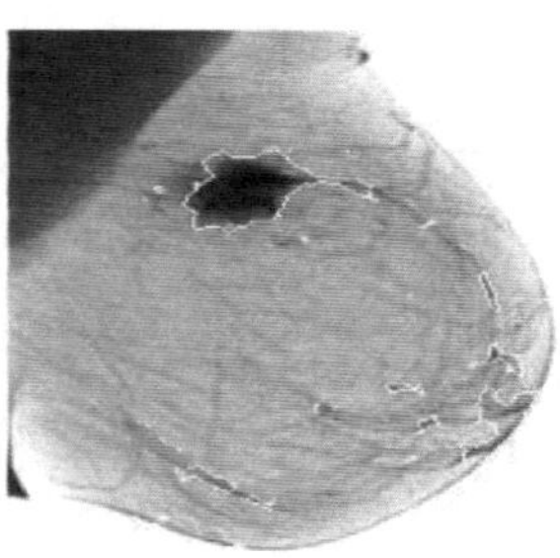

Fig. 4 Microcalcification detection

5. Conclusion

Fractals have applied as a tool for early detection of breast cancer. Physics of fractals with its mathematical formulation are presented. Fractal dimension is used as the measure for detection of abnormality malignancy. Box counting technique is applied to compute the fractal dimension.

6. REFERENCES

[1] Crisan, D.A., Dobrescu, R. ;Planinsic, P., "Mammographic Lesions Discrimination Based on Fractal Dimension as an Indicator". Systems, Signals and Image Processing, 2007, pp. 74-77.

[2] RaduDobrescu, Loretta Ichim, Daniela Crisan. "Diagnosis of breast cancer from mammograms by using fractal measures", Iternational Journal of Medicina Imaging, 2013; 1(2): 32-38. Published online October 30, 2013 (http://www.sicencepublishinggroup.com/j/ijmi)

[3] Napolitano, A.; Ungania, S., Cannata, V., "Fractal Dimension Estimation Methods for Biomedical Images", MATLAB - A fundamental Tool for Scientific Computing and Engineering Applications, vol.3, 2012.

[4] Sedivy, R.M.D., "Fractal Analysis: An Objective Method for Identifying A typical Nuclei", Gynecologic Oncology 75, 1999, pp. 78-83.

Crossover and Mutation Strategies applied in Job Shop Scheduling Problems

J. Vilma Roseline[1], Dr. D. Saravanan[2]
[1] Asst Professor, Department of Mathematics, SRM IST, Ramapuram Campus, Chennai, India, Research Scholar, Bharathiyar University, Coimbatore, India
[2] Professor of Mathematics, Karpaga Vinayaga College of Engineering and Technology, Chennai, India.

*Corresponding author: vilmaroselinej@gmail.com

Abstract: This paper reviews modern genetic algorithm based approaches for solving job shop scheduling problems. This paper elaborates the types of crossover methods such as Partially Mapped crossover (PMX) , Order One Crossover (OX), etc involved in Genetic Algorithm (GA) that are used for Job Shop Scheduling problems. Job Shop Scheduling represents one of the hardest combinatorial optimization problems where number of possible schedules drastically increases with the number of operations and machines. Normal crossover operators will often lead to inadmissible solution . Many specialized combining order or adjacency information from the two parents. The crossover method operators builds an offspring by choosing a subsequence of elements from one parent and preserving the order and position of as many elements as possible from the other parent. A subsequence of elements is selected by choosing two random cut points, which serve as boundaries for the swapping operations. There are few distinct mutation operators widely used for JSS problems such as swap, inversion, insertion (shift) and displacement mutation. Here some modern genetic algorithm-based approaches from the literature are also discussed aswell.

Keywords: genetic operators, job shop scheduling, process planning, makespan, optimization

1. Introduction

Production scheduling represents the process of making the best possible utilization of resources by allocating them over some time period with the purpose of performing some tasks and satisfying certain criteria [1,2]. This type of problem belongs to the group of combinatorial optimization problems and goal is to find best solution within a discrete set of alternative solutions [1]. A significant number of published papers are related to the optimization of production scheduling problems like the ones that will be discussed later. In this paper we are focused on job shop scheduling problem which is the most popular problem of this group. Job shop environment consists of a set of different resources/machines (lathes, milling machines, drilling machines etc.) that are able to perform operations of some jobs where jobs represent parts to be produced [3]. One of the simplest definitions of job shop scheduling (JSS) problem is provided by authors [4]: Given n jobs which have to be processed on m machines, each job consists of a sequence of task operations. Each of these operations requires to be processed without interruption for a given period of time on a given machine (no pre-emption). All machines in the shops are capable of processing at most one job at a time. Tasks of the same job cannot be processed simultaneously & each of them needs to meet each machine only once and the aim is to generate a time schedule to perform the operations on machines. One of the most attractive and mostly considered objectives is to find a scheduling plan that minimizes the makespan, that is, the time needed to complete all the jobs [4]. These assumptions are usually considered in static and deterministic models of classical JSS problems which are widely considered in the literature. During the years, metaheuristics became very popular approaches which are capable of solving instances of problems that are believed to be hard in general, by exploring the usually large solution search space of these instances [5]. Genetic algorithms (GA) are widely known metaheuristics which have been successfully applied to many different optimization problems in last few decades. They belong to the class of evolutionary algorithms which are based on the principles of natural evolution [6] and are very applicable to different combinatorial problems such as job shop scheduling. This paper reviews genetic operators as implementation methods for solving job shop scheduling problems. Many different types of GA components are mentioned and briefly discussed and some modern examples from the literature are analysed. Besides that, integrated process planning and scheduling approach is shortly mentioned and some examples are shown.

2. Job Shop Scheduling

The JSSP can be formulated as follows: For "n" jobs, each one is composed of several operations that must be executed on "m" machines. For each operation uses only one machine for a fixed duration. Each machine can process at most one operation at a time and once an operation starts processing on a given machine, it must complete processing on that same machine without any interruption. The operations of a given job have to be processed in a given set of order [(7)]. The problem's main aim is to schedule the activities or operations of the jobs on the machines such that the total time to finish all jobs, that is the makespan(Cmax), is minimized. The term makespan refers to the cumulative time to complete all the operations of all jobs on all machines. It is a measure of the time period from the starting time of the first operation to the ending time of the lastoperation.
In the literature authors often consider basic assumptions while solving JSS problem [12]:

- Job cannot visit one machine more than once;
- Simultaneous performing of operations not possible;
- Machine can process only one job at current time;
- No specifications of release and due dates;
- No pre-emptions of operations.

During the years, research in the field of JSS problems has expanded towards flexible job shop scheduling problems (FJSS) in which more than one machine for performing each operation are considered. This problem exhibit additional complexity in comparing to the classical JSS. Two different types of FJSS are considered in the literature: Total FJSS, where each operation can be processed on any machine on the shop floor, and partial FJSS, where each operation can be processed on some machine of subset of few existing machines on the shop floor.

3. Genetic Algorithms And Approaches For Improvement

Like the other algorithms in the class of evolutionary algorithms, genetic algorithm is based on population of individuals which assumes that population of possible solutions evolve through iteration process. As population based method, GA is exploration oriented, meaning that algorithm will ensure that all regions of search space are visited [5]. Genetic algorithm was firstly introduced by John Holland in 1975[9] and belongs to the group of evolutionary algorithms, like genetic programming among others. Since then, GA was applied on large number of various types of optimization problems including job shop scheduling. Davis [10] makes the point that although GAs work very well across a range of different problems, they can never be considered the best optimization method for any particular problem. Based on some disadvantages, like slow convergence and lack of exploitation ability (reduced possibility of finding best possible individual in visited regions, i.e. bad local search abilities) [11], there are three different hybridization approaches that improve efficiency of GA [2,10]:

Many authors classify GA representations in two basic ones: direct and indirect. Main distinction is that for direct representations, schedules are directly encoded into chromosomes, while, on the other side, for indirect representations that is not the case. For job shop scheduling problems several GA representations are used. Two main categories are model-based and algorithm based GA representations according to the authors [6,13,14]. Different types of these representations are illustrated on Figure 1.

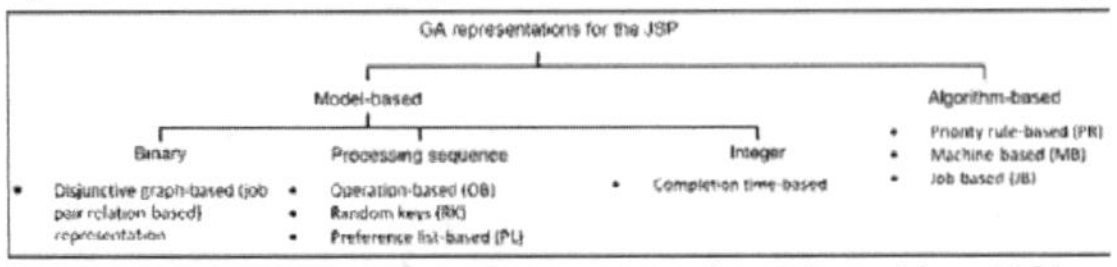

Figure 1. GA types of representations for JSS problem [13]

For model-based representations, the structure of the genotype (configuration of genes) is based on the decision variables of particular job shop mathematical model and the genotype can be directly interpreted to feasible or infeasible solution. If infeasible solution is obtained, it has to be transformed into a feasible one. On the other side, in algorithm based representations genotype used to store information which can be used by an algorithm to generate feasible solutions. Modelbased representation are divided into three groups based on decision variables that are considered; those are binary, processing sequence and integer variables as shown on Figure 1 [6,13]. After evaluating fitness values for each individual in population, standard procedures for selecting best individuals (schedules) are considered. According to author [6] fitness-proportional selection (popular as roulette wheel) and tournament selection are mostly used for job shop problems.

3.1 Crossover

The crossover operates on two chromosomes at a time and generates offspring by combining features of both chromosomes. In general, the performance of the GAs depends, to a great extent, on the performance of the crossover operator used. During the past two decades, various crossover operators have been proposed for literal permutation encodings for JSSP, such as partial-mapped crossover (PMX), order crossover (OX), cycle crossover (CX), position-based crossover, order-based crossover, etc.[(20)]. In this paper, we use a modified OX that proposed by Gao and et al. [(21)], where they used OX in operation sequence; here we used it in job sequence for every machine.

The PMX operator selects two random points and for generating offspring, it copies segment between these points from one parent, and replaces this segment of the second parent. Resulted offspring is called proto offspring which has to be repaired by mappings from parent 1 to parent 2.

Chromosome	Gender	
	Generation t	Generation t+1
C_1	Male	Female
C_2	Female	Male
C_3	Male	Female
C_4	Female	Male
C_5	Male	Female

The modified OX can be described as follows:

Step 1: Select a subsection of job sequence for a machine from one parent at random.

Step 2: Produce offspring by copying the substring of job sequence for a machine into the corresponding positions.

Step 3: Delete the jobs that are already in the substring from the second parent. The resulted sequence of jobs contains jobs that the offspring needs.

Step 4: Place the jobs into the unfixed positions of the offspring from left to right according to the order of the sequence in the second parent. The modified OX for one machine is illustrated in Figure.

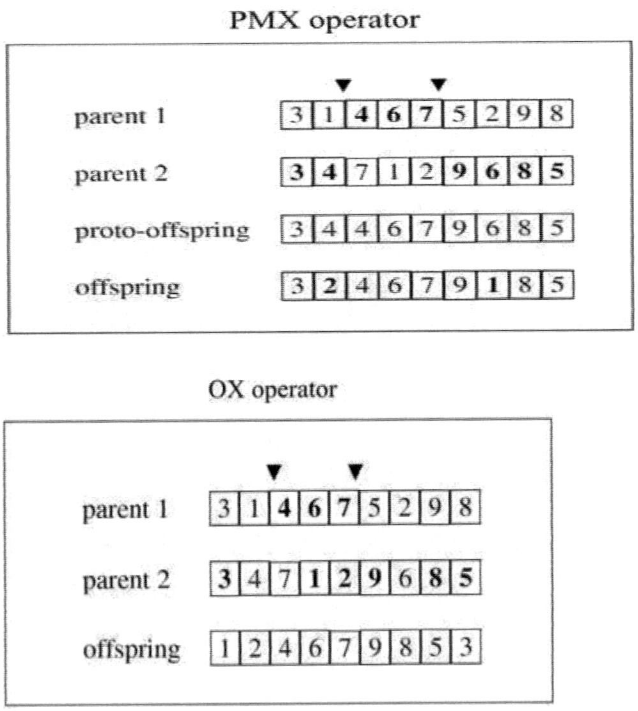

Figure 2. PMX and OX operators of GA [6]

3.2 Sexual Selection

In a standard GA, chromosomes reproduce asexually, that is, any two chromosomes may be the parents in crossover. Gender division and sexual selection inspired a model of gendered GA in which crossover takes place only between chromosomes of opposite sex. Since GA mimics natural evolution, then gender is one of the crucial elements in a GA. Inspired by the nongenetic sex determination system prevalent in some reptile species where sex is determined by the temperature at which the egg is incubated, the population are divided such that the male and female would be selected in an alternate way. The layout of the male and female chromosomes in each generation is different. In other words, when the number of generation is even, the chromosomes and are the male and female, respectively, and vice versa when the number of the generation is odd. For instance, suppose that a population size is five, the layout of males and females from this population is shown inTable 1

3.3 Mutation

Mutation is just used to produce small perturbations on chromosomes in order to maintain the diversity of population. During the last decade, there are several mutation operators have been proposed such as inversion, insertion, displacement, reciprocal exchange mutation, and shift mutation [(21)]. Inversion mutation selects two positions within a chromosome at random and then inverts the substring between these two positions. In this paper we advanced inversion mutation, by choosing two or more positions in the jobs sequence for one machine and invert them as showed in Figure 3. These steps are applied to other machines, to get a new Offspring.

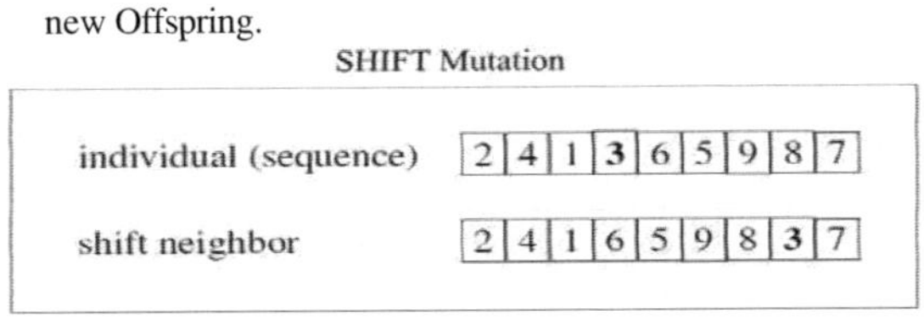

Figure 3. Shift mutation procedure [6]

3.4 New Generation

In this step, the new generation composed by migration initial population with and offspring population. Then we take the best individuals of initial population with generation gap percentage and offspring population.

3.5 Termination Test

The algorithm is terminated either the maximum number of generations is achieved, or when the individuals of the population converge, convergence occurs when all individual positions in the population are identical. In this case, crossover will have no further effect.

4. Conclusion

In the paper the brief overview of genetic algorithms for job shop problems is introduced. This is extending the knowledge of reproduction scenario which we have learnt in all school days. Different types of representations, selection strategies, genetic operators (crossover and mutation) and termination criteria of GA for successful implementation in the search for optimal scheduling plans are also discussed.

References

[1] Kumar, R., Sankha, D.: A Genetic AlgorithmBased Approach for Optimization of Scheduling in Job Shop Environment, Journal of Advanced Manufacturing Systems, Vol. 10, p.p. 223-240, 2011.

[2] Gen, M., Cheng, R.: Genetic Algorithms and Engineering Optimization, John Wiley & Sons, Inc., New York, 2000.

[3] Blazewicz, J., Ecker, K., Pesch, E., Schmidt, G., Węglarz, J.: Handbook on Scheduling: From Theory to Applications, Springer Berlin Heidelberg, p.p. 345-396, Berlin, 2007.

[4] Shafia, M. A., Pourseyed Aghaee, M., Jamili, A.: A new mathematical model for the job shop scheduling problem with uncertain processing times, International Journal of Industrial Engineering Computation, Vol. 2, p.p. 295-306 , 2011.

[5] El-Ghazali, T.: Metaheuristics: From Design to Implementation, John Wiley & Sons, Inc., New Yersey, 2009.

[6] Werner, F.: Genetic Algorithms for Shop Scheduling Problems: A Survey, chapter 8, in: Siarry, P.: Heuristics: Theory and Applications, Nova Science Publishers, Inc., New York, 2013.G. Eason, B. Noble, and I. N. Sneddon, "On certain integrals of Lipschitz-Hankel type involving products of Bessel functions," Phil. Trans. Roy. Soc. London, vol. A247, pp. 529–551, April 1955. *(references)*

[7] 4. Tamilarasi A., Jayasankari S., 2012, Evaluation on GA based Model for solving JSSP, International Journal of Computer Applications, 43(7), 975 – 8887.

[8] Goncalves J.F., Mendes J.J.M., Resende M.G.C., 2005, A hybrid genetic algorithm for the job shops scheduling problem, European Journal of Operational Research, 167, 77–95.K. Elissa, "Title of paper if known," unpublished.

[9] [9] Holland, J.A.: Adaptation in natural and artificial systems, MIT Press, USA, 1975.

[10] Partridge, D.: A review of "Handbook of Genetic Algorithms" from Davis, L.D., Taylor & Francis Group, Vol. 3, p.p. 446-448, 1991.

[11] Beasley, D., Bull, D.R., Martin, R.R.: An Overview of Genetic Algorithms: Part 1, Fundamentals, University Computing, Vol.15, p.p. 58-69, 1993.

[12] Bagchi, T.: Multiobjective Scheduling By Genetic Algorithm, Springer US, New York, 1999. M. Young, The Technical Writer's Handbook. Mill Valley, CA: University Science, 1989.

[13] Grosan, C., Abraham, A.: Hybrid Evolutionary Algorithms: Methodologies, Architectures, and Reviews, Studies in Computational Intelligence, Vol. 75, p.p. 1–17, 2007. [13] Abdelmaguid, T. F.: Representations in Genetic Algorithm for the Job Shop Scheduling Problem: A Computational Study, J. Software Engineering & Applications, Vol. 3, p.p. 1155-1162, December 2010.

[14] Cheng, R., Gen, M., Tsujimura, Y.: A Tutorial Survey of Job-Shop Scheduling Problems using Genetic Algorithms – I. Representation, Computers & Industrial Engineering, Vol. 30, p.p. 983-997, September 1996.

[15] Cheng, R., Gen, M., Tsujimura, Y.: A Tutorial Survey of Job-Shop Scheduling Problems using Genetic Algorithms – II. Hybrid genetic search strategies, Computers & Industrial Engineering, Vol. 36, p.p. 343-364, April 1999.

[16] Maghfiroh, M.F.N., Darmawan, A., Yu, V.F.: Genetic Algorithm for Job Shop Scheduling Problem: A Case Study, International Journal of Innovation, Management and Technology, Vol. 4, February 2013.

[17] Đurđev, M., Milošević, M., Lukić, D., Jovičić, G., Vukman, J.: A review on integrated process planning and production

scheduling approach, 12th International Scientfic Conference MMA 2015-Flexible Technologies, p.p. 117-120 , Novi Sad, 25-26 September, 2015.

[18] Shao, X., Li, X., Gao, L., Zhang, C.: Integration of process planning and scheduling – A modified genetic algorithm-based approach, Computers & Operations Research, Vol. 36, pp. 2082–2096, July 2008.

[19] Zhang, L., Wong, T.N.: An object-coding genetic algorithm for integrated process planning and scheduling, European Journal of Operational Research, Vol. 244, p.p. 434-444, January 2015.

[20] Cheng R., Gen M., Tsujimura Y., 1999, A tutorial survey of job-shop scheduling problems using genetic algorithms, Computers & Industrial Engineering, 36, 343-364.

[21] Gao J., Sun L., Gen M., 2008, A hybrid genetic and variable neighborhood descent algorithm for flexible job shop scheduling problems, Computers & Operations Research, 35, 2892 – 2907.

General Analysis of Two Products Inventory Model – With Production by Two Units System and Sales

S Parvathi[1], M Vijaya Bharathy[2]and A Manikandan[3]

Department of Mathematics,

[1,2,3] SRMIST, Ramapuram Campus, Chennai ,Tamil Nadu, India.

parvathis79@gmail.com, vijirsy87@gmail.com, manimscmphil@gmail.com.

Abstract: In this paper two units system engenders two different products for sale. The two products are engendered piecemeal in pairs. The sale time commences when k dyads are engendered or when the two units system fails whichever occurs first. Postulating the engenderment and sale time of products have general distribution and the failure and rehabilitate rates of the two units are constants, the double Laplace Stieltjes transform of the joint distribution function of the time to sale and sale time are obtained. Their prospect of engenderment time E (T), sales time E(R) is derived by varying the number of engenderment k. To illustrate the application of Numerical results are distribution by giving different values for λ1 λ2 μ.

Mathematics subject classification: 30C45, 30C80.

Keywords: Inventory systems, two different products, seasonal sales.

1 Introduction. Several researchers studied single commodity inventory systems of (s,S) type. Arrow, Karlin and Scart [1] first analysed such inventory systems. Daniel and Ramanarayanan [2] discussed (s, S) inventory system with desultory lead times and unit demand models. Murthy and Ramanarayanan [3, 4, 5, 6] have considered several (s,S) inventory systems. RajaRao.B [7] Life expectancy for setting the clock back to zero property Usha., Eswariprem, Ramanarayanan [8,9,10] have engendered some results in stochastic Analysis of Time to Vital Organs Failure of Gestational Diabetic Person. Wu and Liang-Yuh ouyang [11] studied (Q, R, L) inventory model with defective items. In authentic life inventory models variants of pairs of products are engendered for sales by the companies. For example drug manufacturing companies engender antibiotic and vitamin tablets. Oil refining companies engender fuel oils like petrol and engine oil to reduce friction and auto mobile companies engender engine and body of conveyances. Textiles mills slake the authoritative ordinance for shirting cloth and pant cloth, inner garments and outer garments land so on. In this paper we consider an inventory system in which two different products are engendered in pairs by a two-unit system. The sale time commences when k dyads are engendered a two- unit system fails. The Laplace Stieltjes Transform of distribution function of time to sale and sale time is derived. Their prospect times and numerical examples are presented.

2. Model

Assumptions: The company engenders two different products A & B and at a time only one type is engendered. The engenderment time of product A is arbitrary variable with Cdf $G_P(p)$and that of product B has cdf $G_Q(q)$. Products A & B are engendered in successive manner. The engenderment time P+Q of a dyad has cdf G(P).

The products are engendered by a two unit system which fails when the two units are down and it works when at least one unit is good. Let the probability that either of two units fails during (t, t+Δt) given that the two units are operating at time t, be $\lambda_1\Delta t+o(\Delta t)$ and the probability that one fine-tuned unit fails during (t,t+Δt) given that it is operating at time t, be $\lambda_2\Delta t+o(\Delta t)$. The rehabilitation rate of the failed unit is μ

Sale time commences when k numbers of pairs of products are engendered or when the two- unit system fails.

The products are sold in pairs and the selling time of a dyad is arbitrary variable with Cdf R(q) and the selling time of product A is $R_A(q)$.

Analysis

We note that the probability of n number of pairs produced in (o, t)

$= G_n(t) - G_{n+1}(t)$ where $G_n(t)$ is the Cdf of $\sum_{i=1}^{n}(P_i + Q_i)$. When the nth pair is produced at time P<t during t-P there are two possibilities. After the nth pair production,

(i) the production for product A is over but for B is not over or
(ii) the production for product A is also not over.

Their respective probabilities are given below. We note that the probability of n number of pairs and one production of A is over before t

$$= \int_0^t g_n(x) \int_0^{t-x} g_P(u)\overline{G_Q}(t-x-u)dudx \quad (1)$$

$= P[(\underline{n} + 1)$ number of A products and n number of B products are produced during (0,t)]. Here $\bar{G}(x)$ = 1 - G(P)

Here $g_n(x)$ is the pdf of $\sum_{i=1}^{n}(P_i + Q_i)$ are produced in (0,t). Also the probability of n number of pairs in (0, t) and the production of A is not completed before t is

$$= \int_0^t g_n(x)\overline{G_P}(t-x)dx. \quad (2)$$

Since the selling time starts when k pairs are produced or when the two unit system failed, T- the time to start sales is given by

T = min (time to produce k pairs, the time at which the two- unit system fails)

To find the distribution function of the time to fail time to failure of the two units system, we need the functions $P_{0,0}(u)$=P (At time u the two units system is working, The system does not fail during (0,u) / at time 0 the two units of the system are working). $P_{0,1}(u)$=P (at time u one unit is under repair, system does not fail during (0,u) $|$ (at time 0 the units are working). $P_{0,2}(u)$ be the pdf of time to failure of the two units system, $P_{0,2}(u)$=p(the two –unit system fails during (u,u+du), it does not fail during (0,u) $|$ the two units are working at time 0) .

We now calculate the P(.) functions. $P_{0,0}(P)$ satisfies the following,

$$P_{0,0}(P) = e^{-\lambda_1 x} + \int_0^x \lambda_1 e^{-\lambda_1 u} P_{1,0}(P-u)du$$

Here $P_{1,0}(P)$ = P(At time P the two units are working /at time 0 one unit is in failed state)

$$P_{1,0}(P) = \int_0^x \mu e^{-\mu u} e^{-\lambda_2 u} e^{-\lambda_1 (x-u)}du + \int_0^x \int_0^v \mu e^{-\mu u} e^{-\lambda_2 u} \lambda_1 e^{-\lambda_1 (v-u)} du P_{1,0}(P-v)dv.$$

The first term is the probability that the failed unit is repaired at u and no unit of the system fails during (0, P) and the second term is the probability that the failed unit is repaired at u and other unit does not fail during (0,u), a unit fails at v>u and at P all the two units are working. Laplace transforms of the above equations give,

$$P^*_{0,0}(s) = (\lambda_2+\mu+s)/[s^2+s(\lambda_1+\lambda_2+\mu)+ \lambda_1 \lambda_2] \quad (3)$$

Here* indicates Laplace transform.

Using a similar argument we find

$P_{0,1}(x) = \int_0^x \lambda_1 e^{-\lambda_1 u} e^{-\mu(x-u)} e^{-\lambda_2(x-u)} du + \int_0^x \int_0^v \lambda_1 e^{-\lambda_1 u} \mu e^{-\mu(v-u)} e^{-\lambda_2(v-u)} du P_{0,1}(P-v)dv$.

Taking Laplace Transform $P^*_{0,1}(s) = \lambda_1/[s^2+s(\lambda_1+\lambda_2+\mu)+\lambda_1\lambda_2]$ (4)

The failure density $p_{0,2}(P)$ satisfies

$P_{0,2}(x) = \int_0^x \lambda_1 e^{-\lambda_1 u} p_{1,2}(P-u)du$

where $p_{1,2}(x)\,dx$= p(the two units system fails during (x, x+dx) at time 0 one unit of the system is under repair)

$$P_{1,2}(x) = \lambda_2 e^{-\lambda_2 x} e^{-\mu x} + \int_0^x e^{-\lambda_2 u} \mu e^{-\mu u} P_{0,2}(P-u)du$$

By Laplace transformation,

$$P^*_{0,2}(s) = \lambda_1 \lambda_2/[s^2+s(\lambda_1+\lambda_2+\mu)+\lambda_1\lambda_2] \quad (5)$$

Equation (3) – (5) can be inverted easily β

$P_{0,0}(t) = (\frac{1}{2}) e^{-\alpha t}[e^{\beta t} + e^{-\beta t}] + \left(\frac{1}{4\beta}\right)(\lambda_2 - \lambda_1 + \mu) e^{-\alpha t}[e^{\beta t} - e^{-\beta t}]$

$P_{0,1}(t) = (\frac{\lambda_1}{2\beta}) e^{-\alpha t}[e^{\beta t} - e^{-\beta t}]$

$P_{0,2}(t) = (\frac{\lambda_1\lambda_2}{2\beta}) e^{-\alpha t}[e^{\beta t} - e^{-\beta t}]$

Here $a = (\lambda_1+\lambda_2+\mu)/2$

and $b = (\frac{1}{2})\sqrt{(\lambda_1 - \lambda_2)^2 + (\mu)^2 + 2\mu(\lambda_1 + \lambda_2)}$

$P_{0,0}(t)+P_{0,1}(t)$ = survival function of two unit system.

$$= e^{-(\alpha-\beta)t}[\frac{1}{2}+\frac{\lambda_2}{4\beta}-\frac{\lambda_1}{4\beta}+\frac{\mu}{4\beta}] + e^{-(\alpha+\beta)t}[\frac{1}{2}-\frac{\lambda_2}{4\beta}+\frac{\lambda_1}{4\beta}-\frac{\mu}{4\beta}]$$

$$= c_1 e^{-(\alpha-\beta)t} + c_2 e^{-(\alpha+\beta)t}$$

$p_{0,2}(t)$ = pdf of two unit system

$$= \frac{\lambda_1\lambda_2}{2b}[e^{-(\alpha-\beta)t} - e^{-(\alpha+\beta)t}]$$

$$= c_3 [e^{-(\alpha-\beta)t} - e^{-(\alpha+\beta)t}$$

The pdf of T is

$$f_T(t) = \sum_{i=0}^{k-1}\int_0^t g_i(x)\int_0^{t-x} g_P(u)\bar{G}(t-x-u)dudx + \sum_{i=0}^{k-1}\int_0^t g_i(x)\overline{G_P}(t-x)dx. \quad (6)$$

$$f_T(t) = g_k(t)\,(c_1 e^{-(\alpha-\beta)t} + c_2 e^{-(\alpha+\beta)t}) + c_3 (e^{(\alpha-\beta)t} - e^{-(\alpha+\beta)t})$$
$$\{\sum_{i=0}^{k-1}\int_0^t g_i(x)\int_0^{t-x} g_P(u)\bar{G}(t-x-u)dudx + \sum_{i=0}^{k-1}\int_0^t g_i(x)\overline{G_P}(t-x)dx\}.$$

The first term of the right side of (6) is the part of the pdf that the time to produce k number of pairs is t and the two units system has not failed upto time t. One part of the second term is the part of the pdf that the two unit system fails at time t, the time to produce i number of A and B pairs is P, the $(i+1)^{th}$ A product is produced at time P + u and the production of the $(i+1)^{th}$ B product is not over during (0, t-P-u) for $0 \le i \le k-1$. The other part of the second term is part of the pdf that the two units system fails at time t, the time to produce i number of A and B pairs is P and the $(i+1)^{th}$ production of product A is not over during (0, t-P), $0 \le i \le k-1$

This gives the joint pdf of time to start sales T and total sales time of pairs R as follows considering the sales time of the pairs.

$f_{T,R}(x,y) = g_k(x)(c_1 e^{-(\alpha-\beta)x} + c_2 e^{-(\alpha+\beta)x})\, r_k(y) + c_3 [e^{-(\alpha-\beta)x} - e^{-(\alpha+\beta)x}]$

$$\left[\sum_{i=0}^{k-1}\int_0^x g_i(u)\int_0^{x-u} g_P(v)\overline{G_Q}(x-u-v)dvdu\int_0^y r_i(w)r_A(y-w)dw + \sum_{i=0}^{k-1}\int_0^x g_i(x)\overline{G_P}(x-u)du\, r_i(y)\right] \quad (7)$$

The double Laplace transform of the pdf is given by

$$f_{T,R}^*(\varepsilon,\eta) = \int_0^\infty\int_0^\infty e^{-\varepsilon x}e^{-\eta y}f_{T,R}(x,y)dxdy$$

$$f_{T,R}^*(\varepsilon,\eta) = \int_0^\infty\int_0^\infty e^{-\varepsilon x-\eta y}g_k(x)(c_1\, e^{-(\alpha-\beta)x} + c_2\, e^{-(\alpha+\beta)x}r_k(y)dxdy + \int_0^\infty\int_0^\infty e^{-\varepsilon x-\eta y}\, c_3(e^{-(\alpha-\beta)x} - e^{-(\alpha+\beta)x})\sum_{i=0}^{k-1}\int_0^x g_i(u)\int_0^{x-u} g_P(v)\overline{G_Q}(x-u-v)dvdu\int_0^y r_i(w)r_A(y-w)dwdxdy + \int_0^\infty\int_0^\infty e^{-\varepsilon x-\eta y}\, c_3(\, e^{-(\alpha-\beta)x} - e^{-(\alpha+\beta)x})\sum_{i=0}^{k-1}\int_0^x g_i(u)\overline{G_P}(x-u)du\, r_i(y)dxdy$$

We get

$$f_{T,R}^*(\varepsilon,\eta) = c_1\, g^{*k}(\varepsilon+\alpha-\beta)r^{*k}(\eta) + c_2\, g^{*k}(\varepsilon+\alpha+\beta)r^{*k}(\eta) + c_3\sum_{i=0}^{k-1} g^{*i}(\varepsilon+\alpha-\beta)\, g_P^*((\varepsilon+\alpha-\beta))\overline{G_Q^*}((\varepsilon+\alpha-\beta)\varepsilon+\alpha-\beta)r^{*i}(\eta)r_A^*(\eta) - c_3\sum_{i=0}^{k-1} g^{*i}(\varepsilon+\alpha+\beta)\, g_P^*((\varepsilon+\alpha-\beta))\overline{G_Q^*}((\varepsilon+\alpha-\beta))r^{*i}(\eta)r_A^*(\eta) + c_3\sum_{i=0}^{k-1} g^{*i}(\varepsilon+\alpha-\beta)\,\overline{G_P^*}((\varepsilon+\alpha-\beta))r^{*i}(\eta) - c_3\sum_{i=0}^{k-1} g^{*i}(\varepsilon+\alpha+\beta)\,\overline{G_Q^*}(\varepsilon+\alpha+\beta)r^{*i}(\eta)$$

$$f_{T,R}^*(\varepsilon,\eta) = c_1\, g^{*k}((\varepsilon+\alpha-\beta))r^{*k}(\eta) + c_2\, g^{*k}(\varepsilon+\alpha+\beta)r^{*k}(\eta) + c_3\left[\frac{1-\left(g^*((\varepsilon+\alpha-\beta))r^*(\eta)\right)^k}{1-g^*(\varepsilon+\alpha-\beta)r^*(\eta)}\right]\left[g_P^*(\varepsilon+a-b)\overline{G_Q^*}\left(((\varepsilon+\alpha-\beta))\right)r_A^*(\eta) + \overline{G_P^*}(\varepsilon+\alpha-\beta)\right] - c_3\left[\frac{1-(g^*(\varepsilon+\alpha+\beta)r^*(\eta))^k}{1-g^*(\varepsilon+\alpha+\beta)r^*(\eta)}\right]\left[g_P^*(\varepsilon+\alpha+\beta)\overline{G_Q^*}\left(((\varepsilon+\alpha+\beta))\right)r_A^*(\eta) + \overline{G_P^*}(\varepsilon+\alpha+\beta)\right] \quad (8)$$

The Laplace transform of T is

$$f_{T,R}^{*}(\varepsilon,0)=c_1\mathrm{g}^{*\mathrm{k}}\big((\varepsilon+\alpha-\beta)\big)+c_2\,\mathrm{g}^{*\mathrm{k}}\big((\varepsilon+\alpha+\beta)\big)+c_3\left[\frac{1-\big(\mathrm{g}^{*}((\varepsilon+\alpha-\beta))\big)^{\mathrm{k}}}{1-\mathrm{g}^{*}((\varepsilon+\alpha-\beta))}\right]\Big[\mathrm{g}_{\mathrm{P}}^{*}\big((\varepsilon+\alpha-\beta)\big)\overline{\mathrm{G}_{\mathrm{Q}}^{*}}\Big(\big((\varepsilon+\alpha-\beta)\big)\Big)+\overline{\mathrm{G}_{\mathrm{P}}^{*}}\big((\varepsilon+\alpha-\beta)\big)\Big]-c_3\left[\frac{1-\big(\mathrm{g}^{*}(\varepsilon+\alpha+\beta)\big)^{\mathrm{k}}}{1-\mathrm{g}^{*}((\varepsilon+\alpha+\beta))}\right]\Big[\mathrm{g}_{\mathrm{P}}^{*}(\varepsilon+\alpha+\beta)\overline{\mathrm{G}_{\mathrm{Q}}^{*}}(\varepsilon+\alpha+\beta)+\overline{\mathrm{G}_{\mathrm{P}}^{*}}\big((\varepsilon+\alpha+\beta)\big)\Big] \quad (9)$$

Section 3 : Production Time E(T), Sales Time E(R),

On differentiation of equation (6) we get, α

$\frac{\partial}{\partial\varepsilon}f_{T,R}^{*}(0,0)=-\mathrm{E(T)}$ and we obtain

$$\begin{aligned}E(T)=&-c_1\,\mathrm{g}^{*\mathrm{k}-1}(\alpha-\beta)\mathrm{g}^{*\prime}(\alpha-\beta)-c_2\,\mathrm{g}^{*\mathrm{k}-1}(\alpha+\beta)\mathrm{g}^{*\prime}(\alpha+\beta)\\&+c_3\left[\frac{\mathrm{k}\,\mathrm{g}^{*\mathrm{k}-1}(\alpha-\beta)\mathrm{g}^{*\prime}(\alpha-\beta)}{1-\mathrm{g}^{*}(a-b)}\right]\left[\mathrm{g}_{\mathrm{P}}^{*}(\alpha-\beta)\overline{\mathrm{G}_{\mathrm{Q}}^{*}}(\alpha-\beta)+\overline{\mathrm{G}_{\mathrm{P}}^{*}}(\alpha-\beta)\right]\\&-c_3\left[\frac{1-\mathrm{g}^{*\mathrm{k}}(\alpha-\beta)}{\big(1-\mathrm{g}^{*}(\alpha-\beta)\big)^2}\right]\mathrm{g}^{*\prime}(\alpha-\beta)\left[\mathrm{g}_{\mathrm{P}}^{*}(\alpha-\beta)\overline{\mathrm{Q}}(\alpha-\beta)+\overline{\mathrm{G}_{\mathrm{P}}^{*}}(\alpha-\beta)\right]\\&-c_3\left[\frac{1-\mathrm{g}^{*\mathrm{k}}(\alpha-\beta)}{\big(1-\mathrm{g}^{*}(\alpha-\beta)\big)}\right]\left[\mathrm{g}_{\mathrm{P}}^{*\prime}(\alpha-\beta)\overline{\mathrm{Q}}(\alpha-\beta)+\mathrm{g}_{\mathrm{P}}^{*}(\alpha-\beta)\overline{\mathrm{G}_{\mathrm{Q}}^{*}}'(\alpha-\beta)\right.\\&\left.+\overline{\mathrm{G}_{\mathrm{P}}^{*}}'(\alpha-\beta)\right]\\&-c_3\left[\frac{\mathrm{k}\,\mathrm{g}^{*\mathrm{k}-1}(a+b)\mathrm{g}^{*\prime}(a+b)}{1-\mathrm{g}^{*}(a-b)}\right]\left[\mathrm{g}_{\mathrm{P}}^{*}(\alpha+\beta)\overline{\mathrm{G}_{\mathrm{Q}}^{*}}(a+b)+\overline{\mathrm{G}_{\mathrm{P}}^{*}}(\alpha+\beta)\right]\\&+c_3\left[\frac{1-\mathrm{g}^{*\mathrm{k}}(\alpha+\beta)}{\big(1-\mathrm{g}^{*}(\alpha+\beta)\big)^2}\right]\mathrm{g}^{*\prime}(\alpha+\beta)\left[\mathrm{g}_{\mathrm{P}}^{*}(a+b)\overline{\mathrm{G}_{\mathrm{Q}}^{*}}(\alpha+\beta)+\overline{\mathrm{G}_{\mathrm{P}}^{*}}(\alpha+\beta)\right]\\&+c_3\left[\frac{1-\mathrm{g}^{*\mathrm{k}}(\alpha+\beta)}{\big(1-\mathrm{g}^{*}(\alpha+\beta)\big)^2}\right]\left[\mathrm{g}_{\mathrm{P}}^{*\prime}(\alpha+\beta)\overline{\mathrm{G}_{\mathrm{Q}}^{*}}(\alpha+\beta)+\mathrm{g}_{\mathrm{P}}^{*}(\alpha+\beta)\overline{\mathrm{G}_{\mathrm{Q}}^{*}}'(\alpha+\beta)\right.\\&\left.+\overline{\mathrm{G}_{\mathrm{P}}^{*}}'(\alpha+\beta)\right]\end{aligned} \quad (10)$$

Similarly we note $\frac{\partial}{\partial\eta}f_{T,R}^{*}(0,0)=-\mathrm{E(R)}$ and we obtain

$$\mathrm{E(R)}=c_1\,\mathrm{g}^{*\mathrm{k}}(\alpha-\beta)\mathrm{kE(R_1)}+c_2\,\mathrm{g}^{*\mathrm{k}}(\alpha-\beta)\mathrm{kE(R_1)}-c_3\mathrm{kE(R_1)}\left[\frac{\mathrm{g}^{*\mathrm{k}}(\alpha-\beta)}{1-\mathrm{g}^{*}(\alpha-\beta)}\right]\left[\mathrm{g}_{\mathrm{P}}^{*}(\alpha-\beta)\mathrm{G}_{\mathrm{Q}}^{-*}(\alpha-\beta)+\overline{\mathrm{G}_{\mathrm{P}}^{*}}(\alpha-\beta)\right]+c_3\mathrm{E(R_1)}\left[\frac{1-\mathrm{g}^{*\mathrm{k}}(\alpha-\beta)}{\big(1-\mathrm{g}^{*}(\alpha-\beta)\big)^2}\right]\mathrm{g}^{*}(\alpha-\beta)\left[\mathrm{g}_{\mathrm{P}}^{*}(\alpha-\beta)\mathrm{G}_{\mathrm{Q}}^{-*}(\alpha-\beta)+\overline{\mathrm{G}_{\mathrm{P}}^{*}}(\alpha-\beta)\right]+c_3\left[\frac{1-\mathrm{g}^{*\mathrm{k}}(\alpha-\beta)}{1-\mathrm{g}^{*}(\alpha-\beta)}\right]\left[\mathrm{g}_{\mathrm{P}}^{*}(\alpha-\beta)\mathrm{G}_{\mathrm{Y}}^{-*}(\alpha-\beta)(\mathrm{E}(\mathrm{R_y}))\right]+c_3\mathrm{kE(R_1)}\left[\frac{\mathrm{g}^{*\mathrm{k}}(\alpha+\beta)}{1-\mathrm{g}^{*}(\alpha+\beta)}\right]\left[\mathrm{g}_{\mathrm{P}}^{*}(\alpha+\beta)\mathrm{G}_{\mathrm{Q}}^{-*}(\alpha+\beta)+\overline{\mathrm{G}_{\mathrm{P}}^{*}}(\alpha+\beta)\right]-c_3\mathrm{E(R_1)}\left[\frac{1-\mathrm{g}^{*\mathrm{k}}(\alpha+\beta)}{\big(1-\mathrm{g}^{*}(\alpha+\beta)\big)^2}\right]\mathrm{g}^{*}(\alpha+\beta)\left[\mathrm{g}_{\mathrm{P}}^{*}(\alpha+\beta)\mathrm{G}_{\mathrm{Q}}^{-*}(\alpha+\beta)+\overline{\mathrm{G}_{\mathrm{P}}^{*}}(a+b)\right]-c_3\left[\frac{1-\mathrm{g}^{*\mathrm{k}}(\alpha+\beta)}{1-\mathrm{g}^{*}(\alpha+\beta)}\right]\left[\mathrm{g}_{\mathrm{P}}^{*}(\alpha+\beta)\mathrm{G}_{\mathrm{Q}}^{-*}(\alpha+\beta)(\mathrm{E(y)})\right] \quad (11)$$

Section: 4 : Special case of this Model

We now consider the special case in which P & Y re exponential random variables with parameters α and β respectively, this gives

$$g_P^*(\alpha-\beta)=\frac{r}{r+\alpha-\beta} \qquad g_P^{*\prime}(\alpha-\beta)=\frac{-r}{(r+\alpha-\beta)^2}$$
$$\overline{G_Q^*}(\alpha-\beta)=\frac{1}{s+\alpha-\beta} \qquad \overline{G_Q^*}'(\alpha-\beta)=\frac{-1}{(s+\alpha-\beta)^2}$$
$$\overline{G_P^*}(\alpha-\beta)=\frac{1}{r+\alpha-\beta} \qquad \overline{G_P^*}'(\alpha-\beta)=\frac{-1}{(r+\alpha-\beta)^2} \tag{12}$$

Using (7), (8) & (9) we find E(T) & E(R), when P & Y re exponentials with parameters α & β as follows.

$$E(T)=k\left(\frac{rs}{(r+\alpha-\beta)(s+\alpha-\beta)}\right)^k\left(\frac{r+s+2\alpha-2\beta}{(r+\alpha-\beta)(s+\alpha-\beta)}\right)\left(\frac{-\lambda_1}{2\beta}\right)+\left[1-\left(\frac{rs}{(r+\alpha-\beta)(s+\alpha-\beta)}\right)^k\right]\left(\frac{\alpha+\beta}{(\alpha-\beta)2\beta}\right)+k\left(\frac{rs}{(r+\alpha+\beta)(s+\alpha-\beta)}\right)^k\left(\frac{r+s+2\alpha-2\beta}{(r+\alpha-\beta)(s+\alpha-\beta)}\right)\left(\frac{\lambda_1}{2\beta}\right)-\left[1-\left(\frac{rs}{(r+\alpha+\beta)(s+\alpha+\beta)}\right)^k\right]\left(\frac{\alpha-\beta}{(\alpha+\beta)2\beta}\right) \tag{13}$$

$$E(R)=kE(R_1)\left(\frac{rs}{(r+\alpha-\beta)(\beta+\alpha-\beta)}\right)^k\left(\frac{-\lambda_1}{2\beta}\right)+\left(\frac{\alpha+\beta}{(\alpha-\beta)2\beta}\right)\left(\frac{r}{(\alpha-\beta)(r+s+\alpha-\beta)}\right)\frac{\beta\big(E(R_1)+(\alpha-\beta)E(R_A)\big)}{(\alpha-\beta)}$$
$$\left[\left[1-\left(\frac{rs}{(r+\alpha-\beta)(s+\alpha-\beta)}\right)^k\right]\right]+kE(R_1)\left(\frac{rs}{(r+\alpha+\beta)(s+\alpha+\beta)}\right)^k\left(\frac{\lambda_1}{2\beta}\right)-\left[1-\left(\frac{rs}{(r+\alpha+\beta)(s+\alpha+\beta)}\right)^k\right]\left(\frac{\alpha-\beta}{(\alpha+\beta)2\beta}\right)\left(\frac{r}{(\alpha+\beta)(r+s+\alpha+\beta)}\right)\left(\frac{s\big(E(R_1)+(\alpha+\beta)E(R_A)\big)}{(\alpha+\beta)}\right) \tag{14}$$

Where E(R) is the sales time of pair of products E (R_A) is the sale time of product A

3. Numerical Example

To illustrate the applications of the above result we give different values for λ1, λ_2 μ, and k and we obtain E(T), E(R)in the following table. For $E(R_1) = 10$ and E(R) = 20

	b	**k**	E(T)	E (R)
10	8.062	1	0.451925	0.092711
10	8.062	2	0.699881	0.513884
10	8.062	3	0.838209	0.771805
10	8.062	4	0.914601	0.879229
10	8.062	5	0.956512	0.917767
10	8.062	6	0.979385	0.930559

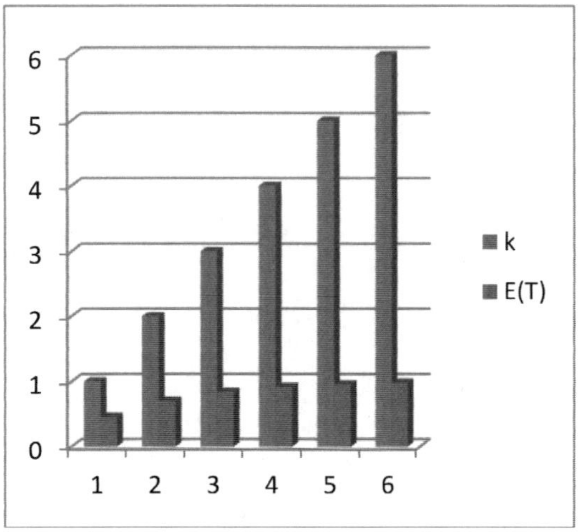

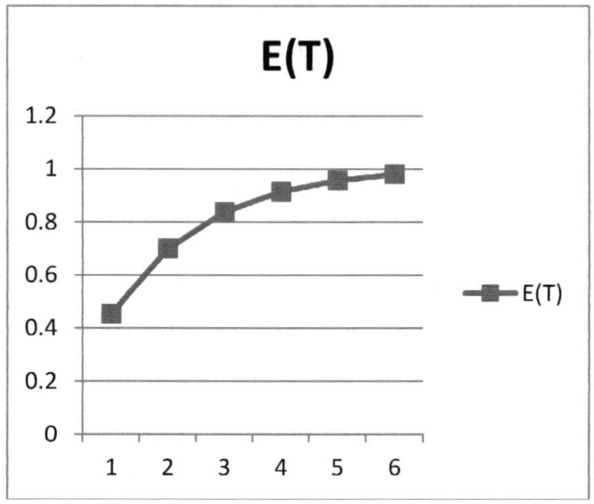

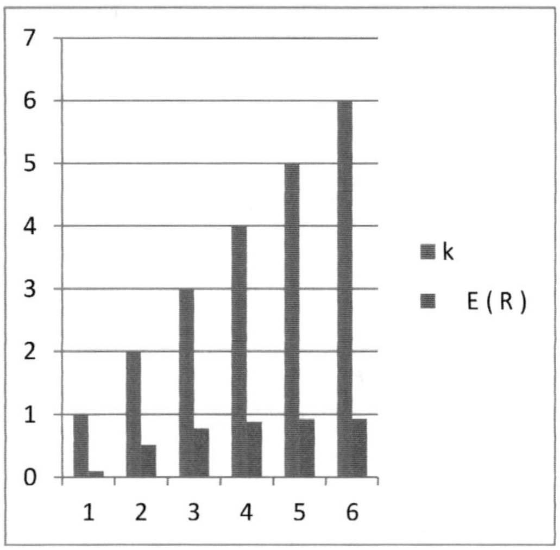

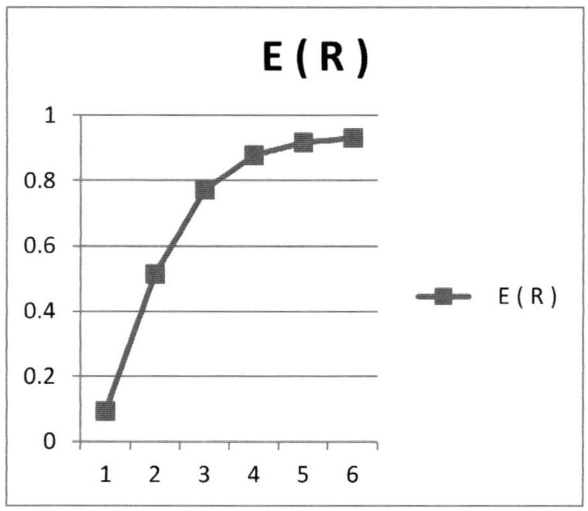

Keeping α=4, &β=6 and varying k values, Production E (T) increases and sales E (R) increases, in 1 & 2 dimensional graph. This graph insist that while season starts, the sales increases and production also increases.

	α=4, E(T)	α=5, E(T)	α=6, E(T)	α=7, E(T)	α=8, E(T)	α=9, E(T)
k						
1	0.452	0.417	0.392	0.372	0.357	0.344
2	0.700	0.660	0.629	0.604	0.584	0.567
3	0.838	0.804	0.776	0.753	0.733	0.717
4	0.915	0.889	0.866	0.847	0.831	0.816
5	0.957	0.938	0.922	0.907	0.893	0.882
6	0.979	0.967	0.955	0.944	0.934	0.925

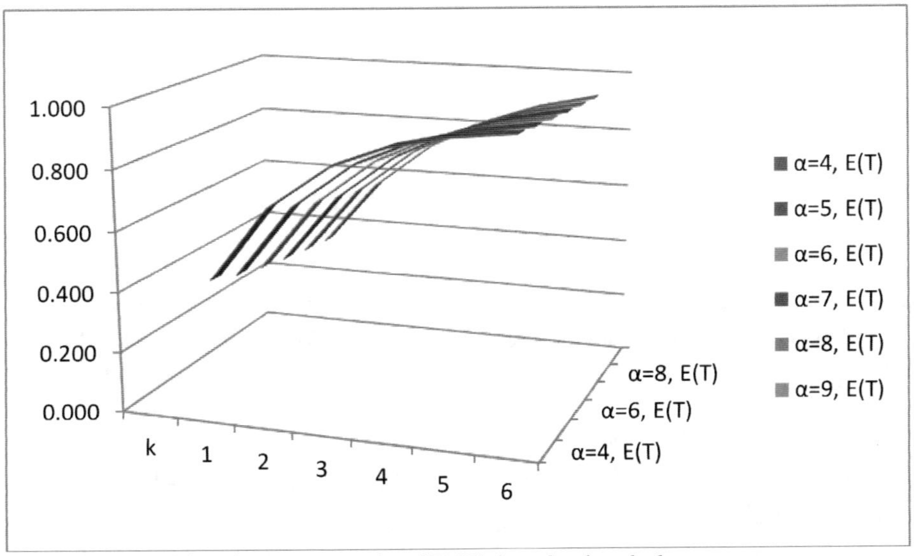

Varying α and k , the 3-d graph of E(T) is obtained above

	α=0.4, E (R)	α=0.5, E (R)	α=0.6, E (R)	α=0.7, E (R)	α=0.8, E (R)	α=0.9, E (R)
k						
1	0.093	0.111	0.128	0.144	0.159	0.172
2	0.514	0.524	0.507	0.468	0.412	0.344
3	0.772	0.848	0.881	0.872	0.828	0.752
4	0.879	1.014	1.107	1.156	1.162	1.125
5	0.918	1.086	1.221	1.320	1.378	1.393
6	0.931	1.115	1.274	1.406	1.504	1.564

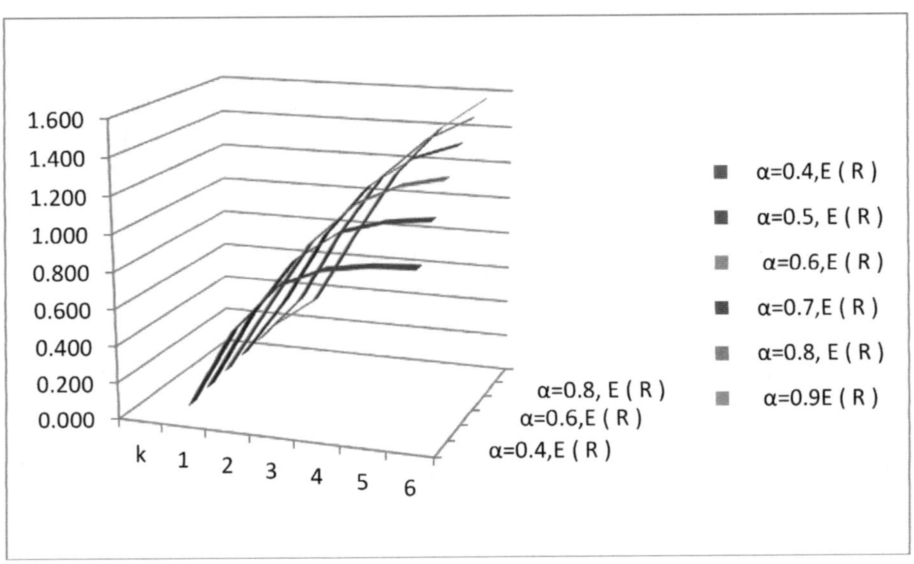

Varying α and k, the 3-d graph of E(R) is obtained above.

References

[1] Arrow.K., Karlin.S., and Scarf.H (1958.), "Studies in the Mathematical Theory of Inventory and Production", Standard University Press., Standard, California,

[2] Daniel.J.K., and Ramanarayanan. R (1988), "An (S, s) inventory system with rest periods to the server", Naval Research Logistics, 35 pp-119–123.

[3] Murthy.S., R.Ramanarayana ((2008), "One ordering and two ordering levels Inventory systems units with SCBZ lead times", International Journal of Pure and Applied Mathematics Vol. 47,3, pp- 427-447

[4] Murthy. S., and Ramanarayanan. R (2008), "Two ordering levels inventory system with different lead times and rest time to the server", Inter. J. Of Applied Math., Vol.21 (2), pp-265-280.

[5] Murthy. S., and Ramanarayanan. R (2008), Two (s, S) inventories with perishable units, The Journal of Modern Mathematics and Statistics, Vol.2 (3) pp-102–108.

[6] Murthy.S., R.Ramanarayana (2008), "Inventory system exposed to Calamity with SCBZ arrival property", The journal of Modern Mathematics and Statistics, Vol.2, 3, pp-109-119

[7] RajaRao.B (1998), "Life expectancy for setting the clock back to zero property", Mathematical Biosciences, pp-251-27

[8]Usha.K., Eswariprem, R. Ramanarayanan (2011), "Stochastic Analysis of Time to Vital Organs Failure of Gestational Diabetic Person", Vol. 5, No.10,

[9]Usha.k, Nithyapriya.N, and R.Ramanarayana (2012), "Cumulative demand inventory system" Inter. J. Of.contemp. Math Sciences, Vol.7, pp- 125-135

[10] Usha.k., Nithyapriya.N, and R.Ramanarayana (2011)," Inventory system with defective suppliers and financial support for customer, Inter. J. Of.contemp. Math Sciences", Vol 6, No 29. Pp-1397 -1410)

[11] Wu,Ouyang Liang-Yuh (2001), "(Q, r, L) inventory model with defective items", Computer and Industrial Engineering, Vol.39, pp-173-185.

12]Usha.K., Eswariprem, R. Ramanarayanan (2010), "Stochastic Analysis Time to Vital Organs Failure of Gestational Diabetic Person", Vol. 1, No.3,pp-287-301

[13]Usha.K., R. Ramanarayanan (2011), "General Analysis of of Systems in which a 2-unit system is a subsystem", Vol. 12, No.4,pp-629-637

[14] Ramanarayanan (1977), "General Analysis 1-out of 2: F Systems exposed to cumulative damage processes"., Vol. 8, No.4,pp-237-245

Journal of Physics: Conference Series **1377** (2019) 012033 doi:10.1088/1742-6596/1377/1/012033

APPLICATION OF CUBIC TRANSITION MATRICES AND PRODUCT FUZZY SOFT TRANSITIONMATRICES

G. Shyamsundar,Assistant Professor,
Department of Mathematics,
SRM Institute of Science and Technology,
Ramapuram Campus, Chennai 600 089.
email Id: sundarshyam1915@gmail.com

INTRODUCTION

In this chapter an application of cubic transition matrix in agricultural planning is provided. This chapter is concluded with a decision theory on product fuzzy soft matrices.

A DECISIONMAKING FOR AGRICULTURAL PLANNING USING CUBIC TRANSITION MATRICES

This section provides an algorithm for decision making to find an optimum solution on agriculture planning for a given data using cubic transition matrices as a tool.

The methodology involves seven steps in the algorithm and concludes that there is an increase in production level of crops.

In many countries the role of farmers in agricultural planning is highly remarkable. The pressure on farmers almost in all parts of the World is to produce more agricultural products and to have more income on it. But in recent years it is really a challenging job for the farmers to do so. Though the farmers face problems from several directions to increase the yield, the major crisis involved in them is due to lack of labors and monsoon failures.

The real time experience of developed countries shows the transfer of labor force from agriculture to non-agriculture; in particular to the industrial sector. Under these circumstances, higher growth in agriculture assumes greater importance and is a matter of concern for policy planners and research scholars in recent times.

In India, there are many farmers having small cultivation agricultural lands on their own. By the climatic condition and the present risk factors, farmers cultivate different crops as per the market need. For crop cultivation, it comprises mainly three levels, namely

1. Sowing of seeds in the land
2. Growth period of the crop
3. Yield period of the crop

Now a days in every period farmers face the major problem in the form of lack of labors. No one can able to utilize the power of labors fruitful. Here we present an algorithm to calculate the lack of labors. From this decision making process the farmers can able to analyze the lack of labors then switch over their crop in to other to maximize the profit.

Algorithm for optimization of labor power in agriculture

Step 1: Form the fuzzy square matrix for lack of labors over that particular crop

Step 2: Divide the square matrix in to number of 2x2 matrices like

$$A_{11} = \begin{bmatrix} a_{11} & a_{12} \\ a_{21} & a_{22} \end{bmatrix}, A_{12} = \begin{bmatrix} a_{12} & a_{13} \\ a_{22} & a_{23} \end{bmatrix}, \dots A_{1n} = \begin{bmatrix} a_{11} & a_{1n} \\ a_{21} & a_{2n} \end{bmatrix}$$

$$B_{21} = \begin{bmatrix} a_{21} & a_{22} \\ a_{31} & a_{32} \end{bmatrix}, B_{22} = \begin{bmatrix} a_{22} & a_{23} \\ a_{32} & a_{33} \end{bmatrix}, \dots B_{2n} = \begin{bmatrix} a_{21} & a_{2n} \\ a_{31} & a_{3n} \end{bmatrix}$$

$$\dots$$

$$K_{m1} = \begin{bmatrix} a_{m-1,1} & a_{m-1,2} \\ a_{m1} & a_{m2} \end{bmatrix}, K_{m2} = \begin{bmatrix} a_{m-1,2} & a_{m-1,3} \\ a_{m2} & a_{m3} \end{bmatrix}, \dots K_{mn} = \begin{bmatrix} a_{m-1,1} & a_{m-1,n} \\ a_{m1} & a_{mn} \end{bmatrix}$$

Step 3: Calculate the evaluation matrix as$\|A\| = \lambda(x)$.

Step 4: Evaluation process

Calculate the determinant using the operations in cubic transition matrix.

Step 5: Select the cubic transition matrix for the evaluation as minimum diagonal length.

Step 6: Find the inner product of cubic transition matrix as$a_{11}.a_{22} \oplus a_{12}.a_{21}$

Step 7: The Utilization factor is 1000x evaluation.

The following example justifies the above algorithm.

Example 5.1

Suppose there are three varieties of crops namely crop 1, crop 2, crop3 or simply 1, 2,3 with which three of labors namely I, II, III are cultivating them. The data is given below.

$$\begin{array}{c} \text{Crops} \\ \\ [\text{Labors}] \begin{array}{c} I \\ II \\ III \end{array} \end{array} \begin{array}{c} \;1 \qquad 2 \qquad 3 \\ \begin{bmatrix} 0.6 & 0.3 & 0.1 \\ 0.5 & 0.6 & 0.3 \\ 0.5 & 0.5 & 0.7 \end{bmatrix}. \end{array}$$

Our attempt is to quote a technique for the minimization of labors and to increase the profit using the above algorithm. We now construct the fuzzy square matrix from the above problem as follows.

Step-1 to Step-6.

$$\begin{array}{c} \text{Crops} \\ \\ [\text{Labors}] \begin{array}{c} I \\ II \\ III \end{array} \end{array} \begin{array}{c} 1 \qquad 2 \qquad 3 \\ \begin{bmatrix} 0.6 & 0.3 & 0.1 \\ 0.5 & 0.6 & 0.3 \\ 0.5 & 0.5 & 0.7 \end{bmatrix}. \end{array}$$

We now divide into sub matrices and proceed to find evaluation matrix, cubic transition matrix and its inner product as follows..

Let $A = \begin{bmatrix} 0.6 & 0.3 \\ 0.5 & 0.6 \end{bmatrix}$

$$||A|| = \begin{bmatrix} 0.6 & 0.3 \\ 0.5 & 0.6 \end{bmatrix} \begin{bmatrix} 0.6 & 0.3 \\ 0.5 & 0.6 \end{bmatrix}$$
$$= 0.6 \oplus 0.5 \oplus 0.5 \oplus 0.6 = 0.2$$

Let $B = \begin{bmatrix} 0.6 & 0.3 \\ 0.5 & 0.6 \end{bmatrix}$

$$||B|| = \begin{bmatrix} 0.3 & 0.1 \\ 0.6 & 0.3 \end{bmatrix} \begin{bmatrix} 0.3 & 0.1 \\ 0.6 & 0.3 \end{bmatrix}$$
$$= 0.9 \oplus 0.6 \oplus 0.6 \oplus 0.9 = 0.$$

Let $C = \begin{bmatrix} 0.6 & 0.3 \\ 0.5 & 0.6 \end{bmatrix}$

$$||C|| = \begin{bmatrix} 0.6 & 0.1 \\ 0.5 & 0.3 \end{bmatrix} \begin{bmatrix} 0.6 & 0.1 \\ 0.5 & 0.3 \end{bmatrix}$$
$$= 0.6 \oplus 0.5 \oplus 0.5 \oplus 0.9 = 0.5$$

Let $D = \begin{bmatrix} 0.6 & 0.3 \\ 0.5 & 0.6 \end{bmatrix}$

$$||D|| = \begin{bmatrix} 0.5 & 0.6 \\ 0.5 & 0.5 \end{bmatrix} \begin{bmatrix} 0.5 & 0.6 \\ 0.5 & 0.5 \end{bmatrix}$$
$$= 0.5 \oplus 0 \oplus 0 \oplus 0.5 = 0$$

Let $E = \begin{bmatrix} 0.6 & 0.3 \\ 0.5 & 0.7 \end{bmatrix}$

$$||E|| = \begin{bmatrix} 0.6 & 0.3 \\ 0.5 & 0.7 \end{bmatrix} \begin{bmatrix} 0.6 & 0.3 \\ 0.5 & 0.7 \end{bmatrix}$$
$$= 0.6 \oplus 0.5 \oplus 0.5 \oplus 0.9 = 0.5$$

Let $F = \begin{bmatrix} 0.5 & 0.3 \\ 0.5 & 0.7 \end{bmatrix}$

$$||F|| = \begin{bmatrix} 0.5 & 0.3 \\ 0.5 & 0.7 \end{bmatrix} \begin{bmatrix} 0.5 & 0.3 \\ 0.5 & 0.7 \end{bmatrix}$$
$$= 0.5 \oplus 0.5 \oplus 0.5 \oplus 0.9 = 0.4$$

Let $G = \begin{bmatrix} 0.6 & 0.3 \\ 0.5 & 0.5 \end{bmatrix}$

$$||G|| = \begin{bmatrix} 0.6 & 0.3 \\ 0.5 & 0.5 \end{bmatrix} \begin{bmatrix} 0.6 & 0.3 \\ 0.5 & 0.5 \end{bmatrix}$$
$$= 0.6 \oplus 0.5 \oplus 0.5 \oplus 0.5 = 0.1$$

Let $H = \begin{bmatrix} 0.3 & 0.1 \\ 0.5 & 0.7 \end{bmatrix}$

$$||H|| = \begin{bmatrix} 0.3 & 0.1 \\ 0.5 & 0.7 \end{bmatrix} \begin{bmatrix} 0.3 & 0.1 \\ 0.5 & 0.7 \end{bmatrix}$$
$$= 0.9 \oplus 0.5 \oplus 0.5 \oplus 0.9 = 0.8$$

Let $I = \begin{bmatrix} 0.6 & 0.1 \\ 0.5 & 0.7 \end{bmatrix}$

$\|I\| = \begin{bmatrix} 0.6 & 0.1 \\ 0.5 & 0.7 \end{bmatrix} \begin{bmatrix} 0.6 & 0.1 \\ 0.5 & 0.7 \end{bmatrix}$

$=0.6 \oplus 0.5 \oplus 0.5 \oplus 0.9 = 0.5$

Then the evaluation matrix is $\begin{bmatrix} 0.2 & 0 & 0.5 \\ 0 & 0.5 & 0.4 \\ 0.1 & 0.8 & 0.5 \end{bmatrix}$

Evaluation $= 0.2(0.5\sim0.2) \ominus 0 \oplus 0.5(0\sim0.5)$

$= 0.2(0.5\sim0.2) \ominus 0 \oplus 0.5(0\sim0.5)$.

$= 0.2(0.3) \oplus 0.5(0.5)$.

$= 0.1$.

We select$\lambda = 0.1$ in an internal cubic transition matrix having minimum diagonal length.

The conclusion matrix $\mathbb{C} = \begin{bmatrix} 1 & 0 \\ 0.9 & 0.1 \end{bmatrix}$ $=0.1 \oplus 0 = 0.1$

Then we conclude that the utilization factor is 10%

Now we minimize the labors.

$$\begin{array}{c} \\ \\ [Labors] \end{array} \begin{array}{c} \\ I \\ II \\ III \end{array} \overset{\text{Crops}}{\begin{array}{ccc} 1 & 2 & 3 \\ 0.5 & 0.2 & 0.1 \\ 0.4 & 0.5 & 0.3 \\ 0.4 & 0.4 & 0.6 \end{array}}.$$

Let $A = \begin{bmatrix} 0.5 & 0.2 \\ 0.4 & 0.5 \end{bmatrix}$

$\|A\| = \begin{bmatrix} 0.5 & 0.2 \\ 0.4 & 0.5 \end{bmatrix} \begin{bmatrix} 0.5 & 0.2 \\ 0.4 & 0.5 \end{bmatrix}$

$=0.5 \oplus 0.8 \oplus 0.8 \oplus 0.5 = 0.6$

Let $B = \begin{bmatrix} 0.2 & 0.1 \\ 0.5 & 0.3 \end{bmatrix}$

$\|B\| = \begin{bmatrix} 0.2 & 0.1 \\ 0.5 & 0.3 \end{bmatrix} \begin{bmatrix} 0.2 & 0.1 \\ 0.5 & 0.3 \end{bmatrix}$

$=0.4 \oplus 0.5 \oplus 0.5 \oplus 0.9 = 0.3$.

Let $C = \begin{bmatrix} 0.5 & 0.1 \\ 0.4 & 0.3 \end{bmatrix}$

$\|C\| = \begin{bmatrix} 0.5 & 0.1 \\ 0.4 & 0.3 \end{bmatrix} \begin{bmatrix} 0.5 & 0.1 \\ 0.4 & 0.3 \end{bmatrix}$

$=0.5 \oplus 0.4 \oplus 0.4 \oplus 0.9 = 0.2$

Let $D = \begin{bmatrix} 0.4 & 0.5 \\ 0.4 & 0.4 \end{bmatrix}$

$\|D\| = \begin{bmatrix} 0.4 & 0.5 \\ 0.4 & 0.4 \end{bmatrix} \begin{bmatrix} 0.4 & 0.5 \\ 0.4 & 0.4 \end{bmatrix}$

$=0.6 \oplus 0 \oplus 0 \oplus 0.6 = 0.2$

Let $E = \begin{bmatrix} 0.5 & 0.3 \\ 0.4 & 0.6 \end{bmatrix}$

$\|E\| = \begin{bmatrix} 0.5 & 0.3 \\ 0.4 & 0.6 \end{bmatrix} \begin{bmatrix} 0.5 & 0.3 \\ 0.4 & 0.6 \end{bmatrix}$

$=0.5 \oplus 0.2 \oplus 0.2 \oplus 0.6 = 0.5$

Let $F = \begin{bmatrix} 0.4 & 0.3 \\ 0.4 & 0.6 \end{bmatrix}$

$\|F\| = \begin{bmatrix} 0.4 & 0.3 \\ 0.4 & 0.6 \end{bmatrix} \begin{bmatrix} 0.4 & 0.3 \\ 0.4 & 0.6 \end{bmatrix}$

$=0.6 \oplus 0.2 \oplus 0.2 \oplus 0.6 = 0.6$

Let $G = \begin{bmatrix} 0.5 & 0.2 \\ 0.4 & 0.4 \end{bmatrix}$

$\|G\| = \begin{bmatrix} 0.5 & 0.2 \\ 0.4 & 0.4 \end{bmatrix} \begin{bmatrix} 0.5 & 0.2 \\ 0.4 & 0.4 \end{bmatrix}$

$=0.5 \oplus 0.8 \oplus 0.8 \oplus 0.6 = 0.7$

Let $H = \begin{bmatrix} 0.2 & 0.1 \\ 0.4 & 0.6 \end{bmatrix}$

$\|H\| = \begin{bmatrix} 0.2 & 0.1 \\ 0.4 & 0.6 \end{bmatrix} \begin{bmatrix} 0.2 & 0.1 \\ 0.4 & 0.6 \end{bmatrix}$
$=0.4 \oplus 0.4 \oplus 0.4 \oplus 0.6 = 0.8$

Let $I = \begin{bmatrix} 0.5 & 0.1 \\ 0.4 & 0.6 \end{bmatrix}$

$\|I\| = \begin{bmatrix} 0.5 & 0.1 \\ 0.4 & 0.6 \end{bmatrix} \begin{bmatrix} 0.5 & 0.1 \\ 0.4 & 0.6 \end{bmatrix}$
$=0.5 \oplus 0.4 \oplus 0.4 \oplus 0.6 = 0.9$

Then the evaluation matrix is $\begin{bmatrix} 0.6 & 0.3 & 0.2 \\ 0.2 & 0.5 & 0.6 \\ 0.7 & 0.8 & 0.9 \end{bmatrix}$

Evaluation $= 0.6(0.5 \sim 0.8) \ominus 0.3(0.8 \ominus 0.2) \oplus 0.2(0.6 \sim 0.5)$
$= 0.6(0.3) \ominus 0.3(0.6) \oplus 0.2(0.1).$
$= 0.8 \sim (0.8) \oplus 0.2.$
$= 0.2.$

We select$\lambda = 0.2$ as an internal cubic transition matrix having minimum diagonal length.

The conclusion matrix $\mathbb{C} = \begin{bmatrix} 0.2 & 0.8 \\ 0.4 & 0.6 \end{bmatrix}$ $=0.2 \oplus 0.2 = 0.4$

Then we conclude that the rate of utilization is40%.

Again we minimize the labors at most possible level.

$$\begin{array}{c} \text{Crops} \\ \begin{array}{cc} & \begin{array}{ccc} 1 & 2 & 3 \end{array} \\ [Labours] \begin{array}{c} I \\ II \\ III \end{array} & \begin{bmatrix} 0.1 & 0.1 & 0.1 \\ 0.1 & 0.3 & 0.1 \\ 0.1 & 0.1 & 0.1 \end{bmatrix} \end{array} \end{array}.$$

Let $A = \begin{bmatrix} 0.1 & 0.1 \\ 0.1 & 0.3 \end{bmatrix}$

$\|A\| = \begin{bmatrix} 0.1 & 0.1 \\ 0.1 & 0.3 \end{bmatrix} \begin{bmatrix} 0.1 & 0.1 \\ 0.1 & 0.3 \end{bmatrix}$ $=0.1 \oplus 0.1 \oplus 0.1 \oplus 0.9 = 0.2$

Let $B = \begin{bmatrix} 0.1 & 0.1 \\ 0.3 & 0.1 \end{bmatrix}$

$\|B\| = \begin{bmatrix} 0.1 & 0.1 \\ 0.3 & 0.1 \end{bmatrix} \begin{bmatrix} 0.1 & 0.1 \\ 0.3 & 0.1 \end{bmatrix}$ $=0.1 \oplus 0.3 \oplus 0.3 \oplus 0.1 = 0.8.$

Let $C = \begin{bmatrix} 0.1 & 0.1 \\ 0.1 & 0.1 \end{bmatrix}$

$\|C\| = \begin{bmatrix} 0.1 & 0.1 \\ 0.1 & 0.1 \end{bmatrix} \begin{bmatrix} 0.1 & 0.1 \\ 0.1 & 0.1 \end{bmatrix}$ $=0.1 \oplus 0.1 \oplus 0.1 \oplus 0.1 = 0.4$

Let $D = \begin{bmatrix} 0.1 & 0.3 \\ 0.1 & 0.1 \end{bmatrix}$

$\|D\| = \begin{bmatrix} 0.1 & 0.3 \\ 0.1 & 0.1 \end{bmatrix} \begin{bmatrix} 0.1 & 0.3 \\ 0.1 & 0.1 \end{bmatrix}$ $=0.1 \oplus 0.3 \oplus 0.3 \oplus 0.1 = 0.8$

Let $E = \begin{bmatrix} 0.3 & 0.1 \\ 0.1 & 0.1 \end{bmatrix}$

$\|E\| = \begin{bmatrix} 0.3 & 0.1 \\ 0.1 & 0.1 \end{bmatrix} \begin{bmatrix} 0.3 & 0.1 \\ 0.1 & 0.1 \end{bmatrix}$ $=0.9 \oplus 0.1 \oplus 0.1 \oplus 0.1 = 0.2$

Let $F = \begin{bmatrix} 0.1 & 0.1 \\ 0.1 & 0.1 \end{bmatrix}$

$\|F\| = \begin{bmatrix} 0.1 & 0.1 \\ 0.1 & 0.1 \end{bmatrix} \begin{bmatrix} 0.1 & 0.1 \\ 0.1 & 0.1 \end{bmatrix}$ $=0.1 \oplus 0.1 \oplus 0.1 \oplus 0.1 = 0.4$

Let $G = \begin{bmatrix} 0.1 & 0.1 \\ 0.1 & 0.1 \end{bmatrix}$

$\|G\| = \begin{bmatrix} 0.1 & 0.1 \\ 0.1 & 0.1 \end{bmatrix} \begin{bmatrix} 0.1 & 0.1 \\ 0.1 & 0.1 \end{bmatrix}$ $=0.1 \oplus 0.1 \oplus 0.1 \oplus 0.1 = 0.4$

Let $H = \begin{bmatrix} 0.1 & 0.1 \\ 0.1 & 0.1 \end{bmatrix}$

$\|H\| = \begin{bmatrix} 0.1 & 0.1 \\ 0.1 & 0.1 \end{bmatrix} \begin{bmatrix} 0.1 & 0.1 \\ 0.1 & 0.1 \end{bmatrix}$ $=0.1 \oplus 0.1 \oplus 0.1 \oplus 0.1 = 0.4$

Let $\quad I = \begin{bmatrix} 0.1 & 0.1 \\ 0.1 & 0.1 \end{bmatrix}$

$\|I\| = \begin{bmatrix} 0.1 & 0.1 \\ 0.1 & 0.1 \end{bmatrix} \begin{bmatrix} 0.1 & 0.1 \\ 0.1 & 0.1 \end{bmatrix} = 0.1 \oplus 0.1 \oplus 0.1 \oplus 0.1 = 0.4$

Then the evaluation matrix is $\begin{bmatrix} 0.2 & 0.8 & 0.4 \\ 0.8 & 0.2 & 0.4 \\ 0.4 & 0.4 & 0.4 \end{bmatrix}$

$$\begin{aligned} \text{Evaluation} &= 0.2(0.8{\sim}0.6) \ominus 0.8(0.2{\sim}0.6) \oplus 0.4(0.2{\sim}0.8) \\ &= 0.2(0.2) \ominus 0.8(0.4) \oplus 0.4(0.6). \\ &= 0.4{\sim}(0.2) \oplus 0.4. \\ &= 0.6. \end{aligned}$$

We select$\lambda = 0.6$ as an internal cubic transition matrix having minimum diagonal length.

The conclusion matrix $\mathbb{C} = \begin{bmatrix} 0.1 & 0.9 \\ 0.7 & 0.3 \end{bmatrix} = 0.3 \oplus 0.3 = 0.6$

Step-7

We conclude that the rate of utilization is60%. Thus this result holds good and yields 60% profit for above fuzzy matrix A with given labors and crops. One can increase the profit level by choosing the appropriate crop with limited labors to increase the production and maximize their profit.

2 PRODUCT FUZZY SOFT MATRICES

The second section introduces the notion of product fuzzy soft transition matrix. Also this section provides an algorithm for decision making about the lack of agriculture works and migration of farmers from their villages due to monsoon failure using product fuzzy soft transition matrix as a tool.

Definition 2.1

Let $(\mathcal{M}, \mathcal{A}) = \mathcal{S}_{ij}$ and $(\mathcal{M}, \mathcal{B}) = \mathcal{R}_{ij}$ be two fuzzy soft transition matrices. Then the transition multiplication $\mathcal{S}_{ij}$ and $\mathcal{R}_{ij}$is called as product fuzzy soft matrix (or) product fuzzy soft transition matrix.

Algorithm 2.1

The algorithm for farmers job lacking and migration involves the following steps.

Step 1: Select a subset $\mathcal{A}$ with two parameters from the transition matrix.

Step 2: Choose the transition matrices from orthogonal transition matrices having maximum and next maximum elements for required parameters.

Step 3: Select another subset $\mathcal{B}$ with two parameters from transition matrix.

Step 4: Choose the transition matrices from the same orthogonal transition matrices having successive maximum elements for required parameters.

Step 5: Obtain the square matrix$\mathcal{S}_{ij}$ having elements that are orthogonal transition matrices for the subset $\mathcal{A}$.

Step 6: Select another square matrix $\mathcal{R}_{ij}$ having elements that are orthogonal transition matrices for the subset $\mathcal{B}$.

Step 7: Reduce the matrices$\mathcal{S}_{ij}$ and $\mathcal{R}_{ij}$ in two steps corresponding to orthogonal transition sets.

Step 8: Calculate the transition multiplication $\mathcal{K}_{ij} = \mathcal{S}_{ij}\mathcal{R}_{ij}$ having elements in orthogonal transition matrices.

Step 9: Obtain the required result from the reduced evaluation matrix $\mathcal{K}_{ij}$.

In India, monsoon rainfall is not up to the mark during the recent years. Due to this situation, farmer's life style tends to below poverty level. Some of the farmers migrate to cities from their own villages and proceed for another job. Later some farmers return back to village to proceed with their own job but suddenly they migrate to town. We try to represent this situation by means of an example and justify the above algorithm using soft transition matrices as a tool.

Example 2.2

$$P = \begin{array}{c} \\ J \\ M \end{array} \begin{array}{c} \begin{array}{cc} J & M \end{array} \\ \begin{bmatrix} 0.6 & 0.4 \\ 0.2 & 0.8 \end{bmatrix} \end{array}$$

This matrix describes the following phenomenon of the farmers and their jobs.

$J \rightarrow J$- Farmers continuing their own job are 60%.
$J \rightarrow M$- Farmers lost their own job then migrating in to other jobs are 40%.
$M \rightarrow J$- Farmers return to their village and like to comeback their own job are 20%.
$M \rightarrow M$- Home returned farmers again migrate from their village 80%.

Step 1:
Define $\mathcal{A} = \{JM_1, JM_2\}$

Step 2:
Now we select the transition matrix $\begin{bmatrix} 0.6 & 0.4 \\ 0.2 & 0.8 \end{bmatrix}$.

The orthogonal transition matrices of $\begin{bmatrix} 0.6 & 0.4 \\ 0.2 & 0.8 \end{bmatrix}$ is same that of $\begin{bmatrix} 0.1 & 0.9 \\ 0.2 & 0.8 \end{bmatrix}$.

Now we first select the orthogonal transition matrices having entry 0.9 in JM.

$$\alpha = \begin{bmatrix} 0.1 & 0.9 \\ 0.2 & 0.8 \end{bmatrix}, \begin{bmatrix} 0.1 & 0.9 \\ 0.7 & 0.3 \end{bmatrix}.$$

Then we select the orthogonal transition matrices having entry 0.9 in JM in the next orthogonal set of matrices which is orthogonal to $\begin{bmatrix} 0.1 & 0.9 \\ 0.3 & 0.7 \end{bmatrix}$.

$$\beta = \begin{bmatrix} 0.1 & 0.9 \\ 0.3 & 0.7 \end{bmatrix}, \begin{bmatrix} 0.1 & 0.9 \\ 0.8 & 0.2 \end{bmatrix}.$$

Then we select the orthogonal transition matrices having entry 0.9 in JM in the next orthogonal set of matrices which is orthogonal to $\begin{bmatrix} 0.1 & 0.9 \\ 0.1 & 0.9 \end{bmatrix}$.

$$\gamma = \begin{bmatrix} 0.1 & 0.9 \\ 0.4 & 0.6 \end{bmatrix}, \begin{bmatrix} 0.1 & 0.9 \\ 0.9 & 0.1 \end{bmatrix}.$$

Then we select the orthogonal transition matrices having entry 0.8 in JM in the next orthogonal set of matrices which is orthogonal to $\begin{bmatrix} 0.1 & 0.9 \\ 0.2 & 0.8 \end{bmatrix}$.

$$\delta = \begin{bmatrix} 0.2 & 0.8 \\ 0.9 & 0.1 \end{bmatrix}, \begin{bmatrix} 0.2 & 0.8 \\ 0.4 & 0.6 \end{bmatrix}.$$

Similarly we can able to collect next four set of matrices

$$\alpha_1 = \begin{bmatrix} 0.1 & 0.9 \\ 0.2 & 0.8 \end{bmatrix}, \begin{bmatrix} 0.1 & 0.9 \\ 0.7 & 0.3 \end{bmatrix}, \beta_1 = \begin{bmatrix} 0.1 & 0.9 \\ 0.3 & 0.7 \end{bmatrix}, \begin{bmatrix} 0.1 & 0.9 \\ 0.8 & 0.2 \end{bmatrix},$$
$$\gamma_1 = \begin{bmatrix} 0.1 & 0.9 \\ 0.4 & 0.6 \end{bmatrix}, \begin{bmatrix} 0.1 & 0.9 \\ 0.9 & 0.1 \end{bmatrix} \text{ and } \delta_1 = \begin{bmatrix} 0.2 & 0.8 \\ 0.9 & 0.1 \end{bmatrix}, \begin{bmatrix} 0.2 & 0.8 \\ 0.4 & 0.6 \end{bmatrix}.$$

Step 3:
Define $\mathcal{B} = \{MM_1, MM_2\}$

Step 4:
Now we select the transition matrix $\begin{bmatrix} 0.6 & 0.4 \\ 0.2 & 0.8 \end{bmatrix}$.

The orthogonal transition matrices of $\begin{bmatrix} 0.6 & 0.4 \\ 0.2 & 0.8 \end{bmatrix}$ is same that of $\begin{bmatrix} 0.1 & 0.9 \\ 0.2 & 0.8 \end{bmatrix}$.

Now we first select the orthogonal transition matrices having entry 0.9 in MM.

$$\alpha = \begin{bmatrix} 0.8 & 0.2 \\ 0.1 & 0.9 \end{bmatrix}, \begin{bmatrix} 0.3 & 0.7 \\ 0.1 & 0.9 \end{bmatrix}.$$

Then we select the orthogonal transition matrices having entry 0.9 in MM in the next orthogonal set of matrices which is orthogonal to $\begin{bmatrix} 0.1 & 0.9 \\ 0.3 & 0.7 \end{bmatrix}$.

$$\beta = \begin{bmatrix} 0.7 & 0.3 \\ 0.1 & 0.9 \end{bmatrix}, \begin{bmatrix} 0.2 & 0.8 \\ 0.1 & 0.9 \end{bmatrix}.$$

Then we select the orthogonal transition matrices having entry 0.9 in MM in the next orthogonal set of matrices which is orthogonal to $\begin{bmatrix} 0.1 & 0.9 \\ 0.1 & 0.9 \end{bmatrix}$.

$$\gamma = \begin{bmatrix} 0.4 & 0.6 \\ 0.1 & 0.9 \end{bmatrix}, \begin{bmatrix} 0.9 & 0.1 \\ 0.1 & 0.9 \end{bmatrix}.$$

Then we select the orthogonal transition matrices having entry 0.8 in MM in the next orthogonal set of matrices which is orthogonal to $\begin{bmatrix} 0.1 & 0.9 \\ 0.2 & 0.8 \end{bmatrix}$.

$$\delta = \begin{bmatrix} 0.1 & 0.9 \\ 0.2 & 0.8 \end{bmatrix}, \begin{bmatrix} 0.6 & 0.4 \\ 0.2 & 0.8 \end{bmatrix}.$$

Similarly we can able to collect next four set of matrices

$$\alpha_1 = \begin{bmatrix}0.9 & 0.1\\0.3 & 0.7\end{bmatrix}, \begin{bmatrix}0.4 & 0.6\\0.3 & 0.7\end{bmatrix}, \beta_1 = \begin{bmatrix}0.9 & 0.1\\0.2 & 0.8\end{bmatrix}, \begin{bmatrix}0.4 & 0.6\\0.2 & 0.8\end{bmatrix},$$

$$\gamma_1 = \begin{bmatrix}0.3 & 0.7\\0.2 & 0.8\end{bmatrix}, \begin{bmatrix}0.8 & 0.2\\0.2 & 0.8\end{bmatrix} \text{and } \delta_1 = \begin{bmatrix}0.1 & 0.9\\0.2 & 0.8\end{bmatrix}, \begin{bmatrix}0.6 & 0.4\\0.2 & 0.8\end{bmatrix}.$$

Step 5:

We construct the fuzzy soft transition matrix $\mathcal{S}_{ij}$ having elements from step2 $\mathcal{S}_{ij} =$

$$\begin{bmatrix} \alpha = \begin{bmatrix}0.1 & 0.9\\0.2 & 0.8\end{bmatrix}, \begin{bmatrix}0.1 & 0.9\\0.7 & 0.3\end{bmatrix} & \beta = \begin{bmatrix}0.1 & 0.9\\0.3 & 0.7\end{bmatrix}, \begin{bmatrix}0.1 & 0.9\\0.8 & 0.2\end{bmatrix} \\ \gamma = \begin{bmatrix}0.1 & 0.9\\0.4 & 0.6\end{bmatrix}, \begin{bmatrix}0.1 & 0.9\\0.9 & 0.1\end{bmatrix} & \delta = \begin{bmatrix}0.2 & 0.8\\0.9 & 0.1\end{bmatrix}, \begin{bmatrix}0.2 & 0.8\\0.4 & 0.6\end{bmatrix} \\ \alpha_1 = \begin{bmatrix}0.1 & 0.9\\0.2 & 0.8\end{bmatrix}, \begin{bmatrix}0.1 & 0.9\\0.7 & 0.3\end{bmatrix} & \beta_1 = \begin{bmatrix}0.1 & 0.9\\0.3 & 0.7\end{bmatrix}, \begin{bmatrix}0.1 & 0.9\\0.8 & 0.2\end{bmatrix} \\ \gamma_1 = \begin{bmatrix}0.1 & 0.9\\0.4 & 0.6\end{bmatrix}, \begin{bmatrix}0.1 & 0.9\\0.9 & 0.1\end{bmatrix} & \delta_1 = \begin{bmatrix}0.2 & 0.8\\0.9 & 0.1\end{bmatrix}, \begin{bmatrix}0.2 & 0.8\\0.4 & 0.6\end{bmatrix} \end{bmatrix}$$

Step 6:

We construct the fuzzy soft transition matrix $\mathcal{T}_{ij}$ having elements from step4 $\mathcal{T}_{ij} =$

$$\begin{bmatrix} \alpha = \begin{bmatrix}0.8 & 0.2\\0.1 & 0.9\end{bmatrix}, \begin{bmatrix}0.3 & 0.7\\0.1 & 0.9\end{bmatrix} & \beta = \begin{bmatrix}0.7 & 0.3\\0.1 & 0.9\end{bmatrix}, \begin{bmatrix}0.2 & 0.8\\0.1 & 0.9\end{bmatrix} \\ \gamma = \begin{bmatrix}0.4 & 0.6\\0.1 & 0.9\end{bmatrix}, \begin{bmatrix}0.9 & 0.1\\0.1 & 0.9\end{bmatrix} & \delta = \begin{bmatrix}0.1 & 0.9\\0.2 & 0.8\end{bmatrix}, \begin{bmatrix}0.6 & 0.4\\0.2 & 0.8\end{bmatrix} \\ \alpha_1 = \begin{bmatrix}0.9 & 0.1\\0.3 & 0.7\end{bmatrix}, \begin{bmatrix}0.4 & 0.6\\0.3 & 0.7\end{bmatrix} & \beta_1 = \begin{bmatrix}0.9 & 0.1\\0.2 & 0.8\end{bmatrix}, \begin{bmatrix}0.4 & 0.6\\0.2 & 0.8\end{bmatrix} \\ \gamma_1 = \begin{bmatrix}0.3 & 0.7\\0.2 & 0.8\end{bmatrix}, \begin{bmatrix}0.8 & 0.2\\0.2 & 0.8\end{bmatrix} & \delta_1 = \begin{bmatrix}0.1 & 0.9\\0.2 & 0.8\end{bmatrix}, \begin{bmatrix}0.6 & 0.4\\0.2 & 0.8\end{bmatrix} \end{bmatrix}$$

Step 7:

First step reduction matrix obtains by multiplying the matrices in their corresponding sets.

$$\mathcal{S}_{ij} = \begin{bmatrix} \alpha = \begin{bmatrix}0.4 & 0.6\\0.8 & 0.2\end{bmatrix} & \beta = \begin{bmatrix}0.3 & 0.7\\0.9 & 0.1\end{bmatrix} \\ \gamma = \begin{bmatrix}0.2 & 0.8\\0.8 & 0.2\end{bmatrix} & \delta = \begin{bmatrix}0.6 & 0.4\\0.2 & 0.8\end{bmatrix} \\ \alpha_1 = \begin{bmatrix}0.1 & 0.9\\0.7 & 0.3\end{bmatrix} & \beta_1 = \begin{bmatrix}0.2 & 0.8\\0.6 & 0.4\end{bmatrix} \\ \gamma_1 = \begin{bmatrix}0.8 & 0.2\\0.2 & 0.8\end{bmatrix} & \delta_1 = \begin{bmatrix}0.4 & 0.6\\0.8 & 0.2\end{bmatrix} \end{bmatrix}$$

Second step reduction matrix obtains by multiplying the matrices in their corresponding sets.

$$\mathcal{S}_{ij} = \begin{bmatrix} \alpha\alpha_1 = \begin{bmatrix}0.6 & 0.4\\0.2 & 0.8\end{bmatrix} & \beta\beta_1 = \begin{bmatrix}0.8 & 0.2\\0.4 & 0.6\end{bmatrix} \\ \gamma\gamma_1 = \begin{bmatrix}0.2 & 0.8\\0.8 & 0.2\end{bmatrix} & \delta\delta_1 = \begin{bmatrix}0.6 & 0.4\\0.2 & 0.8\end{bmatrix} \end{bmatrix}$$

First step reduction matrix obtains by multiplying the matrices in their corresponding sets.

$$\mathcal{T}_{ij} = \begin{bmatrix} \alpha = \begin{bmatrix}0.6 & 0.4\\0.2 & 0.8\end{bmatrix} & \beta = \begin{bmatrix}0.7 & 0.3\\0.1 & 0.9\end{bmatrix} \\ \gamma = \begin{bmatrix}0.2 & 0.8\\0.8 & 0.2\end{bmatrix} & \delta = \begin{bmatrix}0.4 & 0.6\\0.8 & 0.2\end{bmatrix} \\ \alpha_1 = \begin{bmatrix}0.9 & 0.1\\0.3 & 0.7\end{bmatrix} & \beta_1 = \begin{bmatrix}0.8 & 0.2\\0.4 & 0.6\end{bmatrix} \\ \gamma_1 = \begin{bmatrix}0.8 & 0.2\\0.2 & 0.8\end{bmatrix} & \delta_1 = \begin{bmatrix}0.4 & 0.6\\0.8 & 0.2\end{bmatrix} \end{bmatrix}$$

Second step reduction matrix obtains by multiplying the matrices in their corresponding sets.

$$\mathcal{T}_{ij} = \begin{bmatrix} \alpha\alpha_1 = \begin{bmatrix}0.6 & 0.4\\0.2 & 0.8\end{bmatrix} & \beta\beta_1 = \begin{bmatrix}0.8 & 0.2\\0.4 & 0.6\end{bmatrix} \\ \gamma\gamma_1 = \begin{bmatrix}0.2 & 0.8\\0.8 & 0.2\end{bmatrix} & \delta\delta_1 = \begin{bmatrix}0.4 & 0.6\\0.8 & 0.2\end{bmatrix} \end{bmatrix}$$

Step 8: $\mathcal{K}_{ij} = \mathcal{S}_{ij} * \mathcal{T}_{ij}$

$$= \begin{bmatrix} \alpha\alpha_1 = \begin{bmatrix} 0.6 & 0.4 \\ 0.2 & 0.8 \end{bmatrix} & \beta\beta_1 = \begin{bmatrix} 0.8 & 0.2 \\ 0.4 & 0.6 \end{bmatrix} \\ \gamma\gamma_1 = \begin{bmatrix} 0.2 & 0.8 \\ 0.8 & 0.2 \end{bmatrix} & \delta\delta_1 = \begin{bmatrix} 0.6 & 0.4 \\ 0.2 & 0.8 \end{bmatrix} \end{bmatrix} * \begin{bmatrix} \alpha\alpha_1 = \begin{bmatrix} 0.6 & 0.4 \\ 0.2 & 0.8 \end{bmatrix} & \beta\beta_1 = \begin{bmatrix} 0.8 & 0.2 \\ 0.4 & 0.6 \end{bmatrix} \\ \gamma\gamma_1 = \begin{bmatrix} 0.2 & 0.8 \\ 0.8 & 0.2 \end{bmatrix} & \delta\delta_1 = \begin{bmatrix} 0.4 & 0.6 \\ 0.8 & 0.2 \end{bmatrix} \end{bmatrix}$$

$$\mathcal{K}_{ij} = \begin{bmatrix} \alpha\alpha_1 = \begin{bmatrix} 0.4 & 0.6 \\ 0.8 & 0.2 \end{bmatrix} & \beta\beta_1 = \begin{bmatrix} 0.2 & 0.8 \\ 0.6 & 0.4 \end{bmatrix} \\ \gamma\gamma_1 = \begin{bmatrix} 0.8 & 0.2 \\ 0.2 & 0.8 \end{bmatrix} & \delta\delta_1 = \begin{bmatrix} 0.6 & 0.4 \\ 0.2 & 0.8 \end{bmatrix} \end{bmatrix}$$

Now we compute $\langle \begin{bmatrix} 0.4 & 0.6 \\ 0.8 & 0.2 \end{bmatrix}, \begin{bmatrix} 0.4 & 0.6 \\ 0.8 & 0.2 \end{bmatrix} \rangle = 0.6 + 0.8 + 0.8 + 0.4 = 0.6,$

$\langle \begin{bmatrix} 0.2 & 0.8 \\ 0.6 & 0.4 \end{bmatrix}, \begin{bmatrix} 0.2 & 0.8 \\ 0.6 & 0.4 \end{bmatrix} \rangle = 0.4 + 0.8 + 0.8 + 0.6 = 0.6,$

$\langle \begin{bmatrix} 0.8 & 0.2 \\ 0.2 & 0.8 \end{bmatrix}, \begin{bmatrix} 0.8 & 0.2 \\ 0.2 & 0.8 \end{bmatrix} \rangle = 0.4 + 0.4 + 0.4 + 0.4 = 0.6,$

$\langle \begin{bmatrix} 0.6 & 0.4 \\ 0.2 & 0.8 \end{bmatrix}, \begin{bmatrix} 0.6 & 0.4 \\ 0.2 & 0.8 \end{bmatrix} \rangle = 0.6 + 0.8 + 0.8 + 0.4 = 0.6.$

Step 9: The evaluation matrix is $\begin{bmatrix} 0.6 & 0.6 \\ 0.6 & 0.6 \end{bmatrix}$.

Hence we conclude that due to monsoon stoppage 60% of farmers were migrating to other jobs for the given data.

REFERENCES

1. Adke, SR & Manjunath, SM (1984), An Invitation To Finite Markov Process, Wiley Eastern, NewDelhi.
2. Amussen, SA (1987),Applied Probability and Queues, Wiley, Newyork.
3. Bhat, UN (1984), Elements of Applied Stochastic Processes, Second edition, Wiley, New York.
4. Cagman, N& Enginoglu,S(2010),'Soft matrix theory and its decision making', Computers and Mathematics with Applications, vol. 59, no. 10, pp. 3308-3314.
5. Cagman, N & Enginoglu, S(2011),Fuzzy Soft Matrix Theory and Application in decision making, Iranian Journal of Fuzzy Systems, vol. 8, no.3 , pp.137-147.
6. Chung, KL (1979), Elementary Probability theory with Stochastic Processs, Third edition, Springer Verlag, New York.
7. Isaacson, DL & Madsen, RW (1976), Markov Chains-Theory and Applications, Wiley, Newyork.
8. Jun,YB, Kim,CS& Yang,KO (2012), Cubic Sets,Annals of Fuzzy Mathematics and Informatics, vol. 4, no. 1, pp.83- 98
9. Karlin, S & Taylor, HM (1975), A First Course in Stochastic Processes, Academic Press, Newyork.
10. Kim,KH, (1982), Boolean Matrix Theory and Applications, Marcel Dekker, Newyork.
11. Klir,GJ & Folger,TA (1988), Fuzzy sets, Uncertainty and Information Prentice -Hall.
12. Lawless, JF (1982), Statistical Models and Methods of Lifetime Data, Wiley, Newyork.

13. Maji, PK, Biswas, R & Roy A R(2001), Intuitionistic fuzzy soft sets, Journal of Fuzzy Mathematics, vol. 9, no.3, pp. 677 – 691.
14. Majumdar, P &Samanta,SK 2010, Generalized fuzzy soft sets, Computers and Mathematics with Applications , vol.59, no.4, pp. 1425 – 1432.
15. Medhi, J (1982), Stochastic Processes, New Age International (P) Ltd, NewDelhi.
16. Molodtsov, DA, (1999),'Soft set theory—First results'. Computers & Mathematics With Applications, vol. **37** , no. 4, pp.19–31.
17. Sheldon M. Ross (1983), Stochastic Processes, John wiley, NewYork.
18. Shyamal, AK & Pal, M (2006),Interval-valued fuzzy matrices,The Journal of Fuzzy Mathematics,vol.14, no.3, pp.582-594.
19. Srinivasan, SK & Mehata, KM (1981), Probability and Random Processes, Second edition, Tata-McGraw Hill, NewDelhi.
20. Timothy Brown,C, Kais Hamza & Aihua Xia (1998), on the variance to mean ratio for random variables from markov chains and point processes, Journal of Applied Probability vol.35, no.2,pp.303-312.
21. Zaheh, LA (1975), The concept of linguistic variable and its application to approximate reasoning I, Information Sciences, vol.8, pp. 199-249.
22. Zimmermann, H, J (1991), Fuzzy Set theory and its Applications, Second edition, Allied Publishers Limited, New Delhi.

Generalisation of Compatible Ideals with respect to g-open sets

K. Bhavani[1] and R.Sumathy[2]
[1] and [2] SRM Institute of Science and Technology, Ramapuram, Chennai
E-mail: bhavanidhurairaj@gmail.com

Abstract.
In this paper, we define g-compatible ideal with respect to g-open sets and characterize g-local functions in codense ideal topological spaces. Also, we establish some properties of g-local functions with the compatibility of $\mathcal{I}$ with τ with respect to g-open sets and the relative topology

Keywords and Phrases: Ideal topological space, g-local functions, g-open sets, Compatible ideals, ψ_g-operator and Relative topology.

AMS 2010 Subject classification: 54 A 05, 54 A 10.

1. Introduction and Preliminaries.

An *ideal* $\mathcal{I}$ on X [7] is a collection of subsets of X satisfying the following: (i) If $A \in \mathcal{I}$ and $B \subset A$, then $B \in \mathcal{I}$, and (ii) if $A \in \mathcal{I}$ and $B \in \mathcal{I}$, then $A \cup B \in \mathcal{I}$. A topological space (X, τ) together with an ideal $\mathcal{I}$ is called an *ideal topological space* and is denoted by $(X, \tau, \mathcal{I})$. For each subset A of X, $A^\star(\mathcal{I}, \tau) = \{x \in X \mid U \cap A \notin \mathcal{I}$ for every open set U containing $x\}$ is called the *local function* of A [7] with respect to $\mathcal{I}$ and τ. We simply write $A^\star$ instead of $A^\star(\mathcal{I}, \tau)$ in case there is no chance for confusion. We often use the properties of the local function stated in Theorem 2.3 of [5] without mentioning it. Moreover, $cl^\star(A) = A \cup A^\star$ [9] defines a Kuratowski closure operator for a topology $\tau^\star$, which is finer than τ. An ideal $\mathcal{I}$ is a *boundary ideal* [9] or a *codense ideal* [4] if $\tau \cap \mathcal{I} = \{\emptyset\}$. A subset A of a topological space (X, τ) is said to be g- closed [8], if $cl(A) \subset U$ whenever $A \subset U$ and U is open. The complement of a g- closed set is called a g- open set [8]. The collection of all g-open sets in a topological space (X, τ) is denoted by τ_g. The g-closure of A denoted by $cl_g(A)$ [1], defined as the intersection of all g-closed sets containing A and the g-interior of A denoted by $int_g(A)$, defined as the union of all g-open sets contained in A. For every $A \in \wp(X)$, $A^\star(\mathcal{I}, \tau_g) = \{x \in X \mid U \cap A \notin \mathcal{I}$ for every g-open set U containing $x\}$ is called the *g-local function* of A[1] with respect to $\mathcal{I}$ and τ_g and is denoted by $A^\star_g$. Also, $cl^\star_g(A) = A \cup A^\star_g$[1] is a Kurotowski closure operator for a topology $\tau^\star_g = \{X - A \mid$

$cl_g^\star(A) = A\}$ [1] on X which is finer than τ_g. An operator $\psi_g : \wp(X) \to \wp(X)$ is defined by $\psi_g(A) = \{x \in X \mid$ there exists a g-open set U_x containing x, such that $U_x - A \in \mathcal{I}\}$ for every $A \in \wp(X)$[2].

Given an ideal topological space $(X, \tau, \mathcal{I})$, $\mathcal{I}$ is said to be *compatible* with τ [5], denoted by $\mathcal{I} \sim \tau$, if for every subset A of X and for each $x \in A$, there exists an open set U of x such that $U \cap A \in \mathcal{I}$, then $A \in \mathcal{I}$. Compatible ideals are termed as *supercompact* ideals in [9]. If Y is a nonempty subset of an ideal topological space $(X, \tau, \mathcal{I})$, then $(Y, \tau_Y, \mathcal{I}_Y)$ is an ideal topological subspace of X where $\mathcal{I}_Y = \{I \cap Y \mid I \in \mathcal{I}\}$ [3] is an ideal on Y, the restriction of $\mathcal{I}$ to Y and $\tau_Y = \{U \cap Y \mid U \in \tau\}$ is the relative topology on Y.

The following lemmas will be useful in the sequel.

Lemma 1.1. [5, Lemma 6.6] Let $(X, \tau, \mathcal{I})$ be an ideal topological space and $A \subset X$. Then $(\tau_A)^\star(\mathcal{I}_A) = \tau^\star(\mathcal{I})_A$.

Lemma 1.2. [6] Let $(X, \tau, \mathcal{I})$ be an ideal topological space with $B \subseteq A \subseteq X$. Then $B^\star(\mathcal{I}_A, \tau_A) = B^\star(\mathcal{I}, \tau) \cap A$ where τ_A is the relative topology on A.

Lemma 1.3. [6] Let $(X, \tau, \mathcal{I})$ be an ideal topological space and $B \subset A \subset X$. Then $cl_A^\star(B) = cl^\star(B) \cap A$ where $cl_A^\star$ is the closure operator in $(A, \tau_A^\star)$.

Lemma 1.4. [1] Let $(X, \tau, \mathcal{I})$ be an ideal topological space and let A, B be subsets of X. Then the following hold.
(a) $\emptyset_g^\star = \emptyset$.
(b) $A \subset B$ implies $A_g^\star \subset B_g^\star$.
(c) $A_g^\star \subset A^\star$.
(d) $A_g^\star \subset cl_g(A) \subset cl(A)$.
(e) $A_g^\star \cup B_g^\star = (A \cup B)_g^\star$.
(f) $(A_g^\star)_g^\star \subset A_g^\star$.
(g) $(A_g^\star - B_g^\star) \subset (A - B)_g^\star - B_g^\star = (A - B)_g^\star$. (h) $(A - A_g^\star) \cap (A - A_g^\star)_g^\star = \emptyset$.

Lemma 1.5. [2] Let $(X, \tau, \mathcal{I})$ be an ideal topological space and let A, B be subsets of X. Then the following hold.
(a) If $A \subset X$, then $\psi_g(A)$ is g-open.
(b) If $U \in \tau_g$, then $U \subset \psi_g(U)$. Also, If $U \in \tau_g^\star$, then $U \subset \psi_g(U)$.
(c) If $A \subset X$, then $\psi_g(A) \subset \psi_g(\psi_g(A))$.
(d) If $A \subset X$, then $\psi_g(A) = \psi_g(\psi_g(A))$ if and only if $((X - A)_g^\star)_g^\star = (X - A)_g^\star$.
(e) If $A \subset X$ and $I \in \mathcal{I}$, then $\psi_g(A - I) = \psi_g(A) = \psi_g(A \cup I)$.

Journal of Physics: Conference Series **1377** (2019) 012034 doi:10.1088/1742-6596/1377/1/012034

2. g-Compatible Ideals

Definition 2.1. Let $(X, \tau, \mathcal{I})$ be an ideal topological space. $\mathcal{I}$ is said to be *g-compatible* with τ denoted by $\tau \sim_g \mathcal{I}$, if the following condition is satisfied for each subset $A \subset X$: if for every $x \in A$, there exists a g-open set U such that $U \cap A \in \mathcal{I}$, then $A \in \mathcal{I}$.

The following Theorem 2.2 and Theorem 2.3 give some characterizations of g-compatibility of $\mathcal{I}$ with τ_g.

Theorem 2.2. Let $(X, \tau, \mathcal{I})$ be an ideal topological space. Then the following are equivalent.
(a) $\tau \sim_g \mathcal{I}$.
(b) For every $A \subset X$, $A \cap A_g^\star = \emptyset$, then $A \in \mathcal{I}$.
(c) For every $A \subset X$, $A - A_g^\star \in \mathcal{I}$.
(d) For every $\tau_g^\star$-closed subset A of X, $A - A_g^\star \in \mathcal{I}$.
(e) For every $A \subset X$, if A contains no nonempty subset B with $B \subset B_g^\star$, then $A \in \mathcal{I}$.

Proof. $(a) \Rightarrow (b)$. Suppose that $A \cap A_g^\star = \emptyset$ for every $A \subset X$. If $x \in A$, then $x \notin A_g^\star$. Then there exists a g-open set U containing x such that $U \cap A \in \mathcal{I}$. Since $\tau \sim_g \mathcal{I}$, $A \in \mathcal{I}$.

$(b) \Rightarrow (c)$. Let $A \subset X$. Since $A - A_g^\star \subset (A_g^\star)^c$, $(A - A_g^\star) \cap A_g^\star = \emptyset$. Therefore by Lemma 1.4(h), $(A - A_g^\star) \cap (A - A_g^\star)^\star = \emptyset$. Then by hypothesis, $A - A_g^\star \in \mathcal{I}$.

$(c) \Rightarrow (d)$. The proof is clear.
$(d) \Rightarrow (a)$. Let $A \subset X$ and suppose that for every $x \in A$, there exists a g-open set U such that $U \cap A \in \mathcal{I}$. This implies that $x \notin A_g^\star$ and so $A \cap A_g^\star = \emptyset$. Since $cl_g^\star(A) = A \cup A_g^\star$ is $\tau_g^\star$-closed, by hypothesis $(A \cup A_g^\star) - (A \cup A_g^\star)_g^\star \in \mathcal{I}$ and so $(A \cup A_g^\star) - (A_g^\star \cup (A_g^\star)_g^\star) \in \mathcal{I}$. Since $(A_g^\star)_g^\star \subset A_g^\star$, $(A \cup A_g^\star) - A_g^\star \in \mathcal{I}$ and so $A \in \mathcal{I}$ since $A \cap A_g^\star = \emptyset$. Therefore, $\tau \sim_g \mathcal{I}$.
$(c) \Rightarrow (e)$. Let $A \subset X$ and suppose that A contains no nonempty subset B with $B \subset B_g^\star$. By (c), $A - A_g^\star \in \mathcal{I}$ and so $(A - A_g^\star)_g^\star = \emptyset$. Then $(A - A_g^\star)_g^\star \subset A_g^\star$. Since $A \cap A_g^\star = A - (A - A_g^\star)$, $(A \cap A_g^\star)_g^\star = (A - (A - A_g^\star))_g^\star \supset A_g^\star - (A - A_g^\star)_g^\star = A_g^\star$ by Lemma 1.4 (g). Therefore, $(A \cap A_g^\star)_g^\star = A_g^\star$ and so $A_g^\star = \emptyset$ which implies that $A \cap A_g^\star = \emptyset$. Therefore, $A - A_g^\star = A \in \mathcal{I}$.

$(e) \Rightarrow (d)$. Let $A \subset X$. Since $(A - A_g^\star) \cap (A - A_g^\star)_g^\star = \emptyset$, $A - A_g^\star$ contains no nonempty subset B with $B \subset B_g^\star$. Therefore, $A - A_g^\star \in \mathcal{I}$.

Theorem 2.3. Let $(X, \tau, \mathcal{I})$ be an ideal topological space. Then the following are equivalent and are implied by $\tau \sim_g \mathcal{I}$.
(a) For every $A \subset X$, $A \cap A_g^\star = \emptyset$, then $A_g^\star = \emptyset$.
(b) For every $A \subset X$, $(A - A_g^\star)_g^\star = \emptyset$,
(c) For every $A \subset X$, $(A \cap A_g^\star)_g^\star = A_g^\star$.

Proof. $(a) \Rightarrow (b)$. Since $(A-A_g^\star)\cap(A-A_g^\star)_g^\star=\emptyset$ by Lemma 1.4(h), $(A-A_g^\star)_g^\star=\emptyset$ by (a). $(b) \Rightarrow (c)$. Let $A \subset X$ such that $(A-A_g^\star)_g^\star=\emptyset$. Since $A\cap A_g^\star = A-(A-A_g^\star)$, $(A\cap A_g^\star)_g^\star = (A-(A-A_g^\star))_g^\star \supset A_g^\star-(A-A_g^\star)_g^\star = A_g^\star$ and so $A_g^\star \subset (A\cap A_g^\star)_g^\star$. Also, $(A\cap A_g^\star)_g^\star \subset A_g^\star$ which implies that $(A\cap A_g^\star)_g^\star = A_g^\star$.

$(c) \Rightarrow (a)$. Suppose that $A\cap A_g^\star=\emptyset$ for every subset A of X. Let $A\subset X$ such that $(A\cap A_g^\star)_g^\star = A_g^\star$. Then $A_g^\star=(A\cap A_g^\star)_g^\star=\emptyset_g^\star=\emptyset$.

Theorem 2.4. Let $(X,\tau,\mathcal{I})$ be an ideal topological space and $\tau \sim_g \mathcal{I}$. A set A is $\tau_g^\star$-closed if and only if it is the union of a set which is g-closed and a set in $\mathcal{I}$.

Proof. Let A be $\tau_g^\star$-closed subset of X. Then $A_g^\star\subset A$ and so $A=(A-A_g^\star)\cup A_g^\star$ where $A_g^\star$ is a g-closed set. Since $\tau\sim_g\mathcal{I}$, $A-A_g^\star\in\mathcal{I}$. Conversely, let $A=B\cup I$ where B is a g-closed set and $I\in\mathcal{I}$. Then $A_g^\star=(B\cup I)_g^\star=B_g^\star\cup I_g^\star=B_g^\star\subset B\subset A$, since $I_g^\star=\emptyset$. Therefore, $A_g^\star\subset A$ which implies that A is $\tau_g^\star$-closed.

Theorem 2.5. Let $(X,\tau,\mathcal{I})$ be an ideal topological space. Then $\tau\sim_g\mathcal{I}$ if and only if $\psi_g(A)-A\in\mathcal{I}$ for every $A\subset X$.

Proof. Suppose that $\tau\sim_g\mathcal{I}$ and $A\subset X$. Let $x\in\psi_g(A)-A$. Then $x\notin A$ and $x\notin(X-A)_g^\star$ if and only if $x\notin A$ and there exists a g-open set U containing x such that $U-A\in\mathcal{I}$ if and only if there exists a g-open set U containing x such that $x\in U-A\in\mathcal{I}$. Therefore, $x\in(U-A)\cap(\psi_g(A)-A)=U\cap(\psi_g(A)-A)$. Since $U\cap(\psi_g(A)-A)\subset U-A\in\mathcal{I}$, $U\cap(\psi_g(A)-A)\in\mathcal{I}$ by heredity property of ideal. Then by hypothesis, $\psi_g(A)-A\in\mathcal{I}$.

Conversely, suppose that for each $x\in A$, there exists a g-open set U containing x such that $U\cap A\in\mathcal{I}$ where $A\subset X$. Now consider $\psi_g(X-A)=\{x \mid \text{there exists } U\in\tau_g(x) \text{ such that } U\cap A\in\mathcal{I}\}$. Then we have $A\subset\psi_g(X-A)-(X-A)\in\mathcal{I}$ by hypothesis and so $A\in\mathcal{I}$ by heredity. Hence $\tau\sim_g\mathcal{I}$.

As a consequence of the above Theorem 2.5, we have the following Corollary 2.6.

Corollary 2.6. Let $(X,\tau,\mathcal{I})$ be an ideal topological space with $\tau\sim_g\mathcal{I}$. Then $\psi_g(\psi_g(A))=\psi_g(A)$ for every $A\subset X$.
Proof. By Lemma 1.5(c), it follows that $\psi_g(A)\subset\psi_g(\psi_g(A))$. Since $\tau\sim_g\mathcal{I}$, $\psi_g(A)\subset A\cup I$ for some $I\in\mathcal{I}$ by Theorem 2.5. Then $\psi_g(\psi_g(A))\subset\psi_g(A\cup I)=\psi_g(A)$ by Lemma 1.4(h). Therefore, $\psi_g(\psi_g(A))=\psi_g(A)$ for every $A\subset X$.

Theorem 2.7. Let $(X,\tau,\mathcal{I})$ be an ideal topological space with $\tau\sim_g\mathcal{I}$ and $A\subset X$. Then $\psi_g(A)=\bigcup\{\psi_g(U)\mid U\in\tau_g,\ \psi_g(U)-A\in\mathcal{I}\}$.

Proof. Let $\Phi_A = \bigcup\{\psi_g(U) \mid U \in \tau_g,\ \psi_g(U) - A \in \mathcal{I}\}$ and $x \in \psi_g(A)$. Then there exists a g-open set U containing x, such that $U - A \in \mathcal{I}$. By Lemma 1.5(b), $U \subset \psi_g(U)$ and $\psi_g(U) - A \subset (U - A) \cup (\psi_g(U) - U)$. By Theorem 2.5, $\psi_g(U) - U \in \mathcal{I}$ and hence $\psi_g(U) - A \in \mathcal{I}$. Therefore, $x \in \Phi_A$ and so $\Phi_A \supset \psi_g(A)$. Suppose that $x \in \Phi_A$. Then $x \in \psi_g(U)$ for some g-open set U containing x. Since $U \subset \psi_g(U)$, $U - A \subset \psi_g(U) - A \in \mathcal{I}$ which implies that $U - A \in \mathcal{I}$ by heredity. Therefore $x \in \psi_g(A)$ and so $\psi_g(A) \supset \Phi_A$. Hence $\Phi_A = \psi_g(A)$.

Theorem 2.8. Let $(X, \tau, \mathcal{I})$ be an ideal topological space with $\tau \sim_g \mathcal{I}$. If N is a nonempty g-open subset of $A^\star_g \cap \psi_g(A)$, then $N - A \in \mathcal{I}$ and $N \cap A \notin \mathcal{I}$.
Proof. If $N \subset A^\star_g \cap \psi_g(A)$, then $N - A \subset \psi_g(A) - A \in \mathcal{I}$ by Theorem 2.5 and hence $N - A \in \mathcal{I}$ by heredity. Since $N \in \tau_g - \{\emptyset\}$ and $N \subset A^\star_g$, $N \cap A \notin \mathcal{I}$.

Theorem 2.9. Let $(X, \tau, \mathcal{I})$ be an ideal topological space and $A \subset Y \subset X$. If Y is g-open in X, then $A^\star_g(\mathcal{I}_Y, \tau_{gY}) = A^\star_g(\mathcal{I}, \tau_g) \cap Y$ where $\tau_{gY} = \tau_g \mid Y$.

Proof. Let $x \notin A^\star_g(\mathcal{I}, \tau_g) \cap Y$. Then either $x \in Y$ or $x \notin Y$.
Case 1. Suppose that $x \notin Y$. Since $A^\star_g(\mathcal{I}_Y, \tau_{gY}) \subset Y$, $x \notin A^\star_g(\mathcal{I}_Y, \tau_{g\|Y})$.
Case 2. Suppose that $x \in Y$. Since $x \notin A^\star_g(\mathcal{I}, \tau_g)$, there exists a g-open set V in X containing x, such that $V \cap A \in \mathcal{I}$. Since $x \in Y$ and Y is g-open in X, $Y \cap V \in \tau_g$ such that $(Y \cap V) \cap A \in \mathcal{I}$ and so $(Y \cap V) \cap A \in \mathcal{I}_Y$. Consequently, $x \notin A^\star_g(\mathcal{I}_Y, \tau_{g|Y})$. Hence $A^\star_g(\mathcal{I}_Y, \tau_{gY}) \subset A^\star_g(\mathcal{I}, \tau_g) \cap Y$. To prove the converse, consider $x \notin A^\star_g(\mathcal{I}_Y, \tau_{gY})$. Then for some g-open set V in (Y, τ_Y) containing x, there exists $U \in \tau_g$ such that $V = U \cap Y$ and so $(U \cap Y) \cap A \in \mathcal{I}_Y$. Since $A \subset Y$, $U \cap A \in \mathcal{I}_Y \subset \mathcal{I}$ gives $U \cap A \in \mathcal{I}$ for some g-open set U containing x. Therefore, $x \notin A^\star_g(\mathcal{I}, \tau_g)$. Therefore, $A^\star_g(\mathcal{I}_Y, \tau_{gY}) = A^\star_g(\mathcal{I}, \tau_g) \cap Y$.

REFERENCES

[1] K. Bhavani, g-Local Functions, *J. Adv. Stud. Topol.*, 5(1)(2013), 1 - 5.
[2] K. Bhavani, ψ_g-Operator in Ideal Topological Spaces, *J. Adv. Stud. Topol.*, 5(1)(2013), 47 - 49.
[3] J. Dontchev, On Hausdorff Spaces via Topological Ideals and $\mathcal{I}$-irresolute Functions, *Annals of the NewYork academy of Sciences, General Topology and Applications*, 767(1995), 28 - 38.
[4] J. Dontchev, M. Ganster and D. Rose, Ideal Resolvability, Topology and applications, 93(1)1999 : 1-16, doi : 10.1016 /S0166-8641 (97) 00257-5.
[5] D. Jankovic and T. R. Hamlett, New Topologies from Old via Ideals, *Amer. Math. Monthly*, 97 (4) (1990), 295 - 310.
[6] D. Jankovic and T. R. Hamlett, Compatible Extensions of Ideals, *Boll. Un. Mat. Ital.*, (7) 6 - B (1992), 453 - 465.
[7] K. Kuratowski, Topology, Vol. I, Academic Press, New York, 1966.

[8] N. Levine, Generalized closed sets in topology, Rend. Circ. Mat. Palermo, 19(1970), 89 - 96.
[9] R. Vaidyanathaswamy, The localization theory in Set Topology, Proc. Indian Acad. Sci., 20 (1945), 51 - 61.

Journal of Physics: Conference Series **1377** (2019) 012035 doi:10.1088/1742-6596/1377/1/012035

SOME SUBCLASSES OF ANALYTIC FUNCTIONS WITH RESPECT TO (j, k)-SYMMETRICAL POINTS

K. Ramachandran
Assistant Professor,Department of Mathematics, SRM University,Ramapuram, Chennai-600048, Tamilnadu, India.
E-mail: bharaniram22@gmail.com

R. Vanaja
Assistant Professor,Department of Mathematics, SRM University,Ramapuram, Chennai-600048, Tamilnadu, India.
E-mail: rvanajamaths@gmail.com

K. Ramamoorthy
Assistant Professor,Department of Mathematics, LVCE,Walajapad , Kanchipuram, Tamilnadu, India.
E-mail: mailtoramamoorthy@gmail.com

Abstract. In this paper,the author introduces a new subclasses of analytic functions with respect to (j,k)-symmetric points and investigate various inclusion properties for these classes.Integral representation for functions in these classes and some interesting applications.

1. Introduction

Let $\mathcal{A}$ be a class of functions analytic in an open unit disc $\mathcal{U} = \{z \in \mathcal{C} : |z| < 1\}$ and is of the form

$$f(z) = z + \sum_{n=0}^{\infty} a_n z^n$$

. Also let $\mathcal{S}$ be the subclass of $\mathcal{A}$ consisting of all functions which of the form

$$f(z) = z - \sum_{n=2}^{\infty} a_n z^n \qquad (a_n \geq 0)$$

On A Class Of Univalent Starlike Functions

Let $\mathcal{S}(\lambda,\alpha)$ be the subclass of $\mathcal{S}$ consisting of functions $f(z)$ which satisfy the following inequality

$$Re\left(\frac{\frac{zf'(z)}{f(z)}}{\lambda\frac{zf'(z)}{f(z)}+(1-\lambda)}\right)>\alpha \qquad (z\in\mathcal{U}),$$

for some $\alpha(0\leq\alpha<1)$ and $\lambda(0\leq\lambda<1)$ and let $\mathcal{C}(\lambda,\alpha)$ be the subclass of $\mathcal{S}$ consisting of functions $f(z)$ which satisfy the following inequality

$$Re\left(\frac{1+\frac{zf''(z)}{f'(z)}}{1+\lambda\frac{zf''(z)}{f'(z)}}\right)>\alpha \qquad (z\in\mathcal{U}),$$

for some $\alpha(0\leq\alpha<1)$ and $\lambda(0\leq\lambda<1)$.The classes $\mathcal{S}(\lambda,\alpha)$ and $\mathcal{C}(\lambda,\alpha)$ were first introduced and investigated by Altinas and Owa, [1]then were studied by Aouf et al. [2] Let the functions $f(z)$ and $g(z)$ be members of $\mathcal{A}$. We say that the function f is subordinate to g (or g is superordinate to f), written $f\prec g$, if there exists a Schwarz function w analytic in U, with $w(0)=0$ and $|w(z)|<1$ and such that $f(z)=g(w(z))$. In particular, if g is univalent, then $f\prec g$ if and only if $f(0)=g(0)$ and $f(U)\subset g(U)$.

Motivated by the concept introduced by K. Sakaguchi in [11], recently several subclasses of analytic functions with respect to k-symmetric points were introduced and studied by various authors. More prominently, Wang et. al. [12] introduced class $S_s^{(k)}(\varphi)$ of functions $f\in\mathcal{A}$ subject to satisfying the condition

$$\frac{zf'(z)}{f_k(z)}\prec\varphi(z) \quad (z\in U),$$

where $\varphi(z)\in\mathcal{P}$, $k\geq1$ is fixed positive integer and $f_k(z)$ is defined by the equality

$$f_{j,k}(z)=\frac{1}{k}\sum_{\nu=0}^{k-1}\varepsilon^{-\nu}f(\varepsilon^{\nu}z).$$

Similarly, $C_s^{(k)}(\varphi)$ denote the class of functions in $\mathcal{S}$ satisfying the condition

$$\frac{(zf'(z))'}{f_k'(z)}\prec\varphi(z) \quad (z\in U),$$

where $\varphi(z)\in\mathcal{P}$, $k\geq1$ is fixed positive integer.

Liczberski and Połubiński in [7] introduced the notion $(j,\,k)$ symmetrical function $(k=2,\,3,\,\ldots;\,j=0,\,1,\,\ldots,\,k-1)$, which is a generalization of even, odd and k-symmetrical functions. A function $f\in\mathcal{A}$ is said to be $(j,\,k)$-symmetrical if for each $z\in\mathcal{U}$

$$f(\varepsilon z)=\varepsilon^{j}f(z), \tag{1}$$

$$(k=1,\,2,\,\ldots;\,j=0,\,1,\,2,\,\ldots(k-1)),$$

where $\varepsilon = \exp(2\pi i/k)$. The family of $(j,\, k)$-symmetrical functions will be denoted by $\mathcal{F}_k^j$. We observe that $\mathcal{F}_2^1$, $\mathcal{F}_2^0$ and $\mathcal{F}_k^1$ are well-known families of odd functions, even functions and k-symmetrical functions respectively. It was further proved in [7] that each function defined on a symmetrical set can be uniquely represented as the sum of an even function and an odd function.

Also let $f_{j,k}(z)$ be defined by the following equality

$$f_{j,\,k}(z) = \frac{1}{k}\sum_{\nu=0}^{k-1}\frac{f(\varepsilon^{\nu}z)}{\varepsilon^{\nu pj}}, \tag{2}$$

$$(f \in \mathcal{A}_p;\, k = 1,\, 2,\, \ldots;\, j = 0,\, 1,\, 2,\, \ldots(k-1)).$$

Motivated by the classes $\mathcal{S}(\lambda, \alpha)$ and $\mathcal{C}(\lambda, \alpha)$, We now introduced and investigated the following two subclasses of $\mathcal{A}$ with respect to (jk)symmetric points and obtain some interesting results.

We now define the following:

Definition 1.1 *Let $f \in \mathcal{A}$ is in the class $\mathcal{P}_{jk}(\lambda, \alpha)$ if it satisfies the following inequality*

$$Re\left(\frac{\frac{z(f(z))^{(m+1)}}{(f_{jk}(z)^{(m)}}}{\lambda(\frac{z(f(z))^{(m+1)}}{(f_{jk}(z))^{(m)}}) + (1-\lambda)}\right) > \alpha \qquad (z \in \mathcal{U}, m \in \mathbb{N} \cup 0)$$

where $0 \le \lambda < 1$,$0 \le \alpha < 1$, $k \ge 1$ is a fixed positive integer, $f_{jk}(z) \ne 0$ in $\mathcal{U}$ and a function $f^{(m)}(z) \in \mathcal{A}$ is in the class $\mathcal{Q}_{j,k}(\lambda, \alpha)$ if and only if $zf^{(m+1)}(z) \in \mathcal{P}_{j,k}(\lambda, \alpha)$.

2. INTEGRAL REPRESENTATIONS

Theorem 2.1 *Let $f \in \mathcal{P}_{j,\,k}(\lambda,\, \alpha)$, then we have*

$$f_{jk}^{(m)}(z) = z\exp\left\{\frac{1}{k}\sum_{v=0}^{k-1}\int_0^{\varepsilon^{v}z}\frac{2(1-\alpha)w(t)}{t\left[1-\lambda-(1+\lambda-2\alpha\lambda)w(t)\right]}\right\}$$

where $f \in \mathcal{P}_{j,\,k}(\lambda,\, \alpha)$ defined by equality (2), $w(z)$ is analytic in $\mathcal{U}$ with $w(0) = 0$ and $|w(z)| < 1$.

Proof. Let $f \in \mathcal{P}_{j,\,k}(\lambda,\, \alpha)$. In view of the equivalent subordination condition proved by Kuroki and Owa in [6] for the class $f \in P_{j,\,k}(\lambda,\, \alpha)$, we have

$$\frac{\frac{z(f(z))^{(m+1)}}{(f_{jk}(z)^{(m)}}}{\lambda(\frac{z(f(z))^{(m+1)}}{(f_{jk}(z))^{(m)}}) + (1-\lambda)} \prec \frac{1+(1-2\alpha)z}{1-z} \qquad (z \in \mathcal{U}) \tag{3}$$

$$\frac{\frac{z(f(z))^{(m+1)}}{(f_{jk}(z)^{(m)}}}{\lambda(\frac{z(f(z))^{(m+1)}}{(f_{jk}(z))^{(m)}}) + (1-\lambda)} = \frac{1+(1-2\alpha)w(z)}{1-w(z)} \qquad (z \in \mathcal{U}) \tag{4}$$

This yields

$$\frac{z(f(z))^{(m+1)}}{(f_{jk}(z))^{(m)}} = \frac{(1-\lambda)[1+(1-2\alpha)w(z)]}{1-\lambda-(1+\lambda-2\alpha\lambda)w(z)} \qquad (z \in \mathcal{U}) \tag{5}$$

where $w(z)$ is analytic in U and $w(0) = 0$, $|w(z)| < 1$. Substituting z by $\varepsilon^v z$ in the equality (3) respectively$(v = 0, 1, 2, \ldots k-1, \varepsilon^k = 1)$, we have

$$\frac{\varepsilon^v z(f(\varepsilon^v z))^{(m+1)}}{(f_{jk}(\varepsilon^v z))^{(m)}} = \frac{(1-\lambda)[1+(1-2\alpha)w(\varepsilon^v z)]}{1-\lambda-(1+\lambda-2\alpha\lambda)w(\varepsilon^v z)} \tag{6}$$

Using $f_{j,k}(\varepsilon^v z) = \varepsilon^{vpj} f_{j,k}(z)$ can be rewritten in the form

$$\frac{\varepsilon^{v(m+1)-vj} z(f(\varepsilon^v z)^{(m+1)}}{(f_{jk}(z))^{(m)}} = \frac{(1-\lambda)[1+(1-2\alpha)w(\varepsilon^v z)]}{1-\lambda-(1+\lambda-2\alpha\lambda)w(\varepsilon^v z)}$$

Let $v = 0, 1, 2, \ldots k-1$ in respectively and summing them, we get

$$\frac{z(f_{jk}^{(m+1)}(z))}{f_{jk}^{(m)}(z)} = \frac{1}{k}\sum_{v=0}^{k-1} \frac{(1-\lambda)[1+(1-2\alpha)w(\varepsilon^v z)]}{1-\lambda-(1+\lambda-2\alpha\lambda)w(\varepsilon^v z)}$$

On simplifying and integrating, we get

$$\log\left(\frac{f_{jk}^{(m)}(z)}{z}\right) = \frac{1}{k}\sum_{v=0}^{k-1}\int_0^{\varepsilon^v z} \frac{2(1-\alpha)w(t)}{t[1-\lambda-(1+\lambda-2\alpha\lambda)w(t)]}\,dt \tag{7}$$

The difficulty to integrate the term with presence of the first order pole at the origin, has been avoided by integrating from z_0 to z with $z_0 \neq 0$ and then let $z_0 \to 0$. Further simplifying (7), we get

$$f_{jk}^{(m)}(z) = z\exp\left\{\frac{1}{k}\sum_{v=0}^{k-1}\int_0^{\varepsilon^v z} \frac{2(1-\alpha)w(t)}{t[1-\lambda-(1+\lambda-2\alpha\lambda)w(t)]}\,dt\right\}.$$

This completes the proof of theorem.

Theorem 2.2 *Let $f \in \mathcal{P}_{jk}(\lambda, \alpha)$, then we have*

$$f(z) = \int_0^z\int_0^z\cdots\int_0^z \exp\left\{\frac{1}{k}\sum_{v=0}^{k-1}\frac{2(1-\alpha)w(t)}{t[1-\lambda-(1+\lambda-2\alpha\lambda)w(t)]}\right\}$$
$$\times\left[\frac{(1-\lambda)[1+(1-2\alpha)w(\zeta)]}{1-\lambda-(1+\lambda-2\alpha\lambda)w(\zeta)}\right]d\zeta\ldots dz\,dz.$$

Where w(z) is analytic in $\mathcal{U}$ with $w(0) = 0$ and $|w(z)| < 1 \qquad (z \in \mathcal{U})$

Proof. Suppose $f \in \mathcal{P}_{jk}(\lambda, \alpha)$.It follows

$$\frac{z(f(z))^{(m+1)}}{(f_{jk}(z))^{(m)}} = \frac{(1-\lambda)[1+(1-2\alpha)w(z)]}{1-\lambda-(1+\lambda-2\alpha\lambda)w(z)}$$

Or equivalently,

$$f^{(m+1)}(z) = \exp\left\{\frac{1}{k}\sum_{v=0}^{k-1}\frac{2(1-\alpha)w(t)}{t[1-\lambda-(1+\lambda-2\alpha\lambda)w(t)]}\right\} \times \left[\frac{(1-\lambda)[1+(1-2\alpha)w(z)]}{1-\lambda-(1+\lambda-2\alpha\lambda)w(z)}\right]$$

Integrating the above expression $m+1$ times, we have

$$f(z)=\int_0^z\int_0^z\cdots\int_0^z \exp\left\{\frac{1}{k}\sum_{v=0}^{k-1}\frac{2(1-\alpha)w(t)}{t[1-\lambda-(1+\lambda-2\alpha\lambda)w(t)]}\right\}$$

$$\times\left[\frac{(1-\lambda)[1+(1-2\alpha)w(\zeta)]}{1-\lambda-(1+\lambda-2\alpha\lambda)w(\zeta)}\right]d\zeta\ldots dz\,dz.$$

Corollary 2.3 *If $f\in\mathcal{Q}_{jk}(\lambda,\alpha)$ then the integral representation of $f_{jk}^{(m)}(z)$ is given by*

$$f_{jk}^{(m)}(z)=\int_0^z \exp\left\{\frac{1}{k}\sum_{v=0}^{k-1}\int_0^{\varepsilon^v\zeta}\frac{2(1-\alpha)w(t)}{t[1-\lambda-(1+\lambda-2\alpha\lambda)w(t)]}dt\right\}d\zeta.$$

Where $w(z)$ is analytic in $\mathcal{U}$ with $w(0)=0$ and $|w(z)|<1 \qquad (z\in\mathcal{U})$

Corollary 2.4 *If $f\in\mathcal{Q}_{jk}(\lambda,\alpha)$,then*

$$\frac{z(f(z))^{(m+1)}}{(f_{jk}(z))^{(m)}}=\frac{(1-\lambda)[1+(1-2\alpha)w(z)]}{1-\lambda-(1+\lambda-2\alpha\lambda)w(z)}$$

then the integral representation of $f(z)$ is given by

$$f(z)=\int_0^z\int_0^z\cdots\int_0^z\frac{1}{\xi}\int_0^\xi \exp\left\{\frac{1}{k}\sum_{v=0}^{k-1}\int_0^{\varepsilon^v\zeta}\frac{2(1-\alpha)w(t)}{t[1-\lambda-(1+\lambda-2\alpha\lambda)w(t)]}dt\right\}$$

$$\left[\frac{(1-\lambda)[1+(1-2\alpha)w(\zeta)]}{1-\lambda-(1+\lambda-2\alpha\lambda)w(\zeta)}\right]d\zeta d\xi\ldots dz.$$

Where $w(z)$ is analytic in $\mathcal{U}$ with $w(0)=0$ and $|w(z)|<1 \qquad (z\in\mathcal{U})$.

3. Convolution conditions

We provide the convolution conditions for the classes $\mathcal{P}_{jk}(\lambda,\alpha)$ and $\mathcal{Q}_{jk}(\lambda,\alpha)$. Let f, g $\in\mathcal{A}$ Where $f(z)$ is given by

$$f(z)=z+\sum_2^\infty a_nz^n$$

and g(z) is defined by

$$g(z)=z+\sum_2^\infty c_nz^n$$

then the hadamard product $f*g$ is defined by

$$(f*g)(z)=z+\sum_2^\infty a_nc_nz^n=(g*f)(z)$$

.

On A Class Of Univalent Starlike Functions

Remark 3.1 *For a case of $m = 0$ in the definition , a function $f \in \mathcal{P}_{jk}(\lambda, \alpha)$ if and only if*

$$\frac{1}{z}\left\{f * \left\{\frac{z}{(1-z)^2}\{(1-e^{i\theta}) - \lambda[1+(1-2\alpha)e^{i\theta}]\} - (1-\lambda)[1+(1-2\alpha)e^{i\theta}]h(z)\right\}\right\} \neq 0$$

$\forall z \in \mathcal{U}$ and $0 \leq \theta < 2\pi$, Where $h(z)$ is given by

$$h(z) = \frac{1}{k}\sum_{v=0}^{k-1}\frac{z}{1-\varepsilon^{vj}z} \qquad (z \in \mathcal{U})$$

result proved by zhi gang wang and di song.

Remark 3.2 *For a case of $m = 0$ in the definition , a function $f \in \mathcal{Q}_{jk}(\lambda, \alpha)$ if and only if*

$$\frac{1}{z}\left\{f * \left\{z\left\{\frac{z}{(1-z)^2}\{(1-e^{i\theta}) - \lambda[1+(1-2\alpha)e^{i\theta}]\} - (1-\lambda)[1+(1-2\alpha)e^{i\theta}]h(z)\right\}'\right\}\right\} \neq 0$$

$\forall z \in \mathcal{U}$ and $0 \leq \theta < 2\pi$, Where $h(z)$ is given by

$$h(z) = \frac{1}{k}\sum_{v=0}^{k-1}\frac{z}{1-\varepsilon^{vj}z} \qquad (z \in \mathcal{U})$$

result proved by zhi gang wang and di song.

Lemma 3.1 *[8] Let $-1 \leq B_2 \leq B_1 < A_1 \leq A_2 \leq 1$, then we have*

$$\frac{1+A_1z}{1+B_1z} \prec \frac{1+A_2z}{1+B_2z}$$

4. Inclusion Relationships

We give some inclusion relationships for the classes $\mathcal{P}_{jk}(\lambda, \alpha)$ and $\mathcal{Q}_{jk}(\lambda, \alpha)$.

Theorem 4.1 *Let $0 \leq \lambda_1 \leq \lambda_2 < 1$,and $0 \leq \alpha_1 \leq \alpha_2 < 1$ then we have*

$$\mathcal{P}^s_{jk}(\lambda_1, \alpha_1) \subset \mathcal{P}^s_{jk}(\lambda_2, \alpha_2)$$

Proof. Suppose $f(z) \in \mathcal{P}^s_{jk}(\lambda_2, \alpha_2)$,we have

$$\frac{z(f(z))^{(m+1)}}{(f_{jk}(z))^{(m)}} \prec \frac{(1-\lambda_2)[1+(1-2\alpha_2)z]}{1-\lambda_2-(1+\lambda_2-2\alpha_2\lambda_2)z}$$

since $0 \leq \lambda_1 \leq \lambda_2 < 1$,and $0 \leq \alpha_1 \leq \alpha_2 < 1$,then we have
Thus by lemma 3.1.,we have

$$\frac{z(f(z))^{(m+1)}}{(f_{jk}(z))^{(m)}} \prec \frac{(1-\lambda_2)[1+(1-2\alpha_2)z]}{1-\lambda_2-(1+\lambda_2-2\alpha_2\lambda_2)z} \prec \frac{(1-\lambda_1)[1+(1-2\alpha_1)z]}{1-\lambda_1-(1+\lambda_1-2\alpha_1\lambda_1)z}$$

ie $f(z) \in \mathcal{P}^s_{jk}$.this means that

$$\mathcal{P}^s_{jk}(\lambda_2, \alpha_2) \subset \mathcal{P}^s_{jk}(\lambda_1, \alpha_1)$$

.

similarly ,for the class $\mathcal{Q}^s_{jk}(\lambda, \alpha)$ we have

Corollary 4.2 *Let* $0 \leq \lambda_1 \leq \lambda_2 < 1$,*and* $0 \leq \alpha_1 \leq \alpha_2 < 1$ *then we have*

$$\mathcal{Q}^s_{jk}(\lambda_1, \alpha_1) \subset \mathcal{Q}^s_{jk}(\lambda_2, \alpha_2)$$

5. References

[1] O.Altinas and S.Owa, on subclasses of univalent functions with negative coefficients,pusan kyongnam math J.4(1988),41-56.
[2] M.K.Aouf ,H.M.Hossen and A.Y.Lashin,convex subclass of starlike functions,Kyung-pook amth.J.40(2000),287-297.
[3] Teodor Bulboacă, Differential subordinations and superordinations. Recent result, House of Science Book Publ., Cluj-Napoca, 2005.
[4] A. W. Goodman, *Univalent functions. Vol. I*, Mariner, Tampa, FL, 1983.
[5] I. Graham and G. Kohr, *Geometric function theory in one and higher dimensions*, Dekker, New York, 2003.
[6] K. Kuroki and S. Owa, Notes on new class for certain analytic functions, RIMS Kokyuroku, **1772** (2011) pp. 21-25.
[7] P. Liczberski and J. Połubiński, On (j, k)-symmetrical functions, Math. Bohem. **120** (1995), no. 1, 13–28.
[8] M.S.Liu,on a subclass of p-valent close-to-convex functions of order β and type α,J.Math.study 30(1997),102-104.
[9] S. S. Miller and P. T. Mocanu, Subordinants of differential superordinations, Complex Var. Theory Appl. **48** (2003), no. 10, 815–826.
[10] M. A. Nasr and M. K. Aouf, Starlike function of complex order, J. Natur. Sci. Math. **25** (1985), no. 1, 1–12.
[11] K. Sakaguchi, On a certain univalent mapping, J. Math. Soc. Japan **11** (1959), 72–75.
[12] Z.-G. Wang, C.-Y. Gao and S.-M. Yuan, On certain subclasses of close-to-convex and quasi-convex functions with respect to k-symmetric points, J. Math. Anal. Appl. **322** (2006), no. 1, 97–106.

Functions of Bounded Radius Rotation Problems

S.Prema[1,*] P.Sumathi[2]

[1,2] Asst. Professor, Department of Mathematics, SRM IST, Ramapuram Campus, Chennai, India.
*Corresponding author: premnehaa@gmail.com

Abstract: The classes $V_k(\alpha)$, $R_k(\alpha)$, $V_{k1}(\alpha)$ and $R_{k1}(\alpha)$ of analytic function, We establish a relation between the functions of bounded boundary and bounded radius rotations.

Keywords: Analytic, Starlike, Convex, Subordination, Analytic functions with bounded radius and boundary rotations.

Mathematics Subject Classification: 30C45.

1 Introduction:

Let A be the class of functions f of the form $f(z)=z+\sum_{n=2}^{\infty}a_n z^n$ (1.1)

Which are analytic in the unit disc E={z: |z|<1}.

The definition of subordination is $f\in$A is subordinate to g$\in$A, written as $f\prec g$,there exists Schwarz function w(z) with w(0)=0 and |w(z)|<1 ($z\in$E) such that f(z)=g(w(z)), In particular, when g is univalent, then the above subordination is equivalent to f(0)'=g(0) and f(E)$\subseteq$g(E), for any two analytic functions

$$f(z)=\sum_{n=0}^{\infty}a_n z^n, \qquad g(z)=\sum_{n=0}^{\infty}b_n z^n \qquad (z\in E) \qquad (1.2)$$

then convolution

$$(f*g)(z)=\sum_{n=0}^{\infty}a_n b_n z^n \qquad (z\in E) \qquad (1.3)$$

We denote by $s^*(\alpha)$, $c(\alpha)$, $(0\le\alpha<1)$ the classes of starlike and convex functions of order α, respectively defined by

$$s^*(\alpha)=\left\{f\in A:\mathrm{Re}\frac{zf'(z)}{f(z)}>\alpha,\ \ z\in E\right\}$$

$$c(\alpha)=\left\{f\in A;\ zf'(z)\in s^*(\alpha),\ \ z\in E\right\} \qquad (1.4)$$

For $\alpha=0$, we have the well know classes of starlike and convex univalent functions denoted s^* and c, respectively.

Let $P_k(\alpha)$ be the class of functions P(z) analytic in the unit disc E satisfying the properties P(0)=1 and $\int_0^{2\pi}\left|\operatorname{Re}\frac{P(z)-\alpha}{1-\alpha}\right|d\theta \le k\pi$, (1.5)

Where $z = re^{i\theta}$, $k \ge 2$, and $0 \le \alpha < 1$, using Herglotz – Stieltijes formula

$$P(z) = \left(\frac{k}{4}+\frac{1}{2}\right)P_1(z) - \left(\frac{k}{4}-\frac{1}{2}\right)P_2(z), \quad z \in E \tag{1.6}$$

Where $P(\alpha)$ is the class of functions with real part greater then α and $P_i \in P(\alpha)$, for i=1,2 we define the following classes

$$R_k(\alpha) = \left\{ f : f \in A\,and\, \frac{[(1-t)(\lambda z^2 f''(z) + zf'(z)]}{\lambda z(f'(z) - tf'(z)) + (1-\lambda)(f(z) - tf(z))} \in P_k(\alpha), 0 \le \lambda < 1 \right\}, \quad t>0$$

$$V_k(\alpha) = \left\{ f : f \in A\,and\, \frac{[(1-t)(\lambda z^2 f''(z) + zf'(z)]'}{\lambda z(f''(z) - tf''(z)) + [(1-\lambda)(f'(z) - tf'(z)]'} \in P_k(\alpha), 0 \le \lambda < 1 \right\}, \quad t>0 \tag{1.7}$$

$$R_{k1}(\alpha) = \left\{ f : f \in A\,and\, \frac{(1-t)[\gamma\lambda z^3 f'''(z) + (2\gamma\lambda + \gamma + \lambda)z^2 f''(z) + zf'(z)]}{\gamma\lambda z^2(f''(z) - tf''(z)) + (\lambda-\gamma)z(f'(z) - tf'(z) + (1-\lambda+\gamma)(f(z) - tf(z))} \in P_k(\alpha), \right\}$$

$$0 \le \lambda \le \gamma < 1 \quad , t > 0. \tag{1.8}$$

$$V_{k1}(\alpha) = \left\{ f : f \in A\,and\, \frac{(1-t)[\gamma\lambda z^3 f'''(z) + (2\gamma\lambda + \gamma + \lambda)z^2 f''(z) + zf'(z)]'}{\gamma\lambda z^2(f''(z) - tf''(z)) + (\lambda-\gamma)z(f'(z) - tf'(z) + [(1-\lambda+\gamma)(f(z) - tf(z)]'} \in P_k(\alpha), \right\}$$

$$0 \le \lambda \le \gamma < 1 \quad , t > 0. \tag{1.9}$$

We note that

$$f \in V_k(\lambda) \Leftrightarrow zf' \in R_k(\lambda) \qquad f \in V_{k1}(\lambda) \Leftrightarrow zf' \in R_{k1}(\lambda) \tag{1.10}$$

For $\lambda = 0, t = 0, \gamma = 0$ we obtain the well known classes R_k,R_{k1},V_k,V_{k1} of analytic functions with bounded radius and bounded boundary rotations , respectively. These classes are studied by Noor [3] in more details, also it can easily by seen $R_k^{\,2}(\alpha) = S^*(\alpha)$ and $V_2(\alpha) = c(\alpha)$. Goel[6] proved that $f \in c(\alpha)$ implies that $f \in s^*(\alpha)$ and $f \in S^*(\alpha)$

Where $\beta = \beta(\alpha) = \begin{cases} \dfrac{4^{\alpha}(1-2\alpha)}{4-2^{2\alpha+1}} & ,\alpha \ne \dfrac{1}{2} \\ \dfrac{1}{2}\ln 2 & ,\alpha = \dfrac{1}{2} \end{cases}$

and the result is sharp, In this paper, we prove the result of Goel[6] for the classes $V_k(\alpha)$, $R_k(\alpha)$, $V_{k1}(\alpha)$ and $R_{k1}(\alpha)$ by using convolution and Subordination techniques.

2 Preliminary Results

We need the following results to obtain our results.

Lemma 2.1: Let $u = u_1 + iu_2$, $v = v_1 + iv_2$ and $\Psi(u,v)$ be a complex valued function satisfying the conditions

(i) $\Psi(u,v)$ is continuous in a domain $D \subset C^2$.

(ii) $(1,0) \in D$ and $\operatorname{Re}\psi(1,0) > 0$, whenever $(iu_2, v_1) \le 0$ when ever $(iu_2, v_1) \in D \, and \, v_1 \le -\left(\frac{1}{2}\right)\left(1+u_2^2\right)$

If $h(z) = 1 + c_1 z + \dots\dots$ is a function analytic in E. Such that $(h(z), zh'(z)) \in D$ and $\operatorname{Re}\psi(h(z), zh'(z)) > 0$ for $z \in E$, then Re h (z)>0 in E.

3 Main Results:

Theorem 3.1

Let $f \in V_k(\alpha)$, *then* $f \in R_k(\alpha)$, *where*

$$\beta = \frac{1}{4}(2\alpha - (1+\lambda)(1-t) + 2\alpha(1-t)\lambda) + \sqrt{\begin{matrix}(1+\lambda)^2 + 4\alpha^2 + 4\alpha^2\lambda^2 - 4\alpha(1+\lambda) + 8\alpha^2\lambda - 4\alpha t(1+\lambda) \\ + 8 - 10t - 8\lambda + 8\lambda t + t^2(1+\lambda^2).\end{matrix}}$$

Proof: Let

$$\frac{(1-t)(\lambda z^2 f''(z) + zf'(z))}{\lambda z(f(z) - tf(z)) + (1-\lambda)(f(z) - tf(z))} = (1-\beta)P(z) + \beta \qquad (3.1)$$

$$= (1-\beta)\left[\left(\frac{k}{4}+\frac{1}{2}\right)P_1(z) - \left(\frac{k}{4}-\frac{1}{2}\right)P_2(z)\right] + \beta \qquad (3.2)$$

P(z) is analytic in E with P(0)=1 then

$$\frac{\left[(1-t)\left[\lambda z^2 f''(z) + zf'(z)\right]\right]'}{\lambda z\left[f'(z) - tf'(tz)\right] + (1-\lambda)\left[f(z) - tf(z)\right]} = (1-\beta)p(z) + \beta + \frac{(1-t)\begin{bmatrix}\lambda z^2 f''(z) + ztf'(z) + (1-\beta)p'(z) \\ + (1-\lambda)z + \lambda zf'(tz)\end{bmatrix}}{\lambda z\left[f'(z) - tf'(z)\right] + (1-\lambda)(1-\beta)p(z) + \beta} \qquad (3.3)$$

$$\frac{1}{(1-\alpha)}\left[\frac{\left[(1-t)\left[\lambda z^2 f''(z) + zf'(z)\right]\right]'}{\lambda z\left[f'(z) - tf'(tz)\right] + (1-\lambda)\left[f(z) - tf(z)\right]} - \alpha\right] =$$

$$\frac{1}{(1-\alpha)}\left[(1-\beta)p(z) + \beta - \alpha + \frac{(1-t)\left[\lambda z^2 f''(z) + z\lambda f'(z) + (1-\lambda)z\right](1-\beta)p'(z)}{\lambda z\left[f'(z) - tf'(z)\right] + (1-\lambda)\left[(1-\beta)p(z) + \beta\right]}\right]$$

$$= \frac{\beta-\alpha}{1-\alpha} + \frac{1-\beta}{1-\alpha}\left[p(z) + \left(\frac{\left(\frac{1}{(1-\beta)}\right)(L+M)p'(z)}{(N+Q)(p(z) + \frac{\beta}{1-\beta}}\right)\right] \qquad (3.4)$$

Where $L = (1-t)\lambda z^2 f''(z)$

$M = (1-t)\lambda zf'(z) + (1-\lambda)z$

$N = \lambda z\left[f'(z) - tf'(tz)\right]$

$Q = (1-\lambda)t$

Since $f \in V_k(\alpha)$, *it implies that*

$$=\frac{\beta-\alpha}{1-\alpha}+\frac{1-\beta}{1-\alpha}\left[p(z)+\frac{\left(\frac{1}{1-\beta}\right)(L+M)p'(z)}{(N+Q)\left(p(z)+\frac{\beta}{(1-\beta)}\right)}\right]\in p_{k,} \qquad z\in E \qquad (3.5)$$

We define $\varphi_{a,b}(z)=\frac{1}{1+b}\frac{z}{(1-z)\alpha}+\frac{b}{1+b}\frac{z}{(1-z)^{1+a}}$ *(3.6)*

with $a=\frac{1}{1-\beta}$, $b=\frac{\beta}{1-\beta}$ *by using (3.1) with convolution, see[5], we have, that*

$$\frac{\varphi_{a,b}(z)}{z}*p(z)=\left(\frac{k}{4}+\frac{1}{2}\right)\left[\frac{\varphi_{a,b}(z)}{z}*p_1(z)\right]-\left(\frac{k}{4}-\frac{1}{2}\right)\left[\frac{\varphi_{a,b}(z)}{z}*p_2(z)\right]$$

Implies

$$p(z)+\frac{a(L+M)p'(z)}{(N+Q-R)(p(z)+b)}=\left(\frac{k}{4}+\frac{1}{2}\right)\left(p_1(z)+\frac{a(L+M)p_1'(z)}{(N+Q)(p_1(z)+b}\right)-\left(\frac{k}{4}+\frac{1}{2}\right)\left(p_2(z)+\frac{a(L+M)p_2'(z)}{(N+Q)(p_2(z)+b}\right) \qquad (3.7)$$

Thus from (3.5) and (3.7) we have

$$\frac{\beta-\alpha}{1-\alpha}+\frac{(1-\beta)}{(1-\alpha)}\left[p_i(z)+\frac{a(L+M)p_i'(z)}{(N+Q)(p_i(z)+b}\right]\in p,\ i=1, \qquad (3.8)$$

We now from functional $\psi(u,v)$ *by choosing* $u=p_i(z), v=zp_i'(z)$ *in(3.8) and note that the first two conditions of lemma:2.1 are likely satisfied, we check the condition as follows*

$$\mathrm{Re}[\psi(iu_2,v_1)]=\frac{1}{1-\alpha}\left[(\beta-\alpha)+\mathrm{Re}\left(\frac{(1-t)\lambda z^2 f''(z)p_i'(z)+(1-t)f'(z)v_1}{\left(iu_2+\frac{\beta}{1-\beta}\right)[\lambda z(f''(z)-tf''(tz)]+(1-\lambda)[f'(z)-f'(tz)]'}\right)\right]$$

$$=\frac{1}{1-\alpha}\left[(\beta-\alpha)+\frac{\left((1-t)\lambda z^2 f''(z)p_i'(z)+(1-t)f'(z)v_1\right)\left(\frac{\beta}{(1-\beta}\right)}{\left(iu_2+\frac{\beta}{1-\beta}\right)\left(\begin{array}{c}\lambda z(f''(z)+tf'(z))+(1-\lambda)\\(f(z)-tf(z))\end{array}\right)'}\right]$$

$$\le\frac{1}{1-\alpha}\left((\beta-\alpha)-\frac{\frac{1}{2}\left((1+u_2^2)(1-t)f'(z)+(1-t)\lambda z^2 f''(z)\right)\left(\frac{\beta}{1-\beta}\right)}{\left(u_2^2+\left(\frac{\beta}{1-\beta}\right)^2\right)\left((f(z)-tf(z))+(1-t)\lambda z^2 f''(z)\right)}\right)$$

$$= \frac{2(\beta-\alpha) - \left((1+u_2^2)(1-t)f'(z) + (1-t)\lambda z^2 f''(z)\right)\left(\frac{\beta}{1-\beta}\right)}{2\left(u_2^2 + \left(\frac{\beta}{1-\beta}\right)^2\right)\left((f(z) - tf(z) + (1-t)\lambda z^2 f''(z)\right)} \qquad (3.9)$$

$$= \frac{A + Bu_2^2}{2c}, \qquad 2C > 0,$$

Where

$$A = \frac{\beta}{(1-\beta)^2}\left[2(\beta-\alpha)\beta[(f(z) - f(tz)] + (1-t)\,\lambda z^2\ f''(z)\,]' - (1-t)\,[f'(z) + \lambda z^2\ f''(z)\,]\,(1-\beta)\right]$$

$$B = \frac{1}{(1-\beta)}\ [2(\beta-\alpha)(1-\beta)\ [(f(z) - f(tz)] + (1-t)\,\lambda z^2\ f''(z)\,]'' - \beta(1-t)\,[f'(z) + \lambda z^2\ f''(z)\,]$$

$$C = \left[u_2^2 + \left(\frac{\beta}{1-\beta}\right)^2\right]\ (1-\alpha)\ [(f(z) - f(tz)] + (1-t+\lambda)\,\lambda z^2\ f''(z)\,]' \quad (3.10)$$

The right hand side of (3.9) is negative if $A \le 0$ *we have*

$$\beta = \beta(\alpha) = \frac{1}{4}(2\alpha - (1+\lambda)(1-t) + 2\alpha(1-t)\lambda) + \sqrt{\begin{array}{l}(1+\lambda)^2 + 4\alpha^2 + 4\alpha^2\lambda^2 - 4\alpha(1+\lambda) + 8\alpha^2\lambda - 4\alpha t(1+\lambda)\\ +8 - 10t - 8\lambda + 8\lambda t + t^2(1+\lambda^2).\end{array}}$$

Theorem 3.2

Let $f \in V_{k1}(\alpha)$, *then* $f \in R_{k1}(\alpha)$, *where*

$$\beta = \frac{1}{4}(2\alpha - (1+\lambda)(1+\gamma)(1-t) + 2\alpha(1-t)\gamma\lambda) + \sqrt{\begin{array}{l}(1+\lambda+\gamma)^2 + 4\alpha^2\lambda^2\gamma + 6\alpha^2\gamma^2\lambda^2 - 4\alpha(1+\gamma)(1+\lambda)\\ +16\alpha^2\lambda\gamma - 4\alpha t(1+\lambda)(1+\gamma) + 8(1+\lambda)(1+\gamma) - 11t - 12\gamma\lambda\\ +t^2(1+\gamma^2)(1+\lambda^2).\end{array}}$$

Reference

[1] B.Pinchuk, "Functions of bounded boundary rotation, "Israel journal of Mathematics, vol. 10, no. 1, pp.6-16, 1971.

[2]K.S.Padmanabhan and R.Parvathan, "Properties of a class of functions with bounded boundary rotation, "Annales Polomici Mathematici, vol.31, no.3, pp.311-323, 1975.

[3] K.I. Noor, "Some properties of certain analytic function,"journal of natural geometry, vol .7, no.1, pp.11-20, 1995

[4]K.I.Noor,"on some subclasses of function with bounded radius and bounded boundary rotation,"panamerican mathematical journal, vol.6, no.1, pp.75-81, 1996

[5]K.L.Noor,"on analytic function related to certain family of integral operators,"journal of inequalities in pure and applied mathematics, vol.7, no.2, article 69, 6 pages, 2006.

[6]R.N.Goel,"functions starlike and convex of order α , "Journal of the London Mathematical society.Second Series, vol.s2-9, no.1, pp.128-130, 1974

[7]D.A.Brannan,"on functions of bounded boundary rotation .I," proceedings of the Edinburgh mathematical society .series II, vol.16, pp.339-347, 1969

[8]D.Pinchuk,"on starlike and convex function order a," duke mathematical journal, vol.35, pp.721-734, 1968

[9]S.S.Miller,"Differential in equalities and caratheodory functions,"bulletin of the American mathematical society, vol.81, pp.79-81, 1975.

[10]S.S.Miller and P.T.Mocanu,differential subordinations :theory and application ,vol.225of monographs and text bookes in pure and applied mathematicals , marcel dekker ,new york ,NY,USA,2000

Journal of Physics: Conference Series **1377** (2019) 012036 doi:10.1088/1742-6596/1377/1/012036

[11]Z.Nahari conformal mappings, Dover, New [11] Z.Nahari conformal mappings, Dover, New York, NY, USA, 1952
[12]I.S.Jack,"functions starlike and convex of order a,"journal of the London mathematical society .second series, vol.3, pp.469-474, 1971
[13]T.H.MacGregor,"A subordination for convex function of order a," journal of the London mathematical society .second series, vol.9, pp.530-536, 1975.

TOTAL NEIGHBOURHOOD PRIME LABELING OF CERTAIN GRAPHS

S.Sripriya
Assistant Professor, Department of Mathematics, SRMIST ,Ramapuram, Chennai-600048, Tamilnadu, India.
E-mail: ssripriya78@gmail.com

K. Ramachandran
Assistant Professor, Department of Mathematics, SRMIST,Ramapuram, Chennai-600048, Tamilnadu, India.
E-mail: bharaniram22@gmail.com

Keywords:Neighbourhood prime labelling,Total neighbourhood prime labeling
Subject classification: 05C78

Abstract. In this paper,we investigate the total neighbourhood prime labelling of Lily graphs, Umbrella graph, Friendship graph F_n, Generalised friendship graph $Fr_n^{(4)}$,$\bar{Fr}_n^{(3)}$ and $\bar{Fr}_n^{(4)}$.

1. INTRODUCTION

Graphs in which vertices and edges are assigned values subject to certain conditions is called graph labeling.It has various applications in real life problems.Graph labeling is one of the mathematical model for addressing communication network addressing system,circuit design,coding theory,missile guidance codes,and radio astronomy problems etc.The notion of prime labeling is introduced by Roger Entringer and was discussed in paper Tout et. al, [3].
Many researchers have published many results in prime labeling.
S.K.Patel and N.P.Srimali [4] introduced the notion of neighbourhood prime labeling and proved many results.Rajesh kumar and Mathew varkey [6] introduced total neighbourhood prime labeling and proved paths and cycles are total neighborhood prime graphs.Motivated by this work N.P.Shrimali et.al.,[7] also proved that the comb,disjoint union of paths,disjoint union of sunlet graphs,disjoint union of wheel graphs, subdivision of bistar are total neighbourhood prime graphs.
In this paper we study total neighbourhood prime labeling of lily graph,umbrella graph and different friendship graphs. Throughout the paper ,all the graphs considered are

finite,simple,undirected and no isolated vertices.For terminology and notations we follow [1]and for latest survey in graph labeling we refer to gallian 2018 [2].

Definition 1.1 *Let G be a graph with m vertices and p edges.A bijection ϕ from $V(G)$ into $\{1,2,3.....m\}$ is called neighbourhood prime labeling if for each vertex $v \in V(G)$, with degree of V greater than 1, $gcd\{\phi(u) : u \in N(v)\} = 1$.A graph which admits neighbourhood prime labeling is called a neighbourhood prime graph.*

Definition 1.2 *Let G be a graph with m vertices and p edges.A bijection ϕ from $V(G) \cup E(G)$ into $\{1,2,3.....m+p\}$ is said to be total neighbourhood prime labeling if for each vertex of degree atleast two, the labeling on its adjacent vertices are relatively prime and labeling on the incident edges are also relatively prime.*

2. Main Results

2.1. Lily Graph

Definition 2.1 *The lily graph L_n , $n \geq 2$ is obtained by two star graphs $2K_{1,n}$, $n \geq 2$ joining two path graphs $2P_n$, $n \geq 2$ with a common vertex.*

Theorem 2.1 *Lily graph L_n,$n \geq 3$ admits total neighbourhood prime labeling where n is any positive integer.*
Proof. *Let G be a lily graph L_n,$n \geq 3$.*
Let $u_1, u_2, ...u_{2n}, v_1, v_2, ...v_{2n-1}$ be the vertices of L_n and $e_i = \{v_iv_{i+1}\}$, $e_i' = \{v_nu_i\}$ be the edges of L_n.
Here $|V(G)| = 4n-1$, $|E(G)| = 4n-2$ and $|V(G) \cup E(G)| = 8n-3$. Define a labeling f from $V(G) \cup E(G)$ to $\{1,2,......8n-3\}$ as follows
$f(u_i) = i,\ 1 \leq i \leq 2n$
$f(v_i) = \left\{ 4\,n - \frac{i}{2} - \frac{1}{2} \quad i = 1,3,....2n-1 \right.$
$3n - \frac{i}{2} \quad i = 2,4,........2n-2$
$f(v_iv_{i+1}) = 4n+i-1 \quad ,1 \leq i \leq 2n-2$
$f(v_nu_i) = 6n+i-3 \quad ,1 \leq i \leq 2n$
To prove that f is a total neighbourhood prime labeling
Note that u_i ,$1 \leq i \leq 2n$ and v_1, v_{2n-1} are vertices of degree one.

- *Case 1: For $v = v_i$,$2 \leq i \leq 2n-2$,$d(v)$=2 the labels of neighbourhood vertices of v_i are consecutive integers and labels of neighbourhood edges of v_i are also consecutive integers.*
  Therefore $gcd\{f(x), x \in N_V(v_i)\} = 1$ and $gcd\{f(e), e \in N_E(v_i)\} = 1$.

- *Case 2: For $v = v_n$,$d(v) > 2$ the labels of neighbourhood vertices of v_n are consecutive integers and labels of neighbourhood edges of v_n are also consecutive*

integers.
Therefore $gcd\{f(x), x \in N_V(v_n)\} = 1$ *and* $gcd\{f(e), e \in N_E(v_n)\} = 1$.

Hence f is a total neighbourhood prime labeling.

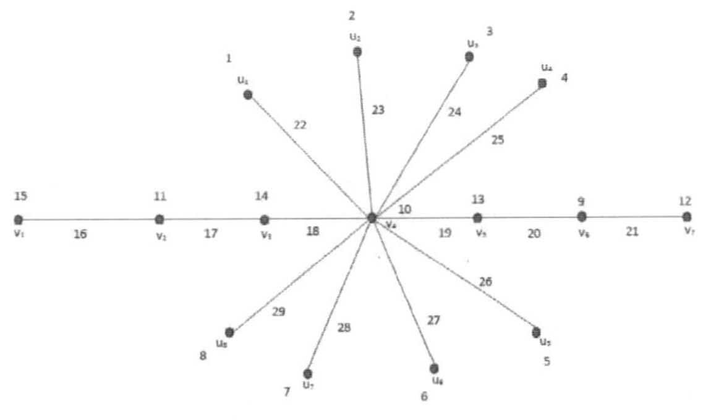

Figure 1. lily graph

2.2. Umbrella Graph

Definition 2.2 *An umbrella graph U(n) is obtained by connecting a path graph* P_n *and a central vertex of a fan graph* F_{n+1}.

Theorem 2.2 *Umbrella graph* U_n,$n \geq 3$ *admits total neighbourhood prime labeling except for* $2n + 1 \equiv 0(mod15)$ *where n is any positive integer.*
Proof. *Let* U_n *be an umbrella graph* $n \geq 3$.*Let* $u_1, u_2, ...u_n, v_1, v_2, ...v_{n+1}$ *be the vertices of* U_n *and* $\{v_iv_{i+1}\}$,$1 \leq i \leq n$,$\{u_iu_{i+1}\}$,$1 \leq i \leq n-1$, $\{v_iu_n\}$,$1 \leq i \leq n+1$ *be the edges of* U_n.
Here $|V(G)| = 2n + 1$, $|E(G)| = 3n$ *and* $|v(G) \cup E(G)| = 5n + 1$.
Define a labeling f from $V(G) \cup E(G)$ *to* $\{1, 2, ...|V(G) \cup E(G)|\}$ *as follows*
The vertex labelings are

- *case 1: when n is odd*
  $f(v_i) = i,\ 1 \leq i \leq n + 1$
  $f(u_{2i-1}) = \frac{3n}{2} + i + \frac{1}{2},\ 1 \leq i \leq \lfloor \frac{n+1}{2} \rfloor$
  $f(u_{2i}) = n + i + 1,\ 1 \leq i \leq \lfloor \frac{n}{2} \rfloor$

- *case 2: when n is even*
  $f(v_i) = i,\ 1 \leq i \leq n + 1$
  $f(u_{2i-1}) = \frac{3n}{2} + i + 1,\ 1 \leq i \leq \lfloor \frac{n}{2} \rfloor$
  $f(u_{2i}) = n + i + 1,\ 1 \leq i \leq \lfloor \frac{n}{2} \rfloor$
  and the edge labelings are
  $f(u_iu_{i+1}) = 2n + i + 1,\ 1 \leq i \leq n - 1$

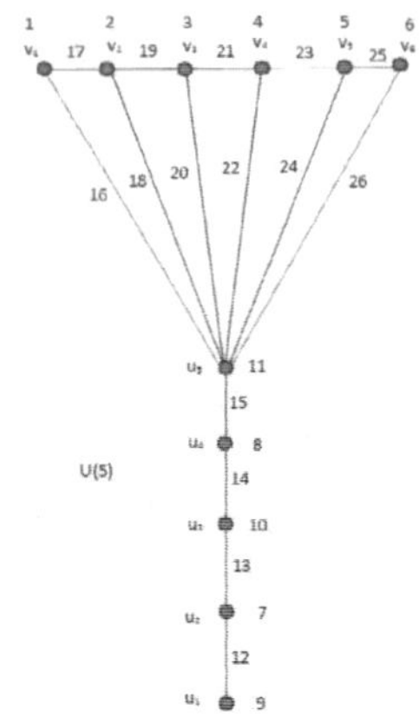

Figure 2. U(5)graph

$f(v_iv_{i+1}) = 3n + 2i,\ 1 \leq i \leq n$
$f(v_iu_n) = 3n + 2i - 1,\ 1 \leq i \leq n + 1$

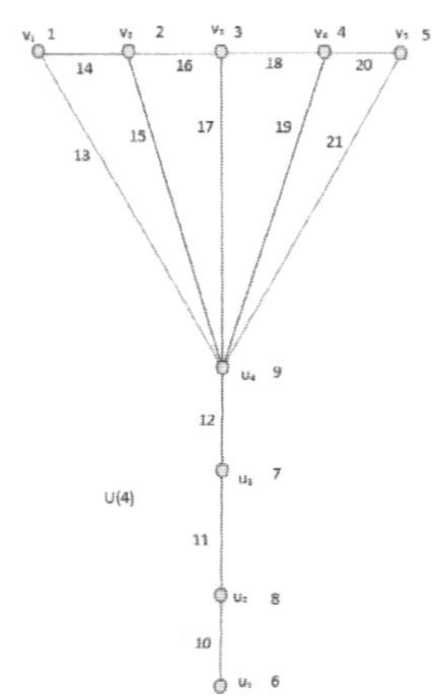

Figure 3. U(4)graph

To prove that f is a total neighbourhood prime labeling

For $v = u_i$, $2 \leq i \leq n$,$d(v) \geq 2$, the labels of neighbourhood vertices contains consecutive integers therefore $gcd\{f(x), x \in N_V(u_i)\} = 1$ and labels of neighbourhood edges contains two consecutive integers therefore $gcd\{f(d), d \in N_E(u_i)\} = 1$.

For $v = v_i$, $1 \leq i \leq n + 1$, $d(v) \geq 2$,the labels of neighbourhood vertices of each vertex v_i are $2i, 2n + 1$ or $i - 1, i + 1, 2n + 1$ or $n - 1, 2n + 1$. From this it is clear that the gcd of the labeling of the neighbourhood vertices of each vertex is one. Also labels of *neighbourhood edges contains consecutive integers, therefore* $gcd\{f(d), d \in N_E(v_i)\} = 1$. *hence f is a total neighbourhood prime labeling.*

2.3. Friendship Graph

Definition 2.3 *A friendship graph F_n is a collection of n triangles with a common vertex 2n+1 vertices and 3n edges.(i.e., one point union of n copies of cycles C_3.)*

Theorem 2.3 *Friendship graph F_n is a total neighbourhood prime labeling for all positive integers n .*

Proof.

Let F_n be a friendship graph with n copies of cycle C_3. Let u be a apex vertex and other vertices in the triangles be $u_1, u_2, ...u_n$ and $v_1, v_2, ...v_n$. The edges in the triangle are uu_i, uv_i, u_iv_i, where $1 \leq i \leq n$.

Define a bijection f from $V(F_n) \cup E(F_n)$ to { 1,2,...5n+1} as follows

$f(u) = 1$

$f(u_i) = 2i, \quad 1 \leq i \leq n$

$f(v_i) = 2i + 1, \quad 1 \leq i \leq n$

$f(uu_i) = 2n + 3i - 1, \quad 1 \leq j \leq n$

$f(uv_i) = 2n + 3i + 1, \quad 1 \leq j \leq n$

$f(u_iv_i) = 2n + 3i, \quad 1 \leq j \leq n$

To prove F_n is total neighbourhood prime labeling.

- *Let $v = u$ then $N_V(v)$ contains consecutive integers and $N_E(v)$ contains atleast two consecutive integers .So $gcd[N_V(v)]$ and $gcd[N_E(v)]$ are one.*

- *For $v = u_i$ the $N_V(u_i)$ contains 1. Therefore $gcd[N_V(u_i)] = 1$ and $N_E(u_i)$ contains two consecutive integers, so $gcd[N_E(u_i)]$ is one.*

- *For $v = v_i$ the $N_V(v_i)$ contains 1. Therefore $gcd[N_V(v_i)] = 1$ and $N_E(v_i)$ contains two consecutive integers, so $gcd[N_E(v_i)]$ is one.*

Hence it is total neighbourhood prime labeling.

2.4. Friendship Graph

Definition 2.4 *A generalised friendship graph or a flower $Fr_n^{(4)}$ is a planar undirected graph with 3n+1 vertices and 4n edges constructed by joining n copies of the cycle graph C_4 with common vertex.*

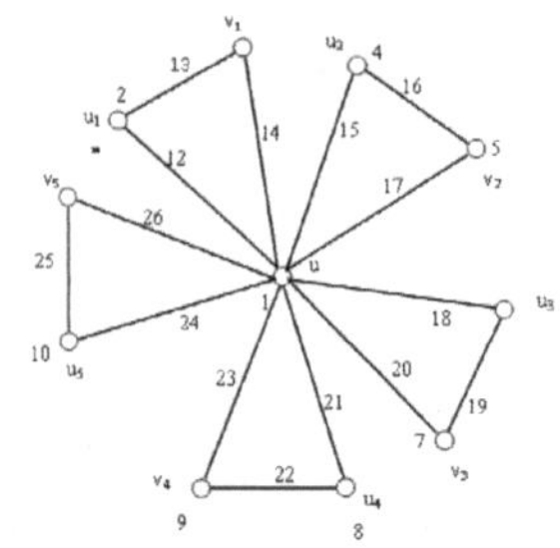

Figure 4. Friendship Graph

Theorem 2.4 *Friendship graph $Fr_n^{(4)}$ is a total neighbourhood prime labeling.*

Proof. *Let G be a friendship graph.Here $|V(G)| = 3n+1$,$|E(G)| = 4n$ and $|V(G) \cup E(G)| = 7n+1$.*

Define bijection $f : V(G) \cup E(G) \rightarrow \{1, 2, ...7n+1\}$ as follows

$f(u_0) = 1$

$f(u_i) = i+1, \quad 1 \leq i \leq n$

$f(v_i) = 5n+2i, \quad 1 \leq i \leq n$

$f(w_i) = 5n+2i+1 \quad 1 \leq i \leq n$

$f(u_i v_i) = n+4i-1, \quad 1 \leq i \leq n$

$f(u_i w_i) = n+4i, \quad 1 \leq i \leq n$

$f(u_0 v_i) = n+4i-2, \quad 1 \leq i \leq n$

$f(u_0 w_i) = n+4i+1, \quad 1 \leq i \leq n$

- *If $v = u_0$,$d(u_0) > 1$ then $N_V(v_i)$ contains consecutive integers, so $gcd\{f(x)/x \in N_V(v) = 1\}$ and $N_E(v)$ contains two consecutive integers. Therefore $gcd\{f(e)/e \in N_E(v)$ is one.*

- *If $v = v_i$,$1 \leq i \leq n$,$d(v)$=2,$N_V(v_i)$ contains one.Therefore $gcd\{f(x)/x \in N_V(v_i)\} = 1$. $N_E(v_i)$ contains two consecutive integers.So gcd $f(e)$ $e \in N_E(v_i) = 1$.*

- *If $v = w_i$,$1 \leq i \leq n$,$d(v)$=2,$N_V(w_i)$ contains one. $N_E(v_i)$ contains two consecutive integers .Therefore $gcdN_V(w_i)$ and $gcdN_E(w_i) = 1$.*

- *If $v = u_i$, $1 \leq i \leq n$,$d(u_i) > 2$ then $N_V(u_i)$ contains consecutive integers and $N_V(u_i) = 1$ and $[N_E(u_i)]$ contains two consecutive integers, therefore they are relatively prime.*

Hence it is a total neighbourhood prime labelling for all n.

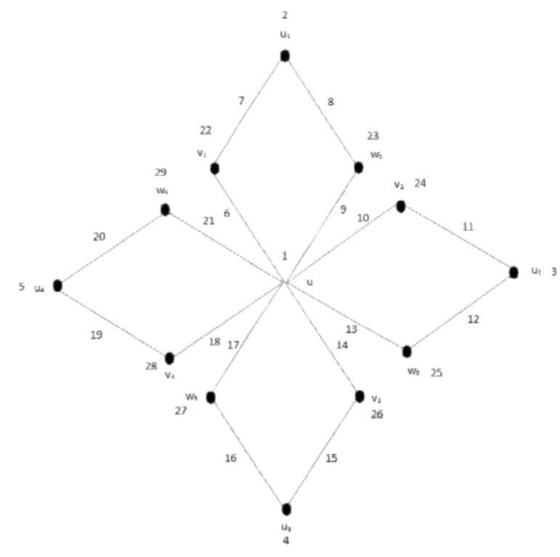

Figure 5. Flower Graph

2.5. Friendship Graph F3

Definition 2.5 *A friendship graph $\bar{Fr}_n^{(3)}$ is a planar undirected graph with 3n+1 vertices and 5n edges constructed by joining n copies of the fan graph F_3 with a common vertex.*

Theorem 2.5 *Friendship graph $\bar{Fr}_n^{(3)}$ is a total neighbourhood prime labeling for all positive integers n .*

Proof.

Let $\bar{Fr}_n^{(3)}$ be a graph with common vertex u and other vertices be $u_1, u_2, ...u_n$, $v_1, v_2, ...v_n$ and $w_1, w_2, ...w_n$, where $1 \leq i \leq n$ .Let the edges be $uu_i, uv_i, uw_i, u_iv_i, v_iw_i$, where $1 \leq i \leq n$ of $\bar{Fr}_n^{(3)}$.

Define a bijection f from $V(\bar{Fr}_n^{(3)} \cup E(\bar{Fr}_n^{(3)})$ to { 1,2,3,......8n+1} as follows

$f(u) = 1$
$f(u_i) = 3i - 1, \quad 1 \leq i \leq n$
$f(v_i) = 3i, \quad 1 \leq i \leq n$
$f(w_i) = 3i + 1, \quad 1 \leq i \leq n$
$f(uu_i) = 3n + 5i - 3, \quad 1 \leq i \leq n$
$f(uv_i) = 3n + 5i - 1, \quad 1 \leq i \leq n$
$f(uw_i) = 3n + 5i + 1, \quad 1 \leq i \leq n$
$f(u_iv_i) = 3n + 5i - 2, \quad 1 \leq i \leq n$
$f(v_iw_i) = 3n + 5i, \quad 1 \leq i \leq n$

To prove $\bar{Fr}_n^{(3)}$ is a total neighbourhood prime labeling.

Let $v = u$,$d(u) > 2$ then $N_V(v)$ contains consecutive integers and $N_E(v)$ contains atleast two consecutive integers $gcd[N_V(v)]$ and $gcd[N_E(v)]$ are one.

For $v = u_i$, $N_V(u_i)$ contain 1. Therefore $gcd[N_V(u_i)] = 1$ and $N_E(u_i)$ contains two

consecutive integers $gcd[N_E(u_i)]$ *is one.*
For $v = v_i$, $N_V(v_i)$ *contain 1. Therefore* $gcd[N_V(v_i)] = 1$ *and* $N_E(v_i)$ *contains two consecutive integers* $gcd[N_E(v_i)]$ *is one.*
For $v = w_i$, $N_V(w_i)$ *contain 1. Therefore* $gcd[N_V(w_i)] = 1$ *and* $N_E(w_i)$ *contains two consecutive integers* $gcd[N_E(w_i)]$ *is one.*
Hence it is total neighbourhood prime labeling.

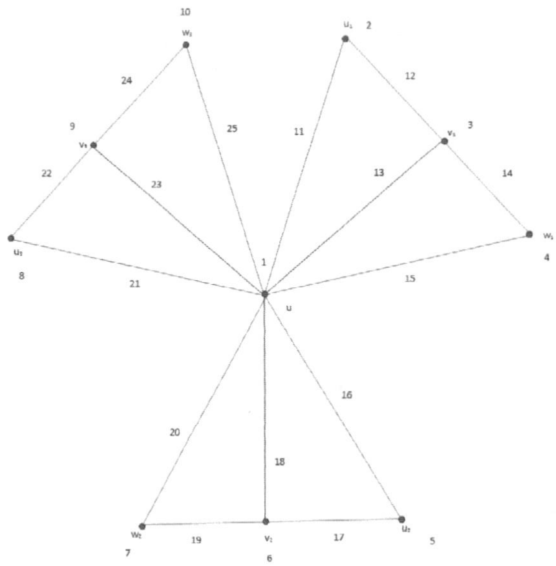

Figure 6. Friendship Graph

2.6. Friendship graph $\bar{Fr}_n^{(4)}$

Definition 2.6 *A friendship graph* $\bar{Fr}_n^{(4)}$ *is a planar undirected graph with 4n+1 vertices and 7n edges constructed by joining n copies of the fan graph* F_4 *with a common vertex.*

Theorem 2.6 *Friendship graph* $\bar{Fr}_n^{(4)}$ *is a total neighbourhood prime labeling for all positive integers n .*
Proof. *Let* $\bar{Fr}_n^{(4)}$ *be a planar undirected graph with* $4n + 1$ *vertices and* $7n$ *edges constructed by joining n copies of the fan graph* F_4 *with a common vertex. Let u be a apex vertex and other vertices are* $u_1, u_2, ...u_n$, $v_1, v_2, ...v_n$, $w_1, w_2, ...w_n$ *and* $z_1, z_2, ...z_n$ *and the edges be* $uu_i, uv_i, uw_iuz_i, u_iv_i, v_iw_i, w_iz_i$, *where* $1 \leq i \leq n$ *of* $\bar{Fr}_n^{(4)}$.

Define a bijection f from $V(\bar{Fr}_n^{(4)}) \cup E(\bar{Fr}_n^{(3)})$ *to* { *1,2,..., 11n+1*} *as follows*
$f(u) = 1$
$f(u_i) = 4i - 2, \quad 1 \leq i \leq n$
$f(v_i) = 4i - 1, \quad 1 \leq i \leq n$

$f(w_i) = 4i, \quad 1 \le i \le n$
$f(z_i) = 4i + 1, \quad 1 \le i \le n$
$f(uu_i) = 4n + 7i - 5, \quad 1 \le i \le n$
$f(uv_i) = 4n + 7i - 3, \quad 1 \le i \le n$
$f(uw_i) = 4n + 7i - 1, \quad 1 \le i \le n$
$f(uz_i) = 4n + 7i + 1, \quad 1 \le i \le n$
$f(u_iv_i) = 4n + 7i - 4, \quad 1 \le i \le n$
$f(v_iw_i) = 4n + 7i - 2, \quad 1 \le i \le n$
$f(w_iz_i) = 4n + 7i, \quad 1 \le i \le n$

To prove $\bar{Fr}_n^{(4)}$ is a total neighbourhood prime labeling.

Let $v = u$,$d(u) > 2$ then $N_V(v)$ contains consecutive integers and $N_E(v)$ contains atleast two consecutive integers $gcd[N_v(v)]$ and $gcd[N_e(V)]$ are one.

For $v = u_i$ the $N_V(u_i)$ contains 1. Therefore $gcd[N_v(u_i)] = 1$ and $N_E(u_i)$ contains two consecutive integers.So $gcd[N_E(u_i)]$ is one.

For $v = v_i$ the $N_V(v_i)$ contains 1. Therefore $gcd[N_v(v_i)] = 1$ and $N_E(v_i)$ contains two consecutive integers.So $gcd[N_E(v_i)]$ is one.

For $v = w_i$ the $N_V(w_i)$ contains 1. Therefore $gcd[N_v(w_i)] = 1$ and $N_E(w_i)$ contains two consecutive integers.So $gcd[N_E(w_i)]$ is one.

For $v = z_i$ the $N_V(z_i)$ contains 1. Therefore $gcd[N_v(z_i)] = 1$ and $N_E(z_i)$ contains two consecutive integers.So $gcd[N_E(z_i)]$ is one.

Hence it is total neighbourhood prime labeling.

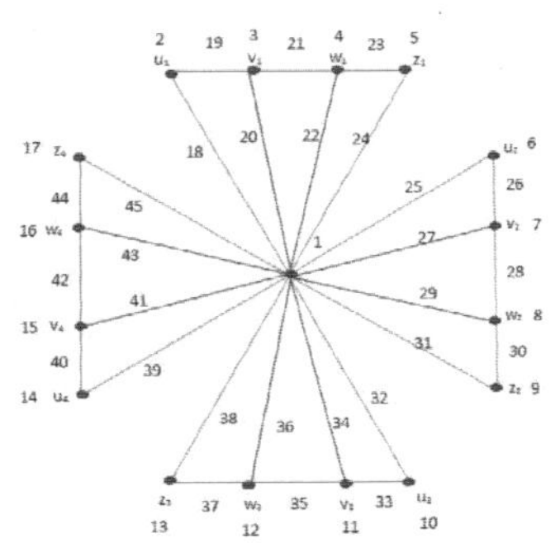

Figure 7. Friendship Graph

labelFr4

[1] F. Harary, Graph Theory, Addison Wesley, Reading, MA, 1969.
[2] Joseph A.Gallian,A dynamic survey of graph labeling the electronic journal of combinatorics, 17[2018].
[3] Tout A., Dabboucy A.N, Hawalla k," prime labeling of graphs", Nat.Acad.Sci. Letters,11[1982, 365-368.]
[4] S.K.Patel, N.P. shrimali," Neighbourhood prime labeling",International journal of mathematics and soft computing, Vol.5, No.2(2015), 135-143.
[5] Mathew varkey T.K, Rajesh kumar T.J,"A note on neighbourhood prime labeling", International journal of mathematics combinatrics, Vol.4(2016), 161-167.

Journal of Physics: Conference Series **1377** (2019) 012037 doi:10.1088/1742-6596/1377/1/012037

[6] Rajesh kumar T.J, Mathew varkey T.K, "A note on total neighbourhood prime labeling", International journal of mathematics combinatrics, 4(2018).
[7] N.P.Shrimali, parul B. pandya, "Total neighbourhood prime labeling of some graphs", International journal of scientific research in mathematical and statistical sciences, vol(5), Dec 2018, 157- 163.

SOLUTION OF MULTI OBJECTIVE TRANSPORTATION PROBLEM UNDER FUZZY ENVIRONMENT

R Gowthami[a] and K Prabakaran[b *]
Department of Mathematics,
SRM Institute of Science and Technology,
Kattankulathur, Chennai - 603203, INDIA
***Corresponding author:* [b] prabakak2@srmist.edu.in;**
[a]gowthami1111997@gmail.com

ABSTRACT
The crisp transportation problem is one of the special case of linear programming problems. It is explained with the hypothesis that the cost parameters are indicated inaccurate way. But in the present world, the transportation parameters may be uncertain due to many uncontrolled factors; this gives rise to fuzzy environment. Here, we proposed a method for the solution of fully fuzzy multi objective transportation problems involving triangular fuzzy numbers. We solve the fully fuzzy multi objective transportation problems without converting it into crisp form. A numerical example is discussed to demonstrate the theory formulated in this article.

Key words: Multi-objective transportation problem, Triangular fuzzy numbers, parametric form, Fuzzy ranking, Fuzzy arithmetic.

1. INTRODUCTION

The transportation problem is particular case of linear programming problems which handle with the allocation of a product from variety of sources of supply to variety of destinations of demand in such a way that the total transportation cost is minimum. Decision making depends exclusively on a single condition is insufficient whenever the decision–making process handled with the complex organizational situation. So, we must understand the presence of numerous criteria that lead to the improvement of multi-criteria decision making. To formulate and solve the transportation problem, the judgment parameters such as supply, demand, unit transportation cost, time, etc. are known precisely. But in real life situations, the data obtainable is of vague in nature and there is an inexactness degree of vagueness or uncertainty present in the problem under consideration. The uncertainty and inexactness can be tackle by the concept of fuzzy sets can be used as an important decision making tool. These imprecise data may be represented by fuzzy numbers. A fuzzy set introduced by Zadeh[19] in 1965 is applied to tackle such uncertain environments.

Most of the authors have given their ideas to solve the different types of fuzzy transportation models to save time and money. Abd El-Wahed[2] gave an interactional fuzzy goal programming approach for multi-objective transportation problems in 2006. Sudhakar and Navaneetha Kumar[16] used the zero-suffix method for solving two stage multi objective fuzzy transportation problems in 2010. Amit Kumar and Anila Gupta[3] projected a innovative technique for solution of linear multi objective transportation problems involving fuzzy parameters in 2012. Khan and Das [7] worked on fuzzy multi objective optimization with mention to multi objective transportation problems in 2014. In 2016, Muruganandam and Srinivasan[12] obtained an optimal solution for a multi objective two stage fuzzy transportation problem. Sheema et.al[15] worked on multi objective fractional transportation problem in fuzzy environment in 2017. Here, we solve the multi-objective fuzzy transportation problem based on new algorithm without converting it to a classical one.

2. PRELIMINARIES

In this section we recall the basic concepts and the results of triangular fuzzy number and their arithmetic operations.

Definition 2.1

A fuzzy set $\tilde{a}$ defined on the set of real numbers R is said to be a fuzzy number if its membership function $\tilde{a}: R \to [0,1]$ has the following characteristics:

(i) $\tilde{a}$ is convex, i.e. $\tilde{a}(\lambda x_1 + (1-\lambda)x_2) \geq \min\{\tilde{a}(x_1), \tilde{a}(x_2)\}$, for all $x_1, x_2 \in R$ and $\lambda \in [0,1]$

(ii) $\tilde{a}$ is normal i.e. there exists an $x \in R$ such that $\tilde{a}(x) = 1$.

(iii) $\tilde{a}$ is Piecewise continuous.

Definition 2.2

A fuzzy number $\tilde{a}$ on R is said to be a triangular fuzzy number (TFN) if its membership function $\tilde{a}: R \to [0,1]$ has the following characteristics:

$$\tilde{a}(x) = \begin{cases} \dfrac{x - a_1}{a_2 - a_1}, & \text{for } a_1 \leq x \leq a_2 \\ \dfrac{a_3 - x}{a_3 - a_2}, & \text{for } a_2 \leq x \leq a_3 \\ 0, & \text{elsewhere} \end{cases}$$

We denote this triangular fuzzy number by $\tilde{a} = (a_1, a_2, a_3)$. We use $F(R)$ to denote the set of all triangular fuzzy numbers. Also if $m = a_2$ represents the modal value or midpoint, $\alpha = (a_2 - a_1)$ represents the left spread and $\beta = (a_3 - a_2)$ represents the right spread of the triangular fuzzy number $\tilde{a} = (a_1, a_2, a_3)$, then the triangular fuzzy number $\tilde{a}$ can be represented by the triplet $\tilde{a} = (\alpha, m, \beta)$. i.e. $\tilde{a} = (a_1, a_2, a_3) = (\alpha, m, \beta)$.

Definition 2.3

A triangular fuzzy number $\tilde{a} \in F(R)$ can also be represented as a pair $\tilde{a} = (\underline{a}, \overline{a})$ of functions $\underline{a}(r)$ and $\overline{a}(r)$ for $0 \leq r \leq 1$ which satisfies the following requirements:

(i) $\underline{a}(r)$ is a bounded monotonic increasing left continuous function.

(ii) $\overline{a}(r)$ is a bounded monotonic decreasing left continuous function.

(iii) $\underline{a}(r) \leq \overline{a}(r)$, $0 \leq r \leq 1$

Definition 2.4

For an arbitrary triangular fuzzy number $\tilde{a} = (\underline{a}, \overline{a})$, the number $a_0 = \left(\dfrac{\underline{a}(1) + \overline{a}(1)}{2}\right)$ is said to be a location index number of $\tilde{a}$. The two non-decreasing left continuous functions $a_* = (a_0 - \underline{a})$, $a^* = (\overline{a} - a_0)$ are called the left fuzziness index function and the right fuzziness index function respectively. Hence every triangular fuzzy number $\tilde{a} = (a_1, a_2, a_3)$ can also be represented by $\tilde{a} = \left(a_0, a_*, a^*\right)$.

2.5 Ranking of triangular Fuzzy Numbers

Many different approaches for the ranking of fuzzy numbers have been proposed in the literature. Abbasbandy and Hajjari[1] proposed a new ranking method based on the left and the right spreads at some α -levels of fuzzy numbers.

For an arbitrary triangular fuzzy number $\tilde{a} = (a_1, a_2, a_3) = \left(a_0, a_*, a^*\right)$ with parametric form $\tilde{a} = \left(\underline{a}(r), \overline{a}(r)\right)$, we define the magnitude of the triangular fuzzy number $\tilde{a}$ by

$$\mathrm{Mag}(\tilde{a}) = \frac{1}{2}\left(\int_0^1 (\underline{a} + \overline{a} + a_0)\ f(r)\ dr\right)$$
$$= \frac{1}{2}\left(\int_0^1 (a^* + 4a_0 - a_*)\ f(r)\ dr\right).$$

where the function $f(r)$ is a non-negative and increasing function on $[0,1]$ with $f(0)=0$, $f(1)=1$ and $\int_0^1 f(r)\ dr = \frac{1}{2}$. The function $f(r)$ can be considered as a weighting function. In real life applications, $f(r)$ can be chosen by the decision maker according to the situation. In this paper, for convenience we use $f(r) = r$.

For any two triangular fuzzy numbers $\tilde{a} = \left(a_0, a_*, a^*\right)$ and $\tilde{b} = \left(b_0, b_*, b^*\right)$ in F(R), we define the ranking of $\tilde{a}$ and $\tilde{b}$ by comparing the Mag ($\tilde{a}$) and Mag ($\tilde{b}$) on R as follows:

(i) $\tilde{a} \succeq \tilde{b}$ if and only if Mag($\tilde{a}$) $\geq$ Mag($\tilde{b}$)

(ii) $\tilde{a} \preceq \tilde{b}$ if and only if Mag($\tilde{a}$) $\leq$ Mag($\tilde{b}$)

(iii) $\tilde{a} \approx \tilde{b}$ if and only if Mag($\tilde{a}$) = Mag($\tilde{b}$)

2.6 Arithmetic operation on triangular Fuzzy Numbers

Ming Ma et al.[11] have proposed a new fuzzy arithmetic based upon both location index and fuzziness index functions. The location index number is taken in the ordinary arithmetic, whereas the fuzziness index functions are considered to follow the lattice rule which is least upper bound in the lattice L. That is for a, b∈L we define $a \vee b = \max\{a, b\}$ and $a \wedge b = \min\{a, b\}$.

For arbitrary triangular fuzzy numbers $\tilde{a} = \left(a_0, a_*, a^*\right)$ and $\tilde{b} = \left(b_0, b_*, b^*\right)$ and $* = \{+, -, \times, \div\}$, the arithmetic operations on the triangular fuzzy numbers are defined by $\tilde{a} * \tilde{b} = \left(a_0 * b_0, a_* \vee b_*, a^* \vee b^*\right)$. In particular for any two triangular fuzzy numbers $\tilde{a} = \left(a_0, a_*, a^*\right)$ and $\tilde{b} = \left(b_0, b_*, b^*\right)$, we define

(i) Addition : $\tilde{a} + \tilde{b} = (a_0, a_*, a^*) + (b_0, b_*, b^*) = (a_0 + b_0, \max\{a_*, b_*\}, \max\{a^*, b^*\})$

(ii) Subtraction : $\tilde{a} - \tilde{b} = (a_0, a_*, a^*) - (b_0, b_*, b^*) = (a_0 - b_0, \min\{a_*, b_*\}, \min\{a^*, b^*\})$

(iii) Multiplication : $\tilde{a} \times \tilde{b} = (a_0, a_*, a^*) \times (b_0, b_*, b^*) = (a_0 \times b_0, \max\{a_*, b_*\}, \max\{a^*, b^*\})$

(iv) Division: $\tilde{a} \div \tilde{b} = (a_0, a_*, a^*) \div (b_0, b_*, b^*) = (a_0 \div b_0, \min\{a_*, b_*\}, \min\{a^*, b^*\})$

3 MODEL REPRESENTATION

In real life situations, usually every organizer wants to achieve multiple goals simultaneously while making transportation of goods. So MOTP developed by researchers to achieve multiple goals. Like classical transportation problem, in MOTP, quantity $\tilde{x}_{ij}$ is to be transported from sources i (i=1,2,…,m) to destinations j (j=1,2,…,n) with cost $\tilde{c}_{ij}$, where$\tilde{c}_{ij}$can be transportation cost, total delivery time, energy consumption or minimizing transportation risk etc. The k objectives $\tilde{z}_1$,$\tilde{z}_2$,……,$\tilde{z}_k$ are to be associated with transportation cost, total delivery time, cost of damage and/or cost of security etc. So MOTP (η) can be represented mathematically as follows:

$$\text{Min } \tilde{Z}_1 \approx \sum_{i=1}^{m}\sum_{j=1}^{n}\tilde{c}_{ij}{}^{1}\,\tilde{x}_{ij}$$

$$\vdots$$

$$\text{Min}\tilde{Z}_k \approx \sum_{i=1}^{m}\sum_{j=1}^{n}\tilde{c}_{ij}{}^{k}\,\tilde{x}_{ij},$$

Subject to

$\sum_{i=1}^{m}\tilde{x}_{ij} \approx \tilde{a}_i$, $\tilde{a}_i \geq 0$, (i=1,2,…,m)

$\sum_{j=1}^{n}\tilde{x}_{ij} \approx \tilde{b}_j$, $\tilde{b}_j \geq 0$,(j=1,2,….,n)

$\tilde{x}_{ij} \succeq 0$, for all i,j and $\sum_{i=1}^{m}\tilde{a}_i \approx \sum_{j=1}^{n}\tilde{b}_j$

Where $\tilde{c}_{ij}{}^{r}$co-efficient of the rth objective,

$\tilde{a}_i$ supply amount of the item for consumption at source i$(\tilde{S}_i)$,

$\tilde{b}_j$ demand of the item for consumption at destination j $(\tilde{D}_j)$,

and $\tilde{a}_i \succeq 0$, $\tilde{b}_j \succeq 0$.

4 NEW ALGORITHM FOR PROPOSED METHOD

Step 1: Formulate the given multi objective Transportation problem into the parametric form.
Step 2: Identity maximum row cost ($\tilde{\alpha}$) as $\tilde{\alpha}_i{}^{r}$= max ($\tilde{c}_{ij}{}^{r}$), for fixed i, $1 \leq j \leq n$and $1 \leq r \leq k$, maximum column cost ($\tilde{\beta}$) as $\tilde{\beta}_j{}^{r}$= max ($\tilde{c}_{ij}{}^{r}$), for fixed j, $1 \leq i \leq m$ and $1 \leq i \leq$m and $1 \leq r \leq k$, where

$\tilde{\alpha}$={$\tilde{\alpha}_1{}^{1}$,…,$\tilde{\alpha}_1^{r}$;….;$\tilde{\alpha}_m{}^{1}$,……,$\tilde{\alpha}_m{}^{r}$} and

$\tilde{\beta}$={$\tilde{\beta}_1^1$,. …, $\tilde{\beta}_1^k$;, ….; $\tilde{\beta}_n^1$, … …, $\tilde{\beta}_n^k$}.

Step 3: Choose Q= $\max_{1 \leq i \leq m, 1 \leq j \leq n}$ ($\tilde{\alpha}_i{}^{r}$,$\tilde{\beta}_j{}^{r}$), ∀r.

Step 4: Choose the cell (C) having Q as one of its objective value. If there exist more than one cell (C)then select the one, which has maximum cost for another objectives.

Step 5: Choose the cell containing $\min(\sum_{i=1}^{m} \tilde{c}_{ij}{}^{r}, for\ fixed\ j)$ in corresponding row or column of cell choosen in Step 4. If tie occurs, then choose the one to which highest allocation is possible.

Step 6: Make maximum possible allocation to the cell selected in Step 5 and cross out the row or column for which supply / demand is satisfied.

Step 7: Repeat the procedure of Step 3 to Step 6 for remaining sources and destinations until whole supply or demand requirements are not met.

5 NUMERICAL EXAMPLE

Example 5.1: Consider a fully multi-objective fuzzy transportation problem the data for the fuzzy cost and time together with the supply and demand are measured to be triangular fuzzy numbers and is given below

Table 5.1:Multi-objective fuzzy Transportation problem

	D1	D2	D3	D4	Supply
S1	(1,1.5,2) (3,4,5)	(1,2,3) (2,4,6)	(5,7,9) (2,3,4)	(4,6,8) (1,3,5)	(7,8,9)
S2	(1,1.5,2) (4,5,6)	(7,8.5,10) (7,8,9)	(2,4,6) (7,8.5,10)	(3,4,5) (9,10,11)	(17,19,21)
S3	(7,8,9) (4,6,8)	(7,9,11) (1,2,3)	(3,4,5) (3,4.5,6)	(5,6,7) (1,1.5,2)	(16,17,18)
Demand	(10,11,12)	(2,3,4)	(13,14,15)	(15,16,17)	

Table 5.2 : Fuzzy transportation problem in parametric form

Sources	D1	D2	D3	D4	Supply
S1	(1.5,0.5-0.5r,0.5-0.5r) (4,1-r,1-r)	(2,1-r,1-r) (4,2-2r,2-2r)	(7,2-2r,2-2r) (3,1-r,1-r)	(6,2-2r,2-2r) (3,2-2r,2-2r)	(8,1-r,1-r)
S2	(1.5,0.5-0.5r,0.5-0.5r) (5,1-r,1-r)	(8.5,1.5-1.5r,1.5-1.5r) (8,1-r,1-r)	(4,2-2r,2-2r) (8.5,1.5-1.5r,1.5-1.5r)	(4,1-r,1-r) (10,1-r,1-r)	(19,2-2r,2-2r)
S3	(8,1-r,1-r) (6,2-2r,2-2r)	(9,2-2r,2-2r) (2,1-r,1-r)	(4,1-r,1-r) (4.5,1.5-	(6,1-r,1-r) 1.5,0.5-	(17,1-r,1-r)

			1.5r,1.5-1.5r)	0.5r,0.5-0.5r)	
Demand	(11,1-r,1-r)	(3,1-r,1-r)	(14,1-r,1-r)	(16,1-r,1-r)	

Table 5.3 : Fuzzy transportation problem after giving the allocation

Sources	D1	D2	D3	D4	Supply
S1	(1.5,0.5-0.5r,0.5-0.5r) (4,1-r,1-r)	(2,1-r,1-r) (4,2-2r,2-2r) **(3,1-r,1-r)**	(7,2-2r,2-2r) (3,1-r,1-r) **(5,1-r,1-r)**	(6,2-2r,2-2r) (3,2-2r,2-2r)	(8,1-r,1-r)
S2	(1.5,0.5-0.5r,0.5-0.5r) (5,1-r,1-r) **(11,2-2r,2-2r)**	(8.5,1.5-1.5r,1.5-1.5r) (8,1-r,1-r)	(4,2-2r,2-2r) (8.5,1.5-1.5r,1.5-1.5r) **(8,1-r,1-r)**	(4,1-r,1-r) (10,1-r,1-r)	(19,2-2r,2-2r)
S3	(8,1-r,1-r) (6,2-2r,2-2r)	(9,2-2r,2-2r) (2,1-r,1-r)	(4,1-r,1-r) (4.5,1.5-1.5r,1.5-1.5r) **(1,1-r,1-r)**	(6,1-r,1-r) 1.5,0.5-0.5r,0.5-0.5r) **(16,1-r,1-r)**	(17,1-r,1-r)
Demand	(11,1-r,1-r)	(3,1-r,1-r)	(14,1-r,1-r)	(16,1-r,1-r)	

Total cost = (2,1-r,1-r)(3,1-r,1-r) + (7,2-2r,2-2r) (5,1-r,1-r)+(1.5,0.5-0.5r,0.5-0.5r) (11,2-2r,2-2r) +(4,2-2r,2-2r) (8,1-r,1-r) +(4,1-r,1-r) (1,1-r,1-r)+ (6,1-r,1-r) (16,1-r,1-r)

=(189.5,2-2r,2-2r) , $\forall r \in [0,1]$

If r=0 then total cost = (187.5,189.5,191.5)

Total time = (4,2-2r,2-2r)(3,1-r,1-r)+(3,1-r,1-r)(5,1-r,1-r)+ (5,1-r,1-r) (11,2-2r,2-2r)

+(8.5,1.51.5r,1.5-1.5r) (8,1-r,1-r)+(4.5,1.51.5r,1.5-1.5r)(1,1-r,1-r) +(1.5,0.5-0.5r,0.5-0.5r)(16,1-r,1-r)

= (178.5,2-2r,2-2r) , $\forall r \in [0,1]$

If r=0 then total time = (176.5,178.5,180.5)

Example 5.2:

Consider an example of unbalanced fully multi-objective fuzzy transportation problem the data for the fuzzy cost and time together with the supply and demand are measured to be triangular fuzzy numbers and is given below

Table 5.4: Multi-objective fuzzy Transportation problem

Source	D1	D2	D3	D4	Supply
S1	(0.6,0.8,0.9) (0.1,0.3,0.9)	(2.4,3,3.5) (1,1.5,2)	(1.8,2.1,2.3) (1,1.4,1.6)	(1.5,1.8,2) (0.7,0.9,1)	(17.2,18,19.2)
S2	(1.1,1.3,1.4) (0.4,0.6,0.7)	(3.1,3.6,4.1) (1.5,1.8,2.1)	(1.4,1.6,1.7) (0.8,0.9,1.1)	(2.2,2.5,2.7) (1,1.4,1.8)	(22.6,24,25.8)
S3	(1.5,1.8,2) (0.6,0.9,1.1)	(3.1,3.5,3.9) (1.4,1.8,2.2)	(2.1,2.4,2.7) (1.3,1.5,1.7)	(0.8,1,1.1) (0.2,0.3,0.4)	(12.6,13,13.6)
Demand	(11.2,12,12,6)	(5.4,6,6.4)	(14.8,16,16.8)	(19,20,20.8)	

Table 5.5: Fuzzy transportation problem in parametric form

Source	D1	D2	D3	D4	Supply
S1	(0.8,0.2-0.2r,0.1-0.1r) (0.3,0.2-0.2r,0.6-0.6r)	(3,0.6-0.6r,0.5-0.5r) (1.5,0.5-0.5r,0.5-0.5r)	(2.1,0.3-0.3r,0.2-0.2r) (1.4,0.4-0.4r,0.2-0.2r)	(1.8,0.3-0.3r,0.2-0.2r) (0.9,0.2-0.2r,0.1-0.1r)	(18,0.8-0.8r,1.2-1.2r)
S2	(1.3,0.2-0.2r,0.1-0.1r) (0.6,0.2-0.2r,0.1-0.1r)	(3.6,0.5-0.5r,0.5-0.5r) (1.8,0.3-0.3r,0.3-0.3r)	(1.6,0.2-0.2r,0.1-0.1r) (0.9,0.1-0.1r,0.2-0.2r)	(2.5,0.3-0.3r,0.2-0.2r) (1.4,0.4-0.4r,0.4-0.4r)	(24,1.4-1.4r,1.8-1.8r)
S3	(1.8,0.3-0.3r,0.2-0.2r) (0.9,0.3-0.3r,0.2-0.2r)	(3.5,0.4-0.4r,0.4-0.4r) (1.8,0.4-0.4r,0.4-0.4r)	(2.4,0.3-0.3r,0.3-0.3r) (1.5,0.2-0.2r,0.2-0.2r)	(1,0.2-0.2r,0.1-0.1r) (0.3,0.1-0.1r,0.1-0.1r)	(13,0.4-0.4r,0.6-0.6r)
Demand	(12,0.8-0.8r,0.6-0.6r)	(6,0.6-0.6r,0.4-0.4r)	(16,1.2-1.2r,0.8-0.8r)	(20,1-r,0.8-0.8r)	

Journal of Physics: Conference Series **1377** (2019) 012038 doi:10.1088/1742-6596/1377/1/012038

Table 5.6 : Balanced Fuzzy transportation problem in parametric form

Source	D1	D2	D3	D4	D5	Supply
S1	(0.8,0.2-0.2r,0.1-0.1r) (0.3,0.2-0.2r,0.6-0.6r)	(3,0.6-0.6r,0.5-0.5r) (1.5,0.5-0.5r,0.5-0.5r)	(2.1,0.3-0.3r,0.2-0.2r) (1.4,0.4-0.4r,0.2-0.2r)	(1.8,0.3-0.3r,0.2-0.2r) (0.9,0.2-0.2r,0.1-0.1r)	(0,0-r,0-r) (0,0-r,0-r)	(18,0.8-0.8r,1.2-1.2r)
S2	(1.3,0.2-0.2r,0.1-0.1r) (0.6,0.2-0.2r,0.1-0.1r)	(3.6,0.5-0.5r,0.5-0.5r) (1.8,0.3-0.3r,0.3-0.3r)	(1.6,0.2-0.2r,0.1-0.1r) (0.9,0.1-0.1r,0.2-0.2r)	(2.5,0.3-0.3r,0.2-0.2r) (1.4,0.4-0.4r,0.4-0.4r)	(0,0-r,0-r) (0,0-r,0-r)	(24,1.4-1.4r,1.8-1.8r)
S3	(1.8,0.3-0.3r,0.2-0.2r) (0.9,0.3-0.3r,0.2-0.2r)	(3.5,0.4-0.4r,0.4-0.4r) (1.8,0.4-0.4r,0.4-0.4r)	(2.4,0.3-0.3r,0.3-0.3r) (1.5,0.2-0.2r,0.2-0.2r)	(1,0.2-0.2r,0.1-0.1r) (0.3,0.1-0.1r,0.1-0.1r)	(0,0-r,0-r) (0,0-r,0-r)	(13,0.4-0.4r,0.6-0.6r)
Demand	(12,0.8-0.8r,0.6-0.6r)	(6,0.6-0.6r,0.4-0.4r)	(16,1.2-1.2r,0.8-0.8r)	(20,1-r,0.8-0.8r)	(1,0-r,0-r)	

Table 5.7 : Fuzzy transportation problem after giving the allocation

Source	D1	D2	D3	D4	D5	Supply
S1	(0.8,0.2-0.2r,0.1-0.1r) (0.3,0.2-0.2r,0.6-0.6r) **(5,0.8-0.8r,1.2-1.2r)**	(3,0.6-0.6r,0.5-0.5r) (1.5,0.5-0.5r,0.5-0.5r)**(6,0.6-0.6r,1.4-1.4r)**	(2.1,0.3-0.3r,0.2-0.2r) (1.4,0.4-0.4r,0.2-0.2r)	(1.8,0.3-0.3r,0.2-0.2r) (0.9,0.2-0.2r,0.1-0.1r) **(7,1-r,0.8-0.8r)**	(0,0-r,0-r) (0,0-r,0-r)	(18,0.8-0.8r,1.2-1.2r)

S2	(1.3,0.2-0.2r,0.1-0.1r) (0.6,0.2-0.2r,0.1-0.1r) **(7,0.8-0.8r,0.6-0.6r)**	(3.6,0.5-0.5r,0.5-0.5r) (1.8,0.3-0.3r,0.3-0.3r)	(1.6,0.2-0.2r,0.1-0.1r) (0.9,0.1-0.1r,0.2-0.2r) **(16,1.2-1.2r,0.8-0.8r)**	(2.5,0.3-0.3r,0.2-0.2r) (1.4,0.4-0.4r,0.4-0.4r)	(0,0-r,0-r) (0,0-r,0-r) **(1,1.4-1.4r,1.8-1.8r)**	(24,1.4-1.4r,1.8-1.8r)
S3	(1.8,0.3-0.3r,0.2-0.2r) (0.9,0.3-0.3r,0.2-0.2r)	(3.5,0.4-0.4r,0.4-0.4r) (1.8,0.4-0.4r,0.4-0.4r)	(2.4,0.3-0.3r,0.3-0.3r) (1.5,0.2-0.2r,0.2-0.2r)	(1,0.2-0.2r,0.1-0.1r) (0.3,0.1-0.1r,0.1-0.1r) **(13,0.4-0.4r,0.6-0.6r)**	(0,0-r,0-r) (0,0-r,0-r)	(13,0.4-0.4r,0.6-0.6r)
Demand	(12,0.8-0.8r,0.6-0.6r)	(6,0.6-0.6r,0.4-0.4r)	(16,1.2-1.2r,0.8-0.8r)	(20,1-r,0.8-0.8r)	(1,0-r,0-r)	

Total cost = (0.8,0.2-0.2r,0.1-0.1r) (5,0.8-0.8r,1.2-1.2r)+(3,0.6-0.6r,0.5-0.5r) (6,0.6-0.6r,0.4-0.4r)

+(1.8,0.3-0.3r,0.2-0.2r) (7,1-r,0.8-0.8r) +(1.3,0.2-0.2r,0.1-0.1r) (7,0.8-0.8r,0.6-0.6r)

+(1.6,0.2-0.2r,0.1-0.1r) (16,1.2-1.2r,0.8-0.8r)+(0.0-r,0-r) (1,1.4-1.4r,1.8-1.8r)

+(1,0.2-0.2r,0.1-0.1r) (13,0.4-0.4r,0.6-0.6r)

=(82.3,1.4-1.4r,1.8-1.8r), $\forall r \in [0,1]$

If r=0 then total cost = (80.9, 82.3, 84.1)

Total time = (0.3,0.2-0.2r,0.6-0.6r) (5,0.8-0.8r,1.2-1.2r)+(1.5,0.5-0.5r,0.5-0.5r) (6,0.6-0.6r,1.4-1.4r) +(0.9,0.2-0.2r,0.1-0.1r) (7,1-r,0.8-0.8r)+(0.6,0.2-0.2r,0.1-0.1r) (7,0.8-0.8r,0.6-0.6r)

+(0.9,0.1-0.1r,0.2-0.2r) (16,1.2-1.2r,0.8-0.8r)+(0,0-r,0-r) (1,1.4-1.4r,1.8-1.8r)

+(0.3,0.1-0.1r,0.1-0.1r) (13,0.4-0.4r,0.6-0.6r)

= (39.3,1.4-1.4r,1.8-1.8r) , $\forall r \in [0,1]$

If r=0 then total time = (37.9, 39.3, 41.1)

6. CONCLUSION

In the above model, we developed a technique for the solution of multi objective transportation problem. Using the proposed algorithm we discussed numerical examples without changing them into equivalent crisp transportation problem. Here we represented the triangular fuzzy numbers are in terms of location index and fuzziness index functions and a solution is obtained in its parametric form. Additionally the projected technique is more flexible for the Decision maker to come to a decision r $\in [0,1]$ regarding upon the condition.

REFERENCES

[1] Abbasbandy S and Hajjari T 2009 A new approach for ranking of trapezoidal fuzzy numbers, *Computers and Mathematics with Applications,* **57** pp 413–419.
[2] Abd W. F. and El-Wahed 2006 Interactive fuzzy goal programming for multi-objective transportation problems, Omega, **34** pp 158-166.
[3] Anila Gupta and Amit Kumar 2012 A new method for solving linear multi-objective transportation problems with fuzzy parameters, *Applied Mathematical Modelling* **36** pp 1421-1430.
[4] Bellman R E and Zadeh L A 1970 Decision Making in a fuzzy environment, *Management Science*,**17**pp 141-164.
[5] Chanas Sand Kulcha D 1966 A concept of optimal Solution of the transportation with fuzzy cost co-efficient, *Fuzzy Sets and systems*, **82 (9)**pp 299 – 305.
[6] Kaufmann A and Gupta M.M 1985 Introduction to Fuzzy Arithmetic's: Theory and Applications, Van Nostrand Reinhold, New York.
[7] Khan A J and Das D K 2014, Fuzzy multi objective optimization: With reference to multi objective transportation problem, Recent Research in Science and Technology, **6(1)** pp 274-282.
[8] Lakhveer Kaur, MadhuchandaRakshit, & Sandeep signh 2018 A New Approach to solve Multi-objective Transportation Problem,Application ,*Applied Mathematics: An International Journal*,**13**,150-159.
[9] Maxler R T and Arora J S, Survey of multi-objective optimization methods for engineering 2009*Struct Multidisc Optim.,* **26** pp 369 - 395.
[10] Melita E Vinoliah and Ganesan K 2018 Solution to a Multi-objective Fuzzy Transportation Problem-A New Approach,*International Journal of Pure and Applied Mathematics,* **119** pp 385-393.
[11] Ming Ma, Menahem Friedman, Abraham kandel 1999 A new fuzzy arithmetic, *Fuzzy sets and systems*,**108** pp 83-90.
[12] Muruganandam S and Srinivasan R 2016, Optimal Solution for Multi-Objective Two Stage Fuzzy Transportation Problem, *Asian Journal of Research in Social Sciences and Humanities* ,**6(5)** pp 744-752.
[13] Nagoor Gani and Razak K A 2006 Two stage fuzzy transportation problem, *Journal of Physical Sciences*, **10**pp 63 – 69.
[14] Sharif Uddin M, Sushanta K. Roy, and Mesbahuddin Ahmed M 2018 An approach to solve multi-objective transportation problem using fuzzy goal programming and genetic algorithm, *AIP Conference Proceedings* 1978, 470095.
[15] Sheema Sadia, Neha Gupta and Qazi M. Ali 2017 Multiobjective Fractional Transportation Problem in Fuzzy Environment, *International Journal of Mathematical Archive*, **8(12)** pp197-209.
[16] Sudhakar V J 2010 Solving the Multi-Objective Two Stage Fuzzy Transportation Problem by Zero Suffix Method,*Journal of Mathematics Research*,**4** pp 135-140.

Journal of Physics: Conference Series **1377** (2019) 012038 doi:10.1088/1742-6596/1377/1/012038

[17] Thorani Y L P and Ravishankar N 2014 Fuzzy multi-objective transportation model based on new ranking index on generalized LR fuzzy numbers, *Applied Math. Sciences*, **8(138)**pp 6849-6879.

[18] Waiel,F.Abd and El-Wahed 2001 A multi-objective transportation problem under fuzziness, *Fuzzy sets and systems*, **117** pp 27 – 33.

[19] Zadeh L A 1965 Fuzzy sets, *information and control*, **8** pp 338-353.

SOLVING SEQUENCING PROBLEMS UNDER FUZZY ENVIRONMENT

S Srimathi[a] and K Prabakaran[b *]
Department of Mathematics,
SRM Institute of Science and Technology,
Kattankulathur, Chennai – 603203. INDIA
Corresponding author: **[b]prabakak2@srmist.edu.in;**
[a]ssrimathi98@gmail.com

ABSTRACT
A sequencing problem is to select the order in which the vital tasks are to be done to minimize the total elapsed time taken for all the tasks. Generally, in job sequencing problems, the processing times are precisely valued. But in reality, it is observed that the processing times during the performance of the job are indefinite. For that reason, the conception of fuzzy job sequencing problem make available for use an effective outline which is useful for real-life situations with fuzzy processing times. Here, we suggest a new technique for solution of fuzzy sequencing problems involving triangular fuzzy numbers (TFN) without transforming them into equivalent crisp sequencing problem.

Key words: Triangular fuzzy numbers, Fuzzy job sequencing problem, Crisp equivalent problem

1. INTRODUCTION

In this competitive world, government agencies, industrial organizations and construction sectors has to plan their projects in order to make best use of resources and to decrease the overall cost to gain profit. A sequencing problem is to establish the best possible sequence where the finite number of works(n) to be accomplished by the finite number of machines(m) and variety of desirable system like lowest amount of total elapsed time, inactive time, and stock price with the specified environment, the sequence of finite number of machine where each job supposed to be accomplished and the exact time necessary for the finite number of works on individual finite number of machine. Ordering is widely used in manufacture stores where diverse number of products are to be handled over a choice of mixtures of finite number of machine. But in reality, it is observed that the processing times through concert of the job are indefinite. To handle vagueness fuzzy set theory plays a major role as fuzzy set is a best mathematical way for representing unequal exactness.

So the problems on fuzzy sequencing take part in a dynamic role in expressing the ambiguity in actual situation. Chen et al.[4] has studied a fuzzy production system of scheduling under multi-objective. Anew algorithm is identified to get the best possible order of diverse one destination as the manufacture principles and the fuzzy minimum-operator with not sequential connection task as the experiment condition. Adamopoulos et al.[2] considered a fuzzy come close to to a one machine scheduling problem. The condition for scheduling problems is the universal suitable date, the full earlier in time and belatedness and the convenient length of the jobs' dealing out times. Their idea is to launch the continuance of the meting out period, to order the jobs to be performed in the machine and, lastly, to establish the universal appropriate day in a close to best possible way. Amalore Arumica et al. [3] have considered 3-machine, n-jobs flow shop arrangement problem in which the vagueness are involved in dealing out time, arrangement time and shipping time are considered by means of octagonal fuzzy numbers. But here we find the best possible order (sequence) for the fuzzy sequencing problem exclusive of changing to traditional problems.

2 PRELIMINARIES

In this section we recall the basic concepts and the results of triangular fuzzy number and their arithmetic operations.

Definition: 2.1

A fuzzy set $\tilde{a}$ defined on the set of real numbers R is said to be a fuzzy number if its membership function $\tilde{a}: R \rightarrow [0,1]$ has the following characteristics:

(i) $\tilde{a}$ is convex, i.e. $\tilde{a}(\lambda x_1 + (1-\lambda)x_2) \geq \min\{\tilde{a}(x_1), \tilde{a}(x_2)\}$, for all $x_1, x_2 \in R$ and $\lambda \in [0,1]$

(ii) $\tilde{a}$ is normal i.e. there exists an $x \in R$ such that $\tilde{a}(x) = 1$.

(iii) $\tilde{a}$ is Piecewise continuous.

Definition: 2.2

A fuzzy number $\tilde{a}$ on R is said to be a triangular fuzzy number (TFN) if its membership function $\tilde{a}: R \rightarrow [0,1]$ has the following characteristics:

$$\tilde{a}(x) = \begin{cases} \dfrac{x - a_1}{a_2 - a_1}, & \text{for } a_1 \leq x \leq a_2 \\ \dfrac{a_3 - x}{a_3 - a_2}, & \text{for } a_2 \leq x \leq a_3 \\ 0, & \text{elsewhere} \end{cases}$$

We denote this triangular fuzzy number by $\tilde{a} = (a_1, a_2, a_3)$. We use F(R) to denote the set of all triangular fuzzy numbers. Also if $m = a_2$ represents the modal value or midpoint, $\alpha = (a_2 - a_1)$ represents the left spread and $\beta = (a_3 - a_2)$ represents the right spread of the triangular fuzzy number $\tilde{a} = (a_1, a_2, a_3)$, then the triangular fuzzy number $\tilde{a}$ can be represented by the triplet $\tilde{a} = (\alpha, m, \beta)$. i.e. $\tilde{a} = (a_1, a_2, a_3) = (\alpha, m, \beta)$.

Definition: 2.3

A triangular fuzzy number $\tilde{a} \in F(R)$ can also be represented as a pair $\tilde{a} = (\underline{a}, \overline{a})$ of functions $\underline{a}(r)$ and $\overline{a}(r)$ for $0 \leq r \leq 1$ which satisfies the following requirements:

i. $\underline{a}(r)$ is a bounded monotonic increasing left continuous function.
ii. $\overline{a}(r)$ is a bounded monotonic decreasing left continuous function.
iii. $\underline{a}(r) \leq \overline{a}(r)$, $0 \leq r \leq 1$

Definition: 2.4

For an arbitrary triangular fuzzy number $\tilde{a} = (\underline{a}, \overline{a})$, the number $a_0 = \left(\dfrac{\underline{a}(1) + \overline{a}(1)}{2}\right)$ is said to be a location index number of $\tilde{a}$. The two non-decreasing left continuous functions $a_* = (a_0 - \underline{a})$, $a^* = (\overline{a} - a_0)$ are called the left fuzziness index function and the right fuzziness index function respectively. Hence every triangular fuzzy number $\tilde{a} = (a_1, a_2, a_3)$ can also be represented by $\tilde{a} = (a_0, a_*, a^*)$

2.1 Ranking of triangular Fuzzy Numbers

Many different approaches for the ranking of fuzzy numbers have been proposed in the literature. Abbasbandy and Hajjari[1] proposed a new ranking method based on the left and the right spreads at some α -levels of fuzzy numbers.

For an arbitrary triangular fuzzy number $\tilde{a} = (a_1, a_2, a_3) = (a_0, a_*, a^*)$ with parametric form $\tilde{a} = (\underline{a}(r), \overline{a}(r))$, we define the magnitude of the triangular fuzzy number $\tilde{a}$ by

$$\mathrm{Mag}(\tilde{a}) = \frac{1}{2}\left(\int_0^1 (\underline{a} + \overline{a} + a_0)\ f(r)\ dr\right)$$
$$= \frac{1}{2}\left(\int_0^1 (a^* + 4a_0 - a_*)\ f(r)\ dr\right).$$

where the function $f(r)$ is a non-negative and increasing function on $[0,1]$ with $f(0)=0$, $f(1)=1$ and $\int_0^1 f(r)\ dr = \frac{1}{2}$. The function $f(r)$ can be considered as a weighting function. In real life applications, $f(r)$ can be chosen by the decision maker according to the situation. In this paper, for convenience we use $f(r) = r$.

For any two triangular fuzzy numbers $\tilde{a} = (a_0, a_*, a^*)$ and $\tilde{b} = (b_0, b_*, b^*)$ in F(R), we define the ranking of $\tilde{a}$ and $\tilde{b}$ by comparing the Mag ($\tilde{a}$) and Mag ($\tilde{b}$) on R as follows:

(i) $\tilde{a} \succeq \tilde{b}$ if and only if $\mathrm{Mag}(\tilde{a}) \geq \mathrm{Mag}(\tilde{b})$

(ii) $\tilde{a} \preceq \tilde{b}$ if and only if $\mathrm{Mag}(\tilde{a}) \leq \mathrm{Mag}(\tilde{b})$

(iii) $\tilde{a} \approx \tilde{b}$ if and only if $\mathrm{Mag}(\tilde{a}) = \mathrm{Mag}(\tilde{b})$

2.2 Arithmetic operation on triangular Fuzzy Numbers

Ming Ma et al.[11] have proposed a new fuzzy arithmetic based upon both location index and fuzziness index functions. The location index number is taken in the ordinary arithmetic, whereas the fuzziness index functions are considered to follow the lattice rule which is least upper bound in the lattice L. That is for a, b∈ L we define $a \vee b = \max\{a, b\}$ and $a \wedge b = \min\{a, b\}$.

For arbitrary triangular fuzzy numbers $\tilde{a} = (a_0, a_*, a^*)$ and $\tilde{b} = (b_0, b_*, b^*)$ and $* = \{+, -, \times, \div\}$, the arithmetic operations on the triangular fuzzy numbers are defined by $\tilde{a} * \tilde{b} = (a_0 * b_0, a_* \vee b_*, a^* \vee b^*)$. In particular for any two triangular fuzzy numbers $\tilde{a} = (a_0, a_*, a^*)$ and $\tilde{b} = (b_0, b_*, b^*)$, we define

(i) Addition : $\tilde{a} + \tilde{b} = (a_0, a_*, a^*) + (b_0, b_*, b^*) = (a_0 + b_0, \max\{a_*, b_*\}, \max\{a^*, b^*\})$

(ii) Subtraction : $\tilde{a} - \tilde{b} = (a_0, a_*, a^*) - (b_0, b_*, b^*) = (a_0 - b_0, \min\{a_*, b_*\}, \min\{a^*, b^*\})$

(iii) Multiplication : $\tilde{a} \times \tilde{b} = (a_0, a_*, a^*) \times (b_0, b_*, b^*) = (a_0 \times b_0, \max\{a_*, b_*\}, \max\{a^*, b^*\})$

(iv) Division: $\tilde{a} \div \tilde{b} = (a_0, a_*, a^*) \div (b_0, b_*, b^*) = (a_0 \div b_0, \min\{a_*, b_*\}, \min\{a^*, b^*\})$

3. PROPOSED ALGORITHM

Assume that the finite number of jobs (n) to be completed by finite number of machines (m) $m_1, m_2, m_3, \ldots, m_m$ and $\tilde{T}_{ij}$ be the time engaged to finish the ith job by the jth machine.

Journal of Physics: Conference Series **1377** (2019) 012039 doi:10.1088/1742-6596/1377/1/012039

Step 1: Find (i) Least amount of time $\tilde{T}_{il}$ (ii) Least amount of time $\tilde{T}_{im}$ and (iii) greatest amount of time $\tilde{T}_{ik}$ whenever k= 2,3,..., m-1.

Step 2: confirm the choice between alternatives.(i) minimum $\tilde{T}_{il} \geqslant$ maximum $\tilde{T}_{ik}$ when ever k= 2,3,..., m-1.or (ii) least amount of time $\tilde{T}_{im} \geqslant$ greatest amount of time $\tilde{T}_{ik}$ whenever k= 2,3,..., m-1.

Step 3: If earlier step doesn't fulfilled, then the algorithm flops, If its satisfied move to next Step.

Step 4: Transform the finite number of machine crisis into two machine crisis by familiarizing the new artificial machines P and Q condition that, $\tilde{T}_{iP}=\tilde{T}_{i1}+\tilde{T}_{i2}+\cdots+\tilde{T}_{im-1}$ for i= 1,2,3,...,n and $\tilde{T}_{iQ}=\tilde{T}_{i2}+\tilde{T}_{i3}+\cdots+\tilde{T}_{im}$ for i= 1,2,3,...,n.

Step 5: Currently $\tilde{a}_{11}, \tilde{a}_{21}, \tilde{a}_{31}, \ldots \tilde{a}_{n1}$ be required times of finite number of works on Machine P and $\tilde{a}_{11}, \tilde{a}_{21}, \tilde{a}_{31}, \ldots \tilde{a}_{n2}$ be required times of finite number of works on Machine Q.

a. Identify least amount of time $(\tilde{a}_{i1}, \tilde{a}_{i2})$ i= 1,2,3,...,n.
b. If the least amount of time is $\tilde{a}_{i1}$ for i =1progression the lth work initially.

and in case the least amount of time is $\tilde{a}_{m2}$ for i=m progression the mth work at the end .

Step 6: Remove the works previously allocated and replicate (a) and (b) of previous step in anticipation of all the works has allocated.

4 . NUMERICAL EXAMPLES

Example 1: Sequencing problems containing 2 machines and 5 jobs, the processing time are triangular fuzzy numbers.

Jobs	1	2	3	4	5
Machine A	(6,7,8)	(4,5,6)	(2,4,5)	(6,8,12)	(0,2,4)
Machine B	(10,11,13)	(6,7,11)	(10,12,14)	(0,6,7)	(0,1,2)

Solution:

JOBS	MACHINE A	MACHINE B
1	(7,1-r,1-r)	(11,1-r,2-2r)
2	(5,1-r,1-r)	(7,1-r,4-4r)
3	(4,2-2r,1-r)	(12,2-2r,2-2r)
4	(8,2-2r,4-4r)	(6,6-6r,1-r)
5	(2,2-2r,2-2r)	(1,1-r,1-r)

Journal of Physics: Conference Series **1377** (2019) 012039 doi:10.1088/1742-6596/1377/1/012039

The order of jobs as follows

3	2	1	4	5

JOBS	MACHINE A		MACHINE B	
	In time	Out time	In time	Out time
3	(0,0,0)	(4,2-2r,1-r)	(4,2-2r,1-r)	(16,2-2r,2-2r)
2	(4,2-2r,1-r)	(8,2-2r,1-r)	(16,2-2r,2-2r)	(23,2-2r,2-2r)
1	(8,2-2r,1-r)	(15,2-2r,1-r)	(23,2-2r,2-2r)	(34,2-2r,2-2r)
4	(15,2-2r,1-r)	(23,2-2r,1-r)	(34,2-2r,2-2r)	(40,2-2r,2-2r)
5	(23,2-2r,1-r)	(25,2-2r,1-r)	(40,2-2r,2-2r)	(41,2-2r,2-2r)

The inactive time for machine A is (16,2-2r,2-2r) hours

The inactive time for machine B is (4,2-2r,2-2r) hours

The best possible time obligatory to finish all the jobs is (41, 2-2r, 2-2r) hours

The best possible time obligatory to finish all the jobs is (39, 41, 43) hours if we chose r=0

But Srikant Gupta et al. [16] converting into crisp equivalent problem and got the solution 41 hours

Example2: Sequencing problems containing 3 machines and 5 jobs, the processing time are triangular fuzzy numbers.

Jobs	J1	J2	J3	J4	J5
Machine A	(4,5,6)	(6,7,8)	(5,6,7)	(8,9,10)	(4,5,6)
Machine B	(1,2,3)	(1,1,1)	(3,4,5)	(4,5,6)	(2,3,4)
Machine C	(2,3,4)	(5,7,9)	(4,5,6)	(5,6,7)	(6,7,8)

Journal of Physics: Conference Series **1377** (2019) 012039 doi:10.1088/1742-6596/1377/1/012039

SOLUTION:

JOBS	MACHINE A	MACHINE B	MACHINE C
J1	(5,1-r,1-r)	(2,1-r,1-r)	(3,1-r,1-r)
J2	(7,1-r,1-r)	(1,0-r,0-r)	(7,2-2r,2-2r)
J3	(6,1-r,1-r)	(4,1-r,1-r)	(5,1-r,1-r)
J4	(9,1-r,1-r)	(5,1-r,1-r)	(6,1-r,1-r)
J5	(5,1-r,1-r)	(3,1-r,1-r)	(7,1-r,1-r)

The order of job as follows

J2	J5	J4	J3	J1

Jobs	Machine A		Machine B		Machine C	
	In time	Out time	In time	Out time	In time	Out time
J2	(0,0,0)	(7,1-r,1-r)	(7,1-r,1-r)	(8,1-r,1-r)	(8,1-r,1-r)	(15,2-2r,2-2r)
J5	(7,1-r,1-r)	(12,1-r,1-r)	(12,1-r,1-r)	(15,1-r,1-r)	(15,2-2r,2-2r)	(22,2-2r,2-2r)
J4	(12,1-r,1-r)	(21,1-r,1-r)	(21,1-r,1-r)	(26,1-r,1-r)	(26,2-2r,2-2r)	(32,2-2r,2-2r)
J3	(21,1-r,1-r)	(27,1-r,1-r)	(27,1-r,1-r)	(31,1-r,1-r)	(32,2-2r,2-2r)	(37,2-2r,2-2r)
J1	(27,1-r,1-r)	(32,1-r,1-r)	(32,1-r,1-r)	(34,1-r,1-r)	(37,2-2r,2-2r)	(40,2-2r,2-2r)

The inactive time for machine A is (8, 2-2r, 2-2r) hours

The inactive time for machine B is (25, 2-2r, 2-2r) hours

The inactive time for machine C is (52, 2-2r, 2-2r) hours

The best possible time obligatory to finish all the jobs is (40, 1-r, 1-r) hours

The best possible time obligatory to finish all the jobs is (39, 40, 41) hours if we choose r=0

But Kripa et al.[8] got the solution (33,40,47) hours

Example 3: Sequencing problems containing 5 machines and 4 jobs, the processing time are triangular fuzzy numbers

Jobs	Machine A	Machine B	Machine C	Machine D	Machine E
J1	(8,9,10)	(6,7,8)	(3,4.5)	(4,5,6)	(10,11,12)
J2	(7,8,9)	(7,8,9)	(5,6,7)	(6,7,8)	(11,12,13)
J3	(6,7,8)	(5,6,7)	(6,7,8)	(7,8,9)	(9,10,11)
J4	(9,10,11)	(4,5,6)	(4,5,6)	(3,4,5)	(7,8,9)

Solution:

Jobs	Machine A	Machine B	Machine C	Machine D	Machine E
J1	(9,1-r,1-r)	(7,1-r,1-r)	(4,1-r,1-r)	(5,1-r,1-r)	(11,1-r,1-r)
J2	(8,1-r,1-r)	(8,1-r,1-r)	(6,1-r,1-r)	(7,1-r,1-r)	(12,1-r,1-r)
J3	(7,1-r,1-r)	(6,1-r,1-r)	(7,1-r,1-r)	(8,1-r,1-r)	(10,1-r,1-r)
J4	(10,1-r,1-r)	(5,1-r,1-r)	(5,1-r,1-r)	(4,1-r,1-r)	(8,1-r,1-r)

The order of jobs as follows

J1	J3	J2	J4

Jobs	Machine A		Machine B		Machine C		Machine D		Machine E	
	In time	Out time	In time	Out time	In time	Out time	In time	Out time	In time	Out time
J1	(0,0,0)	(9,1-r,1-r)	(9,1-r,1-r)	(16,1-r,1-r)	(16,1-r,1-r)	(20,1-r,1-r)	(20,1-r,1-r)	(25,1-r,1-r)	(25,1-r,1-r)	(36,1-r,1-r)
J3	(9,1-r,1-r)	(16,1-r,1-r)	(16,1-r,1-r)	(22,1-r,1-r)	(22,1-r,1-r)	(29,1-r,1-r)	(29,1-r,1-r)	(37,1-r,1-r)	(37,1-r,1-r)	(47,1-r,1-r)
J2	(16,1-r,1-r)	(24,1-r,1-r)	(24,1-r,1-r)	(32,1-r,1-r)	(32,1-r,1-r)	(38,1-r,1-r)	(38,1-r,1-r)	(45,1-r,1-r)	(47,1-r,1-r)	(59,1-r,1-r)
J4	(24,1-r,1-r)	(34,1-r,1-r)	(34,1-r,1-r)	(39,1-r,1-r)	(39,1-r,1-r)	(44,1-r,1-r)	(45,1-r,1-r)	(49,1-r,1-r)	(59,1-r,1-r)	(67,1-r,1-r)

The inactive time for machine A is (33, 1-r, 1-r) hours

The inactive time for machine B is (41, 1-r, 1-r) hours

The inactive time for machine C is (45, 1-r, 1-r) hours

The inactive time for machine D is (43, 1-r, 1-r) hours

The inactive time for machine E is (26, 1-r, 1-r) hours

The best possible time obligatory to finish all the jobs is (67, 1-r, 1-r) hours

The best possible time obligatory to finish all the jobs is (66, 67, 68) hours if we chose r=0

But K.Kripa et al.[8]got the solution (59,67,75) hours

5. CONCLUSION:

In this article, the time measured as inaccurate numbers in particular triangular fuzzy numbers which will be much convincing and common in nature. We extended the well-known Johnson algorithm to fuzzy numbers to identifying the order exclusive of changing them to conventional sequencing problem. We are solved numerical examples using this projected technique exclusive of changing the specified problem to crisp corresponding problem and the results have been compared. Additionally the fuzzy best possible lengths managed by the projected technique are more reasonable than the existing ones, and decision maker to make a decision r $\in [0,1]$concerning upon the situation.

6. REFERENCES

[1] Abbasbandy S and Hajjari T 2009 A new approach for ranking of trapezoidal fuzzy numbers, *Computers and Mathematics with Applications,***57**pp 413–419.

[2] Adamopoulos .G.I and Pappis .C.P. 1996 "A fuzzy – linguistic approach to a multi-criteria sequencing problem", *European Journal of Operational Research*, **92**,pp 628 - 636.

[3] Amalore Arumica and Felbin C. Kennedy 2017 "Role of octagonal fuzzy numbers in three stage flow shop scheduling problem with setup time and transportation time", *International J. of Math. Sci. & Engg. Appls,***2(2)** , pp. 181-192.

[4] Chen .D, Pan .Y and Xue .J 1984 "A fuzzy production system for multi objective scheduling to an one- machine n-parts problem" *the 23rd IEEE Conference on Decision and Control,***(23)**, pp 1101 - 1102.

[5] Helen R and Sumathi R 2015 Solving single machine scheduling problem using type-2 trapezoidal fuzzy numbers, *Int.Journal of Innovative Research in Science Engineering and Technology*, **4(3)**, pp 1383-1391.

[6] Jansi R and Mohana K 2018 A heptagonal fuzzy number in solving fuzzy sequencing problem, *International journal of Innovative research in technology*,**5(6)**, pp 240-243.

[7] Johnson .S.M 1954 Optimal two and three stage production schedules with setup times included, *Noval Research Logistics Quarterly*, **1**, pp 61 - 68.

[8] Kripa K and Govindarajan R 2016 Fuzzy sequencing problem, *International journal of scientific research***5(5)**, pp 663-671.

[9] Kripa K and Govindarajan R 2016 Fuzzy sequencing problem using generalised triangular fuzzy numbers, *Int.Journal of Engineering Research and Application*, **6(6)**, pp 61-64.

[10] McCahon C S and Lee E S 1990 Job sequencing with fuzzy processing times, *Computers Math. Applic.* **19(7)**, pp 31-41.

[11] Ming Ma, Menahem Friedman, Abraham kandel 1999 A new fuzzy arithmetic, *Fuzzy sets and systems,***108**pp 83-90.

[12] Shanmugasundari M 2017 Fuzzy sequencing problem: A Novel Approach,*International journal of pure and applied mathematics* **113(13)**, pp 56-64.

[13] Selvakumari K and Santhi S 2018 A pentagonal fuzzy number in solving fuzzy sequencing problem, *Internatioanl Journal of mathematics and applications*,**6(2-B)**, pp 207-211.

[14] Selvakumari K and Santhi S 2018 An approach for solving fuzzy sequencing problems with octogonal fuzzy numbers using robust ranking techniques, *International Journal of Mathematics Trends and Technology*, **56(3)**, pp 148-152.

[15] Selvakumari K and Sowmiya G 2017 Fuzzy network problems using job sequencing technique in hexagonal fuzzy numbers, *Int.Journal of Advance Engineering and Research Development*, **4(9)**, pp 116-121.

[16] Srikant Gupta and Irfan Aliand Aquil Ahmed 2018 SAI method for solving job shop sequencing problem under certain and uncertain environment, *Sri lankan Journal of Applied Statistics,* **(18-3)**, pp 167-186.

[17] Zadeh L A 1965 Fuzzy sets, *information and control*, **8**,pp 338-353.

A solution approach to fully fuzzy linear fractional programming problems

T. Loganathan[1] , K. Ganesan[2]

[1,2]Department of Mathematics,
Faculty of Engineering and Technology,
SRM Institute of Science & Technology,
Kattankulathur, Chennai - 603203, India.

loganatt@srmist.edu.in, ganesank@srmist.edu.in.

Abstract: In this paper, we propose a method of solving the fully fuzzy linear fractional programming problems, Express all the parameters and variables are triangular fuzzy numbers. Convert all the triangular fuzzy numbers in their parametric form, we convert the fractional programming problem in to a single objective linear programming problem in parametric form. We put new fuzzy arithmetic and fuzzy ranking, we obtain the optimal solution the given fully fuzzy linear fractional programming problem without converting to its equivalent crisp linear programming problem. A numerical example is provided to illustrate the efficiency of the proposed method.
Keywords: fractional programming, triangular fuzzy numbers, parametric form, fuzzy arithmetic, fuzzy ranking.

1. Introduction

Linear fractional programming is a ratio of two linear functions, is optimized. Its applications is used several fields such as production, financial, corporate, health care etc. Bellman and Zadeh [1] proposed the concept of decision making in a fuzzy environment. Subsequently, Many authors discussed the FLFPP. e.g. Chakraborty and Gupta [3], Li et.al [9], Mehlawat et.al[10], Mitlif et.al [12], Muruganandam et.al [14][15], Nachammai et.al[16], Pop et.al[17], Sanjay Jain et.al [19], Stanojevic et. al [20][21], Das et al. [4][5][6][7] have proposed more methods for solving FLFPP. Safaei et.al [18] solving FLP problems with fuzzy goal and fuzzy constraints in two-dimensional space by geometric approach. Pop and Stancu Minasian [17] represent the variables are triangular fuzzy numbers and solving FFLFP problems. Veeramani et.al [22] changed the FFLFP problem into a bi-objective LPP and obtained the optimal solution.

In this paper, we consider the fully fuzzy linear fractional programming problem (FFLFP). First, the FFLFP problem is reduced into a fully fuzzy linear programming (FFLP) problem and then expressed in its parametric form. We propose a simplex type algorithm for the solution of fully fuzzy linear programming problem without changing to an equivalent crisp problem. In section 2, we give some basic definitions and notations along with some preliminary results to be used in subsequent sections. In section 3, we talk about fully fuzzy linear fractional programming problem and present details of the solutions method to solve it. In section 4, we present a numerical example to illustrate the efficiency of the solution approach developed in this paper.

Journal of Physics: Conference Series **1377** (2019) 012040 doi:10.1088/1742-6596/1377/1/012040

2. Preliminaries

Definition 2.1. A fuzzy set $\tilde{a}$ defined on the set of real numbers R is said to be a fuzzy number, if its membership function $\tilde{a} : R \to [0,1]$ has the following characteristics:

(i) $\tilde{a}$ is convex, (i.e.) $\tilde{a}(\lambda x_1 + (1-\lambda)x_2) \geq \min\{\tilde{a}(x_1), \tilde{a}(x_2)\}, \lambda \in [0,1]$, for all $x_1, x_2 \in R$

(ii) $\tilde{a}$ is normal, (i.e.) there exists an $x \in R$ such that $\tilde{a}(x) = 1$

(iii) $\tilde{a}$ is piecewise continuous.

Definition 2.2. A fuzzy number $\tilde{a}$ on R is a triangular fuzzy number if its membership function $\tilde{a} : R \to [0,1]$ has the following characteristics:

$$\tilde{a}(x) = \begin{cases} \dfrac{x - a_1}{a_2 - a_1}, & a_1 \leq x \leq a_2 \\ \dfrac{a_3 - x}{a_3 - a_2}, & a_2 \leq x \leq a_3 \\ 0, & \text{otherwise} \end{cases}$$

We denote this triangular fuzzy number as $\tilde{a} = (a_1, a_2, a_3)$. We use $F(R)$ to denote the set of all triangular fuzzy numbers defined on R.

Definition 2.3. A triangular fuzzy number $\tilde{a} = (a_1, a_2, a_3) \in F(R)$ can also be represented as an pair $\tilde{a} = (\underline{a}, \overline{a})$ of functions $\underline{a}(r), \overline{a}(r)$, for $0 \leq r \leq 1$ which satisfies the following requirements:

(i) $\underline{a}(r)$ is a bounded monotonic increasing left continuous function.

(ii) $\overline{a}(r)$ is a bounded monotonic decreasing left continuous function.

(iii) $\underline{a}(r) \leq \overline{a}(r)$, $0 \leq r \leq 1$.

It is also represented by $\tilde{a} = (a_0, a_*, a^*)$ where $a_* = (a_0 - \underline{a}), a^* = (\overline{a} - a_0)$ are called the left fuzziness index function and the right fuzziness index function respectively. For an arbitrary triangular fuzzy number $\tilde{a} = (\underline{a}, \overline{a})$ the number $a_0 = \left(\dfrac{\underline{a}(1) + \overline{a}(1)}{2}\right)$ is said to be a location index number of $\tilde{a}$.

2.1. Ranking of Triangular Fuzzy Numbers

Several approaches for the ranking of fuzzy numbers have been proposed in the literature. An efficient approach for comparing the fuzzy numbers is by the use of a ranking function based on their graded means. We define the magnitude of the triangular fuzzy number $\tilde{a}$ by

$$R(\tilde{a}) = \left(\frac{a^* + 4a_0 - a_*}{4}\right) = \left(\frac{\underline{a} + \overline{a} + a_0}{4}\right)$$

Consider any two triangular fuzzy numbers

$\tilde{a} = (a_1, a_2, a_3) = (a_0, a_*, a^*)$ and $\tilde{b} = (b_1, b_2, b_3) = (b_0, b_*, b^*)$ in $F(R)$ we have

$$\text{(i) } \tilde{a} \succeq \tilde{b} \text{ if and only if } R(\tilde{a}) \geq R(\tilde{b})$$
$$\text{(ii) } \tilde{a} \preceq \tilde{b} \text{ if and only if } R(\tilde{a}) \leq R(\tilde{b})$$
$$\text{(iii) } \tilde{a} \approx \tilde{b} \text{ if and only if } R(\tilde{a}) = R(\tilde{b})$$

2.2. Arithmetic Operations of TrFN

Ming Ma et. al.[11], analyzed fuzzy computation based on location index, fuzziness index functions. The location index number used ordinary computation and the fuzziness index functions used the lattice guidelines which is least upper bound in the lattice L. let $\tilde{a}, \tilde{b} \in L$, we derive $\tilde{a} \vee \tilde{b} = \max\{\tilde{a}, \tilde{b}\}$ and $\tilde{a} \wedge \tilde{b} = \min\{\tilde{a}, \tilde{b}\}$. If any two triangular fuzzy number $\tilde{a} = (a_0, a_*, a^*), \tilde{b} = (b_0, b_*, b^*)$ and $\bullet = \{+, -, \times, \div\}$, the computation operations on the fuzzy numbers then

$$\tilde{a} \bullet \tilde{b} \quad = (a_0, a_*, a^*) \bullet (b_0, b_*, b^*) = (a_0 \bullet b_0, \max\{a_*, b_*\}, \max\{a^*, b^*\})$$

3. Fully fuzzy linear fractional programming problem (FFLFPP)

A general FFLFPP is given by

$$\max \tilde{z} = \frac{\sum \tilde{c}_j \tilde{x}_j + \tilde{\alpha}_j}{\sum \tilde{d}_j \tilde{x}_j + \tilde{\beta}_j}$$

$$\text{Subject to } \sum_{j=1}^{n} \tilde{a}_{ij} \tilde{x}_j \preceq \tilde{b}_i, \qquad i = 1, 2, 3, \ldots, m \tag{3.1}$$

$$\text{and } \tilde{x}_j \succeq \tilde{0}, \text{ for all } j = 1, 2, 3, \ldots, n.$$

We assume that, $\tilde{a}_{ij}$, $\tilde{b}_i$, $\tilde{c}_j$, and $\tilde{x}_j$ are TrFN for each $i = 1, 2, 3, \ldots, m$ and $j = 1, 2, 3, \ldots, n$.

The general fully fuzzy linear fractional programming problem is expressed in matrix form as

$$\max \tilde{z} = \frac{\tilde{\mathbf{c}} \tilde{\mathbf{x}} + \tilde{\alpha}}{\tilde{\mathbf{d}} \tilde{\mathbf{x}} + \tilde{\beta}}$$

$$\text{Subject to } \tilde{A} \tilde{\mathbf{x}} \preceq \tilde{\mathbf{b}}, \tag{3.2}$$

$$\text{and } \tilde{\mathbf{x}} \succeq \tilde{0},$$

where $\tilde{A} = (\tilde{a}_{ij})_{(m \times n)}$, $\tilde{\mathbf{x}} = (\tilde{x}_1, \tilde{x}_2, \tilde{x}_3, \ldots, \tilde{x}_n)$, $\tilde{\mathbf{b}} = (\tilde{b}_1, \tilde{b}_2, \tilde{b}_3, \ldots, \tilde{b}_m)$ and $\tilde{\mathbf{c}} = (\tilde{c}_1, \tilde{c}_2, \tilde{c}_3, \ldots, \tilde{c}_n)$ $\tilde{\mathbf{d}} = (\tilde{d}_1, \tilde{d}_2, \tilde{d}_3, \ldots, \tilde{d}_n)$, $\tilde{\alpha}, \tilde{\beta} \in F(R)$.

It is assumed that the denominator is positive for all feasible solution.

3.1. Conversion of fuzzy linear fractional programming problem into fuzzy linear programming problem

We assume that the feasible region $\tilde{S} = \{\tilde{\mathbf{x}} \in F^n(R) : \tilde{A} \tilde{\mathbf{x}} \preceq \tilde{\mathbf{b}}, \tilde{\mathbf{x}} \succeq \tilde{0}\}$ is nonempty and bounded and the denominator $(\tilde{\mathbf{d}} \tilde{\mathbf{x}} + \tilde{\beta}) \succ \tilde{0}$.

If $(\tilde{\mathbf{d}}\,\tilde{\mathbf{x}}+\tilde{\beta})\prec\tilde{0}$, then the condition $\dfrac{\tilde{\beta}(\tilde{A}\,\tilde{\mathbf{x}}-\tilde{\mathbf{b}})}{\tilde{\beta}(\tilde{\mathbf{d}}\,\tilde{\mathbf{x}}+\tilde{\beta})}\preceq\tilde{0}$ will not hold. As a result solution to the LFP cannot be found.

Now we can convert the above LFP into an LP in the following way assuming that $\tilde{\beta}\neq\tilde{0}$.

3.2. Transformation of the objective function

Multiplying both the denominator and the numerator of $\max\ \tilde{z}=\dfrac{\tilde{\mathbf{c}}\,\tilde{\mathbf{x}}+\tilde{\alpha}}{\tilde{\mathbf{d}}\,\tilde{\mathbf{x}}+\tilde{\beta}}$ by $\tilde{\beta}$, where $\tilde{\beta}\neq\tilde{0}$.

We have
$$\tilde{z}=\frac{\tilde{\mathbf{c}}\,\tilde{\mathbf{x}}\tilde{\beta}+\tilde{\alpha}\tilde{\beta}}{\tilde{\mathbf{d}}\,\tilde{\mathbf{x}}\tilde{\beta}+\tilde{\beta}\tilde{\beta}}=\frac{\tilde{\mathbf{c}}\,\tilde{\mathbf{x}}\tilde{\beta}+\tilde{\alpha}\tilde{\beta}}{\tilde{\beta}(\tilde{\mathbf{d}}\,\tilde{\mathbf{x}}+\tilde{\beta})}=\frac{\tilde{\mathbf{c}}\,\tilde{\mathbf{x}}\tilde{\beta}-\tilde{\mathbf{d}}\,\tilde{\mathbf{x}}\tilde{\alpha}+\tilde{\mathbf{d}}\,\tilde{\mathbf{x}}\tilde{\alpha}+\tilde{\alpha}\tilde{\beta}}{\tilde{\beta}(\tilde{\mathbf{d}}\,\tilde{\mathbf{x}}+\tilde{\beta})}$$
$$=\frac{(\tilde{\mathbf{c}}\tilde{\beta}-\tilde{\mathbf{d}}\tilde{\alpha})\tilde{\mathbf{x}}+(\tilde{\mathbf{d}}\,\tilde{\mathbf{x}}+\tilde{\beta})\tilde{\alpha}}{\tilde{\beta}(\tilde{\mathbf{d}}\,\tilde{\mathbf{x}}+\tilde{\beta})}=\left(\tilde{\mathbf{c}}-\tilde{\mathbf{d}}\frac{\tilde{\alpha}}{\tilde{\beta}}\right)\frac{\tilde{\mathbf{x}}}{(\tilde{\mathbf{d}}\,\tilde{\mathbf{x}}+\tilde{\beta})}+\frac{\tilde{\alpha}}{\tilde{\beta}}=\tilde{\mathrm{p}}\tilde{\mathbf{y}}+\tilde{\mathrm{g}}$$

where $\tilde{\mathrm{p}}=\left(\tilde{\mathbf{c}}-\tilde{\mathbf{d}}\dfrac{\tilde{\alpha}}{\tilde{\beta}}\right)$, $\tilde{\mathbf{y}}=\dfrac{\tilde{\mathbf{x}}}{(\tilde{\mathbf{d}}\,\tilde{\mathbf{x}}+\tilde{\beta})}$ and $\tilde{\mathrm{g}}=\dfrac{\tilde{\alpha}}{\tilde{\beta}}$.

Hence $F(y)=\tilde{\mathrm{p}}\tilde{\mathbf{y}}+\tilde{\mathrm{g}}$.

3.3. Transformation of the constraints

From the constraint $\tilde{A}\,\tilde{\mathbf{x}}\preceq\tilde{\mathbf{b}}$, we have, $\tilde{A}\,\tilde{\mathbf{x}}-\tilde{\mathbf{b}}\preceq\tilde{0}\Rightarrow\dfrac{\tilde{\beta}(\tilde{A}\,\tilde{\mathbf{x}}-\tilde{\mathbf{b}})}{\tilde{\beta}(\tilde{\mathbf{d}}\,\tilde{\mathbf{x}}+\tilde{\beta})}\preceq\tilde{0}\Rightarrow\dfrac{(\tilde{A}\tilde{\beta}\,\tilde{\mathbf{x}}-\tilde{\mathbf{b}}\tilde{\beta})}{\tilde{\beta}(\tilde{\mathbf{d}}\,\tilde{\mathbf{x}}+\tilde{\beta})}\preceq\tilde{0}$

$$\Rightarrow\frac{(\tilde{A}\tilde{\beta}\,\tilde{\mathbf{x}}+\tilde{\mathbf{b}}\tilde{\mathbf{d}}\,\tilde{\mathbf{x}}-\tilde{\mathbf{b}}\tilde{\mathbf{d}}\,\tilde{\mathbf{x}}-\tilde{\mathbf{b}}\tilde{\beta})}{\tilde{\beta}(\tilde{\mathbf{d}}\,\tilde{\mathbf{x}}+\tilde{\beta})}\preceq\tilde{0}\Rightarrow\frac{(\tilde{A}\tilde{\beta}\,\tilde{\mathbf{x}}+\tilde{\mathbf{b}}\tilde{\mathbf{d}}\,\tilde{\mathbf{x}})}{\tilde{\beta}(\tilde{\mathbf{d}}\,\tilde{\mathbf{x}}+\tilde{\beta})}-\frac{(\tilde{\mathbf{b}}\tilde{\mathbf{d}}\,\tilde{\mathbf{x}}+\tilde{\mathbf{b}}\tilde{\beta})}{\tilde{\beta}(\tilde{\mathbf{d}}\,\tilde{\mathbf{x}}+\tilde{\beta})}\preceq\tilde{0}$$

$$\Rightarrow\frac{\tilde{\beta}(\tilde{A}+\frac{\tilde{\mathbf{b}}}{\tilde{\beta}}\tilde{\mathbf{d}})\tilde{\mathbf{x}}}{\tilde{\beta}(\tilde{\mathbf{d}}\,\tilde{\mathbf{x}}+\tilde{\beta})}-\frac{\tilde{\mathbf{b}}(\tilde{\mathbf{d}}\,\tilde{\mathbf{x}}+\tilde{\beta})}{\tilde{\beta}(\tilde{\mathbf{d}}\,\tilde{\mathbf{x}}+\tilde{\beta})}\preceq\tilde{0}\Rightarrow\left(\tilde{A}+\frac{\tilde{\mathbf{b}}}{\tilde{\beta}}\tilde{\mathbf{d}}\right)\frac{\tilde{\mathbf{x}}}{(\tilde{\mathbf{d}}\,\tilde{\mathbf{x}}+\tilde{\beta})}-\frac{\tilde{\mathbf{b}}}{\tilde{\beta}}\preceq\tilde{0}\Rightarrow\left(\tilde{A}+\frac{\tilde{\mathbf{b}}}{\tilde{\beta}}\tilde{\mathbf{d}}\right)\frac{\tilde{\mathbf{x}}}{(\tilde{\mathbf{d}}\,\tilde{\mathbf{x}}+\tilde{\beta})}\preceq\frac{\tilde{\mathbf{b}}}{\tilde{\beta}}$$

$\Rightarrow\tilde{G}\tilde{\mathbf{y}}\preceq\tilde{\mathbf{h}}$ where $\tilde{G}=\left(\tilde{A}+\dfrac{\tilde{\mathbf{b}}}{\tilde{\beta}}\tilde{\mathbf{d}}\right)$, $\tilde{\mathbf{y}}=\dfrac{\tilde{\mathbf{x}}}{(\tilde{\mathbf{d}}\,\tilde{\mathbf{x}}+\tilde{\beta})}$ and $\tilde{\mathbf{h}}=\dfrac{\tilde{\mathbf{b}}}{\tilde{\beta}}$.

From the above equations we finally obtain the new LP form of the given LFP as follows:

$$\begin{aligned}&\max\ F(y)=\tilde{\mathrm{p}}\tilde{\mathbf{y}}+\tilde{\mathrm{g}}\\&\text{subject to}\ \tilde{G}\tilde{\mathbf{y}}\preceq\tilde{\mathbf{h}}\\&\text{and}\ \tilde{\mathbf{y}}\succeq\tilde{0}.\end{aligned}\tag{3.3}$$

3.4. Algorithm

Step 1: Convert the objective function in max type, if it is in min type.
Step 2: Transform the FLFPP into an equivalent FLPP.
Step 3: Express the transformed FLPP into its parametric form.

Journal of Physics: Conference Series **1377** (2019) 012040 doi:10.1088/1742-6596/1377/1/012040

Step 4: Solve the parametric form FLPP using Simplex method.

Step 5: If the fuzzy optimum value of the transformed problem is

$$\max{}^{*}F(y)=\tilde{q} \text{ with } \tilde{\mathbf{y}}=(\tilde{y}_1,\tilde{y}_2,\tilde{y}_3,\ldots,\tilde{y}_n).$$

Step 6: Find the value of $\tilde{\mathbf{x}}=(\tilde{x}_1,\tilde{x}_2,\tilde{x}_3,\ldots,\tilde{x}_n)$, using the value of $\tilde{\mathbf{y}}=(\tilde{y}_1,\tilde{y}_2,\tilde{y}_3,\ldots,\tilde{y}_n)$ and $\tilde{\mathbf{x}}=\dfrac{\tilde{\mathbf{y}}\tilde{\beta}}{1-\tilde{\mathbf{d}}\tilde{\mathbf{y}}}$.

Step 7: Substitute the value of $\tilde{\mathbf{x}}=(\tilde{x}_1,\tilde{x}_2,\tilde{x}_3,\ldots,\tilde{x}_n)$ in the objective function of the original fuzzy linear fractional programming problem to get the required optimal solution.

4. Numerical Example

A company makes two products A and B. Added a permanent rate of around 1 rupee to the rate function.

products	Profit(around)/ per unit	Rate (around) / per unit	Row material(pound)	Man-hours(daily)
A	5	5	3	5 hours
B	3	2	5	2 hours
Available			15	10 hours

Compute how many products A and B should be produced in order to maximize the total profit.

Solution. The mathematical formulation of the FFLFPP is:

$$\max \tilde{z}=\frac{\tilde{5}\tilde{x}_1+\tilde{3}\tilde{x}_2}{\tilde{5}\tilde{x}_1+\tilde{2}\tilde{x}_2+\tilde{1}}$$

$$\text{subject to } \tilde{3}\tilde{x}_1+\tilde{5}\tilde{x}_2\leq\widetilde{15}$$

$$\tilde{5}\tilde{x}_1+\tilde{2}\tilde{x}_2\leq\widetilde{10}$$

$$\tilde{x}_1,\tilde{x}_2\geq\tilde{0}.$$

Applying the proposed algorithm, the FFLFPP is transformed into an equivalent FFLPP as

$$\max \tilde{z}=\tilde{5}\tilde{y}_1+\tilde{3}\tilde{y}_2$$

subject to

$$\widetilde{78}\,\tilde{y}_1+\widetilde{35}\,\tilde{y}_2\leq\widetilde{15}$$

$$\widetilde{55}\,\tilde{y}_1+\widetilde{22}\,\tilde{y}_2\leq\widetilde{10}$$

$$\tilde{y}_1,\tilde{y}_2\geq\tilde{0}.$$

$$\text{That is } \max\tilde{z}=(3,5,7)\tilde{y}_1+(2,3,4)\tilde{y}_2$$

$$\text{subject to } (76,78,80)\tilde{y}_1+(34,35,36)\tilde{y}_2\leq(11,15,19)$$

$$(53,55,57)\tilde{y}_1+(21,22,23)\tilde{y}_2\leq(9,10,11)$$

$$\text{and } \tilde{y}_1,\tilde{y}_2\geq\tilde{0}.$$

Express this FFLP in its parametric form as

$$\text{Max } \tilde{z} = (5,\ 2-2r\ ,\ 2-2r)\tilde{y}_1 + (3,\ 1-r,\ 1-r)\tilde{y}_2$$

$$\text{subject to } (78,\ 2-2r\ ,\ 2-2r)\tilde{y}_1 + (35,\ 1-r,\ 1-r)\tilde{y}_2 \le (15,\ 4-4r\ ,\ 4-4r)$$

$$(55,\ 2-2r\ ,\ 2-2r)\tilde{y}_1 + (22,\ 1-r\ ,\ 1-r)\tilde{y}_2 \le (10,\ 1-r\ ,\ 1-r)$$

$$\text{and } \tilde{y}_1, \tilde{y}_2 \ge \tilde{0}.$$

The standard fully fuzzy linear programming (FFLP) problem into its parametric form is

$$\text{Max } \tilde{z} = (5, 2-2r, 2-2r)\tilde{y}_1 + (3,\ 1-r,\ 1-r)\tilde{y}_2 + (1,0,0)\tilde{s}_1 + (1,0,0)\tilde{s}_2$$

$$\text{subject to } (78,\ 2-2r,\ 2-2r)\tilde{y}_1 + (35,\ 1-r,\ 1-r)\tilde{y}_2 + (1,0,0)\tilde{s}_1 = (15,\ 4-4r,\ 4-4r)$$

$$(55,\ 2-2r,\ 2-2r)\tilde{y}_1 + (22,\ 1-r\ ,\ 1-r)\tilde{y}_2 + (1,0,0)\tilde{s}_2 = (10,\ 1-r\ ,\ 1-r)$$

$$\text{and } \tilde{y}_1, \tilde{y}_2, \tilde{s}_1, \tilde{s}_2 \ge \tilde{0}.$$

The initial fuzzy basic feasible solution is $\tilde{s}_1 = (15,\ 4-4r\ ,\ 4-4r)$, $\tilde{s}_2 = (10,\ 1-r\ ,\ 1-r)$.

Table 1: Initial iteration

		$\tilde{c}_j$	(5, 2-2r , 2-2r)	(3, 1-r , 1-r)	(0,0,0)	(0,0,0)
$\tilde{c}_B$	$\tilde{y}_B$	$\tilde{x}_B$	$\tilde{y}_1$	$\tilde{y}_2$	$\tilde{s}_1$	$\tilde{s}_2$
(0,0,0)	$\tilde{s}_1$	(15, 4-4r , 4-4r)	(78, 2-2r , 2-2r)	(35, 1-r , 1-r)	(1,0,0)	(0,0,0)
(0,0,0)	$\tilde{s}_2$	(10, 1-r , 1-r)	(55, 2-2r , 2-2r)	(22, 1-r , 1-r)	(0,0,0)	(1,0,0)
		$\tilde{z}_j$	(0,0,0)	(0,0,0)	(0,0,0)	(0,0,0)
		$(\tilde{z}_j - \tilde{c}_j)$	(-5, 2-2r , 2-2r)	(-3, 1- r , 1-r)	(0,0,0)	(0,0,0)

Here leaving variable is $\tilde{s}_2$ and entering variable is $\tilde{y}_1$.

Table 2: Optimal iteration

		$\tilde{c}_j$	(5, 2-2r, 2-2r)	(3, 1-r , 1-r)	(0,0,0)	(0,0,0)
$\tilde{c}_B$	$\tilde{y}_B$	$\tilde{x}_B$	$\tilde{y}_1$	$\tilde{y}_2$	$\tilde{s}_1$	$\tilde{s}_2$
(3, 1-r , 1-r)	$\tilde{y}_2$	(0.43, 4-4r , 4-4r)	(2.23, 4-4r , 4-4r)	(1,4-4r,4-4r)	(0.03,4-4r,4-4r)	(0,4-4r,4-4r)
(0,0,0)	$\tilde{s}_2$	(0.57, 4-4r , 4-4r)	(5.97, 4-4r , 4-4r)	(0,4-4r,4-4r)	(-0.63,4-4r,4-4r)	(1,4-4r,4-4r)
		$\tilde{z}_j$	(6.69, 4-4r , 4-4r)	(3,4-4r,4-4r)	(0.09,4-4r,4-4r)	(0,4-4r,4-4r)
		$(\tilde{z}_j - \tilde{c}_j)$	(1.69, 4-4r , 4-4r)	(0,4-4r,4-4r)	(0.09,4-4r,4-4r)	(0,4-4r,4-4r)

Since all the $(\tilde{z}_j - \tilde{c}_j) \succeq \tilde{0}$, current solution is optimal.

Hence we get $\tilde{y}_1 \approx \tilde{0}$ and $\tilde{y}_2 \approx (0.43, 4-4r, 4-4r)$

Next, by the proposed algorithm we find the values of $\tilde{x}_1$, $\tilde{x}_2$ using the transformation

$$(\tilde{x}_1,\tilde{x}_2)=\frac{(\tilde{y}_1,\tilde{y}_2)\tilde{\beta}}{1-\tilde{\mathbf{d}}(\tilde{y}_1,\tilde{y}_2)}.$$

That is $(\tilde{x}_1,\tilde{x}_2)=\dfrac{[(0,0,0),(0.43,4-4r,4-4r)](3,1-r,1-r)}{(0,0,0)-[(5,1-r,1-r)(2,1-r,1-r)][(0,0,0),(0.43,4-4r,4-4r)]}$

$$\Rightarrow(\tilde{x}_1,\tilde{x}_2)=\left((0,4-4r,4-4r),(3.07,4-4r,4-4r)\right)$$

$$\Rightarrow \tilde{x}_1=(0,4-4r,4-4r) \text{ and } \tilde{x}_2=(3.07,4-4r,4-4r).$$

Substituting these values in the original linear fractional objective function, we have

$$\max\tilde{z}=\frac{\tilde{5}\tilde{x}_1+\tilde{3}\tilde{x}_2}{\tilde{5}\tilde{x}_1+\tilde{2}\tilde{x}_2+\tilde{1}}=\frac{\tilde{5}(0,4-4r,4-4r)+\tilde{3}(3.07,4-4r,4-4r)}{\tilde{5}(0,4-4r,4-4r)+\tilde{2}(3.07,4-4r,4-4r)+\tilde{1}}=(1.29,\ 4-4r,\ 4-4r)$$

$$\Rightarrow \max\tilde{z}=(1.29,\ 4-4r,\ 4-4r)$$

We get the optimal solution of the given problem as $\max\tilde{z}=(1.29,\ 4-4r,\ 4-4r)$ with $\tilde{x}_1=(0,\ 4-4r,\ 4-4r)$ and $\tilde{x}_2=(3.07,\ 4-4r,\ 4-4r)$.

Table 3: Optimal solution for different values of "r"

Value of r	$\max\tilde{z}$	Value of r	$\max\tilde{z}$	Value of r	$\max\tilde{z}$
$r=0$	$(-2.71,1.29,\ 5.29)$	$r=0.72$	(0.17, 1.29, 2.41)	$r=0.82$	$(0.57,1.29,\ 2.01)$
$r=0.25$	$(-1.71,1.29,\ 4.29)$	$r=0.74$	(0.25, 1.29, 2.33)	$r=0.85$	$(0.69,1.29,\ 1.89)$
$r=0.5$	$(-0.71,1.29,\ 3.29)$	$r=0.75$	$(0.29,1.29,\ 2.29)$	$r=0.90$	$(0.89,1.29,\ 1.69)$
$r=0.68$	(0.01, 1.29, 2.57)	$r=0.78$	$(0.41,1.29,\ 2.17)$	$r=0.95$	$(1.09,1.29,\ 1.49)$
$r=0.70$	(0.09, 1.29, 2.49)	$r=0.80$	$(0.49,1.29,\ 2.09)$	$r=1$	$(1.29,1.29,\ 1.29)$

Table 4: Comparison with other methods

Methods	$\max\tilde{z}$
veeramani et al.	$(0.53,\ 1.27,\ 9)$
Pop et al.	$(0.4,\ 1.28,\ 11)$
Stanojevic-Stancu Minasian method	$(0.4,\ 1.3,\ 11)$
Safaei method	$(0.4,\ 1.28,\ 11)$
Proposed Method	$\max\tilde{z}$ for $r\in[0.68,1]$

5. Conclusion

We have proposed a new method for the solution of fully fuzzy linear FPP without reducing to its equivalent crisp form. From table 3, it is evident that the proposed method gives flexibility to the decision maker to choose his preferred solution whereas the other method does not give such facility to the decision maker. Also from table 4, we see that the proposed method gives vagueness reduced results comparing with other methods.

References

[1]. Bellman, R. E., & Zadeh, L. A. 1970. Decision-making in a fuzzy environment. *Management Science*, **17**(4), B-141.

[2]. M. Borza , A. S. Rambely and M. Saraj, 2012 , Solving linear fractional programming with intervals coefficients in the objective function-A new approach , Applied Mathematical sciences, **6** (69) , 3443-3452

[3]. Chakraborty, M., & Gupta, S, 2002, Fuzzy mathematical programming for multi-objective linear fractional programming problem. Fuzzy sets and systems, **125**(3), 335-342.

[4]. Das, S. K., & Edalatpanah, S. A. 2016, A general form of fuzzy linear fractional programs with trapezoidal fuzzy numbers. International journal of data envelopment analysis and operations research, **2** (1), 16-19.

[5]. Das, S. K., Mandal, T., & Edalatpanah, S. A. 2017, A new approach for solving fully fuzzy linear fractional programming problems using the multi-objective linear programming. RAIRO-Operations Research **51** (1), 285-297.

[6]. SK Das, SA Edalatpanah, and T Mandal, 2018, A proposed model for solving fuzzy linear fractional programming problem: numerical point of view, Journal of Computational Science, **25**,367-375.

[7]. SK Das, T Mandal, 2017, A new model for solving fuzzy linear fractional programming problem with ranking function, Journal of applied research on industrial engineering, **4** (2), 89-96.

[8]. Ganesan, K., & P. Veeramani, V., 2006, Fuzzy linear programs with trapezoidal fuzzy numbers. Ann. Oper. Res. **143**, 305-315.

[9]. D.F. Li, S. Chen, 1996, A fuzzy programming approach to fuzzy linear fractional programming with fuzzy coefficients, J. Fuzzy Math. **4**, 829–834.

[10]. M.K. Mehlawat, S. Kumar, 2012, A solution procedure for a linear fractional programming problem with fuzzy numbers, Adv. Intell. Soft Comput. **130**, 1037–1049.

[11]. Ming Ma, Menahem Friedman and Abraham Kandel, 1999, A new approach fuzzy arithmetic, Fuzzy sets and systems, **108**, 83-90.

[12]. R. J. Mitlif, 2016, Solving fuzzy fractional linear programming problems by ranking function methods, Journal of college of education, **1**, 93-108.

[13]. S. Mohanaselvi and K. Ganesan, 2017, A new approach for solving linear fuzzy fractional transportation problem. International Journal of Civil Engineering and Technology, **8** (8), 1123–1129.

[14]. S. Muruganandam and P. Ambika, 2017, A New Approach for Solving Fully Fuzzy Linear Fractional Programming Problem, Middle-East Journal of Scientific Research **25** (3), 647-652,

[15]. S. Muruganandam and P. Ambika, 2018, Solving fuzzy integer linear fractional programming problem. Global Journal for Research Analysis, **7** (4), 103–105.

[16]. A.Nachammai and P.Thangaraj, 2012, Solving fuzzy linear fractional programming problem using metric distance ranking, Applied Mathematical Sciences, **6**(26), 1275-1285.

[17]. B. Pop, I.M. Stancu-Minasian, 2008, A method of solving fully fuzzified linear fractional programming problems, Journal of applied mathematics and computing, **27**, 227–242.

[18]. N. Safaei, 2014, A new method for solving fully fuzzy linear fractional programming with a triangular fuzzy numbers, App. Math. and Comp. Intel., **3**(1), 273–281

[19]. Sanjay Jain, Adarsh Mangal, and P. R. Parihar, 2011, Solution of fuzzy linear fractional programming problem, OPSEARCH, **48**(2), 129–135
[20]. Stanojevic, B & Stancu-Minasian, I. M. 2009. On solving fuzzified linear fractional programs, Advanced modeling and optimization, **11**, 503-523.
[21]. Stanojević, B., & Stancu-Minasian, I. M. 2012, Evaluating fuzzy inequalities and solving fully fuzzified linear fractional programs. Yugoslav Journal of operations research, **22**(1), 41-50.
[22]. Veeramani, C., & Sumathi, M. 2016, Solving the linear fractional programming problem in a fuzzy environment: numerical approach, Applied Mathematical Modelling, **40**, 6148–6164.

AN APPROACH FOR SOLVING FUZZYASSIGNMENT PROBLEM USING BRANCH ANDBOUND TECHNIQUE

S.Mohanaselvi[1,a)] and V.Jayashree[2, b)]
[1,2]Department of Mathematics,
Faculty of Engineering & Technology,
SRM Institute of Science and Technology,
Kattankulathur –603 203, Tamilnadu,India
Email: [a)]mohanass@srmist.edu.in;[b)]v.jaya166@gmail.com

Abstract This paper aims to obtain a minimum fuzzy assignment cost for a fuzzy assignment problem by using fuzzy branch and bound technique. Without transforming the fuzzy assignment problem into a corresponding crisp assignment problem a solution procedure to assign different resources to different activities is proposed to attain the fuzzy optimal cost. In this paper for comparing fuzzy numbers a ranking function in terms of parametric form is used. To showcase the proposed method an example is discussed and a comparison is made with a existing method.

Key words: Trapezoidal Fuzzy numbers, Fuzzy assignment problem, Fuzzy branch and bound technique, Fuzzy number ranking

1.Introduction

An assignment problem is a linear programming problem which is solved using Hungarian method where allocation is done on one – one way in which different activities are assigned to different resources so that the total time or cost involved is minimized and the total profit or sale is maximized.Many real life problems can be solved by assignment problemtechnique. The main ideaof assignment problem is to determine the optimum solution for a given problem. In order to solve assignment problemthe decision parameters must be crisp one. But in real life problems we deal with vagueness and imprecise parameters. This impreciseness occurs due to measurement inaccuracy, variation of parameters, computational errors etc. In 1965, Lotfi Zadeh [12] introduced fuzzy sets that give a procedure to deal with uncertainity.

Pandian etal [7] proposed parallel moving method to solve Fuzzy assignment problem and an optimal solution is found. Nagoor Gani etal [5]suggested a ranking method that converts fuzzy assignment problem to crisp assignment problem in LPP form by LINGO 9.0. Dhanasekar etal [2] has solved fuzzy assignment problem without changing the fuzzy cost into crisp cost using Yagers ranking. AnchalChoudhary etal [1] solved fuzzy assignment problem using trapezoidal fuzzy numbers. Soland etal [9] solved generalized assignment problem using branch and bound algorithm by resolving a series of binary knapsack problem to decide the bounds. Narayanamoorthy etal [6] suggested branch and bound technique with trapezoidal fuzzy number to solve fuzzy assignment problem. Srinivas etal [8] proposed bound and bound algorithm to solve triangular and trapezoidal fuzzy number and Robust ranking technique is applied to transform fuzzy numbers to crisp values. Hema etal [4] an optimal solution is obtained without converting fuzzy numbers to crisp numbers by using branch and bound method. Hosseinzadeh etal[3] proposed a method for solving fuzzy linear programming by lexicography method. Thorani etal [10] proposed a new algorithm is proposed for solving fuzzy assignment problem with fuzzy cost. Thangavelu etal [11] proposed a method of converting Generalized trapezoidal fuzzy number to crisp one using a new ranking method.

This paper aims at finding a fuzzy optimal assignment cost to a fuzzy assignment problem with assignment cost represented as trapezoidal fuzzy number(TrFN). This paper is structured as follows. In section 2, the basic definition, ranking function and arithmetic operation of TrFN are reviewed. In section 3, we introduce the fuzzy assignment problem with TrFN and the fuzzy branch and bound algorithm. A numerical example is provided in section 4 to compare the suggested method with a existing method.

2.Preliminaries

In this section we review certain definition, ranking function and arithmetic operation of TrFN.

2.1 Fuzzy set

A fuzzy set is characterized by a unique membership function which maps each element in the universe of discourse X to the unit interval $[0,1]$

$$\tilde{A} = \{(x, \mu_{\tilde{A}}(x)); x \in X\}$$

The degree of membership function of $\tilde{A}$ is defined by $\mu_{\tilde{A}}(x) : X \to [0,1]$.In a fuzzy set $\tilde{A}$, $\mu_{\tilde{A}}(x)$ is called as the membership degree of $x \in X$.

2.2 Trapezoidal Fuzzy number (TrFN)

A TrFN is a four valued argument $\tilde{A} = (a_1, a_2, a_3, a_4)$ and is described by the following membership function

$$\mu_{\tilde{A}}(x) = \begin{cases} \dfrac{x - a_1}{a_2 - a_1} & a_1 \le x \le a_2 \\ 1 & a_2 \le x \le a_3 \\ \dfrac{a_4 - x}{a_4 - a_3} & a_3 \le x \le a_4 \\ 0 & elsewhere \end{cases}$$

2.3 Parametric representation of TrFN

A fuzzy number can also be illustrated in parametric form as (m, α, β) where m is the mean value and α, β are left and right spreads respectively. Hence any TrFN $\tilde{A} = (a_1, a_2, a_3, a_4)$ is represented as $\tilde{A} = (m, \alpha, \beta)$ where $m = \dfrac{a_2 + a_3}{2}; \alpha = a_2 - a_1$ and $\beta = a_4 - a_3$.

2.4Ranking Function for TrFN:

A function $\Re : F(R) \to R$ is defined as a ranking function where $F(R)$ represents the set of all TrFN. A natural order is provided by the ranking function in which each of the trapezoidal fuzzy number is mapped into a real number in R .Let $\tilde{A} = (m, \alpha, \beta)$, then the ranking function of $\tilde{A}$ is defined as

$$\Re(\tilde{A}) = m + \frac{\beta - \alpha}{4}$$

2.5 Arithmetic operation for TrFN

Let $\tilde{A}_1 = (m, \alpha, \beta)$ and $\tilde{A}_2 = (n, \gamma, \delta)$ be any two TrFNs. Based on the parametric representation of TrFN is the arithmetic operations are given by

(i) $\tilde{A}_1 + \tilde{A}_2 = (m + n, \alpha + \gamma, \beta + \delta)$

(ii) $-\tilde{A}_1 = (-m, \beta, \alpha)$

(iii) $\tilde{A}_1 - \tilde{A}_2 = (m - n, \alpha + \delta, \beta + \gamma)$

3. Mathematical Expression of Fuzzy Assignment Problem

Mathematical model of fuzzy assignment problem is given by,

$$\text{Min } Z = \sum_{i=1}^{n} \sum_{j=1}^{n} \tilde{a}_{ij} \tilde{x}_{ij}$$

subject to $\sum_{i=1}^{n} \tilde{x}_{ij} = 1$

$$\sum_{j=1}^{n} \tilde{x}_{ij} = 1$$

Here $\tilde{x}_{ij}$ denotes the assignment of worker i to job j such that

$$\tilde{x}_{ij} = \begin{cases} 1 & \text{if } i^{th} \text{ worker is assigned to } j^{th} \text{ job} \\ 0 & \text{otherwise} \end{cases}$$

and $\tilde{a}_{ij}$ is the assigning cost of jth job i^{th} worker.

Table 1 Data Matrix

	Job1	Job 2	Job 3	... Job j	Job n
Worker 1	$\tilde{a}_{11}$	$\tilde{a}_{12}$	$\tilde{a}_{13}$	$\ldots\tilde{a}_{ij}$	$\tilde{a}_{1n}$
Worker 2	$\tilde{a}_{21}$	$\tilde{a}_{22}$	$\tilde{a}_{23}$	$\ldots\tilde{a}_{ij}$	$\tilde{a}_{2n}$
.	-	-	-	-	-
.	-	-	-	-	-
Worker i	$\tilde{a}_{i1}$	$\tilde{a}_{i2}$	$\tilde{a}_{i3}$	$\ldots\tilde{a}_{ij}$	-
.	-	-	-	-	-
Worker n	$\tilde{a}_{n1}$	$\tilde{a}_{n2}$	$\tilde{a}_{n3}$	$\ldots\tilde{a}_{nj}$	$\tilde{a}_{nn}$

3.1 *Fuzzy Branch and Bound Algorithm*

The fuzzy branch and bound procedure is utilized for resolving fuzzy assignment problem. The first step in the algorithm is to determine the fuzzy upper and the lower bound of the problem solving. The fuzzy lower and upper bound are used to close up the region of solution value. When the bound value is found the value is used to determine which of the sub problem should be eliminated. The feasible solution is represented by leaves of the tree.

There are two ways for computing the fuzzy cost function:

1. For each worker, we select a job with least fuzzy cost from list of non-allocated jobs.
2. For each job, we select a worker with least fuzzy cost for that job from list of non-allocated workers.

   Stepwise procedure of the fuzzy branch and bound algorithm is given below

Step:1 Minimum fuzzy cost is added from each row and fix the fuzzy lower bound for the given problem.

Step:2 Eliminate the fuzzy least cost for assigning first worker to n jobs and form the sub problem for worker 2.
Step:3 Continue by eliminating the worker2, worker 3,……..worker n by assigning the job with fuzzy minimum cost and obtain the fuzzy feasible solution.

4.Numerical example

Let us consider an example discussed by Srinivas [7].A fuzzy assignment problem involving four jobs and four workers is given below whose cost are represented by trapezoidal fuzzy number.

Table 2 Fuzzy assignment problem

Workers	Jobs			
	1	2	3	4
a	(3,5,6,7)	(5,8,11,12)	(9,10,11,15)	(5,8,10,11)
b	(7,8,10,11)	(3,5,6,7)	(6,8,10,12)	(5,8,9,10)
c	(2,4,5,6)	(5,7,10,11)	(8,11,13,15)	(4,6,7,10)
d	(6,8,10,12)	(2,5,6,7)	(5,7,10,11)	(2,4,5,7)

$$\text{Min}\quad (3,5,6,7)x_{11}+(5,8,11,12)x_{12}+(9,10,11,15)x_{13}+(5,8,10,11)x_{14}+(7,8,10,11)x_{21}+(3,5,6,7)x_{22}+ (6,8,10,12)x_{23}+(5,8,9,10)x_{24}+(2,4,5,6)x_{31}+(5,7,10,11)x_{32}+(8,11,13,15)x_{33}+(4,6,7,10)x_{34}+ (6,8,10,12)x_{41}+(2,5,6,7)x_{42}+(5,7,10,11)x_{43}+(2,4,5,7)x_{44}$$

Subject to

$$x_{11}+x_{12}+x_{13}+x_{14}=1;x_{21}+x_{22}+x_{23}+x_{24}=1\ x_{31}+x_{32}+x_{33}+x_{34}=1\ x_{41}+x_{42}+x_{43}+x_{44}=1$$
$$x_{11}+x_{21}+x_{31}+x_{41}=1\ x_{12}+x_{22}+x_{32}+x_{42}=1\ x_{13}+x_{23}+x_{33}+x_{43}=1\ x_{14}+x_{24}+x_{34}+x_{44}=1$$

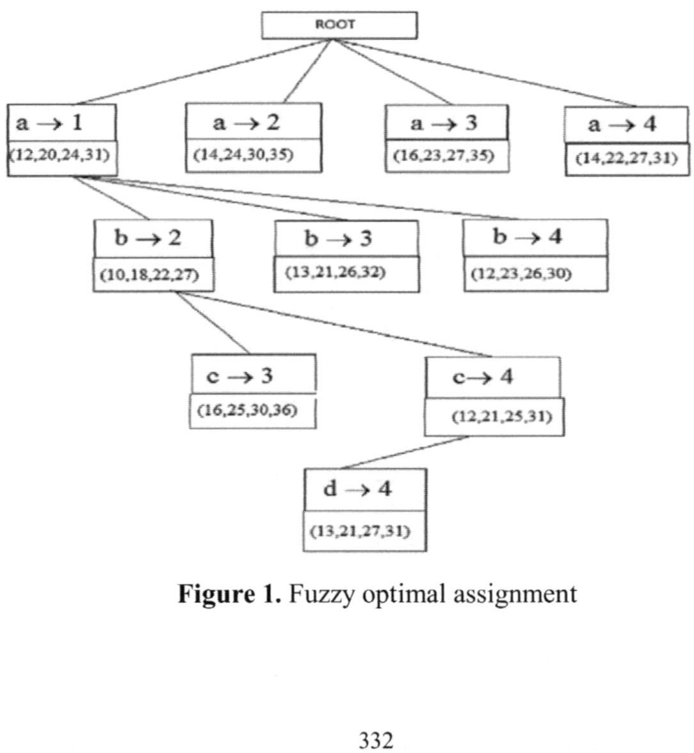

Figure 1. Fuzzy optimal assignment

The fuzzy optimal assignment is given by $a \to 1;\quad b \to 2;\quad c \to 4;\quad d \to 3$. The minimized fuzzy assignment cost is obtained as $(15, 23, 29, 35)$.

Table - 3

Method	Fuzzy assignment cost
Proposed method	$(15, 23, 29, 35)$
Srinivas et al method	$(19, 28, 35, 40)$

Hence we obtained an improved minimized cost.

5.Conclusion

In this paper,a fuzzy branch and bound technique is introduced to resolve fuzzy assignment problem. Without defuzzification procedure we directly solve the fuzzy assignment problem and obtained the optimal assignment cost. A comparison study is made with the existing method to show case the improvement of solution procedure.

References

[1] AnchalChoudhary Jat R N Sharma.S C Sanjay Jain 2016A New Algorithm for solving Fuzzy Assignment problem using Branch and Bound Method *International Journal of Mathematical Archive* **7**(3) 5-11

[2] Dhanasekar S Harikumar K and Sattanathan R 2012 A New approach for solving Fuzzy Assignment Problem *Ultra Scientist* **24**(1)A111 -116

[3] Hosseinzadeh A and Edalatpanah S A 2016 A new Approach for Solving Fully Fuzzy Linear programming by using the Lexicography method *Advances in fuzzy systems ID1538496* 1-6

[4] Muruganandam S Hema K 2017 Solving Fully Fuzzy Assignment Problem using Branch and Bound Technique *Global Journal of Pure and Applied Mathematics* **13**(9) 4515-4522

[5] NagoorGani A and Mohamed V N 2013 Solution of a Fuzzy Assignment Problem by using New ranking method *Intern J Fuzzy Mathematical Archive* **2** 8-16

[6] Narayanamoorthy S and Vidhya P 2017 An adoptable technique to solve Fuzzy Assignment problem *International journal of Pure and Applied mathematics* **117** (14) 199-207

[7] Pandian P and Kavitha K 2012 A new method for solving Fuzzy Assignment Problems *Annals of Pure and Applied Mathematics* **1**(1) 69 – 83

[8] Srinivas B Ganesan G 2015 Method for solving Branch and Bound Techniques for Assignment problems using Triangular and Trapezoidal fuzzy numbers *International Journal in Management and social science* **3** (3) 167- 176

[9] TerryRoss G Richard Soland M 1975 A Branch and Bound Algorithm for the generalized Assignment Problem *Mathematical programming* 8 91 – 103

[10] Thorani Y L P and Ravi Shankar N 2013 Fuzzy Assignment Problem with Generalized Fuzzy Numbers *Applied Mathematical Sciences* **7** (71) 3511 – 3537

[11] Thangavelu K Uthra G Umamageswari R M 2016 Solution of Fuzzy Assignment Problem with Ranking of Generalized Trapezoidal Fuzzy Numbers *International journal of pure and applied mathematics* **106** (6) 9 –16

[12] Zadeh L A 1965 Fuzzy sets *Information Control* 8 338-353

A FUZZY DYNAMIC PROGRAMMING APPROACH TO FUZZY LEAST COST ROUTE PROBLEM

S.Mohanaselvi[1,a)] and S.Suparna Mondal[2, b)]
[1,2]Department of Mathematics,
Faculty of Engineering & Technology,
SRM Institute of Science and Technology,
Kattankulathur –603 203, Tamilnadu,India
Email: [a)]mohanass@srmist.edu.in;[b)]riyankarinku2003998@gmail.com

Abstract This paper focuses on obtaining a solution to generalized trapezoidal fuzzy least cost route problem using fuzzy dynamic programming approach. Using the fuzzy forward and backward recursive equations the fuzzy optimal solution is developed by without converting them into an equivalent crisp problem. Using appropriateranking and arithmetic operation in terms of parametric form the minimized fuzzy cost is attained. To explain the procedure an example is illustrated.
Key words: Generalized trapezoidal Fuzzy numbers, Fuzzy dynamic programming, Fuzzy recursive equation

1.Introduction

Dynamic Programming is a mathematical procedure of optimization using multistage decision progression. Dynamic Programming was established by Richard Bellman [2] in 1953. It is a general algorithm design method for solving problems formulated as repetitions with overlapping sub instances. Zadeh[12] in 1965 developed fuzzy set theory which has become an significant device in dealing roughness and imprecisin in real world problems. In 1970 Bellman etal[3] applied fuzzy set theory in decision making problems. The technique of solving optimization problems in dynamic Programming involving fuzzy parameters is known fuzzy dynamic programming. Kacprzyk[7] [8] [9] have stated the applications and developments in the field of fuzzy dynamic programming.

.Hussien etal[5] have approached a new method using fuzzy dynamic programming for solving multiple criteria resources allocation problem. Baldwin etal[4] have worked on dynamic programming with fuzzy systems and fuzzy environment for multistage problem is expressed and solved by a procedure of fuzzy interpolation. It is an extension of Bellman etal model[3]. Nagalakshmi etal[10] [11] have proposed a paper on solving a fuzzy optimal subdivision problem using fuzzy least cost route problem by using generalized trapezoidal number. Hosseinzadeh etal[10] have approached a new technique for resolving fully fuzzy linear programming by using the lexicography way.

This paper aims at finding afuzzy optimal solution to a fuzzy least cost route problem with transportation cost represented as generalized trapezoidal fuzzy number (GTrFN). The paper is organized as follows. In section 2, the basic definition, ranking function and arithmetic operation of GTrFN are reviewed. In section 3, we introduce the fuzzy least cost route problem with GTrFN and the algorithm to find the minimized fuzzy cost using fuzzy dynamic programming. A numerical example is also provided in section 4 to explain the proposed method.

2. Preliminaries

In this section we review some basic definition, ranking function and arithmetic operation of GTrFN.

2.1Fuzzy set

A fuzzy set is characterized by a unique membership function which maps each element of the universe of discourse X to the unit interval $[0,1]$.

$\tilde{A} = \{(x, \mu_{\tilde{A}}(x)); x \in X\}$

The degree of membership function of $\tilde{A}$ is defined by $\mu_{\tilde{A}}(x): X \to [0,1]$. In a fuzzy set $\tilde{A}$, $\mu_{\tilde{A}}(x)$ is called the membership degree of $x \in X$.

2.2 Generalized trapezoidal Fuzzy number (GTrFN)

A fuzzy set $\tilde{A}$ defined on the universal set of real numbers R , is said to be GTrFN if its membership function has the following characteristics:

(i) $\mu_{\tilde{A}}(x): R \to [0,1]$ is continuous

(ii) $\mu_{\tilde{A}}(x) = 0$ for all $x \in (-\infty, a_1) \cup (a_4, \infty)$

(iii) $\mu_{\tilde{A}}(x)$ is strictly increasing on $[a_1, a_2]$ and strictly decreasing on $[a_3, a_4]$

(iv) $\mu_{\tilde{A}}(x) = w$ for all $x \in [a_2, a_3]$ where $0 < w \le 1$

A generalized trapezoidal fuzzy number is a four valued argument $\tilde{A} = (a_1, a_2, a_3, a_4; w)$ where w is the height of the GTrFN, $0 < w \le 1$ and is defined by the following membership function ,

$$\mu_{\tilde{A}}(x) = \begin{cases} w\dfrac{x-a_1}{a_2-a_1} & a_1 \le x \le a_2 \\ 1 & a_2 \le x \le a_3 \\ w\dfrac{a_4-x}{a_4-a_3} & a_3 \le x \le a_4 \\ 0 & elsewhere \end{cases}$$

2.3 Parametric representation of GTrFN

Any fuzzy number can also be represented in parametric form as $(m, \alpha, \beta; w)$ where m is the mean value and α, β are left and right spreads respectively and $0 < w \le 1$. Hence any GTrFN $\tilde{A} = (a_1, a_2, a_3, a_4; w)$ can be represented as $\tilde{A} = (m, \alpha, \beta; w)$ where $m = \dfrac{a_2 + a_3}{2}; \alpha = a_2 - a_1$ and $\beta = a_4 - a_3$.

2.4 Ranking Function for GTrFN

A function $\Re: F(R) \to R$ is defined as a ranking function where $F(R)$ represents the set of all GTrFN. The function maps each GTrFN into a real line R, where a natural order exists. Let $\tilde{A} = (m, \alpha, \beta; w)$, then the ranking function of $\tilde{A}$ is defined as

$$\Re(\tilde{A}) = w\left[m + \frac{\beta - \alpha}{4}\right]$$

2.5 Arithmetic operation for GTrFN

Let $\tilde{A} = (m,\alpha,\beta;w_1)$ and $\tilde{B} = (n,\gamma,\delta;w_2)$ be any two GTrFNs. Based on the parametric representation of GTrFNs , the arithmetic operations are defined as follows:

(i) $\tilde{A}+\tilde{B} = (m+n,\alpha+\gamma,\beta+\delta;\min(w_1,w_2))$

(ii) $-\tilde{A} = (-m,\beta,\alpha;w_1)$

(iii) $\tilde{A} - \tilde{B} = (m-n,\alpha+\delta,\beta+\gamma;\min(w_1,w_2))$

3. Fuzzy least cost route problem

The fuzzy least cost route problem deals with allocation of properties among available routes with the purpose of attaining a minimum cost of transportation from one city to another. Suppose if a firm has to transport its goods from city A_1 to city A_n. The fuzzy transportation costs are given along the lines which connects the nodes. Each node represents each city. It is required to determine the least cost route connecting the cities A_1 and A_n.The problem is divided to sub-stages. In each stage decision will be made to select a path out of many alternatives. For example, at stage 1 one of the two paths A_1A_2 and A_1A_3 is to be selected and at stage 2 one of the four paths A_3A_4, A_3A_5, A_2A_5 and A_2A_6 is to be selected. The optimal policy will consist of a set of paths connecting A_1 with A_n. Hence a fuzzy least cost route problem consists of some basic steps such as stages, states at each stages and decision replacements at each stage.

3.1 General fuzzy dynamic programming algorithm

The steps given below are followed for using fuzzy dynamic programming methodology to solve fuzzy least cost route problem

*step1:*Detect the problem fuzzy decision variables and state the fuzzy objective function to be optimized under definite restrictions, if any.

*step 2:*The given problem is divided into sub problems . At each stage classify the fuzzy state variables and write down the transformation function as a function of the fuzzy state variable and fuzzy decision variable at the next stage.

*step 3:*Then by the use of general fuzzy recursive relationship we obtain optimum decisions for the problem. Usage of fuzzy forward or backward recursive formula depends upon the problem.

If d_n is the decision variables that define the immediate destinations and s_n is the state variables describes a specific city at any stage. C_{s_n,d_n} is the cost associated with the state variable s_n and the d_n decision variable for the current nth stage. $\tilde{f}_n(s_n,d_n)$ is the minimum total distance for the last n stages, given that salesman is in state s_n and selects d_n as immediate destination. $\tilde{f}_n^*(s_n)$ is the optimal path (minimum cost) when the salesman is in state with n more stages to go for reaching the final stage.Fuzzy forward recursive equation for one stage return is given by

$$\tilde{f}_1 = \tilde{c}_1(s_1,d_1)$$

and the optimal value of $\tilde{f}$ under the state variable s_1 can be obtained by selecting a suitable decision variable d_1. That is

$$\tilde{f}_1^*(s_1) = \min_{d_1}\{\tilde{c}_1(s_1,d_1)\}$$

The range of d_1 is determined by s_1, but s_1 is determined by what has happened in stage 2. Then in stage 2the return function will take the form

$$\tilde{f}_2(s_2,d_2)=c_{s_2,d_2}+\tilde{f}_1^*(s_1)$$

The fuzzy recursive equation is given by

$$\tilde{f}_n^*(s_n)=\underset{d_n}{\text{Min}}\left\{C_{s_n,d_n}+\tilde{f}_{n-1}^*(d_n)\right\}$$

step4: Construct suitable tables to show the essential values of the return function at each stage.

*step 5:*Decide the overall optimum decisions and its value. More than one such optimal policy is found.

4. Illustrative example

Let us consider the following numerical example figure 1 of a company that wants to transport its goods from city A_1 to A_{10}. The optimal route connecting the routes A_1 and A_{10} is to be determined.

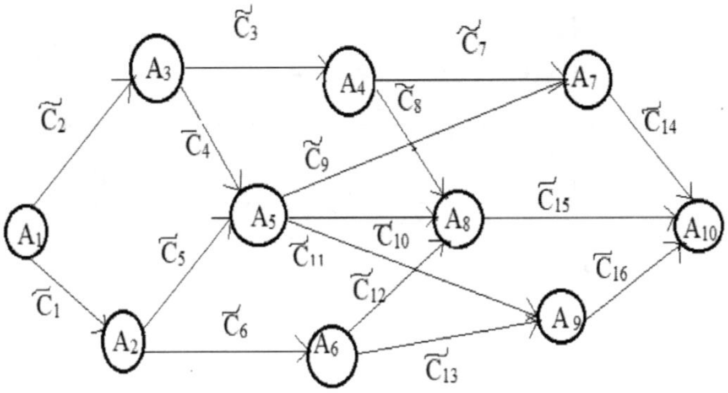

Figure 1. Fuzzy least cost route problem

The cost of transportation table1is taken as GTrFNs as given below

Table 1:GTrFN cost of transportation

Cost variables	Values	Cost variables	Values
$\tilde{C}_1$	$(3,4,5,6;0.1)$	$\tilde{C}_9$	$(4,5,6,7;0.2)$
$\tilde{C}_2$	$(5,6,7,8;0.3)$	$\tilde{C}_{10}$	$(6,7,8,9;0.3)$
$\tilde{C}_3$	$(3,4,5,6;0.2)$	$\tilde{C}_{11}$	$(4,5,6,7;0.2)$
$\tilde{C}_4$	$(2,3,4,5;0.1)$	$\tilde{C}_{12}$	$(7,8,9,10;0.3)$
$\tilde{C}_5$	$(1,2,3,4;0.1)$	$\tilde{C}_{13}$	$(9,10,11,12;0.3)$
$\tilde{C}_6$	$(5,6,7,8;0.3)$	$\tilde{C}_{14}$	$(6,7,8,9;0.3)$

$\tilde{C}_7$	(2,3,4,5;0.1)	$\tilde{C}_{15}$	$(1,2,3,4;0.1)$
$\tilde{C}_8$	$(5,6,7,8;0.3)$	$\tilde{C}_{16}$	$(5,6,7,8;0.3)$

The problem is divided into four stages. To solve the given problem we define problem stages, fuzzy decision variables, fuzzy state variables, return function and transition function.Before proceeding we convert this problem into parametric form $(m,\alpha,\beta;w)$ is given below

Table 2: Parametric representation of cost of transportation

Cost variables	Values	Cost variables	Values
$\tilde{C}_1$	$(4.5,1,1;0.1)$	$\tilde{C}_9$	$(5.5,1,1;0.2)$
$\tilde{C}_2$	$(6.5,1,1;0.3)$	$\tilde{C}_{10}$	$(7.5,1,1;0.3)$
$\tilde{C}_3$	$(4.5,1,1;0.2)$	$\tilde{C}_{11}$	$(5.5,1,1;0.2)$
$\tilde{C}_4$	$(3.5,1,1;0.1)$	$\tilde{C}_{12}$	$(8.5,1,1;0.3)$
$\tilde{C}_5$	$(2.5,1,1;0.1)$	$\tilde{C}_{13}$	$(10.5,1,1;0.3)$
$\tilde{C}_6$	$(6.5,1,1;0.3)$	$\tilde{C}_{14}$	$(7.5,1,1;0.3)$
$\tilde{C}_7$	$(3.5,1,1;0.1)$	$\tilde{C}_{15}$	$(2.5,1,1;0.1)$
$\tilde{C}_8$	$(6.5,1,1;0.3)$	$\tilde{C}_{16}$	$(6.5,1,1;0.3)$

Working backward in stages from A_{10} to A_1, we determine the minimum fuzzy cost to node A_{10} in stage4, from node A_7, A_8 and A_9.These results are given in table3

TABLE 3: stage-4

State s_1	Decisions d_1	$\tilde{f}_1(s_1,d_1)=c_1(s_1,d_1)$	Minimum distance $\tilde{f}_1^*(s_1)$	d_1
	A_7	$(7.5,1,1;0.3)$	$(7.5,1,1;0.3)$	A_{10}
	A_8	$(2.5,1,1;0.3)$	$(2.5,1,1;0.3)$	A_{10}
	A_9	$(6.5,1,1;0.3)$	$(6.5,1,1;0.3)$	A_{10}

Similarly using the fuzzy backward recursive equations stage by stage table(4)(5)(6), we obtain the fuzzy optimal route.

Table 4: stage-3

States s_2	Decisions d_2	$\tilde{f}_2(s_2,d_2)=c_{s_2,d_2}+\tilde{f}_1^*(s_1)$			Minimum distance $\tilde{f}_2^*(s_2)$	d_2
		A_7	A_8	A_9		
	A_4	$(11,2,2;0.1)$	$(9,2,2;0.1)$	-	$(9,2,2;0.1)$	A_8
	A_5	$(13,2,2;0.2)$	$(10,2,2;0.1)$	$(12,2,2;0.2)$	$(10,2,2;0.1)$	A_8
	A_6	-	$(11,2,2;0.1)$	$(17,2,2;0.3)$	$(11,2,2;0.1)$	A_8

Table 5: stage-2

States s_3	Decisions d_3	$\tilde{f}_3(s_3,d_3)=c_{s_3,d_3}+\tilde{f}_2^*(s_2)$			Minimum distance $\tilde{f}_3^*(s_3)$	d_3
		A_4	A_5	A_6		
	A_3	$(13.5,3,3;0.1)$	$(13.5,2,2;0.1)$	-	$(13.5,2,2;0.1)$	A_4 (or) A_5
	A_2	-	$(12.5,3,3;0.1)$	$(17.5,3,3;0.1)$	$(12.5,3,3;0.1)$	A_5

Table 6:stage-2

States s_4	Decisions d_4	$\tilde{f}_4(s_4,d_4)=c_{s_4,d_4}+\tilde{f}_3^*(s_3)$		Minimum distance $\tilde{f}_4^*(s_4)$	d_4
		A_3	A_4		
	A_1	$(20,3,3;0.1)$	$(17,4,4;0.1)$	$(17,4,4;0.1)$	A_2

We obtain the fuzzy optimum route using backward fuzzy recursive equationsas $A_{10}\rightarrow A_8\rightarrow A_5\rightarrow A_2\rightarrow A_1$The minimum fuzzy cost connecting from A_1 to A_{10} is found as

$\tilde{C}_{15}+\tilde{C}_{10}+\tilde{C}_5+\tilde{C}_1=(17,4,4;0.1)$.

5.Conclusion

In this paper a solution procedure to generalized trapezoidal fuzzy least cost route problem using fuzzy dynamic programming approach is proposed. Using the fuzzy backward recursive equations the fuzzy optimum solution is developed by without converting them into an equivalent crisp problem. To explain the procedure an example is illustrated.

References

[1] Amit Kumar Pushpinder Singh Parampreet Kaur and AmarpreetKaur 2010 A New Approach for Ranking of GTrFN *International Journal of Computer Electrical Automation Control and Information and Engineering***4**(8)299-302

[2] Bellman R E 1953 The Theory of Dynamic Programming *An Address Delivered before the Summer Meeting of the Society in Laramie*

[3] Bellman R E and Zadeh L A 1970 Decision-Making in a Fuzzy Environment *Management sci* **17**141-164[2]

[4] Baldwin J F and B W Pilswoth 1994 Dynamic Programming for Fuzzy Systems with Fuzzy Environment *J Math Anal Appl***85** 1-23

[5] Hussien M L Abo-sinna M A 1995 A Fuzzy Dynamic Approach to the Multi-Criteria Resource Allocation Problem *Fuzzy sets and systems***69** 115-124

[6] Hosseinzadeh A Edalatpanah S A 2016 A New Approach for Solving Fully Fuzzy Linear Programming by using the Lexicography Method *Advances in Fuzzy SystemsArticle ID 1-6.*

[7] Kacprzyk J 1994Fuzzy Dynamic Programming–basics issues in M Delgado etal Eds *Fuzzy Optimization Recent Advances Physica Heidelberg* 321-331
[8] Kacprzyk J Esogbue A O 1996 Fuzzy Dynamic Programming Main Developments and Applications*Fuzzy sets and system***81**31-45
[9] Kacprzy J 1997 Multistage Fuzzy Control A Model-Based Approach to *Fuzzy Control and DecisionMaking Wiley New York USAand Engineering***4**(8)
[10] Nagalakshmi T and Uthra G 2017 An Approach of Finding an Optimal Solution for a Fuzzy Least Cost Route Problem by using Generalized Trapezoidal Fuzzy Numbers *Advances in Fuzzy Mathematics ISSN 0973-533X* **12**(3) 737-745
[11] Nagalakshmi T and Uthra G 2018 On a Solving Fuzzy Optimal Subdivision Problem using Fuzzy Dynamic Programming*International journal of mathematics trends and technology IJMTT* **56**244-251
[12] Zadeh L A 1965 Fuzzy Sets *Information and control***8** 338-353

A NEW APPROACH FOR SOLVING INTEGER INTERVAL TRANSPORTATION PROBLEM WITH MIXED CONSTRAINTS

G. Keerthana[a] and G. Ramesh[b*]

Department of Mathematics, SRM Institute of science and Technology,Kattankulathur,Chennai-603203,INDIA
[*]*corresponding author*:[b]**rameshapg@gmail.com**
[a]**keerthanagurunathan@gmail.com**

Abstract:
In this paper, we suggest an interval Transportation problem where the source and destination parameters are chosen as interval numbers. The main objective of the problem is to minimize the total transportation cost and total delivery time. Algorithm is proposed (VAM and MODI) to determine initial allocation to the basic cells and the corresponding interval transportation cost. The optimal solution is obtained without converting the problem to its classical version. Numerical example is obtained which will be easy to understand the solution and procedure of the proposed method.

Keywords: Interval numbers, Interval Transportation problem, Interval VAM method, Interval MODI Method.

1.Introduction

Transportation problem is one of the most important and special type of linear programming problem. where the main objective of the problem is to minimize the transportation cost of distributing a product from a number of sources to a number of destinations, In actual life we cannot predict the shipping cost precisely due to unpredictable and unchangeable life conditions, so shipping cost will be high,therefore normal classical problem will not give a proper solution, hence it can be solved by using Intervals which gives the solution more precisely. Many researchers have proposed different types of interval transportation problem. A. Akilbasha et al. [1] studied about integer interval transportation problem in rough nature. P. Pandian et al. [2] proposed optimal solution is obtained by using intervals by non-fuzzy method. A. Akilbasha et al. [3] Introduced Fully rough integer interval transportation problem is solved by using rough slice sum method. which gives the appropriate solution for rough nature. A.Akilbashaet al. [1,3]studied solution is obtained by using intervals in rough environment. Source and destination is considered to be uncertain and rough cost are assigned in model proposed by Subhakanta Dash et al. [4]. Pradipkundu et al. [5] proposed solid transportation problem unit transportation cost with all parameters are considered as rough variables. Li et al. [6] proposed a Fuzzy compromise programming approach to a multi objective transportation problem. G.Ramesh et al. [7] proposed A method for solving interval number linear programming problem without converting it to a classical linear programming problem. K.Ganesan [8] proposed Arithmetic operations on interval numbers. That is discussed to solvethe system of interval linear equations Sundaresan et al. [9] (2013)Resource Management Techniques OR. Ramne Moore et al. [10] Introduction to Interval Analysis (1-15).

2. Preliminaries

The aim of this section is to present some notations and results which are useful for our further consideration

2.1. Interval numbers

Let $\tilde{a}=[a_1,a_2]=\{x\in R: a_1\le x\le a_2 \text{ and } a_1,a_2\in R\}$ be an interval on the real line R. If $\tilde{a}=a_1=a_2=a$, then $\tilde{a}=[a,\ a]=a$ is a real number (or a degenerate interval).The mid-point and width (or half-width) of an interval number $\tilde{a}=[a_1,\ a_2]$ can be defined as $m(\tilde{a})=\left(\frac{a_1+a_2}{2}\right)$ and $w(\tilde{a})=\left(\frac{a_2-a_1}{2}\right)$. The interval number ã can be expressed in terms of its midpoint and width as $\tilde{a}=[a_1,a_2]=\langle m(\tilde{a}),w(\tilde{a})\rangle$.

2.2.Ranking of Interval Numbers

Sengupta and Pal [11] proposed a simple and efficient index by comparing any two intervals on IR through decision maker's satisfaction.

Definition 2.1. Let $\preceq$ be an extended order relation between the interval numbers $\tilde{a}=[a_1,\ a_2]$, $\tilde{b}=[b_1,\ b_2]$ in IR, then for $m(\tilde{a})<m(\tilde{b})$, we construct a premise $(\tilde{a}\preceq\tilde{b})$ which implies that ã is inferior to $\tilde{b}$ (or $\tilde{b}$ is superior to ã).

An *acceptability function* $A_{\preceq}: IR\times IR\to[0,\infty)$ is defined as:

$$A_{\preceq}(\tilde{a},\tilde{b})=A(\tilde{a}\preceq\tilde{b})=\frac{m(\tilde{b})-m(\tilde{a})}{w(\tilde{b})+w(\tilde{a})},\quad \text{where } w(\tilde{b})+w(\tilde{a})\neq 0.$$

$A_{\prec}$ be interpreted as the grade of acceptability of the the first interval number to be inferior to the second interval number. For any two interval numbers ã and $\tilde{b}$ in IR either $A(\tilde{a}\preceq\tilde{b})\ge 0$ (or $A(\tilde{b}\succeq\tilde{a})\succeq 0$ (or) $A(\tilde{a}\preceq\tilde{b})=0$ (or) $A(\tilde{b}\succeq\tilde{a})=0$ (or) $A(\tilde{a}\preceq\tilde{b})+A(\tilde{b}\preceq\tilde{a})=0$. $A(\tilde{a}\preceq\tilde{b})=0$ and $A(\tilde{b}\preceq\tilde{a})=0$, then we say that interval numbers ã and $\tilde{b}$ are equivalent (non-inferior to each other) and we denote it by $\tilde{a}\approx\tilde{b}$. Also if $A(\tilde{a}\preceq\tilde{b})\ge 0$, then $\tilde{a}\preceq\tilde{b}$ and if $A(\tilde{b}\preceq\tilde{a})\ge 0$, then $\tilde{b}\preceq\tilde{a}$.

2.3 A New Interval Arithmetic

Ming Ma et al. have proposed a new fuzzy arithmetic based on both location index and fuzziness index function . The location index number is taken in the ordinary arithmetic, whereas the fuzziness index functions are considered to follow the lattice rule which exist as the least upper bound and greatest lower bound in the lattice L. That is for $a,b\in L$ we define $a\vee b=\max\{a,b\}$ and $a\wedge b=\min\{a,b\}$. Let us define for any two intervals $\tilde{b}=[b_1,\ b_2]\in IR$ and for $*\in\{+,-,\cdot,\div\}$, the arithmetic operations on ã and $\tilde{b}$ are defined as

(i). Addition : $\tilde{a}+\tilde{b}=\langle m(\tilde{a}),w(\tilde{a})\rangle+\langle m(\tilde{b}),w(\tilde{b})\rangle=\langle m(\tilde{a})+m(\tilde{b}),\ \max\{w(\tilde{a}),w(\tilde{b})\}\rangle$.

(ii). Subtraction : $\tilde{a}-\tilde{b}=\langle m(\tilde{a}),w(\tilde{a})\rangle-\langle m(\tilde{b}),w(\tilde{b})\rangle=\langle m(\tilde{a})-m(\tilde{b}),\ \max\{w(\tilde{a}),w(\tilde{b})\}\rangle$.

(iii). Multiplication : $\tilde{a}\times\tilde{b}=\langle m(\tilde{a}),w(\tilde{a})\rangle\times\langle m(\tilde{b}),w(\tilde{b})\rangle=\langle m(\tilde{a})\times m(\tilde{b}),\ \max\{w(\tilde{a}),w(\tilde{b})\}\rangle$.

(iv). Division : $\tilde{a}\div\tilde{b}=\langle m(\tilde{a}),w(\tilde{a})\rangle\div\langle m(\tilde{b}),w(\tilde{b})\rangle=\langle m(\tilde{a})\div m(\tilde{b}),\max\{w(\tilde{a}),w(\tilde{b})\}\rangle$, provided $m(\tilde{b})\neq\tilde{0}$.

3.Main Results

Let us consider a fully interval transportation problem with m sources and n destinations involving numbers. Let $\tilde{a}_i \succeq \tilde{0}$ be the availability at source i and $\tilde{b}_j$, ($\tilde{b}_j \succeq \tilde{0}$) be the requirement at destination j. Let $\tilde{c}_{ij}$ ($\tilde{c}_{ij} \succeq \tilde{0}$) be the unit interval transportation cost from source i to destination j. Let x_{ij} denote the number of interval units to be transported from source i to destination j. Then the problem is to find a feasible way of transporting the available amount at each source to satisfy the demand at each destination so that the total interval transportation cost is minimized.

3.1. Mathematical formulation of interval transportation problem

The mathematical model of fully interval integer transportation problem is as follows

$$\text{Minimize } \tilde{Z} \approx \sum_{i=1}^{m}\sum_{j=1}^{n} \tilde{c}_{ij}\tilde{x}_{ij}$$

$$\text{subject to } \sum_{j=1}^{n} \tilde{x}_{ij} \approx \tilde{a}_i,\ i = 1,2,3,\ldots,m$$

$$\sum_{i=1}^{m} \tilde{x}_{ij} \approx \tilde{b}_j,\ j = 1,2,3,\ldots,n \qquad (3.1)$$

$$\sum_{i=1}^{m} \tilde{a}_i \approx \sum_{j=1}^{n} \tilde{b}_j,\ \text{where } i = 1,2,3,\ldots,m;\ j = 1,2,3,\ldots,n \text{ and } \tilde{x}_{ij} \succeq \tilde{0} \text{ for all i and j}$$

And$\tilde{a}_i$,$\tilde{b}_j\tilde{c}_{ij}$, $\tilde{x}_{ij}$ in IR, where $\tilde{c}_{ij}$ is the interval unit transportation cost from i^{th} source to the jth destination. The main objective is to minimize the total fuzzy transportation cost, In this paper the fuzzy transportation problem is solved by interval version of VAM and MODI method**.** This interval transportation problem is explicitly represented by the following interval transportation table

Definition 3.1. A set of non-negative allocations $\tilde{x}_{ij}$ which satisfies (in the sense equivalent) the row and the column restrictions is known as fuzzy feasible solution.

Definition 3.2. An interval feasible solution to an interval transportation problem with m sources and n destinations is said to be an interval basic feasible solution if the number of positive allocations are (m+n-1). If the number allocations in an interval basic solution is less than (m+n-1), it is called degenerate interval basic feasible solution.

Definition 3.3. An interval feasible solution is said to be interval optimal solution if it minimizes the total interval transportation cost .

3.2. Basic Theorems

Theorem 3.1. (Existence of an interval feasible solution)

		Destination				
		1	2	…	N	Supply
Sources	1	$\tilde{c}_{11}$	$\tilde{c}_{12}$	…	$\tilde{c}_{1n}$	$\tilde{a}_1$
	2	$\tilde{c}_{21}$	$\tilde{c}_{22}$	….	$\tilde{c}_{2n}$	$\tilde{a}_2$
	⋮	⋮	⋮	⋮	⋮	⋮
	M	$\tilde{c}_{m1}$	$\tilde{c}_{m2}$	….	$\tilde{c}_{mn}$	$\tilde{a}_n$
	Demand	$\tilde{b}_1$	$\tilde{b}_2$	…	$\tilde{b}_m$	

The necessary and sufficient condition for the existence of an interval feasible solution to the fully interval transportation problem (3.1)

$$\sum_{i=1}^{m}\tilde{a}_i \approx \sum_{j=1}^{n}\tilde{b}_j,\ i=1,2,3,...,m \text{ and } j=1,2,3,...,n$$

(Total supply $\approx$ Ttotal demand). (3.2)

Proof: (Necessary condition)

Let there exist an interval feasible solution to the fully interval transportation problem

$$\text{Minimize } \tilde{Z} \approx \sum_{i=1}^{m}\sum_{j=1}^{n}\tilde{c}_{ij}\tilde{x}_{ij}$$

$$\text{subject to } \sum_{j=1}^{n}\tilde{x}_{ij} \approx \tilde{a}_i, i=1,2,3,...,m.$$

$$\sum_{i=1}^{m}\tilde{x}_{ij} \approx \tilde{b}_j, j=1,2,3,...,n.$$

$$\text{and } \tilde{x}_{ij} \succeq \tilde{0} \text{ for all i and j.}$$

(3.3)

From $\sum_{j=1}^{n}\tilde{x}_{ij} \approx \tilde{a}_i, (i=1,2,3,...,m)$, we have $\sum_{i=1}^{m}\sum_{j=1}^{n}\tilde{x}_{ij} \approx \sum_{i=1}^{m}\tilde{a}_i$ (3.4)

Also from $\sum_{i=1}^{m}\tilde{x}_{ij} \approx \tilde{b}_j$, $(j=1,2,3,...,n)$, we have $\sum_{j=1}^{n}\sum_{i=1}^{m}\tilde{x}_{ij} \approx \sum_{j=1}^{n}\tilde{b}_j$ (3.5)

From equations (3.4) and (3.5), we have $\sum_{i=1}^{m}\tilde{a}_i \approx \sum_{j=1}^{n}\tilde{b}_j$

(Sufficient condition)

Since all $\tilde{a}_i$ and $\tilde{b}_j$ are positive, $\tilde{x}_{ij}$must be all positive.Therefore equation (3.2) yields a feasible solution.

Theorem 3.2.The dimension of the basis of an (m×n) fully interval transportation are (m+n-1).That is a fully interval transportation problem has only (m+n-1) independent structural constraints and its basic feasible solution has only (m+n-1) positive components.

Proof: Consider a fully interval transportation problem with m sources and n destinations,

$$\text{Minimize } \tilde{Z} \approx \sum_{i=1}^{m}\sum_{j=1}^{n}\tilde{c}_{ij}\tilde{x}_{ij}$$

$$\text{subject to } \sum_{j=1}^{n}\tilde{x}_{ij} \approx \tilde{a}_i, i=1,2,3,...,m.$$

$$\sum_{i=1}^{m}\tilde{x}_{ij} \approx \tilde{b}_j, j=1,2,3,...,n.$$

$$\text{and } \tilde{x}_{ij} \succeq \tilde{0} \text{ for all i and j.}$$

Let us consider the fully interval transportation has m rows (supply constraint equations) and n columns (demand constraint equations). There are totally (m+n) constraint equations.

This is due to the condition that $\sum_{i=1}^{m} \tilde{a}_i \approx \sum_{j=1}^{n} \tilde{b}_j$ which is the last requirement constraint. Therefore one of (m+n) constraints can always be derived from the remaining (m+n-1). Thus there exists only (m+n-1) independent constraints and its basic feasible solution has only (m+n-1) positive components

Theorem 3.3. The values of the interval basic feasible solution are all differences between the partial sum of $\tilde{a}_i$ and the partial $\tilde{b}_j$ that is $\tilde{x}_{ij} = \pm \sum_{i=1}^{m} \tilde{r}_i \tilde{a}_i \pm \sum_{j=1}^{n} \tilde{s}_j \tilde{b}_j$, where $\tilde{r}_i$ and $\tilde{s}_j$ are either $\tilde{1}=[1,1]$ or $\tilde{0}=[0,0]$

Theorem 3.4. The fully interval transportation problem has a triangular basis.

Proof: We note that every equation has a basic variable; otherwise, the equation cannot be satisfied for $\tilde{a}_i \neq 0,\ \tilde{b}_j \neq 0$. Suppose every row and column equations has atleast two basic variables, since there are m rows and n columns, the total number of basic variables in row equations and column equations will be at least 2m and 2n respectively. Suppose if the total number of basic variables is D, then obviously $D \geq 2m,\ D \geq 2n$.

Case (i). If $m < n$, then $m+n < n+n \Rightarrow m+n < 2n \Rightarrow m+n < 2n < D \Rightarrow m+n < D$.

Case(ii). If $m > n$, then $m+n > m+n \Rightarrow m+m > m+n \Rightarrow 2m > m+n$
$\Rightarrow m+n < 2m \Rightarrow m+n < 2m \leq D \Rightarrow m+n < D$.

Case(iii). If $m = n$, then $m+n = m+n \Rightarrow m+m = m+n \Rightarrow 2m = m+n$
$\Rightarrow m+n = 2m \Rightarrow m+n = 2m < D \Rightarrow m+n < D$.

Thus in all cases $D \geq m + n$. But the number of basic variables in a fully interval transportation problem is (m+n-1) which is a contradiction.

Therefore there is atleast one equation, either row or column having only one basic variable. Let the rth equation have only one basic variable. Let $\tilde{x}_{rs}$ be the only basic variable in the rth row and s^{th} column, then $\tilde{x}_{rs} \approx \tilde{a}_r$. Eliminate r^{th} row from the system of equations and substitute $\tilde{x}_{rs} \approx \tilde{a}_r$ in s^{th} column equation and replace $\tilde{b}_s$ by $\tilde{b}_s' \approx \tilde{b}_s - \tilde{a}_r$. After eliminating the rth row, the system has (m+n-2) linearly independent constraints. Hence the number of basic variables is (m+n-2).

Repeat the process and conclude that in the reduced system of equations there is an equation which has only one basic variable. But if this is in s^{th} column equation, then it will have two basic variables.

This concludes that in our original system of equations, there is an equation which has atleast two basic variables. Continue the process to get the theorem.

4. PROPOSED ALGORITHMS

4.1. Interval Version of Vogel's Approximation Method (IVAM)

Step 1. Express the interval transportation problem in the transportation table.

Step 2. Express the all interval parameters supply, demand and unit transportation cost in the transportation problem in terms of midpoint and half width. That is in the form of $\tilde{a} = [a_1, a_2] = \langle m(\tilde{a}), w(\tilde{a}) \rangle$.

Journal of Physics: Conference Series **1377** (2019) 012043 doi:10.1088/1742-6596/1377/1/012043

Step 3. In the interval Transportation table, determine the row penalties for each row by subtracting the lowest cost from the next lowest cost of that row. In the similar way calculate the column penalties for each column. Write down the row penalties aside each row and the column penalties below each column.

Step 4. Identify the column or row with largest fuzzy penalty. In case of tie, break the tie arbitrarily. Select a cell with minimum fuzzy cost in the selected column (or row, as the case may be) and assign the maximum units possible by considering the demand and supply position corresponding to the selected cell.

Step 5. Delete the column/row for which the supply and demand requirements are met.

Step 6. Continue steps 1 to 3 for the resulting fuzzy transportation tableuntil the supply and demand of all sources and destinations have been met.

4.2. Interval Version of MODI Method

Step 1: Start with interval basic feasible solution consisting of (m+n–1) allocations in independent positions.

Step 2: Determine the values of (m+n) dual variables $\tilde{u}_i$ $(i=1,2,...,m)$ and $\tilde{v}_j$ $(j=1,2,...,n)$ for all the rows and columns such that for each occupied cell (i, j), the following condition is satisfied $\tilde{c}_{ij} = \tilde{u}_i + \tilde{v}_j$.

Step 3: Compute the opportunity cost $\tilde{d}_{ij}$ for each unoccupied cell (i, j) by using the formula:

$\tilde{d}_{ij} = \tilde{c}_{ij} - (\tilde{u}_i + \tilde{v}_j)$ for all i and j.

Step 4: Check the sign of the opportunity cost $\tilde{d}_{ij}$ for each unoccupied cell (i, j) and conclude that

(i) If all $\tilde{d}_{ij} \succeq \tilde{0}$, then the current solution is optimal.

(ii)If at least one $\tilde{d}_{ij} \prec \tilde{0}$, then the current solution is not optimal and an improved solution can be obtained.
In this case, the unoccupied cell with the most negative value of $\tilde{d}_{ij}$ is considered for the new transportation schedule.

Step 5: Construct aclosed path (loop) for the unoccupied cell (i, j) having largest negative opportunity cost. Mark a (+) sign in this cell and move along the rows (or columns) to find an occupied cell. Mark a (-) sign in this cell and find out another occupied cell. Repeat the process and mark the occupied cells with (+) and (-) signs alternatively. Close the path back to the selected unoccupied cell.

Step 6: Select the smallest quantity amongst the cells marked with (-) sign. Allocate this value to the unoccupied cell of the loop and add and subtract it in the occupied cells as per their signs. Thus an improved solution is obtained by calculating the total transportation cost by this method.

Step 7: Test the revised solution further for optimality. The procedure terminates when all $\tilde{d}_{ij} \succeq \tilde{0}$, for unoccupied cells.

5.Numerical Examples

Example 5.1. Consider the following interval transportation problem discussed by Akilbasha and Natarajan [3]. They proposed a new method namely, roughslice sum method for finding an optimal solution for interval transportation problems where transportation cost, supply and demand are intervals.

Table 5.1. Interval Transportation table

	D_1	D_2	D_3	Supply
S_1	[3,4]	[7,9]	[6,8]	[6,7]
S_2	[7,8]	[5,7]	[3,5]	[8,10]
S_3	[9,10]	[11,13]	[4,6]	[11,13]
Demand	[9,10]	[12,14]	[7,9]	

A.Akilbasha et al. obtained the optimum interval transportation cost of this problem as [120,195]. Let us consider the same problem and solve by VAM and MODI method. Let us define all the interval parameters as $\tilde{a}=[a_1,a_2]$ in terms of midpoint and width $\tilde{a}=[a_1,a_2]=\langle m(\tilde{a}), w(\tilde{a})\rangle$. Now the given interval transportation problem is as follows

Table 5.2. Interval Transportation table

	D_1	D_2	D_3	Supply
S_1	⟨3.5,0.5⟩	⟨8,1⟩	⟨7,1⟩	⟨6.5,0.5⟩
S_2	⟨7.5,0.5⟩	⟨6,1⟩	⟨4,1⟩	⟨9,1⟩
S_3	⟨9.5,0.5⟩	⟨12,1⟩	⟨5,1⟩	⟨12,1⟩
Demand	⟨9.5,0.5⟩	⟨13,1⟩	⟨8,1⟩	

Table :5.3.Interval transportation table

To balance the above problem, now we have to add one row

	D_1	D_2	D_3	Supply
S_1	⟨3.5,0.5⟩	⟨8,1⟩	⟨7,1⟩	⟨6.5,0.5 ⟩
S_2	⟨7.5,0.5 ⟩	⟨6,1⟩	⟨4,1⟩	⟨9,1⟩
S_3	⟨9.5,0.5⟩	⟨12,1⟩	⟨5,1⟩	⟨12,1⟩
S_4	⟨0,0⟩	⟨0,0⟩	⟨0,0⟩	⟨3,0⟩
Demand	⟨9.5,0.5⟩	⟨13,1⟩	⟨8,1⟩	⟨30.5,1⟩

Applying the Interval version of Vogel's Approximation Method (IVAM), the initial interval basic feasible solution is obtained as follows

Table :5.4.Interval transportation table

	D_1	D_2	D_3	Supply	θ
S_1	⟨3.5,0.5⟩	⟨8,1⟩	⟨7,1⟩	⟨6.5,0.5⟩	⟨3.5,1⟩
S_2	⟨7.5,0.5⟩	⟨6,1⟩	⟨4,1⟩	⟨9,1⟩	⟨2,1⟩
S_3	⟨9.5,0.5⟩	⟨12,1⟩	⟨5,1⟩	⟨12,1⟩	⟨4.5,1⟩
S_4	⟨0,0⟩	⟨0,0⟩ ⟨3,0⟩	⟨0,0⟩	⟨3,0⟩	⟨0,0⟩
Demand	⟨9.5,0.5⟩	⟨10,1⟩	⟨8,1⟩		
θ	⟨3.5,0.5⟩	⟨ 6,1⟩	⟨4,1⟩		

Table :5.5.Interval transportation table

	D_1	D_2	D_3	Supply	θ
S_1	⟨3.5,0.5⟩	⟨8,1⟩	⟨7,1⟩	⟨6.5,0.5⟩	⟨3.5,1⟩
S_2	⟨7.5,0.5⟩	⟨6,1⟩	⟨4,1⟩	⟨9,1⟩	⟨2,1⟩
S_3	⟨9.5,0.5⟩	⟨12,1⟩	⟨5,1⟩ ⟨8,1⟩	⟨4,1⟩	⟨4.5,1⟩
Demand	⟨9.5,0.5⟩	⟨10,1⟩	⟨8,1⟩		
θ	⟨4,0.5⟩	⟨2,1⟩	⟨1,1⟩		

Table :5.6.Interval transportation table

	D_1	D_2	Supply	θ
S_1	⟨3.5,0.5⟩ ⟨6.5,0.5⟩	⟨8,1⟩	⟨6.5,0.5⟩	⟨4.5,1⟩
S_2	⟨7.5,0.5⟩	⟨6,1⟩	⟨9,1⟩	⟨1.5,1⟩
S_3	⟨9.5,0.5⟩	⟨12,1⟩	⟨4,1⟩	⟨2.5,1⟩
Demand	⟨3,0.5⟩	⟨10,1⟩		
θ	⟨4,0.5⟩	⟨2,1⟩		

Table :5.7.Interval transportation table

	D_1	D_2	Supply	θ
S_2	⟨7.5,0.5⟩	⟨6,1⟩ ⟨9,1⟩	⟨9,1⟩	⟨1.5,1⟩
S_3	⟨9.5,0.5⟩	⟨12,1⟩	⟨4,1⟩	⟨2.5,1⟩
Demand	⟨3,0.5⟩	⟨1,1⟩		
θ	⟨2,0.5⟩	⟨6,1⟩		

Table :5.8.Interval transportation table

	D_1	D_2	Supply	θ
S_3	⟨9.5,0.5⟩	⟨12,1⟩ ⟨1,1⟩	⟨3,1⟩	⟨2.5,1⟩
Demand	⟨3,0.5⟩	⟨1,1⟩		
θ	⟨9.5,0.5⟩	⟨12,1⟩		

Table :5.9.Interval transportation table

	D_1	Supply
S_3	⟨9.5,0.5⟩ ⟨3,1⟩	⟨ 3,1⟩
Demand	⟨3,0.5⟩	

Hence the initial basic feasible solution of the interval transportation problem by using interval Vogel's approximation method is

Table : 5.10.Interval transportation table

	D_1	D_2	D_3	Supply
S_1	⟨3.5,0.5⟩ ⟨6.5,0.5⟩	⟨8,1⟩	⟨7,1⟩	⟨6.5,0.5⟩
S_2	⟨7.5,0.5⟩	⟨6,1⟩ ⟨9,1⟩	⟨4,1⟩	⟨9,1⟩
S_3	⟨9.5,0.5⟩ ⟨3,1⟩	⟨12,1⟩ ⟨1,1⟩	⟨5,1⟩ ⟨8,1⟩	⟨12,1⟩
S_4	⟨0,0⟩	⟨0,0⟩ ⟨3,0⟩	⟨0,0⟩	⟨3,0⟩
Demand	⟨9.5,0.5⟩	⟨13,1⟩	⟨8,1⟩	⟨30.5,1⟩

The initial interval transportation cost

= ⟨3.5,0.5⟩⟨6.5,0.5⟩+⟨6,1⟩⟨9,1⟩+⟨9.5,0.5⟩⟨ 3,1⟩+⟨12,1⟩⟨1,1⟩+⟨5,1⟩⟨8,1⟩ +⟨0,0⟩⟨3,0⟩

= ⟨0,0⟩ +⟨40,1⟩+⟨22.75,0.5⟩+⟨54,1⟩+⟨12,1⟩⟨1,1⟩ +⟨28.5,1⟩

= ⟨157.25,1⟩

Now apply the interval version of MODI method for checking the optimality, we have

Table :5.11.Interval transportation table

⟨3.5,0.5⟩ ⟨6.5,0.5⟩	⟨8,1⟩	⟨7,1⟩	u_1=⟨ 0,1⟩
⟨7.5,0.5⟩	⟨6,1⟩ ⟨9,1⟩	⟨4,1⟩	u_2= ⟨0,1⟩
⟨9.5,0.5⟩ ⟨3,1⟩	⟨12,1⟩ ⟨1,1⟩	⟨5,1⟩ ⟨8,1⟩	u_3=⟨6,0.5⟩
⟨0,0⟩	⟨0,0⟩ ⟨3,0⟩	⟨0,0⟩	u_4=⟨-6,1⟩
v_1=⟨3.5,0.5⟩	v_2= ⟨6,1⟩	v_3=⟨-1,1⟩	

Since all $\tilde{d}_{ij} \succeq \tilde{0}$, the current solution is an interval optimal solution. Hence the optimal interval transportation cost is

= ⟨3.5,0.5⟩⟨6.5,0.5⟩+⟨6,1⟩⟨9,1⟩+⟨9.5,0.5⟩⟨ 3,1⟩+⟨12,1⟩⟨1,1⟩+⟨5,1⟩⟨8,1⟩ +⟨0,0⟩⟨3,0⟩

= ⟨157.25, 1⟩

= [156.25, 158.25]

It is to be noted that our solution is very much sharper than the solution obtained by Akilbasha et al.
Example 5.2. Consider another problem discussed by Akilbasha et al.

For this problem, Akilbasha et al. obtained the optimum interval transportation cost as [63,288].

Table :5.12.Interval transportation table

	D_1	D_2	D_3	Supply
S_1	[2,5]	[5,11]	[4,10]	[5,8]
S_2	[6,9]	[3,9]	[1,7]	[6,12]
S_3	[8,11]	[9,15]	[2,8]	[9,15]
Demand	[8,11]	[10,16]	[5,11]	

Let us define all the interval parameters "$\tilde{a} = [a_1, a_2]$ in terms of midpoint and width as $\tilde{a} = [a_1, a_2] = \langle m(\tilde{a}), w(\tilde{a}) \rangle$. "Now the given interval transportation problem will be as follows

Table 5.13. Interval Transportation table

	D_1	D_2	D_3	Supply
S_1	⟨3.5,1.5⟩	⟨8,3⟩	⟨7,3⟩	⟨6.5,1.5⟩
S_2	⟨7.5,1.5⟩	⟨6,3⟩	⟨4,3⟩	⟨9,3⟩
S_3	⟨9.5,1.5⟩	⟨12,3⟩	⟨5,3⟩	⟨12,3⟩
Demand	⟨9.5,1.5⟩	⟨13,3⟩	⟨8,3⟩	

Table :5.14.Interval transportation tableTo balance the above problem, now we have to add one row

	D_1	D_2	D_3	Supply
S_1	⟨3.5,1.5⟩	⟨8,3⟩	⟨7,3⟩	⟨6.5,1.5 ⟩
S_2	⟨7.5,1.5 ⟩	⟨6,3⟩	⟨4,3⟩	⟨9,3⟩
S_3	⟨9.5,1.5⟩	⟨12,3⟩	⟨5,3⟩	⟨12,3⟩
S_4	⟨0,0⟩	⟨0,0⟩	⟨0,0⟩	⟨3,0⟩
Demand	⟨9.5,1.5⟩	⟨13,3⟩	⟨8,3⟩	⟨30.5,3⟩

Applying the Interval version of Vogel's Approximation Method (IVAM), the initial interval basic feasible solution is obtained as follows

Table :5.15.Interval transportation table

	D_1	D_2	D_3	Supply	θ
S_1	⟨3.5,1.5⟩	⟨8,3⟩	⟨7,3⟩	⟨6.5,1.5⟩	⟨3.5,3⟩
S_2	⟨7.5,1.5⟩	⟨6,3⟩	⟨4,3⟩	⟨ 9,3⟩	⟨2,3⟩
S_3	⟨9.5,1.5⟩	⟨12,3⟩	⟨5,3⟩	⟨12,3⟩	⟨4.5,3⟩
S_4	⟨0,0⟩	⟨0,0⟩ ⟨3,0⟩	⟨0,0⟩	⟨ 3,0⟩	⟨0,0⟩
Demand	⟨9.5,1.5⟩	⟨10,3⟩	⟨8,3⟩		
θ	⟨3.5,1.5⟩	⟨ 6,3⟩	⟨4,3⟩		

Table :5.16.Interval transportation table

	D_1	D_2	D_3	Supply	θ
S_1	⟨3.5,1.5⟩	⟨8,3⟩	⟨7,3⟩	⟨6.5,1.5⟩	⟨3.5,3⟩
S_2	⟨7.5,1.5⟩	⟨6,3⟩	⟨4,3⟩	⟨9,3⟩	⟨2,3⟩
S_3	⟨9.5,1.5⟩	⟨12,3⟩	⟨5,3⟩ ⟨8,3⟩	⟨4,3⟩	⟨4.5,3⟩
Demand	⟨9.5,1.5⟩	⟨10,3⟩	⟨8,3⟩		
θ	⟨4,1.5⟩	⟨2,3⟩	⟨1,3⟩		

Table :5.17.Interval transportation table

	D_1	D_2	Supply	θ
S_1	⟨3.5,1.5⟩ ⟨6.5,1.5⟩	⟨8,3⟩	⟨6.5,1.5⟩	⟨4.5,3⟩
S_2	⟨7.5,1.5⟩	⟨6,3⟩	⟨9,3⟩	⟨1.5,3⟩
S_3	⟨9.5,1.5⟩	⟨12,3⟩	⟨4,3⟩	⟨2.5,3⟩
Demand	⟨3,1.5⟩	⟨10,3⟩		
θ	⟨4,1.5⟩	⟨2,3⟩		

Table :5.18.Interval transportation table

	D_1	D_2	Supply	θ
S_2	⟨7.5,1.5⟩	⟨6,3⟩ ⟨9,3⟩	⟨9,3⟩	⟨1.5,3⟩
S_3	⟨9.5,1.5⟩	⟨12,3⟩	⟨4,3⟩	⟨2.5,3⟩
Demand	⟨3,1.5⟩	⟨1,3⟩		
θ	⟨2,1.5⟩	⟨6,3⟩		

Table :5.19.Interval transportation table

	D_1	D_2	Supply	θ
S_3	⟨9.5,1.5⟩	⟨12,3⟩ ⟨1,3⟩	⟨3,3⟩	⟨2.5,3⟩
Demand	⟨3,1.5⟩	⟨1,3⟩		
θ	⟨9.5,1.5⟩	⟨12,3⟩		

Table :5.20.Interval transportation table

	D_1	Supply
S_3	⟨9.5,1.5⟩ ⟨3,3⟩	⟨ 3,3⟩
Demand	⟨3,1.5⟩	

Hence the initial basic feasible solution for the interval transportation problem using interval Vogel's approximation method is

Table :5.21.Interval transportation table

	D_1	D_2	D_3	Supply
S_1	⟨3.5,1.5⟩ ⟨6.5,1.5⟩	⟨8,3⟩	⟨7,3⟩	⟨6.5,1.5⟩
S_2	⟨7.5,1.5⟩	⟨6,3⟩ ⟨9,3⟩	⟨4,3⟩	⟨9,3⟩
S_3	⟨9.5,1.5⟩ ⟨3,3⟩	⟨12,3⟩ ⟨1,3⟩	⟨5,3⟩ ⟨8,3⟩	⟨12,3⟩
S_4	⟨0,0⟩	⟨0,0⟩ ⟨3,0⟩	⟨0,0⟩	⟨3,0⟩
Demand	⟨9.5,1.5⟩	⟨13,3⟩	⟨8,3⟩	⟨30.5,3⟩

The initial interval transportation cost

= ⟨3.5,1.5⟩⟨6.5,1.5⟩+⟨6,3⟩⟨9,3⟩+⟨9.5,1.5⟩⟨ 3,3⟩+⟨12,3⟩⟨1,3⟩+⟨5,3⟩⟨8,3⟩ +⟨0,0⟩⟨3,0⟩

= ⟨0,0⟩ +⟨40,3⟩+⟨22.75,1.5⟩+⟨54,3⟩+⟨12,3⟩⟨1,3⟩ +⟨28.5,3⟩

= ⟨157.25, 3⟩

Now by applying the interval version of MODI method for checking the optimality, we have

Table :5.22.Interval transportation table

⟨3.5,1.5⟩ ⟨6.5,1.5⟩	⟨8,3⟩	⟨7,3⟩	u_1=⟨ 0,3⟩
⟨7.5,1.5⟩	⟨6,3⟩ ⟨9,3⟩	⟨4,3⟩	u_2= ⟨0,3⟩
⟨9.5,1.5⟩ ⟨3,3⟩	⟨12,3⟩ ⟨1,3⟩	⟨5,3⟩ ⟨8,3⟩	u_3=⟨6,1.5⟩
⟨0,0⟩	⟨0,0⟩ ⟨3,0⟩	⟨0,0⟩	u_4=⟨-6,3⟩
v_1=⟨3.5,1.5⟩	v_2= ⟨6,3⟩	v_3=⟨-1,3⟩	

Since all $\tilde{d}_{ij} \succeq \tilde{0}$, the present solution is an interval optimal solution. Hence the optimal interval transportation cost is

$= \langle 3.5,1.5\rangle\langle 6.5,1.5\rangle+\langle 6,3\rangle\langle 9,3\rangle+\langle 9.5,1.5\rangle\langle 3,3\rangle+\langle 12,3\rangle\langle 1,3\rangle+\langle 5,3\rangle\langle 8,3\rangle+\langle 0,0\rangle\langle 3,0\rangle$

$= \langle 157.25, 3\rangle$

$= [154.25, 160.25]$

It is to be noted that our solution is very much sharper than [63,288] the solution obtained by Akilbasha et al.

6. Conclusion

TheTransportation problem with mixed constraints having all parameters as integer intervals is considered in this paper. A method namely VAM and MODI method is proposed to solve fully integer interval transportation problem without converting it to the classical transportation problem so that the total shipping cost is minimum. The above method is a systematic procedure which is both easy to understand and apply in real life. This method is a method provides an optimal solution which is sharper than the other methods.

References

[1]. AAkilbashaG Natarajan and P Pandian 2016Finding an optimal solution of the interval integer transportation problem with rough nature by split and separation method*In.J.of Pure and Applied Math.*

[2]. P Pandian and G Natarajan 2010 A new method of finding an optimal solution of fully interval integer transportation problems*Applied Mathematical Sciences*

[3]. P Pandian 2017 Solving Transportation problem with mixed constraints in Rough Environment*In. J. of Pure and Applied Math.*

[4]. Subhakanta Dash and S.P.Mohanty 2017 Transportation Programming under uncertain Environment. *In.J.of. Eng. Research and Development***7 22-28**

[5]. Pradipkundu, Samarjitkar and Manoranjanmaiti 2013 Some Solid transportation model with crisp and rough costs*In.J. of mathematical and computational sciences.* **vol -7**

[6]. Lushu Li and Lai K K 2000 A fuzzy approach to the multi objective transportation problem*Computers and Operations Research***27 43-57**

[7]. G Ramesh and K Ganesh 2011 Interval Linear Programming with generalized interval arithmetic *International Journal of Scientific & Engineering Research***2 11**

[8]. K Ganesan 2006 On some properties of interval matrices*In.J.of .Mathematical sciences.*

[9]. Sundaresan VGanapathy Subramanian K S and Ganesan K Resource management techniques OR.

[10]. Ramne Moore 1971 Bakerkearfottr and Michael J. Cloud Introduction to interval analysis **(1-15)**

[11]. S K Das A Goswami and S SAlam 1999 Multiobjective transportation problem with interval cost source and destination parameters *European Journal of Operational Research***117 100 – 112**

[12]. Ming Ma Menahem Friedman and Abraham Kandel 1999 A new fuzzy arithmetic Fuzzy sets and systems **108 83-90**

Journal of Physics: Conference Series **1377** (2019) 012044 doi:10.1088/1742-6596/1377/1/012044

Fuzzy project planning and scheduling using critical path techique

T Leelavathy[1]and S Padmadevi[2]

[1]Department of Mathematics,
Faculty of Engineering and Technology,
SRM Institute of Science and Technology,
Kattankulathur-603203, India.
[2] Department of Mathematics, SRM Institute of Science and Technology,
Kattankulathur-603203, India.

E-mail: leelavat@srmist.edu.in and
padmadevi1301@gmail.com

Abstract. In this article, we intend to determine the total duration of completion of the project for fuzzy networking problem with the activity duration represented as trapezoidal fuzzy number using Critical Path technique. A new fuzzy arithmetic operation, ranking method and a precise step by step plan is designed to obtain the total fuzzy duration, utilizing the available resources. Finally the proposed study is discussed with a numerical example.

1. Intoduction

Network Analysis or Network planning and scheduling techniques used for planning, scheduling and controlling large and complex projects. These techniques are based on the representation of the projects as a network of activities. A project involves a large number of interrelated activities that must be completed on or before a specified time limit, in a specified sequence, with specified quantity and minimum cost of using resources such as money, material, facilities and space. Network representations are widely used for problems in areas such as production, distribution, project planning, resource management and financial planning.

A Network is a graphical representation of arrows and nodesprovides a powerful visual and conceptual aid for portraying the relationships between the components of systems and shows the logical sequence of various activities to be performed to achieve project objectives. Planning is a phase which involves listing of tasks or jobs that must be performed to complete a project under consideration. Scheduling is a phase which involves the laying out of the actual activities of the project in a logical sequence of time in which they have to be performed. Control is a phase consists of reviewing the progress of the project whether the actual performance is according to the planned schedule, and finding the reason for difference, if any, between the schedule and performance. Total project time is the time taken to complete a project and found from the sequence of critical activities. The path connecting the first initial node to the very last terminal node of the longest duration in a project network is called critical path.

Critical path method provides the solution for complicated project network when the decision parameters are crisp. But in reality, due to uncertainty of information as well as the variation of management scenario, it is difficult for the decision maker to obtain the exact time duration for the project activities. Hence we cannot use the traditional classical method for solving the project networking problems. Zadeh [10] introduced the concept of fuzzy set in 1965, which plays a vital role in handling such situations effectively. Buckly [1] has applied trapezoidal fuzzy number in linear programming. The basic concepts of fuzzy sets and its application are discussed by Dubois et.al [2]. Kikuchi [3] deals the method of defuzzification of fuzzy numbers. Stefan Chanas et.al [8] Found critical path in the network where the duration of each activity are fuzzy numbers. Revathi et.al [5] found the fuzzy critical path of fuzzy networking problem using trapezoidal fuzzy number. Rajendran et.al [4] used octagonal fuzzy number to find fuzzy critical path in fuzzy project networking problem. Shin-Pin-Chen et.al, [7] proposed fuzzy LPP formulation to find the fuzzy critical path. Stephan Dinagar et.al,[9] obtained fuzzy critical path for the problem with duration of each activities represented by fuzzy number. Selvakumari et.al, [6] solved the fuzzy networking problem with hexagonal fuzzy number using job sequencing technique.

In this paper, we investigate a more realistic problem, namely project scheduling problem with fuzzy duration represented as trapezoidal fuzzy number. The paper is arranged as follows: In section 2, we review the basic definitions, ranking of trapezoidal fuzzy number and arithmetic operation of trapezoidal fuzzy number. In section 3, we proposed an algorithm to find the critical path and maximum duration to complete the project, where the duration of activities are trapezoidal fuzzy number. In section 4, a numerical problem is provided to find the efficiency of the proposed approach.

2. Preliminaries

In this section, we represent the idea about the fuzzy numbers and definitions which involved in the research work. Fuzzy set theory have been defined by Zadeh [10] and the fundamental concepts are given by Dubois and Prade [2] are applied to fuzzy environment.

Definition 2.1

A fuzzy set $\tilde{A}$ in X is a set of ordered pair defined by, $\tilde{A}(x) = \{(x, \mu_{\tilde{A}}(x)) : x \in X, \mu_{\tilde{A}}(x) \in [0,1]\}$ where $\mu_{\tilde{A}}(x)$ is a membership function.

Definition 2.2

A fuzzy set $\tilde{A}$ defined on a set of real number R is said to be a fuzzy number, if its membership function. $\tilde{A} : R \to [0,1]$ has the following characteristic:

1)$\tilde{A}$ is convex

(i.e.,) $\tilde{A}\{\lambda x_1 + (1-\lambda)x_2\} \geq \min\{\tilde{A}(x_1), \tilde{A}(x_2)\}$ for all $x_1, x_2 \in R$ and $\lambda \in [0,1]$

2) $\tilde{A}$ is normal, i.e., there exist an $x \in R$ such that $\tilde{A}(x) = 1$

3) $\tilde{A}$ is piecewise continuous.

Definition 2.3

A fuzzy number $\tilde{A}$ is a trapezoidal fuzzy number denoted by $\tilde{A} = (a_1,a_2,a_3,a_4)$ where a_1,a_2,a_3,a_4 are real numbers and its membership function $\tilde{A}(x)$ is given below,

$$\tilde{A}(x) = \begin{cases} \dfrac{x - a_1}{a_2 - a_1}, & \text{if } a_1 \leq x \leq a_2 \\ 1, & \text{if } a_2 \leq x \leq a_3 \\ \dfrac{a_{4-x}}{a_4 - a_3}, & \text{if } a_3 \leq x \leq a_4 \\ 0, & \text{otherwise} \end{cases}$$

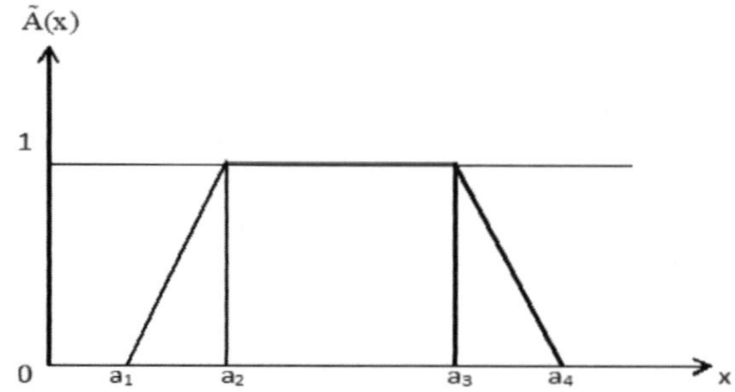

Figure 2.1. The membership function of Trapezoidal Fuzzy Number

In this paper, the trapezoidal fuzzy number is represented as, $\tilde{A} = (a_1,a_2,a_3,a_4) = (m,w,a^*,a_*)$ where $m = \frac{(a_1 + a_3)}{2}$ and $w = \frac{(a_3 - a_2)}{2}$ are the midpoint and width of the core $[a_2,a_3]$ respectively. Also $a_* = (a_2 - a_1)$ and $a^* = (a_4 - a_3)$ represents the left and right spreads respectively of the trapezoidal fuzzy number $\tilde{A} = (a_1, a_2, a_3, a_4)$.

2.1 Ranking of trapezoidal fuzzy nmber

The ranking function $\Re: F(R) \to R$ is defined by its graded mean as $R(\tilde{A}) = \left[\frac{(a_2 + a_3)}{2} + \frac{(a_* - a^*)}{4}\right]$, where T(R) is the set of all trapezoidal fuzzy numbers defined on R. A trapezoidal fuzzy number $\tilde{A} = (a_1,a_2,a_3,a_4)$ in T(R) is said to be positive, if and only if $\Re(\tilde{A}) > 0$ and is denoted by $\tilde{A} \succ \tilde{0}$. for any two trapezoidal fuzzy number $\tilde{A} = (a_1,a_2,a_3,a_4)$ and $\tilde{B} = (b_1, b_2, b_3, b_4)$ in T(R), we have the following comparison:

(i). $\tilde{A} \succeq \tilde{B}$ if and only if $\Re(\tilde{A}) \geq \Re(\tilde{B})$

(ii). $\tilde{A} \preceq \tilde{B}$ if and only if $\Re(\tilde{A}) \leq \Re(\tilde{B})$

(iii). $\tilde{A} \approx \tilde{B}$ if and only if $\Re(\tilde{A}) = \Re(\tilde{B})$

2.2. Arithmetic operations on trapezoidal fuzzy number

For arbitrary trapezoidal fuzzy numbers $\tilde{A} = (\langle m(\tilde{a}), w(\tilde{a})\rangle, a_*, a^*)$ and $\tilde{B} = (\langle m(\tilde{b}), w(\tilde{b})\rangle, a_*, a^*)$ and $* = \{+, -, \times, \div\}$, the arithmetic operations on the trapezoidal fuzzy numbers are defined by,

$$\tilde{A} \circ \tilde{B} = (m(\tilde{a}) \circ m(\tilde{b}), w(\tilde{a}) \vee w(\tilde{b}), a_* \vee b_*, a^* \vee b^*)$$

$$= (m(\tilde{a}) \circ m(\tilde{b}), \max\{w(\tilde{a}), w(\tilde{b})\}, \max\{a_*, b_*\}, \max\{a^*, b^*\})$$

In particular, for any two trapezoidal fuzzy numbers, $\tilde{A} = (\langle m(\tilde{a}), w(\tilde{a})\rangle, a_*, a^*)$ and $\tilde{B} = (\langle m(\tilde{b}), w(\tilde{b})\rangle, b_*, b^*)$ we define,

Addition: $$\tilde{A}+\tilde{B}=\begin{pmatrix} m(\tilde{a})+m(\tilde{b}), \max\{w(\tilde{a}), w(\tilde{b})\}, \\ \max\{a_*, b_*\}, \max\{a^*, b^*\} \end{pmatrix}$$

Subtraction: $$\tilde{A}-\tilde{B}=\begin{pmatrix} m(\tilde{a})-m(\tilde{b}), \max\{w(\tilde{a}), w(\tilde{b})\}, \\ \max\{a_*, b_*\}, \max\{a^*, b^*\} \end{pmatrix}$$

Multiplication: $$\tilde{A}\times\tilde{B}=\begin{pmatrix} m(\tilde{a})\times m(\tilde{b}), \max\{w(\tilde{a}), w(\tilde{b})\}, \\ \max\{a_*, b_*\}, \max\{a^*, b^*\} \end{pmatrix}$$

Division: $$\tilde{A}\div\tilde{B}=\begin{pmatrix} m(\tilde{a})\div m(\tilde{b}), \max\{w(\tilde{a}), w(\tilde{b})\}, \\ \max\{a_*, b_*\}, \max\{a^*, b^*\} \end{pmatrix}$$

3. Network Scheduling and Planning

Any individuals operation, which utilizes resources and has an end and a beginning, is called activity. An event represents the start and the end point of an activity. It is shown by a circle called nodes.

3.1 Fuzzy Critical Path Analysis

The objective of fuzzy critical path analysis is to estimate the total project duration and to assign starting and finishing time of all activities involved in the project. This helps to check the actual progress against the scheduled duration of the project.

The following factors in fuzzy should be known in order to prepare the project scheduling

(1) Total completion time of the project.
(2) Earlier and latest start time of each activity.
(3) Critical activities and critical path.
(4) Float for each activity (i.e), the amount of time by which the completion of a non-critical activity can be delayed, without delaying the total project completion time.

3.2. Notations to represent fuzzy factors

$FE\tilde{S}_{ij}$ = Fuzzy Earliest start time of an activity (i, j)

$FL\tilde{S}_{ij}$ = Fuzzy Latest start time of an activity (i, j)

$FE\tilde{F}_{ij}$ = Fuzzy Earliest finish time of an activity (i, j)

$FL\tilde{F}_{ij}$ = Fuzzy Latest finish time of an activity (i, j)

$F\tilde{t}_{ij}$ = Fuzzy Duration of an activity (i, j)

3.3. Fuzzy Critical Path Algorithm

Step 1: Construct the fuzzy project network with predecessor and successor events.

Step 2: Calculate fuzzy earliest start time, $FE\tilde{S}_{ij}=\max_i\left[FE\tilde{S}_i+\tilde{t}_{ij}\right]$, i = number of preceding nodes.

Step 3: Calculate fuzzy earliest finish time, $FE\tilde{F}_{ij}=\left[FE\tilde{S}_{ij}+\tilde{t}_{ij}\right]$.

Step 4: Calculate fuzzy latest finish time, $FL\tilde{F}_{ij}=\min_j\left[FL\tilde{F}_{ij}-\tilde{t}_{ij}\right]$, j = number of succeeding nodes.

Step 5: Calculate fuzzy latest start time, $FL\tilde{S}_{ij}=\left[FL\tilde{F}_{ij}-\tilde{t}_{ij}\right]$.

Journal of Physics: Conference Series **1377** (2019) 012044 doi:10.1088/1742-6596/1377/1/012044

Step 6: Calculate fuzzy total floating time (i.e), $FT\tilde{F}_{ij} = \left[FL\tilde{F}_{ij} - FE\tilde{F}_{ij}\right]$ or $FT\tilde{F}_{ij} = \left[FL\tilde{S}_{ij} - FE\tilde{S}_{ij}\right]$

Step 7: If $FT\tilde{F}_{ij} = 0$ then those activities are called as fuzzy critical activities.

Step 8: Connect all the fuzzy critical activities from source node to destination node. The path connecting all the fuzzy critical activities is called as fuzzy critical path.

Step 9: The maximum duration to complete the fuzzy network project is obtained by adding all the fuzzy activity duration of the critical activities in the critical path.

4. Numerical example

The proposed technique is illustrated by the following example. Consider fuzzy networking problem were each activity duration is trapezoidal fuzzy number.

Table 1: Fuzzy network problem with fuzzy activity duration as trapezoidal fuzzy number

Activity	1-2	1-3	2-4	2-5	3-4
Fuzzy duration	(25,28,32,35)	(40,55,65,70)	(32,37,43,48)	(35,38,42,45)	(20,25,35,40)
Activity	3-6	4-7	5-7	6-7	
Fuzzy duration	(42,45,49,60)	(60,65,75,85)	(65,75,85,90)	(15,18,22,26)	

Table 2: Trapezoidal fuzzy activity duration of each activity is represented as $\left(\langle m(\tilde{a}), w(\tilde{a})\rangle, a_*, a^*\right)$

Activity	1-2	1-3	2-4	2-5	3-4
Fuzzy duration	<30,2,3,3>	<60,5,15,5>	<40,3,5,5>	<40,2,3,3>	<30,5,5,5>
Activity	3-6	4-7	5-7	6-7	
Fuzzy duration	<47,2,3,11>	<70,5,5,10>	<80,5,10,5>	<20,2,3,4>	

Table 3: Total float for each activity in a Trapezoidal Fuzzy Project Network and Critical Path

Activity	Fuzzy duration	Fuzzy Earliest Start	Fuzzy Earliest finish	Fuzzy Latest Start	Fuzzy Latest Finish	Fuzzy Total Float
(1,2)	<30,2,3,3>	<0,0,0,0>	<30,2,3,3>	<10,5,15,10>	<40,5,15,10>	<10,5,15,10>
(1,3)	<60,5,15,5>	<0,0,0,0>	<60,5,15,5>	<0,5,15,10>	<60,5,15,10>	**<0,5,15,10>**
(2,4)	<40,3,5,5>	<30,2,3,3>	<70,3,5,5>	<50,5,15,10>	<90,5,15,10>	<20,5,15,10>
(2,5)	<40,2,3,3>	<30,2,3,3>	<70,2,3,3>	<40,5,15,10>	<80,5,15,10>	<10,5,15,10>
(3,4)	<30,5,5,5>	<60,5,15,5>	<90,5,15,5>	<60,5,5,10>	<90,5,15,10>	**<0,5,15,10>**
(3,6)	<47,2,3,11>	<60,5,15,5>	<107,5,15,11>	<93,5,15,11>	<140,5,15,10>	<33,5,5,11>
(4,7)	<70,5,5,10>	<90,5,15,5>	<160,5,15,10>	<90,5,15,10>	<160,5,15,10>	**<0,5,15,10>**
(5,7)	<80,5,10,5>	<70,2,3,3>	<150,5,10,5>	<80,5,15,10>	<160,5,15,10>	<10,5,15, 10 >
(6,7)	<20,2,3,4>	<107,5,15,11>	<127,5,15,11>	<140,5,15,10>	<160,5,15,10>	<33,5,15,11>

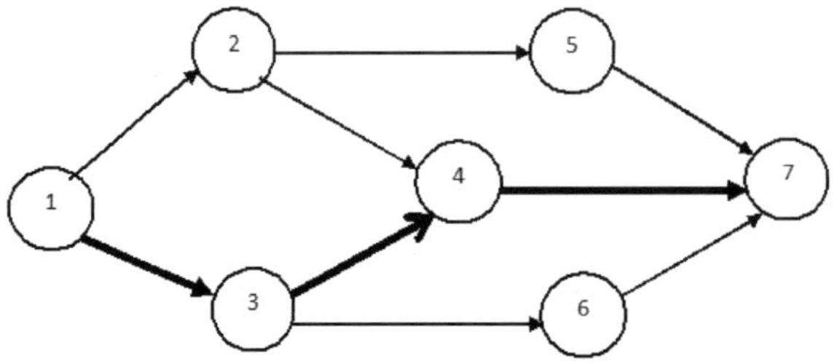

Figure. 4.1. Critical Path for Fuzzy Networking Problem.

4.1. Result

According to the fuzzy total float, the fuzzy critical activities are 1-3, 3-4 and 4-7.

Therefore, connecting all the fuzzy critical activities we get the critical path for fuzzy project network problem as $1 \to 3 \to 4 \to 7$.

Thus, the total duration of completion of the fuzzy networking problem is (160,5, 15,10).

Hence the optimal fuzzy project duration in terms of trapezoidal fuzzy number
$(a_1, a_2, a_3, a_4) = (140,155,165,175)$.

Our results are better comparing with theproject duration obtained by Revathi and Saravanan [6] , where they have converted the fuzzy duration of trapezoidal fuzzy number into crisp number.

5. Conclusion

In this paper, we proposed a new critical path algorithm, to obtain an optimal duration of completion of a fuzzy networking problem with trapezoidal fuzzy number, without converting the problem as a classical networking problem. The numerical illustration clearly shows that, the proposed methodology is more efficient than the existing method and provides a better solution to the fuzzy networking problem.

6. References

[1] Buckley JJ1988 possibility linear programming with triangular fuzzy numbers, *Fuzzy sets and system*, **26**135-138
[2] Dubois D and PradeH1980*Fuzzy sets and system theory and applications*Academic press, Newyork.
[3] Kikuchi S2000 A method to defuzzify the fuzzy numbers, Transportation problem application, *Fuzzy sets and system***11**63-6
[4] Rajendran C andAnandanarayananM2015 Fuzzy Critical Path method using octagonal fuzzy numbers*International Journal of Scientific Engineering and Research***5** issue 10 pp.15-18
[5] Revathi M and Saravanan R 2013Optimal solution for fuzzy networking problem using ranking method, *International Journal of Current Research***5**, Issue 07, pp. 1794-1796
[6] Selvakumari K and Sowmiya G 2017 Fuzzy networking problems using job sequencing technique in hexagonal fuzzy numbers, *International Journal of Advance Engineering and Research Development*. **4** Issue 9
[7] Shih -Pin Chen and Yi-JuHsueh 2008 A simple approach to fuzzy critical path analysis in project network, *Applied Mathematical Modelling***32** 1289-1297
[8] Stefan Chan and Pawel Zibeline 2001 Critical path analysis in the network with fuzzy activities times*Fuzzy sets and systems* 195-204

[9] Stephan Dingar D and Rameshan N 2018 A new approach for fuzzy critical path method using octagonal fuzzy numbers*International Journal of Pure and Applied Mathematics***119** No. 13, 357-364

[10] Zadeh L A1965Fuzzy sets, *Information and control*, **8**, pp.338-353

Fuzzysequencing problemwith uncertain processing time

T. Leelavathy[1]andM. Kowsalya[2]

[1]Department of Mathematics,
Faculty of Engineering and Technology,
SRM Institute of Science and Technology,
Kattankulathur-603203, India.
[2] Department of Mathematics, SRM Institute of Science and Technology,
Kattankulathur-603203, India.

E-mail: leelavat@srmist.edu.in and
kowsalya0318@gmail.com

Abstract. In this paper, we suggested a new approach to obtained an optimal sequence for fuzzy sequencing problem with trapezoidal fuzzy number. Using new fuzzy arithmetic operation and ranking method, an improved algorithm is proposed to obtain the fuzzy total elapsed time for processing all the jobs without converting problem into classical form. The approach is efficiently illustrated with a numerical example.

1. Introduction

A sequencing problem specifies the optimal sequence (order) in which 'n' jobs to be processed by 'm' machine. It determines an appropriate order for a service of jobs to be done on a finite number of service facilities, so as to minimise the total time taken for finishing all the jobs. A sequencing problem could involve, jobs in a manufacturing plant, aircraft waiting for landing and clearance, maintenance scheduling in a factory, programmes to be run on a computer, customer in bank and so-on.

Johnson proposed an exact algorithm to get the minimum make span for two machine problem in 1954. The Johnson algorithm has also been extended to m-machine problem with make span as the objective. But it is difficult to apply those conventional approaches to real world problem.

In real life situation the information available is of imprecise nature and there is an uncertainty present in the problem. In order to tackle this uncertainty the concept of fuzzy sets can be used as an important decision making tool. These imprecise data may be represented by fuzzy numbers. The index of fuzzy set was introduced by Zadeh [11] in 1965.

K. Kripa et al [5] have discussed an algorithm to obtain an optimal sequence of a fuzzy sequencing problem with generalised triangular fuzzy numbers by converting fuzzy sequencing problem into crisp sequencing problem. S. U. Malini et al [7] proposed a ranking to solve fuzzy sequencing problem with octagonal fuzzy number. K. Selvakumari et al [10] framed the fuzzy sequencing problem where processing time taken as octagonal fuzzy number and pentagonal fuzzy number and solved by robust ranking method. LaxminarayanSahoo [6] used yagers ranking index method to fuzzy job sequencing problem, where processing time being taken as trapezoidal fuzzy numbers. R. Jansi et al [3] used Pascal's algorithm to obtain optimal sequence of the fuzzy sequencing problem with trapezoidal fuzzy number.

In this paper, we proposed a new algorithm to find the optimal sequence for the fuzzy sequencing problem with trapezoidal fuzzy number, without converting in to a classical problem. In section 2, we review the basic definitions, ranking of trapezoidal fuzzy number and arithmetic operation of trapezoidal fuzzy number. In section 3, we proposed an improved algorithm to find the optimal sequence and total elapsed time of the sequencing problem with trapezoidal fuzzy number. In section 4, a numerical problem is provided to show the competence of methodology.

2. Preliminaries

In this section, we represent the idea about the fuzzy numbers and definitions which involved in the research work. Fuzzy set theory have been defined by Zadeh [11] and the fundamental concepts are given by Dubois and Prade [1] are applied to fuzzy environment.

Definition 2.1

A fuzzy set $\tilde{A}$ in X is a set of ordered pair defined by, $\tilde{A}(x) = \{(x, \mu_{\tilde{A}}(x)) : x \in X, \mu_{\tilde{A}}(x) \in [0,1]\}$ where $\mu_{\tilde{A}}(x)$ is a membership function.

Definition 2.2

A fuzzy set $\tilde{A}$ defined on a set of real number R is said to be a fuzzy number, if its membership function. $\tilde{A} : R \rightarrow [0,1]$ has the following characteristic:

1) $\tilde{A}$ is convex (i.e.,) $\tilde{A}\{\lambda x_1 + (1-\lambda)x_2\} \geq \min\{\tilde{A}(x_1), \tilde{A}(x_2)\}$ for all $x_1, x_2 \in R$ and $\lambda \in [0,1]$

2) $\tilde{A}$ is normal, i.e., there exist an $x \in R$ such that $\tilde{A}(x) = 1$

3) $\tilde{A}$ is piecewise continuous.

Definition 2.3

A fuzzy number $\tilde{A}$ is a trapezoidal fuzzy number denoted by $\tilde{A} = (a_1,a_2,a_3,a_4)$ where a_1,a_2,a_3,a_4 are real numbers and its membership function $\tilde{A}(x)$ is given below,

$$\tilde{A}(x) = \begin{cases} \dfrac{x-a_1}{a_2-a_1}, & \text{if } a_1 \leq x \leq a_2 \\ 1, & \text{if } a_2 \leq x \leq a_3 \\ \dfrac{a_{4-x}}{a_4-a_3}, & \text{if } a_3 \leq x \leq a_4 \\ 0, & \text{otherwise} \end{cases}$$

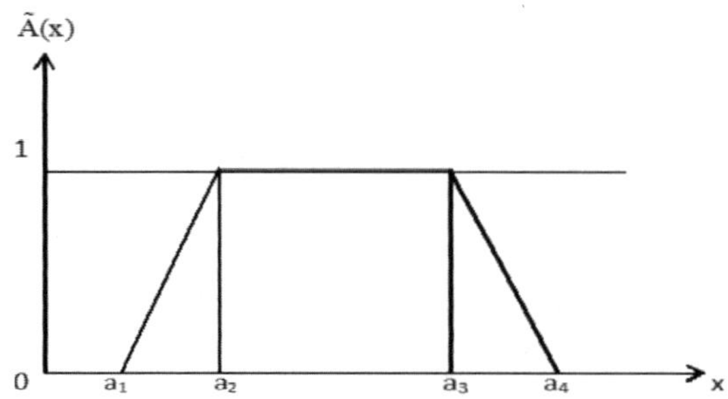

Figure 2.1. The membership function of Trapezoidal Fuzzy Number

In this paper, the trapezoidal fuzzy number is represented as, $\tilde{A} = (a_1,a_2,a_3,a_4) = (m,w,a^*,a_*)$ where $m = \frac{(a_1 + a_3)}{2}$ and $w = \frac{(a_3 - a_2)}{2}$ are the midpoint and width of the core $[a_2,a_3]$ respectively. Also $a_* = (a_2 - a_1)$ and $a^* = (a_4 - a_3)$ represents the left and right spreads respectively of the trapezoidal fuzzy number $\tilde{A} = (a_1, a_2, a_3, a_4)$.

2.1. RANKING ON TRAPEZOIDA L FUZZY NUMBER

The ranking function $\Re:F(R)\to R$ is defined by its graded mean as $R(\tilde{A})=\left[\frac{(a_2+a_3)}{2}+\frac{(a_* - a^*)}{4}\right]$,

where T(R) is the set of all trapezoidal fuzzy numbers defined on R. A trapezoidal fuzzy number $\tilde{A} = (a_1,a_2,a_3,a_4)$ in T(R) is said to be positive, if and only if $\Re(\tilde{A})>0$ and is denoted by $\tilde{A} \succ \tilde{0}$. for any two trapezoidal fuzzy number $\tilde{A} = (a_1,a_2,a_3,a_4)$ and $\tilde{B} = (b_1, b_2, b_3, b_4)$ in T(R), we have the following comparison:

(i). $\tilde{A} \succeq \tilde{B}$ if and only if $\Re(\tilde{A}) \geq \Re(\tilde{B})$

(ii). $\tilde{A} \preceq \tilde{B}$ if and only if $\Re(\tilde{A}) \leq \Re(\tilde{B})$

(iii). $\tilde{A} \approx \tilde{B}$ if and only if $\Re(\tilde{A}) = \Re(\tilde{B})$

2.2. ARITHMATIC OPERATION ON TRAPEZOIDAL FUZZY NUMBER

For arbitrary trapezoidal fuzzy numbers $\tilde{A}=(\langle m(\tilde{a}),w(\tilde{a})\rangle, a_*, a^*)$ and $\tilde{B}=(\langle m(\tilde{b}),w(\tilde{b})\rangle, a_*, a^*)$ and $* = \{+,-,\times,\div\}$, the arithmetic operations on the trapezoidal fuzzy numbers are defined by,

$$\tilde{A}\circ\tilde{B}=\left(m(\tilde{a})\circ m(\tilde{b}), w(\tilde{a})\vee w(\tilde{b}), a_*\vee b_*, a^*\vee b^*\right)$$

$$=\left(m(\tilde{a})\circ m(\tilde{b}), \max\{w(\tilde{a}),w(\tilde{b})\}, \max\{a_*,b_*\}, \max\{a^*,b^*\}\right)$$

In particular, for any two trapezoidal fuzzy numbers, $\tilde{A} = (\langle m(\tilde{a}), w(\tilde{a})\rangle, a_*, a^*)$ and $\tilde{B} = (\langle m(\tilde{b}), w(\tilde{b})\rangle, b_*, b^*)$ we define,

Addition: $$\tilde{A} + \tilde{B} = \begin{pmatrix} m(\tilde{a}) + m(\tilde{b}), \max\{w(\tilde{a}), w(\tilde{b})\}, \\ \max\{a_{*}, b_{*}\}, \max\{a^{*}, b^{*}\} \end{pmatrix}$$

Subtraction: $$\tilde{A} - \tilde{B} = \begin{pmatrix} m(\tilde{a}) - m(\tilde{b}), \max\{w(\tilde{a}), w(\tilde{b})\}, \\ \max\{a_{*}, b_{*}\}, \max\{a^{*}, b^{*}\} \end{pmatrix}$$

Multiplication: $$\tilde{A} \times \tilde{B} = \begin{pmatrix} m(\tilde{a}) \times m(\tilde{b}), \max\{w(\tilde{a}), w(\tilde{b})\}, \\ \max\{a_{*}, b_{*}\}, \max\{a^{*}, b^{*}\} \end{pmatrix}$$

Division: $$\tilde{A} \div \tilde{B} = \begin{pmatrix} m(\tilde{a}) \div m(\tilde{b}), \max\{w(\tilde{a}), w(\tilde{b})\}, \\ \max\{a_{*}, b_{*}\}, \max\{a^{*}, b^{*}\} \end{pmatrix}$$

3. Fuzzy Sequencing Problem

A suitable order in which the number of jobs can be assigned to a finite number of service facilities is called sequencing. The jobs or items are the primary stimulus for sequencing. There should be a certain number of jobs say 'n' to be processed or sequenced. A machine is characterised by a certain processing capability of facility through which a job must pass before it is completed in the shop. There must be certain number of machines say 'k' to be used for processing the jobs. In reality, the duration has vagueness or uncertainty, which is managed by fuzzy set theory Zadeh [11]. The sequencing problem with uncertain processing time is termed as fuzzy sequencing problem. Using fuzzy set theory with arithmetic operations, a new algorithm is proposed to sequence the jobs to be processed in various machines, with minimum total processing time. The assumptions for the classical problem are also applicable for fuzzy sequencing problem.

3.1 Notations

d_{ij}– Processing duration for job i on machine j .

T - Total elapsed time for processing all the jobs.

I_{ij}- Idle time on machine j from the end of job (i-1) to the start of job i.

3.2Proposed Algorithm

Let there be n jobs, each of which is to be processed through m machines $M_1, M_2, \ldots\ldots, M_m$ in the order $M_1, M_2, \ldots\ldots, M_m$. Let $\tilde{d}_{ij}$ be the fuzzy processing time to complete the ith job on j[th] machine.The optimal solution can be obtained if either or both of the following condition hold good,

i) $\min\{\tilde{d}_{1j}\} \geq \max\{\tilde{d}_{ij}\}$; $j = 2,3,\ldots\ldots,m-1$ *and / or*

(ii) $\min\{\tilde{d}_{mj}\} \geq \max\{\tilde{d}_{ij}\}$; $j = 2,3,\ldots\ldots,m-1$

That is the minimum fuzzy processing time on machines M_1 and M_m is as great as the maximum fuzzyprocessing time on any of the remaining (m-1) machines. If either or both these conditions hold, then proceed with the steps of the algorithm.

In particular, we consider n jobs each of which is to be processed through 3 machines M_1, M_2, M_3. The condition for optimality is, minimum fuzzy processing time on machines M_1 and M_3 should be greater than or equal to the maximum fuzzy processing time on machine M_2.

Journal of Physics: Conference Series **1377** (2019) 012045 doi:10.1088/1742-6596/1377/1/012045

The algorithm is summarized as follows:

STEP 1: Observe the fuzzy processing times of the given jobs on all three machines. If either or both the conditions mentioned above hold, then go to step2 otherwise the algorithm fails.

STEP2: Introducing two mythical machines R and L such that, $\tilde{d}_{iR} = \tilde{d}_{i1} + \tilde{d}_{i2} + ... + \tilde{d}_{im-1}$ *for* $i=1,2,.......,n$

$\tilde{d}_{iL} = \tilde{d}_{i2} + \tilde{d}_{i3} + ... + \tilde{d}_{im}$ *for* $i=1,2,.......,n$

STEP 3: Examine the fuzzy processing time $\tilde{T}_{i1}$ *and* $\tilde{T}_{i2}$ *for* $i = 1,2,........,n$ of the transferred two machine problem and identify $\min(\tilde{T}_{i1}, \tilde{T}_{i2})$.

STEP 4: If the minimum fuzzy processing time appears in the first machine then place the job in the first available position in the sequence. If the minimum fuzzy processing time appears in the second machine then place the job at the last position in the sequence. If there is a tie then select the job corresponding to the largest job subscript and place it in the right position in the sequence.

STEP 5: Cross off assigned jobs from the table and repeat steps 3 & 4, until all the jobs have been assigned. Therefore, the obtained sequence is the optimal sequence (order) in which the jobs are to be processed on all the machines.

STEP 6: The total elapsed time to process all the jobs through the machines, is the time when the nth job in a sequence finishes on last machine.

4. Numerical Example

Consider a fuzzy sequencing problem with the processing time as trapezoidal fuzzy numbers are given in the following table.

Table 1: Fuzzy sequencing problem with trapezoidal fuzzy number

Jobs /Machine	M_1	M_2	M_3
1.	(4,6,8,14)	(2,4,5,9)	(2,3,4,7)
2.	(7,8,10,15)	(3,6, 7,8)	(5,9,10,12)
3.	(3,6,7,8)	(0,1,2,5)	(4,6,8,14)
4.	(3,7,8,10)	(1,2,4,5)	(3,6,7,8)
5.	(8,10,12,14)	(2,3,4,7)	(2,4,5,9)

Table 2: Trapezoidal fuzzy number is represented as $\left(m(\tilde{a}), w(\tilde{a}), a_*, a^*\right)$

Jobs/ Machine	M_1	M_2	M_3

1.	(7,1,2,6)	(4.5,0.5,2,4)	(3.5,0.5,1,3)
2.	(9,1,1,5)	(6.5,0.5,3,1)	(9.5,0.5,4,2)
3.	(6.5,0.5,3,1)	(1.5,0.5,1,3)	(7,1,2,6)
4.	(7.5,0.5,4,2)	(3,1,1,1)	(6.5,0.5,3,1)
5.	(11,1,2,2)	(3.5,0.5,1,3)	(4.5,0.5,2,4)

Table 3: Three machine problems is converted into two Machine problem

Jobs	J1	J2	J3	J4	J5
R	(11.5,1,2,(	(15.5,1,3,5	(8,0.5,3,3	(10.5,1,4,	(14.5,1
L	(8,0.5,2,4)	(16,0.5,3,2	(8.5,1,2,6	(9.5,1,3,1	(8,0.5,

Table 4: Fuzzy optimal sequence

J3	J2	J4	J1	J5

Table 5: Fuzzy total elapsed time

Jobs	Machine M_1		Machine M_2		Machine M_3	
	Time In	Time Out	Time In	Time Out	Time In	Time Out
J1	(0, 0 ,0 ,0)	(6.5, 0.5, 3, 1)	(6.5, 0.5, 3, 1)	(8, 0.5, 3, 3)	(8, 0.5, 3, 3)	(15, 1, 3, 6)
J2	(6.5, 0.5, 3, 1)	(15.5,1, 3, 5)	(15.5, 1, 3, 5)	(22, 1, 3, 5)	(22, 1, 3, 5)	(31.5, 1, 4, 5)
J3	(15.5,1, 3, 5)	(23, 1, 4, 5)	(23, 1, 4, 5)	(26, 1, 4, 5)	(31.5, 1, 4, 5)	(38, 1, 4, 5)
J4	(23, 1, 4, 5)	(30, 1, 4, 6)	(30, 1, 4, 6)	(34.5, 1, 4, 6)	(38, 1, 4, 5)	(41.5, 1, 4, 5)
J5	(30, 1, 4, 6)	(41, 1, 4, 6)	(41, 1, 4, 6)	(44.5, 1, 4, 6)	(44.5, 1, 4, 6)	(49, 1, 4, 6)

4.1 Results

Therefore, total elapsed time for processing all jobs through the machines = (49, 1, 4, 6)

The total elapsed time in terms of trapezoidal fuzzy number $(a_1, a_2, a_3, a_4) = (44, 48, 50, 56)$

Idle time on machine M_1 = (8, 1, 4, 6)

Idle time on machine M_2 = (30, 1, 4, 6)

Idle time on machine M_3 = (18, 1, 4, 6)

But the total elapsed time for processing all jobs through the machines obtained by LaxminarayanSahoo [10] is 51 hours, where as we obtain the result as (49, 1, 4, 6), by using the ranking function we get 49.5 hours. Hence the proposed algorithm and the defined fuzzy arithmetic operation reduced the time taken to process all the jobs, which is more efficient than the existing one.

5. Conclusion

In this paper, we consider the processing duration as trapezoidal fuzzy number for a fuzzy sequencing problem. We proposed an improved algorithm, to obtain an optimal sequence and total elapsed time to process all jobs through machines, without converting the problem as a classical sequencing problem. The numerical illustration clearly shows that, the proposed methodology is more efficient than the existing method.

6. Reference

[1] Dubois D. and Prade H. 1980*Fuzzy sets and system theory and applications*, Academic press, Newyork.

[2] George Klir J Bo Yuan 1995 Fuzzy sets and Fuzzy Logic Theory and Application *Prentice Hall*Inc 574p

[3] Jansi and Mohana K 2016 A Heptagonal Fuzzy Number in Solving Fuzzy Sequencing Problem *International Journal of Innovative Research in Technology***5** Issue 6

[4] Kripa Kand Govindarajan R 2016 Fuzzy Sequencing Problem *International Journal of Scientific Research***5** Issue: 5

[5] Kripa K and Govindarajan R 2016 Fuzzy Sequencing Problem using Generalized Triangular Fuzzy Numbers *International Journals of Engineering Research and Application***6** Issue 6 (part-1) pp.61-64

[6] LaxminarayanSahoo 2017 Solving Fuzzy Sequencing Problems with Fuzzy Processing Times *International Journal of Advance Research and Innovative Idea in Education***3** Issue-4

[7] Malini SU and Kalaivani S 2018 Solving Fuzzy Sequencing Problem using Octagonal Fuzzy Number *International Journal of Mathematical Archive***9**(9) 109-114

[8] Selvakumari K And Sowmiya S 2017 Fuzzy Network Problem using Job Sequencing Technique in Hexagonal Fuzzy Number *International Journal of Advance Engineering and Research Development***4**(9).

[9] Selvakumari K and Santhi S 2018 A Pentagonal Fuzzy Number Solving Fuzzy Sequencing Problem *International Journal of Mathematics and its Application***6**(2-B) 207-211

[10] Selvakumari K and Santhi S 2018 An Approach for Solving Fuzzy Sequencing Problems with Octagonal Fuzzy Numbers using Robust Ranking Techniques *International Journal of Mathematics Trends and Technology***56** Number 3

[11] Zadeh L A 1965 *Fuzzy Sets Information and Control***8**pp 338-353

[12] Zimmermann H J 1991 *Fuzzy set Theory and Its Application* Boston: Kulwer

THE HUNGARIAN METHOD FOR THE ASSIGNMENT PROBLEM, WITH GENERALIZED INTERVAL ARITHMETIC AND ITS APPLICATIONS

D Priya[a] and G Ramesh[b] *
Department of Mathematics,
SRM Institute of Science and Technology,
Kattankulathur, Chennai - 603203, INDIA
Corresponding author: **[b] rameshapg@gmail.com;**
[a] priyad.vnb@gmail.com

Abstract: In this paper, we focus on the solution procedure for fully interval assignment problem (FIAP), Hungarian method is considered into account. In consideration method the given FIAP is decayed into Interval Assignment Problem, solving it with existing method and by using its optimal solutions, an optimal interval assignment solution to the given FIAP is obtained. Although the assignment problem can be solved as an ordinary transportation problem or as Linear programming problem, its unique structure can be exploited, resulting in special purpose algorithm, is called Hungarian method. A numerical example is provided to illustrate the solution procedure developed in this paper.

Keywords: Interval numbers, Assignment problems, Hungarian assignment problem.

1. INTRODUCTION

Assignment problem describes that one individual be able to perform one work at a time, to get the optimal solution by maximize the total profit (or) minimize the total cost. Assignment problem can be declared in the form of m x n matrix (c_{ij}) also known as Cost matrix, where c_{ij} is the cost of assigning i^{th} device to j^{th} job as 'n' works to be performed on 'm' machine (one work to one machine). This is also known as Fully interval Assignment problem, it can be solved by using a known process called Hungarian method . Jayalakshmi [6] introduced computation of intervals without using arithmetic operations, as decomposed FIAP into crisp AP using midpoint technique. Sarangam Majumdar [11] proposed interval Assignment method to solve real world linear problems whose cost matrix are in interval form. Ramesh Kumar A et.al [8][9] proposed the application of assignment problem and converting crisp AP into interval forms. For both maximization and minimization of Hungarian method using its extension of intervals was solved by Amutha et.al[1], Deepa et.al[3]introduced a new method known as Best assignment then solving it by intervals, Humayra Dil Afroz et.al[6] proposed a new assignment method comparative with existing method. Whereas Sundaresan et.al and Ramon.E.Moore[12][10] proposes a basic definitions of OR and existing intervals. In real world consumption, there are many values i.e for eg time, ability, etc are in crisp form which does not give a perfect value, In this paper using Interval arithmetic by Hungarian method, the problem can be solved more to get precisely with the basic assumption of one job to one person for minimizes total cost (or) maximizes the profit. The rest of the paper includes review on interval arithmetic are highlighted.

2. Preliminaries

This section includes some notations, notions which results in our further consideration. Let $\tilde{a}=[a_1,a_2]=\{x: a_1 \le x \le a_2,\ x \in R\}$. If $\tilde{a}=a_1=a_2=a$, then $\tilde{a}=[a,a]=a$ is a real number (or a degenerate interval). Let IR= $\{\tilde{a}=[a_1,a_2]: a_1 \le a_2$ and $a_1, a_2 \in R\}$ be the set of all proper intervals and IR $=\{\tilde{a}=[a_1,a_2]: a_1>a_2$ and $a_1,a_2 \in R\}$be the set of all improper intervals on the real line R. We shall use the terms interval and interval number interchangeably. The Midpoint and Width of an interval number $\tilde{a}=[a_1,a_2]$ are defined as $m(\tilde{a}) = \left(\frac{a_1 + a_2}{2}\right)$ and $w(\tilde{a}) = \left(\frac{a_2 - a_1}{2}\right)$. The interval number ã can also be expressed in terms of its midpoint and width as: $\tilde{a} = [a_1, a_2] = \langle m(\tilde{a}), w(\tilde{a})\rangle$.

2.1. A new interval arithmetic

Ming Ma et al. [7] have proposed a new fuzzy arithmetic based upon both fuzziness index and location index function. The fuzziness index functions are in use to follow the lattice rules then they are the least upper bound and greatest lower bound in the lattice L. That is for $a, b \in L$ we define $a \vee b = \max\{a, b\}$ and $a \wedge b = \min\{a, b\}$. whereas the location index number is considered as the ordinary arithmetic which includes basic concepts.
For two intervals $\tilde{a} = [a_1, a_2]$, $\tilde{b} = [b_1, b_2] \in IR$ and $* \in \{+, -, \cdot, \div\}$, the arithmetic operations on ã and $\tilde{b}$ are defined as:

$$\tilde{a} * \tilde{b} = [a_1, a_2] * [b_1, b_2] = \langle m(\tilde{a}), w(\tilde{a})\rangle * \langle m(\tilde{b}), w(\tilde{b})\rangle = \langle m(\tilde{a}) * m(\tilde{b}),\ \max\{w(\tilde{a}), w(\tilde{b})\}\rangle.$$

In Particular

(i). Addition: $\tilde{a} + \tilde{b} = \langle m(\tilde{a}), w(\tilde{a})\rangle + \langle m(\tilde{b}), w(\tilde{b})\rangle = \langle m(\tilde{a}) + m(\tilde{b}),\ \max\{w(\tilde{a}), w(\tilde{b})\}\rangle.$

(ii). Subtraction: $\tilde{a} - \tilde{b} = \langle m(\tilde{a}), w(\tilde{a})\rangle - \langle m(\tilde{b}), w(\tilde{b})\rangle = \langle m(\tilde{a}) - m(\tilde{b}),\ \max\{w(\tilde{a}), w(\tilde{b})\}\rangle.$

(iii). Multiplication: $\tilde{a} \times \tilde{b} = \langle m(\tilde{a}), w(\tilde{a})\rangle \times \langle m(\tilde{b}), w(\tilde{b})\rangle = \langle m(\tilde{a}) \times m(\tilde{b}),\ \max\{w(\tilde{a}), w(\tilde{b})\}\rangle.$

(iv). Division: $\tilde{a} \div \tilde{b} = \langle m(\tilde{a}), w(\tilde{a})\rangle \div \langle m(\tilde{b}), w(\tilde{b})\rangle = \langle m(\tilde{a}) \div m(\tilde{b}), \max\{w(\tilde{a}), w(\tilde{b})\}\rangle,$

provided $m(\tilde{b}) \neq \tilde{0}$.

2.2. Ranking of interval numbers

Sengupta and Pal [2]proposed a simple and efficient index for comparing any two intervals on IR through decision maker's satisfaction.

Defintion 2.2.1: Let $\preceq$ be an extended order relation between the interval numbers $\tilde{a} = [a_1, a_2]$, $\tilde{b} = [b_1, b_2]$ in IR, then for $m(\tilde{a}) < m(\tilde{b})$, we construct a premise ($\tilde{a} \circ \tilde{b}$) which implies that ã is inferior to $\tilde{b}$ (or $\tilde{b}$ is superior to ã).

An acceptability function is defined as: $A_{\preceq} : IR \times IR \rightarrow [0, \infty)$

$$A_{\preceq}(\tilde{a}, \tilde{b}) = A(\tilde{a} \preceq \tilde{b}) = \frac{m(\tilde{b}) - m(\tilde{a})}{w(\tilde{b}) + w(\tilde{a})}, \text{ where } w(\tilde{b}) + w(\tilde{a}) \neq 0.$$

may be interpreted as the grade of acceptability of the the first interval number to be inferior to the second interval number. For any two interval numbers ã and $\tilde{b}$ in IR either $A(\tilde{a} \preceq \tilde{b}) \geq 0$ (or) $A(\tilde{b} \succeq \tilde{a})$ (or) $A(\tilde{a} \preceq \tilde{b}) = 0$ (or) (or) $A(\tilde{a} \preceq \tilde{b}) + A(\tilde{b} \succeq \tilde{a}) = 0$. If $A(\tilde{a} \preceq \tilde{b}) = 0$ and $A(\tilde{b} \succeq \tilde{a}) = 0$, then

Journal of Physics: Conference Series **1377** (2019) 012046 doi:10.1088/1742-6596/1377/1/012046

we say that the interval numbers $\tilde{a}$ and $\tilde{b}$ are equivalent (non-inferior to each other) and we denote it by $\tilde{a} \approx \tilde{b}$. Also if $A(\tilde{a} \preceq \tilde{b}) \geq 0$, then $\tilde{a} \preceq \tilde{b}$ and if $A(\tilde{b} \preceq \tilde{a}) \geq 0$, then $\tilde{b} \preceq \tilde{a}$.

3. MAIN RESULTS

3.1. General Interval Assignment Problem

Let there are m work and n people available with dissimilar skills. If the cost of doing j^{th} work by i^{th} people is c_{ij}. Now the problem is which the work is to be assigned to whom so that the cost of completion of work will be minimum. Mathematically, we express the problem as follows:

Minimize $\tilde{Z}(cost) = \Sigma\, \Sigma\, \tilde{c}_{ij}\, \tilde{x}_{ij}$; i=1,2,…..,n; j=1,2,….,n

Where $\tilde{x}_{ij} = \begin{cases} 1; & \text{if } i^{th} \text{ person is assigned } j^{th} \text{ work} \\ 0; & \text{if } i^{th} \text{ person is not assigned the } j^{th} \text{ work} \end{cases}$

with the restrictions

$\Sigma\, \tilde{x}_{ij} = 1$; j =1,2 ,….,n

i.e., i^{th} person will do only one work

$\Sigma\, \tilde{x}_{ij} = 1$; i = 1,2,….,n

i.e., j^{th} work will be done only by one person

3.2. Interval Hungarian method

In this section an algorithm to solve assignment problem with generalized interval arithmetic using Hungarian method:

Step 1: Find out the mid values of each interval in the cost matrix.

Step 2: Subtract the interval which have smallest mid value in each row from all the entries of its row.

Step 3: Subtract the interval which have smallest mid value from those columns which have no intervals contain zero from all the entries of its column.

Step 4: Draw lines through appropriate rows and columns so that all the intervals contain zero of the cost matrix are covered and the minimum number of such lines is used.

Step 5: Test for optimality (i) If the minimum number of covering lines is equal to the order of the cost matrix, then optimality is reached. (ii) If the minimum number of covering lines is less than the order of the matrix, then go to step 6.

Step 6: Determine the smallest mid values of the intervals which are not covered by any lines. Subtract this entry from all uncrossed elements and add it to the crossing having an interval contain zero. Then go to step 4.

3.3. Tabular form of the problem

The cost matrix of the interval assignment problem is known in the chart

People / Jobs	1	2	3	….j	….n
1	$\tilde{C}_{11}$	$\tilde{C}_{12}$	$\tilde{C}_{13}$	$\tilde{C}_{1j}$	$\tilde{C}_{1n}$
2	$\tilde{C}_{21}$	$\tilde{C}_{22}$	$\tilde{C}_{23}$	$\tilde{C}_{2j}$	$\tilde{C}_{2n}$
. . . i	. . . $\tilde{C}_{i1}$	. . . $\tilde{C}_{i2}$	. . . $\tilde{C}_{13}$	. . . $\tilde{C}_{ij}$	. . . $\tilde{C}_{in}$
. . . n	. . . $\tilde{C}_{n1}$	. . . $\tilde{C}_{n2}$	. . . $\tilde{C}_{n3}$	. . . $\tilde{C}_{nj}$	. . . $\tilde{C}_{nn}$

4. NUMERICAL EXAMPLE

Example 1:

Let us consider an assignment method discussed by Ramesh kumar et al.[8].The problem is solved by interval Hungarian method as follows: A corporation has four salesmen and four open territories available for assignment, that they are not equally rich in their sales potential and also it is estimated that the usual salesman operation in each territory would bring in the following annual Sales:

Territory:	I	II	III	IV
Annual sales (RS):	60,000	50,000	40,000	30,000

It is considered that the four salesmen will differ in ability, that working under the same conditions their yearly is estimated would be proportionately as follows:

Salesman:	A	B	C	D
Proportion:	7	5	5	4

If the criterion is maximum expected total values, the intuitive answer is to assign the best salesman to the richest territory; the next best salesmen to the second richest territory and so on verify this answer by the Hungarian method.

Solution

Step 1: By Constructing the effectiveness of the matrix. By taking 10,000 as one unit as the sales proportion and the maximum sales matrix is obtained as follows:

Table (a) Sales in 10,000 rupees

Sales proportion

		6	5	4	3
		I	II	III	IV
7	A	42	35	28	21
5	B	30	25	20	15
5	C	30	25	20	15
4	D	24	20	16	12

The Value of C_{11} = Sales Proportion × Sales Territory

$$= 7\times6 = 42$$

By applying traditional assignment method Ramesh kumar et al. got an optimal as I,II,III,IV are assign to A,B,C,D with an maximum cost is [95,103]. They converted this assignment problem into interval Hungarian method. Now the cost matrix of the interval Hungarian problem is

Step 2: Crisp entries of Cost matrix

	I	II	III	IV
A	[41,43]	[34,36]	[27,29]	[20,22]
B	[29,31]	[24,26]	[19,21]	[14,16]
C	[29,31]	[24,26]	[19,21]	[14,16]
D	[23,25]	[19,21]	[15,17]	[11,13]

Applying their Assignment method, Ramesh kumar et al.[8] obtained the optimal assignment as A, B, C,D machines are assign I, II, III, IV operators respectively and the optimum assignment cost as [95,103]. After stating that the above assignment is optimal, they claim that the solution is not unique and another optimal solution can be obtained as A,B,C,D are assign to IV, II, III, I respectively and the minimum cost is [89,91]. Hence the result and their conclusion violate the concept of optimality.

Now we shall solve the same interval assignment problem given in the step 2 by applying the method in this paper. Let us express any two interval parameters as

$\tilde{a}$= $[a_1,a_2]$ in terms of midpoint and width as $\tilde{a}=[a_1,a_2]=\langle m(\tilde{a}),w(\tilde{a})\rangle$.

Now all the given interval assignment problem becomes

Step 3: Interval entries of Cost matrix

	I	II	III	IV
A	<42,1>	<35,1>	<28,1>	<21,1>
B	<30,1>	<25,1>	<20,1>	<15,1>
C	<30,1>	<25,1>	<20,1>	<15,1>
D	<24,1>	<20,1>	<16,1>	<12,1>

Step 4: Subtract the smallest interval mid value in each row and column from all the entries of its row and column

	I	II	III	IV
A	<9,1>	<6,1>	<3,1>	<0,1>
B	<3,1>	<2,1>	<1,1>	<0,1>
C	<3,1>	<2,1>	<1,1>	<0,1>
D	<0,1>	<0,1>	<0,1>	<0,1>

Step 5: Interval entries of Cost matrix

	I	II	III	IV
A	<8,1>	<5,1>	<2,1>	<0,1>
B	<2,1>	<1,1>	<0,1>	<0,1>
C	<2,1>	<1,1>	<0,1>	<0,1>
D	<0,1>	<0,1>	<0,1>	<1,1>

Step 6: Interval entries of Cost matrix

	I	II	III	IV
A	<7,1>	<4,1>	<2,1>	<0,1>
B	<1,1>	<0,1>	<0,1>	<0,1>
C	<1,1>	<0,1>	<0,1>	<0,1>
D	<0,1>	<0,1>	<1,1>	<2,1>

The optimal assignment schedule is given by

A→ IV, B→ II, C→ III, D→ I

The optimal assignment cost = <21,1> + <25,1> + <20,1> + <24,1>

= <90, 1>

= [89,91]

It is noted that our solution [89,91] is very much sharper than the solution [95,103] obtained by Ramesh kumar et al[8]

Example 2:

Beta Corporation has four plants each of which can manufacture any one of the four goods Production costs differ from one plant to another as do sales revenue. Given the revenue and cost data below, attain which product each plant should manufacture to maximize profit

Sales revenue (Rs. 000s product)

Plant

	1	2	3	4
A	50	68	49	62
B	60	70	51	74
C	55	67	53	70
D	58	65	54	69

Production cost (Rs. 000s product) plant

	1	2	3	4
A	49	60	45	61
B	55	63	45	69
C	52	62	49	68
D	55	64	48	66

Solution

Step 1: The profit matrix by using production cost sales revenue.

Profit= sales - cost

Profit matrix

	1	2	3	4
A	1	8	4	1
B	5	7	6	5
C	3	5	4	2
D	3	1	6	3

Step 2: Cost matrix with crisp entries"

	1	2	3	4
A	[0,2]	[7,9]	[3,5]	[0,2]
B	[4,6]	[6,8]	[5,7]	[4,6]
C	[2,4]	[4,6]	[3,5]	[1,3]
D	[2,4]	[0,2]	[5,7]	[2,4]

Applying the interval assignment method, Ramesh kumar et al. obtained the optimal solution as A,B,C, D as II, IV, I, III operators respectively and the optimum assignment cost is [18,26]. After stating the above assignment is optimal, they claim that the solution is not unique and another optimal solution can be obtained as I, III, IV, II are assign to A,B,C,D respectively and the minimum cost is [9,11]. Hence their result and their conclusion violate the concept of optimality.

Now we shall solve the same interval assignment problem given in the step 2 by applying the method in this paper. Let us state all the interval parameter

$\tilde{a}$= $[a_1,a_2]$ in terms of midpoint and width as $\tilde{a}=[a_1,a_2]=\langle m(\tilde{a}), w(\tilde{a})\rangle$.

Now all the given interval assignment problem becomes

Step 3: Cost matrix with interval entries

	1	2	3	4
A	<1,1>	<8,1>	<4,1>	<1,1>
B	<5,1>	<7,1>	<6,1>	<5,1>
C	<3,1>	<5,1>	<4,1>	<2,1>
D	<3,1>	<1,1>	<6,1>	<3,1>

Step 4: Subtract the least interval mid value in both row and column from all the entries of its row and column

	1	2	3	4
A	<0,1>	<7,1>	<2,1>	<0,1>
B	<0,1>	<2,1>	<0,1>	<0,1>
C	<1,1>	<3,1>	<1,1>	<0,1>
D	<2,1>	<0,1>	<4,1>	<2,1>

The optimal assignment schedule is given by

A→I, B→III, C→IV, D→II

The optimal assignment cost = <1,1> + <6,1> + <2,1> + <1,1>
= <10, 1>

= [9, 11]

It is noted that our solution [9, 11] is very much sharper than the solution [18, 26] obtained by Ramesh kumar et al.[8]

Example 3

An air line operates seven days a week has time table shown below. Crews must have a least layover (rest) time of 5 hrs, between flights. Obtain the pair of flights that minimizes layover time away from quarters. For any given the crews will e based at the city that effect is the least layover

Journal of Physics: Conference Series **1377** (2019) 012046 doi:10.1088/1742-6596/1377/1/012046

Delhi Jaipur			Jaipur Delhi		
Flight No	Depart	Arrive	Flight No	Depart	Arrive
1	7.00AM	8.00AM	101	8.00AM	9.15AM
2	8.00AM	9.00AM	102	8.30AM	9.45AM
3	1.30PM	2.30PM	103	12.00Noon	1.15PM
4	6.30PM	7.30PM	104	5.30PM	6.45PM

For each pair, state the town where the crews should be based

Solution

Step1: Constructing the table for layover times between flights that the crew is based at Delhi , Consider 15 mins = 1 unit

Flight arrives and Depart from Jaipur, The minimum layover is 5 hrs among flights.

	Flight no 1	Timing	
		Arrives	Depart
101	24 hrs = 96 units	8.00AM	8.00AM
102	24hrs+30 min =98 units	8.00AM	8.30AM
103	24hrs+4 min =112 units	8.00AM	12.00Noon
104	9 hrs+30 min =38 units	8.00AM	5.30PM

	Flight no 2	Timing	
		Arrives	Depart
101	23 hrs = 92 units	9.00AM	8.00AM
102	23hrs+30 min =94 units	9.00AM	8.30AM
103	24hrs+3 min =108 units	9.00AM	12.00Noon
104	8 hrs+30 min =34 units	9.00AM	5.30PM

	Flight no 3	Timing	
		Arrives	Depart
101	17 hrs +30 min = 70 units	2.30PM	8.00AM
102	18 hrs =72 units	2.30PM	8.30AM
103	21hrs+30 min =86 units	2.30PM	12.00Noon
104	24 hrs+3 hrs =108 units	2.30PM	5.30PM

	Flight no 4	Timing	
		Arrives	Depart
101	12 hrs + 30 min = 50 units	7.30PM	8.00AM
102	13hrs =52 units	7.30PM	8.30AM
103	16hrs+30 min =66 units	7.30PM	12.00Noon
104	22 hrs =88 units	7.30PM	5.30PM

Layover times when crew is based on Delhi (Table 1)

Flight No	101	102	103	104
1	96	98	112	38
2	92	94	108	34
3	70	72	86	108
4	50	52	66	88

Flight arrives and Depart from Delhi, The minimum place over is 5 hrs between flights.

	Flight no 101	Timing	
		Arrives	Depart (ND)
1	21 hrs + 45 min = 87 units	9.15AM	7.00AM
2	22hrs+45 min =91 units	9.15AM	8.00AM
3	28hrs+15 min =113 units	9.15AM	1.30PM
4	9 hrs+ 15 min =37 units	9.15AM	6.30PM

	Flight no 102	Timing	
		Arrives	Depart (ND)
1	21 hrs + 15 min = 85 units	9.45AM	7.00AM
2	22hrs+15 min =89 units	9.45AM	8.00AM
3	27hrs+45 min =111 units	9.45AM	1.30PM
4	8 hrs+ 45 min =35 units	9.45AM	6.30PM

	Flight no 103	Timing	
		Arrives	Depart (ND)
1	17 hrs + 45 min = 71 units	1.15PM	7.00AM
2	18hrs+45 min =75 units	1.15PM	8.00AM
3	24hrs+15 min =97 units	1.15PM	1.30PM
4	5 hrs+ 15 min =21 units	1.15PM	6.30PM

	Flight no 104	Timings	
		Arrives	Depart (ND)
1	12 hrs + 15 min = 49 units	6.45PM	7.00AM
2	13hrs+15 min =53 units	6.45PM	8.00AM
3	18hrs+45 min =75 units	6.45PM	1.30PM
4	23 hrs+ 45 min =95 units	6.45PM	6.30PM

Layover times when crew is based on Jaipur (Table 2)

Flight No	101	102	103	104
1	87	85	71	49
2	91	89	75	53
3	113	111	97	75
4	37	35	21	95

Step 2 : Construct the table for minimum layover times among flights with the help of Table 1and 2

Flight No	101	102	103	104
1	87	85	71	38
2	91	89	75	34
3	70	72	86	75
4	37	35	21	88

Step 3: Cost matrix with interval entries

Flight No	101	102	103	104
1	[86,88]	[84,86]	[70,72]	[37,39]
2	[90,92]	[88,90]	[74,76]	[33,35]
3	[69,71]	[71,73]	[85,87]	[74,76]
4	[36,38]	[34,36]	[20,22]	[87,89]

Step 4: Cost matrix with interval entries

Flight No	101	102	103	104
1	<87,1>	<85,1>	<71,1>	<38,1>
2	<91,1>	<89,1>	<75,1>	<34,1>
3	<70,1>	<72,1>	<86,1>	<75,1>
4	<37,1>	<35,1>	<21,1>	<88,1>

Step 5: Subtract the smallest entries in each row and column, we get

Flight No	101	102	103	104
1	<49,1>	<45,1>	<33,1>	<0,1>
2	<57,1>	<53,1>	<41,1>	<0.1>
3	<0,1>	<0,1>	<16,1>	<5,1>
4	<16,1>	<12,1>	<0,1>	<67,1>

<u>Step 6</u>:

Flight No	101	102	103	104
1	<4,1>	<0,1>	<0,1>	<0,1>
2	<12,1>	<8,1>	<8,1>	<0,1>
3	<0,1>	<0,1>	<28,1>	<50,1>
4	<4,1>	<0,1>	<0,1>	<100,1>

Flight 1→102

Flight 2→104

Flight 3→101

Flight 4→103

5. CONCLUSION:

In this paper, we have proposed interval versions of Hungarian method for solving the interval assignment problems without converting them to classical assignment problems. The main objective is to form the solution more precisely, Numerical examples solved and the results obtained are discussed .it is to be noted that the optimal solution obtained by our method is sharper than the solution obtained by others.

6. REFERENCES

[1]. Amutha S Lakshmi S Naramadha S 2017 Method of solving extension of interval in assignment problem *Int.J.Eng-Sci.comp* 7(4)10484-10488

[2]. Atanu Sengupta and Tapan Kumar Pal 2000 Theory and Methodology: On computing interval numbers *European journal of Operational research* PP 28-43

[3]. Deepa S and Ramesh kumar A 2018 New Interval Linear Assignment Problem *IOSR journal of Business and Management* PP 05-08

[4]. Deepa S and Ramesh Kumar A 2018 Solving Interval linear assignment problem using best assignment method, *IOSR journal of Engineering* PP 11-14

[5]. Humayra Dil afroz and Mohammad Anwar Hossen 2017 New proposed method for solving assignment problem and comparative study with the existing method *IOSR journal of mathematics* PP 84-88

[6]. Jayalakshmi M 2017 Solving fully interval assignment problems *IOP Conf series Materials Science And Engineering*

[7]. Ming Ma Menachem Friedman and Abraham kandel 1999 A new fuzzy arithmetic. Fuzzy sets and systems 108:83-90

[8]. Ramesh kumar A and Deepa S 2015 Variations of the interval linear assignment problems, *IOSR journal of Engineering* PP 16-25

[9]. Ramesh kumar A and Deepa S 2016 Solving ONE's interval linear assignment problems *Int.J.Eng-Res App*.6 (10)69-75

[10]. Ramon E.Moore Baker kearfott Michael J.Cloud 1971 Introduction to Interval analysis PP 1-15

[11]. Sarangam majumdar 2013 Interval linear assignment problems *Universal journal of applied mathematics* 1(1) 14-16

Journal of Physics: Conference Series **1377** (2019) 012046 doi:10.1088/1742-6596/1377/1/012046

[12]. Sundaresan V and Ganapathy Subramanian and K S Ganesan K 2013 Resource management techniques (OR)PP 8.1-8.49

Journal of Physics: Conference Series **1377** (2019) 012047 doi:10.1088/1742-6596/1377/1/012047

Fixed Point Theorem in Fuzzy Automata Normed Linear Space

S.Karpagavalli

Department of Mathematics, SRM Institute of Science and Technology, Kattankulathur, Chennai, Tamil Nadu, India.

E-mail: valli2061@gmail.com

V.Visalakshi

Department of Mathematics, SRM Institute of Science and Technology, Kattankulathur, Chennai, Tamil Nadu, India.

E-mail: visalsenthil86@gmail.com

Abstract. The main purpose of this paper is to study the existence of a fixed point in fuzzy automata normed linear space. In particular, fixed point for a self map on compact fuzzy automata normed linear space has been established.

1. Introduction

Zadeh [12] established fuzzy set theory and its logic with vagueness. Any uncertainty can be interpreted by membership function which is consecutively called as a fuzzy set. The problems with ambiguity is mainly concerned with fuzzy set theory.

Fuzzy norm on linear space is introduced first by Katsaras [7] in year 1984. In 1992, Felbin[4] introduced the finite-dimensional fuzzy normed linear space. T.Bag and Samanta [2] aggregated many properties in completeness and compactness on finite dimensional fuzzy normed linear with a restricted choice of t-norm. Cheng and Mordeson [3] introduced fuzzy normed linear space based on the ideas of fuzzy metric space introduced by Karmosil and Michalek [8] have, defined fuzzy normed linear space and aggregated with many properties. The concept of fuzzy metric space introduced by Karmosil and Michalek [8] has been studied with some modifications by George and Veeramani [5]. Ismat Beg, Shaban Sedghi and Nabi Shobe [6] studied the fixed point theorem in fuzzy metric space.

S.P.Tiwari and many authors [9, 10, 11] has studied the algebraic properties of fuzzy automata in lattice ordered monoid and also introduced some topological concepts.
B. Amudhambigai and V. Madhuri [1] introduced fuzzy automata normed linear space also introduced the topological concepts on fuzzy automata normed linear structure. The concept of fuzzy automata normed linear space introduced in [1] influenced us to obtain fixed point of self map satisfying a relation defined on non decreasing function in fuzzy automata normed linear space.

2. Preliminaries

2.1. Fuzzy Set

Let X be a non empty set and $I = [0, 1]$. A fuzzy subset in X is a function with domain X and value in I, that is an element of I^X.

2.2. Fuzzy Normed Linear Space

Let X be a linear space over the field F. A fuzzy subset N on $X \times R$ is called a fuzzy norm on X if and only if $x, y \in X$ and $c \in F$.

(i) $\forall\ t \in R$ with $t \leq 0$, $N(x, t) = 0$
(ii) $\forall\ t \in R, t > 0$, $N(x, t) = 1$ iff $x = 0$
(iii) $\forall\ t \in R, t > 0$ then $N(cx, t) = N(x, \frac{t}{|c|})$ if $c \neq 0$
(iv) $\forall\ s, t \in R,\ x, y \in X$, $N(x + y, t + s) \geq min\,\{N(x, t), N(y, s)\}$
(v) $N(x, .)$ is a non decreasing function of R and $\lim_{t\to\infty} N(x, t)$=1.

The pair (X, N) will be referred to as a fuzzy normed linear space $(FNLS)$.

2.3. Fuzzy Automata Normed Linear Space

Let $N = (Q, X, \delta)$ be a fuzzy automata . A fuzzy automata normed linear space $(FANLS)$ is a 3-tuple (Q, N, T) where Q is non-empty set of states of N and also it is linear space over the field F, T is a t-norm and N is a fuzzy set on $Q \times (0, \infty)$ such that $\forall\ p, q \in Q$ and all $\nu, \mu > 0$ the following condition:

(i) $N(p, \mu) > 0$
(ii) $N(p, \mu) = 1\ \forall\ \mu > 0$ iff $p = 0$
(iii) If $\alpha \neq 0$ then $N(\alpha p, \mu) = N(p, \frac{\mu}{|\alpha|})\ \forall \mu,\alpha \in F$
(iv) $T[N(p, \mu), N(q, \nu)] \leq N(p + q, \mu + \nu)\ \forall \mu, \nu \in F$
(v) $N(p, \cdot)$ is a non-decreasing function F and $\lim_{\mu\to\infty} [N(p, \mu)] = 1$
(vi) Assume that for all $p \neq 0$, $N(p, \cdot)$ is a continuous function of F and strictly increasing on the subset $\mu : 0 < N(p, \mu) < 1$ of F.

3. Fixed Point Theorem on compact FANLS

Let (Q, N, T) be a FANLS and $\phi \neq Y \subseteq Q$. Define $\delta_N(Y, \mu) = \inf\{N(q - r, \mu) : q, r \in Y\}$ $\forall\ \mu > 0$. For an $B_n = \{q_n, q_{n+1}, ...\}$ in a FANLS (Q, N, T). Let $c_n(\mu) = \delta_N(B_n, \mu)$. Then $c_n(\mu)$ is finite $\forall\ n \in N$, $\{c_n(\mu)\}$ is a non-increasing $c_n(\mu) \to c(\mu)$ for some $0 \leq c(\mu) \leq 1$ and also $c_n(\mu) \leq N(q_i - q_j, \mu)\ \forall i, j \geq n$. let $\mathscr{F}$ be the set all continuous function on $\mathbf{F}$: $\mathscr{X} \to [-1, 1]$ where $\mathscr{X}$ is $(0, 1)^3 \times (0, 1)$ such that $\mathbf{F}$ is non-decreasing continuous function such that $\mathbf{F}((t, t, t), s) \leq 0$ it follows that $s \geq \beta(t)$ where β is non decreasing continuous function on [0,1] such that $\beta(s) > s$ for $s \in [0, 1)$.

3.1. Continuous

Let (Q, N, T) be a FANLS N is said to be continuous on $Q \times (0, \infty)$ if $\lim_{n\to} N(q_n - r_n, \mu) = N(q - r, \mu)$ whenever $\{(q_n - r_n, \mu_n)\}$ is a sequence in $Q \times (0, \infty)$ which is converges to a point $(q - r, \mu) \in Q \times (0, \infty)$ that is $\lim_{n\to\infty} N(q_n - q, \mu) = \lim_{n\to\infty} N(r_n - r, \mu) = 1$.

3.2. Convergent

Let (Q, N, T) be a FANLS, let $\{q_n\}$ be a sequence in Q. Then $\{q_n\}$ is said to be convergent if there exist $q \in Q$ such that $\lim_{n\to\infty} N(q_n - q, \mu) = 1\ \forall\ \mu > 0$. Then q is called the limit of the sequence $\{q_n\}$. The limit q of a sequence $\{q_n\}$ is denoted by $q = \lim q_n$.

3.3. Compact
A FANLS in which every sequence has a convergent subsequence is said to be compact FANLS.

3.4. Theorem
Let (Q, N, δ) be a compact FANLS and S is a continuous self map of Q satisfying $\forall\ q - r \in Q$ with $\mathbf{F}(N(q-r,\mu), N(Sq-q,\mu), N(Sq-r,\mu), N(Sq-Sr,\mu)) < 0$ where $\mathbf{F} \in \mathscr{F}$. Then S has atmost one fixed point x in Q such that $S(x) = x$

Proof:

For $n = 1, 2, 3, ...$, S^nQ is Compact and also $S^{n+1}Q \subset S^nQ$. Let $Q_\circ = \bigcap_{n=1}^{\infty} S^nQ$,then $Q_\circ$ is a non empty compact subset of Q and $SQ_\circ = Q_\circ$. It is enough to show that $Q_\circ$ is a Set with one element . Suppose $Q_\circ$ is set with more then one element. Then the function $N : Q^2 \times (0,\infty) \to [-1,1]$ has minimum value. i.e, $\exists\ (q_\circ - r_\circ, \mu) \in Q_\circ^2 \times (0,\infty)$ such that $N(q_\circ - r_\circ, \mu) \leq N(q-r,\mu)\ \forall\ q - r \in Q_\circ$. Since $SQ_\circ = Q_\circ\ \exists\ q_1 - r_1 \in Q_\circ$ such that $Sq_1 = q_\circ$ and $Sr_1 = r_\circ$
$\mathbf{F}(N(q-r,\mu), N(Sq-q,\mu), N(Sq-r,\mu), N(Sq-Sr,\mu)) < 0$
thus We have, by equation(3.1)
$\mathbf{F}(N(q_1-r_1,\mu), N(q_\circ-q_1,\mu), N(q_\circ-r_1,\mu), N(q_\circ-r_\circ,\mu)) < 0$
$\mathbf{F}(N(q_\circ-r_\circ,\mu), N(q_\circ-r_\circ,\mu), N(q_\circ-r_\circ,\mu), N(q_\circ-r_\circ,\mu)) < 0$

$\Rightarrow N(q_\circ - r_\circ,\mu) > N(q_\circ - r_\circ,\mu)$, which contradicts our assumption. Therefore $Q_\circ$ is a set with one element and hence S has a fixed point in Q.

3.5. Corollary
Let (Q, N, δ) be a Compact FANLS, $m \in \mathbf{N}$ and S^m is a continuous self map of Q satisfying $\forall$ $q - r \in Q$.
$\mathbf{F}(N(q-r,\mu), N(S^mq-q,\mu), N(S^mq-r,\mu), N(S^mq-S^mr,\mu)) < 0$ where $\mathbf{F} \in \mathscr{F}$. Then, S has a atmost one fixed point x in Q such that $S(x) = x$

Proof:

S^m has a unique point x in Q. Since $Sx = SS^mx = S^mSx$
Sx is also a fixed point of S^m. By the previous theorem, it follows $Sx = x$.

3.6. Example
Let $Q \subseteq \mathbf{R}$, $Q = \left\{\frac{1}{n}\right\} \mid n \in \mathbf{N} \bigcup \{0\}$. Define $N : Q \times (0,\infty) \to [0,1]$ by $N(p,\mu) = \frac{\mu}{\mu+|p|}\ \forall\ p \in Q$ and $\mu \in [0,\infty)$. Then (Q, N, T) is a compact FANLS. Define a map $S : Q \to Q$ by $S(q) = \frac{q}{2}\ \forall\ q \in Q$ and $\mathbf{F} : \mathscr{X} \to [-1,1]$ where $\mathscr{X}$ is $(0,1)^3 \times (0,1)$ defined by $\mathbf{F}[(u_1,u_2,u_3),u_4] = \beta(min\{u_1,u_2,u_3\}) - u_4$ and non decreasing function β on $[0,1]$ is defined by $\beta(s) = 1 - \sqrt{s}$, $s \in [0,1]$ and $\beta(s) > s$, $s \in (0,1)$. S satisfies all the hypothesis and hence S has atmost one fixed point.

4. Reference

[1] Amudhambigai.B and Madhuri.V, A new view on fuzzy automata normed linear structure space, *Iranian journal of fuzzy system.* (accepted)
[2] Bag .T and Samanta .S, 2003, Finite dimensional fuzzy normed linear space, *The Journal of fuzzy Mathematics*, Vol.**3**, 687-705.
[3] Cheng.S.C and Mordeson.J.N., 1994, Fuzzy linear operators and fuzzy normed linear spaces, *Bulletin of the calcutta mathematical society*, Vol.**86**, pp.429-436.
[4] Felbin .C, 1992, Finite dimensional fuzzy normed linear space, *Fuzzy sets and system*, Vol.**48**, pp. 239-248.
[5] George.A and Veeramani.P,1994, On some result in fuzzy metric space, *Fuzzy set and systems*, Vol.**64**, No.3.
[6] Ismat Beg, Shaban Sedghi and Nabi Shobe,2013, Fixed point theorem in fuzzy metric space, *International journal of analysis*, Article ID 934145.
[7] Katsaras.A.K,1984, Fuzzy topological vector space, *Fuzzy sets and system*, Vol.**12**, No.2, pp.143-154.

[8] Kramosil.A.K and Michakel.J, 1975, Fuzzy metric and statistical metric space, Kybernetica, Vol.**11**, pp.326-334.
[9] Tiwari.S.P and Sharan.S, 2012, Fuzzy automata based on lattice ordered monoids with algebraic and topological aspects, *Fuzzy information and engineering*, Vol.**2**, pp.155-164.
[10] Tiwari.S.P, Singh.A.K,Sharan . S, 2012, Fuzzy automata based on lattice ordered monoid and associated topology,*Journal of uncertain system*, Vol.**2**, pp.155-164.
[11] Tiwari.S.P, Singh .A.K, Sharan.S, Yadev.V.K, 2015, Bifuzzy core of fuzzy automata, *Iranian journal of fuzzy system*, Vol.**12**, pp.63-73.
[12] Zadeh.L.A, 1965, Fuzzy sets, *Information and Control*, Vol.**8**, No.3, pp.338-358.

A SIMPLE APPROACH FOR THE SOLUTION OF FUZZY MULTI OBJECTIVE TRAVELLING SALESMAN PROBLEM

M Pavithra[a], K Ganesan[b*]
Department of Mathematics,
Faculty of Engineering and Technology
SRM Institute of Science and Technology
Kattankulathur, Chennai-603 203, INDIA
***Corresponding author*: [b]ganesank@srmist.edu.in;**
[a]pavithramahendran52@gmail.com

Abstract We propose a simple method for solving multi objective travelling salesman problem whose decision parameters are expressed as triangular fuzzy number. We apply arithmetic operation and ranking for the parametric form of triangular fuzzy number. The proposed gives preference solution to the decision maker without change to classical travelling salesman problem. A numerical example is provided to show the efficiency of the propose method.

Key words: Travelling salesman problem, fuzzy number, fuzzy arithmetic, fuzzy ranking.

1. Introduction

A sales rep regularly should visit various urban communities beginning from his base camp .the separation, time, cost between each pair of urban communities are thought to be known. The issue of finding the most limited separation, least time and least cost if the sales rep begins from his base camp and goes through every city under his locale precisely once and come back to the Headquarters. In this paper, we have utilized Hungarian to determine triangular fuzzy number without changing over them to traditional triangular fuzzy number issue so that shipping cost, time, distance is minimum Bellman and Zadeh[2] first proposed by the concept of a fuzzy decision making.

2.Preliminaries

Definition 2.1. A fuzzy set $\tilde{A}$ defined on X is a collection of ordered pair

$$\tilde{A} = \{(x,\mu_{\tilde{A}}(x)): x\in X\} \text{ where } \mu_{\tilde{A}}(x) \text{ is a membership value of X}$$

Definition 2.2. A fuzzy set $\tilde{a}$ defined on the set of real numbers R is said to be a fuzzy number if its membership function $\tilde{a}: R \to [0,1]$ has the following characteristics

(i) $\tilde{a}$ is convex
(ii) $\tilde{a}$ is normal
(iii) $\tilde{a}$ is Piecewise continuous.

Definition 2.3. A fuzzy number $\tilde{a}$ on R is said to be a triangular fuzzy number (TFN) or linear fuzzy number if its membership function $\tilde{a}: R \to [0,1]$ has the following characteristics:

$$\tilde{a}(x)=\begin{cases}\dfrac{x-a_1}{a_2-a_1}, & \text{for } a_1\le x\le a_2\\ \dfrac{a_3-x}{a_3-a_2}, & \text{for } a_2\le x\le a_3\\ 0, & \text{elsewhere}\end{cases}$$

We denote this triangular fuzzy number by $\tilde{a}=(a_1, a_2, a_3)$. We use $F(R)$ to denote the set of all triangular fuzzy numbers. Also if $m=a_2$ represents the modal value or midpoint, $\alpha=(a_2-a_1)$ represents the left spread and $\beta=(a_3-a_2)$ represents the right spread of the triangular fuzzy number $\tilde{a}=(a_1, a_2, a_3)$, then the triangular fuzzy number $\tilde{a}$ can be represented by the triplet $\tilde{a}=(\alpha,m,\beta)$. i.e. $\tilde{a}=(a_1, a_2, a_3)=(\alpha, m, \beta)$. We use $F(R)$ to denote the set of all triangular fuzzy numbers.

2.1 Ranking of triangular Fuzzy Numbers

Many different approaches for the ranking of fuzzy numbers have been proposed in the literature. Abbasbandy and Hajjari[1] proposed a new ranking method based on the left and the right spreads at some α -levels of fuzzy numbers.

For an arbitrary triangular fuzzy number $\tilde{a}=(a_1, a_2, a_3)=\left(a_0,a_*,a^*\right)$ with parametric form $\tilde{a}=\left(\underline{a}(r), \overline{a}(r)\right)$, we define the magnitude of the triangular fuzzy number $\tilde{a}$ by

$$\begin{aligned}\mathrm{Mag}(\tilde{a})&=\frac{1}{2}\left(\int_0^1(\underline{a}+\overline{a}+a_0)\ f(r)\ dr\right)\\&=\frac{1}{2}\left(\int_0^1(a^*+4a_0-a_*)\ f(r)\ dr\right).\end{aligned}$$

where the function $f(r)$ is a non-negative and increasing function on $[0,1]$ with $f(0)=0$, $f(1)=1$ and $\int_0^1 f(r)\ dr=\frac{1}{2}$. The function $f(r)$ can be considered as a weighting function. In real life applications, $f(r)$ can be chosen by the decision maker according to the situation. In this paper, for convenience we use $f(r)=r$.

For any two triangular fuzzy numbers $\tilde{a}=\left(a_0,a_*,a^*\right)$ and $\tilde{b}=\left(b_0,b_*,b^*\right)$ in F(R), we define the ranking of $\tilde{a}$ and $\tilde{b}$ by comparing the Mag ($\tilde{a}$) and Mag ($\tilde{b}$) on R as follows:

(i) $\tilde{a}\succeq\tilde{b}$ if and only if Mag($\tilde{a}$) $\ge$ Mag($\tilde{b}$)

(ii) $\tilde{a}\preceq\tilde{b}$ if and only if Mag($\tilde{a}$) $\le$ Mag($\tilde{b}$)

(iii) $\tilde{a}\approx\tilde{b}$ if and only if Mag($\tilde{a}$) = Mag($\tilde{b}$)

2.2.Arithmetic operation on triangular Fuzzy Numbers

Ming Ma et al.[7] have proposed a new fuzzy arithmetic based upon both location index and fuzziness index functions. The location index number is taken in the ordinary arithmetic, whereas the fuzziness index functions are considered to follow the lattice rule which is least upper bound in the lattice L. That is for a, b$\in$L we define $a\vee b=\max\{a,b\}$ and $a\wedge b=\min\{a,b\}$.

For arbitrary triangular fuzzy numbers $\tilde{a}=\left(a_0,a_*,a^*\right)$ and $\tilde{b}=\left(b_0,b_*,b^*\right)$ and $*=\{+,-,\times,\div\}$, the arithmetic operations on the triangular fuzzy numbers are defined by $\tilde{a}*\tilde{b}=\left(a_0*b_0,a_*\vee b_*,a^*\vee b^*\right)$. In particular for any two triangular fuzzy numbers $\tilde{a}=\left(a_0,a_*,a^*\right)$ and $\tilde{b}=\left(b_0,b_*,b^*\right)$, we define

(i) Addition : $\tilde{a}+\tilde{b} =(a_0,a_*,a^*)+(b_0,b_*,b^*)=(a_0+b_0,\max\{a_*,b_*\},\max\{a^*,b^*\})$

(ii) Subtraction : $\tilde{a}-\tilde{b} =(a_0,a_*,a^*)-(b_0,b_*,b^*)=(a_0-b_0,\min\{a_*,b_*\},\min\{a^*,b^*\})$

(iii) Multiplication : $\tilde{a}\times\tilde{b}=(a_0,a_*,a^*)\times(b_0,b_*,b^*) =(a_0\times b_0,\max\{a_*,b_*\},\max\{a^*,b^*\})$

(iv) Division: $\tilde{a}\div\tilde{b} =(a_0,a_*,a^*)\div(b_0,b_*,b^*) =(a_0\div b_0,\min\{a_*,b_*\},\min\{a^*,b^*\})$

3.Main result

Fuzzy Multi objective linear programming problem with triangular fuzzy number

A fuzzy multi objective linear programming problem involving triangular fuzzy numbers can be defined as follows:

$$\max \tilde{Z}\approx(\tilde{Z}_1,\tilde{Z}_2,\ldots\ldots,\tilde{Z}_N)^T$$
$$\text{where } \tilde{Z}_1=\sum_{j=1}^{n}\tilde{c}_{1j}\tilde{x}_j,\tilde{Z}_2=\sum_{j=1}^{n}\tilde{c}_{2j}\tilde{x}_j,\ldots\ldots\ldots,\tilde{Z}_N=\sum_{j=1}^{n}\tilde{c}_{Nj}\tilde{x}_j$$

$$\begin{aligned}\text{subject to } &\sum_{j=1}^{n}\tilde{a}_{ij}\tilde{x}_j\preceq\tilde{b}_i \quad \text{for all } i=1,2,\ldots\ldots,m_0\\ &\sum_{j=1}^{n}\tilde{a}_{ij}\tilde{x}_j\succeq\tilde{b}_i \quad \text{for all } i=m_0+1,\ldots\ldots\ldots m \\ \text{and } &\tilde{x}_j\succeq\tilde{0} \quad \text{for all } j=1,2,\ldots\ldots\ldots\ldots n.\end{aligned} \tag{2}$$

where $\tilde{a}_{ij}$, $\tilde{c}_j$, $\tilde{x}_j$, $\tilde{b}_i\in F(R)$, $i=1,2,3,\ldots\ldots.m$ and $j=1,2,3,\ldots\ldots,n.$. If $\tilde{a}_{ij}$, $\tilde{c}_j$, $\tilde{x}_j$ and $\tilde{b}_i$ are represented by location index number, left fuzziness index function and right fuzziness index function respectively, then the above problem can be rewritten as follows:

$$\begin{aligned}&\max \tilde{Z}\approx(\tilde{Z}_1,\tilde{Z}_2,\ldots\ldots,\tilde{Z}_N)\\ &\text{where } \tilde{Z}_1=\sum_{j=1}^{n}\left\langle(c_{1j})_0,(c_{1j})_*,(c_{1j})^*\right\rangle\times\left\langle(x_j)_0,(x_j)_*,(x_j)^*\right\rangle\\ &\tilde{Z}_2=\sum_{j=1}^{n}\left\langle(c_{2j})_0,(c_{2j})_*,(c_{2j})^*\right\rangle\times\left\langle(x_j)_0,(x_j)_*,(x_j)^*\right\rangle\\ &\ldots\ldots\ldots\tilde{Z}_k=\sum_{j=1}^{n}\left\langle(c_{kj})_0,(c_{kj})_*,(c_{kj})^*\right\rangle\times\left\langle(x_j)_0,(x_j)_*,(x_j)^*\right\rangle\\ &\sum_{j=1}^{n}\left\langle(a_{ij})_0,(a_{ij})_*,(a_{ij})^*\right\rangle\times\left\langle(x_j)_0,(x_j)_*,(x_j)^*\right\rangle\preceq\left\langle(b_j)_0,(b_j)_*,(b_j)^*\right\rangle \quad \text{for all } i=1,2,\ldots\ldots,m_0\\ &\sum_{j=1}^{n}\left\langle(a_{ij})_0,(a_{ij})_*,(a_{ij})^*\right\rangle\times\left\langle(x_j)_0,(x_j)_*,(x_j)^*\right\rangle\succeq\left\langle(b_j)_0,(b_j)_*,(b_j)^*\right\rangle \quad \text{for all } i=m_0+1,\ldots\ldots\ldots m\\ &\left\langle(x_j)_0,(x_j)_*,(x_j)^*\right\rangle\succeq\tilde{0} \quad \text{for all } j=1,2,\ldots\ldots\ldots\ldots n.\end{aligned} \tag{3}$$

Practically, it is difficult to reach fuzzy optima for all objective function subject to the given constraints in problem (2) and (3).

3.1. Algorithm to solve travelling salesman problem

Step 1: Express the given multi objective travelling salesman problem in parametric form.
Step 2: Convert the multi objective travelling salesman problem into a single objective travelling salesman problem by giving suitable weight to the objectives.
Step 3: Solve the reduced single objective travelling salesman problem without converting to Crisp form.
Step 4: Optimal solution of the single objective travelling salesman problem is also optimal solution to multi objective travelling salesman problem

4. Numerical example

A travelling salesman has to visit 4 cities in the following multi objective fuzzy travelling salesman problem. He wishes to start from a particular city, visit each city once and then return to his starting point the following matrix display the fuzzy travelling cost matrix our aim is to minimized the cost, time , distance.

Table 1: Cost matrix of the multi objective travelling salesman problem involving triangular fuzzy numbers

	A	B	C	D
A	∞	(18,20,22) (4,5,6) (2,4,6)	(14,15,16) (6,5,4) (3,5,7)	(7,9,11) (1,3,5) (1,2,3)
B	(18,20,22) (3,5,7) (3,4,5)	∞	(29,30,31) (2,5,8) (2,3,4)	(8,10,12) (2,3,4) (1,3,5)
C	(13,15,17) (3,5,7) (4,5,6)	(28,30,32) (4,5,6) (2,3,4)	∞	(18,20,22) (8,10,12) (1,2,3)
D	(9,11,13) (2,3,4) (1,2,3)	(7,10,13) (1,3,5) (2,3,4)	(19,20,21) (9,10,11) (1,2,3)	∞

Table 2: Cost matrix in which triangular fuzzy numbers are expressed in their parametric form

	A	B	C	D
A	∞	(20,2-2r,2-2r) (5,1-r,1-r) (4,2-2r,2-2r)	(15,1-r,1-r) (5,1-r,1-r) (5,2-2r,2-2r)	(9,2-2r,2-2r) (3,2-2r,2-2r) (2,1-r,1-r)
B	(20,2-2r,2-2r) (5,2-2r,2-2r) (4,1-r,1-r)	∞	(30,1-r,1-r) (5,3-3r,3-3r) (3,1-r,1-r)	(10,2-2r,2-2r) (3,1-r,1-r) (3,2-2r,2-2r)
C	(15,2-2r,2-2r) (5,2-2r,2-2r) (5,1-r,1-r)	(30,2-2r,2-2r) (5,1-r,1-r) (3,1-r,1-r)	∞	(20,2-2r,2-2r) (10,2-2r,2-2r) (2,1-r,1-r)
D	(11,2-2r,2-2r) (3,1-r,1-r) (2,1-r,1-r)	(10,3-3r,3-3r) (3,2-2r,2-2r) (3,1-r,1-r)	(20,1-r,1-r) (10,1-r,1-r) (2,1-r,1-r)	∞

Table 3: Cost matrix in which triangular fuzzy numbers are given preferred weights

	A	B	C	D
A	∞	(10,1-r,1-r) (1.5,0.3-0.3r,0.30.3r) (0.8,0.40.4r,0.4-0.4r)	(7.5,0.5-0.5r,0.5-0.5r) (1.5,0.3-0.3r,0.3-0.3r) (1,0.4-0.4r,0.4-0.4r)	(4.5,1-r,1-r) (0.9,0.6-0.6r,0.6-0.6r) (0.4,0.2-0.2r,0.2-0.2r)
B	(10,1-r,1-r) (1.5,0.6-0.6r,0.6-0.6r) (0.8,0.2-0.2r,0.2-0.2r)	∞	(15,0.5-0.5r,0.5-0.5r) (1.5,0.9-0.9r,0.9-0.9r) (0.6,0.2-0.2r,0.2-0.2r)	(5,1-r,1-r) (0.9,0.3-0.3r,0.3-0.3r) (0.6,0.4-0.4r,0.4-0.4r)
C	(7.5,1-r,1-r) (1.5,0.6-0.6r,0.6-0.6r) (1,0.2-0.2r,0.2-0.2r)	(15,1-r,1-r) (1.5,0.3-0.3r,0.3-0.3r) (0.6,0.2-0.2r,0.2-0.2r)	∞	(10,1-r,1-r) (3,0.6-0.6r,0.6-0.6r) (0.4,0.2-0.2r,0.2-0.2r)
D	(5.5,1-r,1-r) (0.9,0.3-0.3r,0.3-0.3r) (0.4,0.2-0.2r,0.2-0.2r)	(5,1.5-1.5r,1.5-1.5r) (0.9,0.6-0.6r,0.6-0.6r) (0.6,0.2-0.2r,0.2-0.2r)	(10,0.5-0.5r,0.5-0.5r) (3,0.3-0.3r,0.3-0.3r) (0.4,0.2-0.2r,0.2-0.2r)	∞

Table 4: Multi objective travelling salesman problem is converted to a single objective travelling salesman problem

	A	B	C	D
A	∞	(12.3,1-r,1-r)	(10,0.5-0.5r,0.5-0.5r)	(5.8,1-r,1-r)
B	(12.3,1-r,1-r)	∞	(17.1,0.9-0.9r,0.9-0.9r)	(6.5,1-r,1-r)
C	(10,1-r,1-r)	(17.1,1-r,1-r)	∞	(13.4,1-r,1-r)
D	(6.8,1-r,1-r)	(6.5,1.5-1.5r,1.5-1.5r)	(13.4,0.5-0.5r,0.5-0.5r)	∞

Table 5: Row reduced Cost matrix

	A	B	C	D
A	∞	(6.5)	(4.2,-0.5+0.5r,-0.5+0.5r)	0
B	(5.8)	∞	(10.6,-0.1+0.1r,-0.1+0.1r)	0
C	0	(7.1)	∞	(3.4)
D	(0.3,-0.5+0.5r,-0.5+0.5r)	0	(6.9,-1+r,-1+r)	∞

Table 6: Column reduced Cost matrix

	A	B	C	D
A	∞	(6.5)	0	0
B	(5.8)	∞	(6.4,0.4-0.4r,0.4-0.4r)	0
C	0	(7.1)	∞	(3.4)
D	(0.3,-0.5+0.5r,-0.5+0.5r)	0	(2.7,-0.5+0.5r,-0.5+0.5r)	∞

The Optimum solution is **A→ D→ B → C → A**

∴ Fuzzy optimal distance = (3,2-2r,2-2r)+(5,3-3r,3-3r)+(5,2-2r,2-2r)+(3,2-2r,2-2r) = (16,3-3r,3-3r)

Fuzzy optimal cost = (9,2-2r,2-2r)+(30,1-r,1-r)+(15,2-2r,2-2r)+(10,3-3r,3-3r) = (64,3-3r,3-3r)

Fuzzy optimal time = (2,1-r,1-r)+(3,1-r,1-r)+(5,1-r,1-r)+(3,1-r,1-r) = (13,1-r,1-r)

The value of r	**The optimal distance** (16,3-3r,3-3r)	**Optimal cost** (64,3-3r,3-3r)	**Optimal time** (13,1-r,1-r)
r=0	(13,16,19)	(61,64,67)	(12,13,14)
r=0.5	(14,16,17)	(62,64,65)	(12,13,13)
r=1	(16,16,16)	(64,64,64)	(13,13,13)

5.Conclusion

In the Travelling Salesman problem all the parameters are triangular fuzzy numbers. In this paper, we have reduced the multi objective travelling salesman problem in to a single objective travelling sales man problem by giving suitable weights for the decision parameters. We have solved the reduced single objective travelling salesman problem involving triangular fuzzy numbers without converting to classical travelling salesman problem so that the distance , cost and time is minimum. The proposed method provide an optimal solution which is sharper than the solution obtained by other methods. It is to be seen that the Decision maker have the versatility of picking r∈[0,1] depending on the condition and his craving by applying the proposed technique.

REFERENCES

[1] Abbasbandy S and Hajjari T 2009 A new approach for ranking of trapezoidal fuzzy numbers *Computers and Mathematics with Applications* **57**pp 413–419

[2] Bellman R E and Zadeh L A 1970 Decision-making in fuzzy environment Management sciences **17**(4) 141-164

[3] Dhanasekar S Hariharan S & Sekar P 2013 Classical Travelling Salesman Problem (TSP) based Approach to Solve Fuzzy TSP using Yager's Ranking. *International Journal of Computer Applications 975* 8887

[4] Hameed I A 2019 March Multi-objective Solution of Traveling Salesman Problem with Time. In *International Conference on Advanced Machine Learning Technologies and Applications* pp **121-132** Springer Cham

[5] Jayamoorthi K Karunanidy D Jayavel A & Ramalingam, S 2017A SURVEY ON MULTI-OBJECTIVE TRAVELLING SALSMAN PROBLEM *IIOAB JOURNAL 8*(2) 223-233

[6] Ma M & Li H 2017 August A hybrid genetic algorithm for solving bi-objective traveling salesman problems In *Journal of Physics: Conference Series* **887**(1)p 012065 IOP Publishing

[7] Ming Ma Menahem Friedman Abraham kandel1999 A new fuzzy arithmetic Fuzzy sets and systems 108 83-90

[8] Mohanaselvi S & Ganesan K 2012 Fully Fuzzy linear programs with triangular fuzzy number *International journal of Mathematical Archieve* ***3***(5) 1838-1845

[9] Remer B & Malikopoulos A A 2019 The multi-objective dynamic traveling salesman problem Last mile delivery with unmanned aerial vehicles assistance *arXiv preprint arXiv 1903 04643*

[10] Ricciardi L A & Vasile M 2019 Solving multi-objective dynamic travelling salesman problems by relaxation

[11] Vandermeulen I Groß R & Kolling A 2019 May Balanced task allocation by partitioning the multiple traveling salesperson problem In *Proceedings of the 18th International Conference on Autonomous Agents and MultiAgent Systems* .International Foundation for Autonomous Agents and Multiagent Systems pp1479-1487

[12] Y J Lai and C L Hwang Fuzzy Multiple Objective Decision Making, Lecture Notes in Economics and Mathematical systems **404** springer Berlin 1994

DECISION MAKERS PRFERENCE SOLUTION FOR A FUZZY MULTI OBJECTIVE ASSIGNMENT PROBLEM

B. Divya, K. Ganesan
Department of mathematics
Faculty of engineering and technology
SRM institute of science and technology
kattankulathur, Chennai-603 203, India
E-mail: divyalathama@gmail.com, ganesank@srmist.edu.in

ABSTRACT: In this paper we propose a novel approach for the solution of fuzzy multi objective assignment problem whose parameters are expressed as triangular fuzzy numbers. We apply arithmetic operations and ranking in the parametric form of triangular fuzzy numbers to solve fuzzy multi objective assignment problem without converting to classical form. The proposed method gives flexibility to the decision maker to his/her preferred fuzzy optimal solution. A numerical example is also given based on this novel method.

Key words: Fuzzy assignment situation, multi objective, fuzzy ranking, fuzzy arithmetic.

1. INTRODUCTION

The Multi-Objective Assignment Problem (MOAP) deals with cost, time, quality etc. The objectives of an assignment problem are to minimize both operating cost and operating time, and to maximize quality simultaneously. Suppose we have to assign n workers to n tasks in such a way that the overall operation cost, labour-time, and quality level are to be optimized. Assignment problems representing real-world situations involve a set of parameters whose values are not known exactly. In that case the parameters of the problem are usually defined by the decision makers (DMs) in an uncertain way. Fuzzy sets introduced by Zadeh [21][21] in 1965 provide a procedure to deal with uncertainty. Bellman and Zadeh [3] proposed the concept of decision-making in a fuzzy environment.

Bao et al. [2] solved a multi-objective assignment problem converting to crisp environment. Kagade and Bajaj [6] discussed multi-objective assignment problem with interval cost. Pramanik and Biswas [15][16] developed a priority based fuzzy goal programming method for generalized trapezoidal fuzzy numbers and applied it for multi-objective assignment problem in which cost and time are considered as a generalized trapezoidal fuzzy numbers. They have also solved a multi-objective assignment problem with fuzzy costs for the case of military affairs. De et.al [4] discussed multi objective assignment problem using interactive fuzzy goal programming approach. Tapkan et al. [14] used the bees algorithm for solving the fuzzy multiple objective generalized assignment problem. Pankaj Gupta et.al [12] developed a fuzzy approach to multi criteria assignment problem using exponential membership functions. Kayvan Saleh [7] discussed multi-objective assignment problem with interval parameters. Anita Ravi et.al [1] solved multi-objective assignment problem using genetic algorithm based hybrid approach. Ventepaka et al.[20] introduced an algorithm for solving multi objective assignment problem using Hungarian algorithm an optimal solution of each objective function by minimizing the resource. Thorani and Ravi Shankar [19] has proposed a fuzzy assignment problem with three parameters fuzzy cost, fuzzy time, fuzzy quality and solved it using their fuzzy ranking method. Leelavathy and Ganesan [8] obtained a compromised solution to multi objective fuzzy assignment problems where all the parameters are trapezoidal fuzzy numbers. Sophia Porchelvi and

Anitha [17] used a fuzzy programming technique with linear membership functions to find the compromise solution of multi-objective assignment problem. Mahbubur Rahman et.al [9] solved multi-objective assignment problem with decision maker's preferences by using genetic algorithm. Melita Vinoliah and Ganesan [10] proposed a different approach for the solution of fuzzy multi objective assignment problems.

This paper aims at finding a fuzzy optimal assignment without reducing it to an equivalent crisp form. The paper is organized as follows. In section 2, the basic definition, ranking function and arithmetic operation of TriFN are reviewed. In section 3, fuzzy multi objective assignment model and formulations are reviewed. A numerical example is also provided in section 4 to compare the proposed method with the existing method.

2. PRELIMINARIES

Definition 2.1. A fuzzy set $\tilde{a}$ characterized on X is an accumulation of requested pair

$\tilde{a} = \{(x, \mu_{\tilde{a}}(x)) : x \in X\}$ where $\mu_{\tilde{a}}(x)$ is a participation estimation of x in $\tilde{a}$.

Definition 2.2. Fuzzy set $\tilde{a}$ is defined on a set R of real number whose membership function $\tilde{a}: R \to [0,1]$ is a fuzzy number under certain conditions:

i. $\tilde{a}: R \to [0,1]$ is normal. (i.e.) height ($\tilde{a}$) = 1
ii. $\tilde{a}: R \to [0,1]$ is convex
iii. $\tilde{a}: R \to [0,1]$ is piecewise continuous.

Definition 2.3. A triangular fuzzy number $\tilde{a} = (a_1, a_2, a_3)$ is defined by its membership function is given by

$$\mu_{\tilde{a}}(x) = \begin{cases} \dfrac{x - a_1}{a_2 - a_1}, & a_1 \le x \le a_2 \\ \dfrac{a_3 - x}{a_3 - a_2}, & a_2 \le x \le a_3 \\ 0, & \text{otherwise} \end{cases}$$

Definition 2.4. A triangular fuzzy number $\tilde{a} = (a_1, a_2, a_3) \in F(R)$ can also be represented as a pair $\tilde{a} = (\underline{a}, \overline{a})$ of functions which satisfies the following requirements:

i. $\underline{a}(r)$ is bounded monotonic increasing left continuous function.
ii. $\overline{a}(r)$ is a bounded monotonic decreasing left continuous function.
iii. $\underline{a}(r) \le \overline{a}(r), 0 \le r \le 1$

Definition 2.5. For an arbitrary triangular fuzzy number $\tilde{a} = (\underline{a}, \overline{a})$, the number $a_0 = \left(\dfrac{\underline{a}(1) + \overline{a}(1)}{2}\right)$ is said to be a location index number of $\tilde{a}$. The two non-decreasing left continuous functions

$a_* = (a_0 - \underline{a})$, $a^* = (\overline{a} - a_0)$ are called the left fuzziness index function and the right fuzziness index function respectively. Hence every triangular fuzzy number $\tilde{a} = (a_1, a_2, a_3)$ can also be represented by $\tilde{a} = (a_0, a_*, a^*)$

2.6. Ranking of triangular fuzzy number

Several approaches for the ranking of fuzzy numbers have been proposed in the literature. An efficient approach for comparing the fuzzy numbers are by use of ranking function based on their graded means. We defined the triangular fuzzy number $\tilde{a}$ by

$$R(\tilde{a}) = \left(\frac{a^* + 4a_0 - a_*}{4}\right) = \left(\frac{\underline{a} + \overline{a} + a_0}{4}\right)$$

For any two triangular fuzzy number $\tilde{a} = (a_1, a_2, a_3) = (a_0, a_*, a^*)$ and $\tilde{b} = (b_1, b_2, b_3) = (b_0, b_*, b^*)$ in $F(R)$

$$\text{(i) } \tilde{a} \succeq \tilde{b} \text{ if and only if } R(\tilde{a}) \geq R(\tilde{b})$$
$$\text{(ii) } \tilde{a} \preceq \tilde{b} \text{ if and only if } R(\tilde{a}) \leq R(\tilde{b})$$
$$\text{(iii) } \tilde{a} \approx \tilde{b} \text{ if and only if } R(\tilde{a}) = R(\tilde{b})$$

2.7. Arithmetic operations of triangular fuzzy numbers

Ming ma et.al. [11] have projected a new fuzzy arithmetic based upon the both position index functions. The position index number is taken in the ordinary arithmetic, whereas the fuzziness index functions are considered to follow the lattice rule which is least upper bound in the lattice L i.e. for $x, y \in L$, $x \vee y = \max\{x, y\}$ and $x \wedge y = \max\{x, y\}$.

The arithmetic operations on triangular fuzzy numbers are

$$\text{(i) Addition} : \tilde{a} + \tilde{b} = (a_0, a_*, a^*) + (b_0, b_*, b^*) = (a_0 + b_0, \max\{a_*, b_*\}, \max\{a^*, b^*\})$$
$$\text{(ii) Subtraction} : \tilde{a} - \tilde{b} = (a_0, a_*, a^*) - (b_0, b_*, b^*) = (a_0 - b_0, \max\{a_*, b_*\}, \max\{a^*, b^*\})$$
$$\text{(iii) Multiplication} : \tilde{a} \times \tilde{b} = (a_0, a_*, a^*) \times (b_0, b_*, b^*) = (a_0 \times b_0, \max\{a_*, b_*\}, \max\{a^*, b^*\})$$
$$\text{(iv) Division} : \tilde{a} \div \tilde{b} = (a_0, a_*, a^*) \div (b_0, b_*, b^*) = (a_0 \div b_0, \max\{a_*, b_*\}, \max\{a^*, b^*\})$$

3. FUZZY MULTI-OBJECTIVE ASSIGNMENT MODEL

Suppose n jobs are to be performed by n persons depending on their efficiency to do the job in one to one basis such that the assignment cost is minimal. An assignment problem with multiple objectives in fuzzy environment is called a multi objective fuzzy assignment problem.

3.1. Mathematical Formulation of Multi Objective Fuzzy Assignment Problem

Consider a multi objective fuzzy assignment problem of assigning 'n' jobs (operations) to 'n' persons (operators) whose cost coefficients are triangular fuzzy numbers

Mathematical model of multi-objective fuzzy assignment problem can be stated as follows:

$$\text{Minimize } Z^k = \sum_{i=1}^{n}\sum_{j=1}^{n} \tilde{c}_{ij}^k \, \tilde{x}_{ij}$$

$$\text{subject to } \sum_{i=1}^{n} \tilde{x}_{ij} = 1, \text{ for } j = 1,2,3,\dots,n$$

$$\sum_{j=1}^{n} \tilde{x}_{ij} = 1, \text{ for } i = 1,2,3,\dots,n$$

$$\text{and} \quad \tilde{x}_{ij} = \begin{cases} 1, & \text{if job i is assigned to person j} \\ 0, & \text{otherwise} \end{cases}$$

where $\tilde{\mathbf{z}}^k = \{\tilde{z}^1, \tilde{z}^2, \dots, \tilde{z}^k\}$ is a vector of k objectives.

3.2. Cost matrix of $(n \times n)$ *multi objective fuzzy assignment problem*

For simplicity let us assume that the multi objective fuzzy assignment problem has three objectives namely cost, time, quality. Let $\tilde{c}_{ij}$ be the fuzzy assignment cost, $\tilde{t}_{ij}$ be the fuzzy assignment time, $\tilde{q}_{ij}$ be the fuzzy assignment quality incurred in assigning the ith person to j^{th} job.

Table 1: Fuzzy cost matrix for multi objective fuzzy assignment problem

	J_1	J_2		J_n
P_1	$\tilde{c}_{11}, \tilde{t}_{11}, \tilde{q}_{11}$	$\tilde{c}_{12}, \tilde{t}_{12}, \tilde{q}_{12}$	…,…,…	$\tilde{c}_{1n}, \tilde{t}_{1n}, \tilde{q}_{1n}$
P_2	$\tilde{c}_{21}, \tilde{t}_{21}, \tilde{q}_{21}$	$\tilde{c}_{22}, \tilde{t}_{22}, \tilde{q}_{22}$	…,…,…	$\tilde{c}_{2n}, \tilde{t}_{2n}, \tilde{q}_{2n}$
	…,…,…	…,…,…	…,…,…	…,…,…
	…,…,…	…,…,…	…,…,…	…,…,…
	…,…,…	…,…,…	…,…,…	…,…,…
P_n	$\tilde{c}_{1n}, \tilde{t}_{1n}, \tilde{q}_{1n}$	$\tilde{c}_{2n}, \tilde{t}_{2n}, \tilde{q}_{2n}$	…,…,…	$\tilde{c}_{nn}, \tilde{t}_{nn}, \tilde{q}_{nn}$

Since the objective functions conflict with each other, a complete fuzzy optimal solution doesn't exist always, so a new solution concept called fuzzy compromise solution is obtained for multi objective fuzzy assignment problem. A Simple method would be to form a composite objective function as the weighted sum of the objectives, where a weight for an objective is proportional to the preference factor assigned to that particular objective.

3.3. Algorithm to solve fuzzy multi objective assignment problem

Step 1: Express the given fuzzy multi objective assignment problem in parametric form.

Step2: Convert the fuzzy multi objective assignment problem into a fuzzy single objective assignment problem by giving a suitable weight to the objectives.

Step3: Solve the fuzzy single objective assignment problem without converting into crisp form.

Step 4: Optimal solution of the fuzzy single objective assignment problem is also a optimal solution to fuzzy multi objective assignment problem.

4. Numerical example

Let us consider an example discussed by Gupta and Mehlawat (2014). A fuzzy assignment problem concerning six jobs and six workers are given below whose cost, time and quality are represented by triangular fuzzy numbers.

Table 2: Cost matrix with triangular fuzzy numbers

Worker(i)	Job (j)					
	Job-1	Job-2	Job-3	Job-4	Job-5	Job-6
Worker-1						
c_{ij}	(4,6,8)	(3,4,6)	(4,5,8)	(6,8,11)	(7,10,14)	(4,6,7)
t_{ij}	(2,4,5)	(16,20,24)	(7,9,12)	(2,3,5)	(5,8,10)	(7,9,12)
q_{ij}	(0,1,3)	(1,3,5)	(0,1,3)	(0,1,3)	(0,1,3)	(3,5,7)
Worker-2						
c_{ij}	(4,6,7)	(4,5,7)	(5,6,9)	(5,8,12)	(6,9,11)	(6,8,11)
t_{ij}	(4,6,9)	(15,18,22)	(6,8,12)	(7,5,10)	(14,17,20)	(6,8,10)
q_{ij}	(1,3,5)	(3,5,7)	(1,3,5)	(3,5,7)	(5,7,9)	(3,5,7)
Worker-3						
c_{ij}	(8,11,14)	(5,7,9)	(2,4,6)	(5,8,12)	(2,3,4)	(3,4,6)
t_{ij}	(2,3,4)	(6,8,10)	(17,20,24)	(5,7,10)	(12,15,18)	(5,7,10)
q_{ij}	(0,1,3)	(5,7,9)	(3,5,7)	(1,3,5)	(3,5,7)	(5,7,9)
Worker-4						
c_{ij}	(7,9,12)	(7,10,12)	(6,8,11)	(4,6,8)	(8,10,12)	(3,4,6)
t_{ij}	(10,12,16)	(10,13,16)	(12,14,18)	(4,6,9)	(7,9,12)	(8,10,14)
q_{ij}	(3,5,7)	(7,9,10)	(1,3,5)	(3,5,7)	(1,3,5)	(1,3,5)
Worker-5						
c_{ij}	(3,4,6)	(4,6,8)	(5,7,10)	(7,9,12)	(6,8,12)	(5,7,10)
t_{ij}	(7,9,12)	(5,8,11)	(5,7,10)	(11,14,18)	(3,5,8)	(7,9,12)
q_{ij}	(1,3,5)	(7,9,10)	(5,7,9)	(3,5,7)	(1,3,5)	(1,3,5)
Worker-6						
c_{ij}	(2,3,4)	(4,5,7)	(8,11,15)	(8,10,13)	(9,12,15)	(6,8,12)
t_{ij}	(14,17,21)	(10,13,16)	(2,3,5)	(3,5,8)	(10,13,17)	(5,7,10)
q_{ij}	(1,3,5)	(1,3,5)	(3,5,7)	(5,7,9)	(3,5,7)	(5,7,9)

Table 3: (Cost matrix) Triangular fuzzy numbers are in their parametric form

	job-1	Job-2	Job-3	Job-4	Job-5	Job-6
Worker-1	(6,2-2r,2-2r) (4,2-2r,2-2r) (1,1-r,1-r)	(4,1-r,1-r) (20,4-4r,4-4r) (3,,2-2r,2-2r)	(6,2-2r,2-2r) (9,2-2r,2-2r) (1,1-r,1-r)	(8,2-2r,2-2r) (3,1-r,1-r) (1,1-r,1-r)	(10,3-3r,3-3r) (10,3-3r,3-3r) (1,1-r,1-r)	(6,2-2r,2-2r) (9,2-2r,2-2r) (5,2-2r,2-2r)
Worker-2	(6,2-2r,2-2r) (6,2-2r,2-2r) (3,2-2r,2-2r)	(5,2-2r,2-2r) (18,3-3r,3-3r) (5,2-2r,2-2r)	(7,2-2r,2-2r) (8,2-2r,2-2r) (3,2-2r,2-2r)	(5,2-2r,2-2r) (7,2-2r,2-2r) (5,2-2r,2-2r)	(9,3-3r,3-3r) (17,3-3r,3-3r) (7,2-2r,2-2r)	(8,3-3r,3-3r) (8,2-2r,2-2r) (5,2-2r,2-2r)
Worker-3	(11,3-3r,3-3r) (3,1-r,1-r) (1,1-r,1-r)	(7,2-2r,2-2r) (8,2-2r,2-2r) (7,2-2r,2-2r)	(4,2-2r,2-2r) (20,3-3r,3-3r) (5,2-2r,2-2r)	(8,3-3r,3-3r) (7,2-2r,2-2r) (3,2-2r,2-2r)	(3,1-r,1-r) (15,3-3r,3-3r) (5,2-2r,2-2r)	(4,1-r,1-r) (7,2-2r,2-2r) (7,2-2r,2-2r)
Worker-4	(9,2-2r,2-2r) (12,2-2r,2-2r) (5,2-2r,2-2r)	(10,3-3r,3-3r) (13,3-3r,3-3r) (9,2-2r,2-2r)	(8,2-2r,2-2r) (14,2-2r,2-2r) (3, 2-2r,2-2r)	(6,2-2r,2-2r) (6,3-3r,3-3r) (5, 2-2r,2-2r)	(10,2-2r,2-2r) (9,2-2r,2-2r) (3,2-2r,2-2r)	(4,1-r,1-r) (10,2-2r,2-2r) (3,2-2r,2-2r)
Worker-5	(4,1-r,1-r) (9,2-2r,2-2r) (3,2-2r,2-2r)	(6,2-2r,2-2r) (8,3-3r,3-3r) (9,2-2r,2-2r)	(7,2-2r,2-2r) (7,3-3r,3-3r) (7,3-3r,3-3r)	(9,2-2r,2-2r) (14,3-3r,3-3r) (5,2-2r,2-2r)	(8,2-2r,2-2r) (5,2-2r,2-2r) (3,2-2r,2-2r)	(7,2-2r,2-2r) (9,2-2r,2-2r) (3,2-2r,2-2r)
Worker-6	(3,1-r,1-r) (17,3-3r,3-3r) (3,2-2r,2-2r)	(5,1-r,1-r) (13,3-3r,3-3r) (3,2-2r,2-2r)	(10,2-2r,2-2r) (5,2-2r,2-2r) (7,2-2r,2-2r)	(10,2-2r,2-2r) (5,2-2r,2-2r) (7,2-2r,2-2r)	(12,3-3r,3-3r) (13,3-3r,3-3r) (5,2-2r,2-2r)	(8,2-2r,2-2r) (7,2-2r,2-2r) (7,2-2r,2-2r)
	c_{ij} t_{ij} q_{ij}	c_{ij} t_{ij} q_{ij}	c_{ij} t_{ij} q_{ij}	c_{ij} t_{ij} q_{ij}	c_{ij} t_{ij} q_{ij}	c_{ij} t_{ij} q_{ij}

Table 4: Triangular fuzzy numbers are given preferred weights

	Job-1	Job-2	Job-3	Job-4	Job-5	Job-6
Worker-1	(3,1-r,1-r) (1.2,0.6-0.6r,0.6-0.6r) (-0.2,0.2-0.2r,0.2-0.2r)	(2,0.5-0.5r,0.5-0.5r) (6,1.2-1.2r,1.2-1.2r) (-0.6,0.4-0.4r,0.4-0.4r)	(3,1-r,1-r) (2.7,0.6-0.6r,0.6-0.6r) (-0.2,0.2-0.2r,0.2-0.2r)	(4,1-r,1-r) (0.9,0.3-0.3r,0.3-0.3r) (-0.2,0.2-0.2r,0.2-0.2r)	(5.5,1.5-1.5r,1.5-1.5r) (2.4,0.9-0.9r,0.9-0.9r) (-0.2,0.2-0.2r,0.2-0.2r)	(3,1-r,1-r) (2.7,0.6-0.6r,0.6-0.6r) (-1,0.4-0.4r,0.4-0.4r)
Worker-2	(3,1-r,1-r) (1.8,0.6-0.6r,0.6-0.6r) (-0.6,0.4-0.4r,0.4-0.4r)	(2.5,1-r,1-r) (5.4,0.9-0.9r,0.9-0.9r) (-1,0.4-0.4r,0.4-0.4r)	(3.5,1-r,1-r) (2.4,0.6-0.6r,0.6-0.6r) (-0.6,0.4-0.4r,0.4-0.4r)	(2.5,1-r,1-r) (2.1,0.6-0.6r,0.6-0.6r) (-1,0.4-0.4r,0.4-0.4r)	(4.5,1.5-1.5r,1.5-1.5r) (5.1,0.9-0.9r-0.9-0.9r) (-1.4,0.4-0.4r,0.4-0.4r)	(4,1.5-1.5r,1.5-1.5r) (2.4,0.6-0.6r,0.6-0.6r) (-1,0.4-0.4r,0.4-0.4r)
Worker-3	(5.5,1.5-1.5r,1.5-1.5r) (0.9,0.3-0.3r,0.3-0.3r) (-0.2,0.2-0.2r,0.2-0.2r)	(3.5,1-r,1-r) (2.4,0.6-0.6r,0.6-0.6r) (-1.4,0.4-0.4r,0.4-0.4r)	(2,1-r,1-r) (6,0.9-0.9r,0.9-0.9r) (-1,0.4-0.4r,0.4-0.4r)	(4,1.5-1.5r,1.5-1.5r) (2.1,0.6-0.6r,0.6-0.6r) (-0.6,0.4-0.4r,0.4-0.4r)	(1.5,0.5-0.5r,0.5-0.5r) (4.5,0.9-0.9r,0.9-0.9r) (-1,0.4-0.4r,0.4-0.4r)	(2,0.5-0.5r,0.5-0.5r) (2.1,0.6-0.6r,0.6-0.6r) (-1.4,0.4-0.4r,0.4-0.4r)
Worker-4	(4.5,1-r,1-r) (3.6,0.6-0.6r,0.6-0.6r) (-1,0.4-0.4r,0.4-0.4r)	(5,1.5-1.5r,1.5-1.5r) (3.9,0.9-0.9r,0.9-0.9r) (-1.8,0.4-0.4r,0.4-0.4r)	(4,1-r,1-r) (4.2,0.6-0.6r,0.6-0.6r) (-0.6,0.4-0.4r,0.4-0.4r)	(3,1-r,1-r) (1.8,0.9-0.9r,0.9-0.9r) (-1,0.4-0.4r,0.4-0.4r)	(5,1-r,1-r) (2.7,0.6-0.6r,0.6-0.6r) (-0.6,0.4-0.4r,0.4-0.4r)	(2,0.5-0.5r,0.5-0.5r) (3,0.6-0.6r,0.6-0.6r) (-0.6,0.4-0.4r,0.4-0.4r)
Worker-5	(2,0.5-0.5r,0.5-0.5r) (2.7,0.6-0.6r,0.6-0.6r) (-0.6,0.4-0.4r,-.4-0.4r)	(3,1-r,1-r) (2.4,0.9-0.9r,0.9-0.9r) (-1.8,0.4-0.4r,0.4-0.4r)	(3.5,1-r,1-r) (2.4,0.9-0.9r,0.9-0.9r) (-1.8,0.4-0.4r,0.4-0.4r)	(4.5,1-r,1-r) (4.2,0.9-0.9r,0.9-0.9r) (-1,0.4-0.4r,0.4-0.4r)	(4,1-r,1-r) (1.5,0.6-0.6r,0.6-0.6r) (-0.6,0.4-0.4r,0.4-0.4r)	(3.5,1-r,1-r) (2.7,0.6-0.6r,0.6-0.6r) (-0.6,0.4-0.4r,0.4-0.4r)

Worker-6	(1.5,0.5-0.5r,0.5-0.5r) (5.1,0.9-0.9r,0.9-0.9r) (-0.6,0.4-0.4r,0.4-0.4r)	(2.5,0.5-0.5r,0.5-0.5r) (3.9,0.9-0.9r,0.9-0.9r) (-0.6,0.4-0.4r,0.4-0.4r)	(5,1-r,1-r) (1.5,0.6-0.6r,0.6-0.6r) (-1.4,0.4-0.4r,0.4-0.4r)	(5,1-r,1-r) (1.5,0.6-0.6r,0.6-0.6r) (-1.4,0.4-0.4r,0.4-0.4r)	(6,1.5-1.5r,1.5-1.5r) (3.9,0.9-0.9r,0.9-0.9r) (-1,0.4-0.4r,0.4-0.4r)	(4,1-r,1-r) (2.1,0.6-0.6r,0.6-0.6r) (-1.4,0.4-0.4r,0.4-0.4r)

Table 5: Fuzzy multi objective assignment problem converted into single objective assignment problem

	Job-1	Job-2	Job-3	Job-4	Job-5	Job-6
W1	(4,1-r,1-r)	(7.4,1.2-1.2r, 1.2-1.2r)	(5.5,1-r, 1-r)	(4.7,1-r,1-r)	(7.9,1.5-1.5r, 1.5-1.5r)	(6.7,1-r,1-r)
W2	(4.2,1-r,1-r)	(6.9,1-r,1-r)	(4.3,1-r, 1-r)	(3.6,1-r,1-r)	(8.2,1.5-1.5r, 1.5-1.5r)	(5.4,1.5-1.5r, 1.5-1.5r)
W3	(6.2,1.5-.5r, 1.5-1.5r)	(4.5,1-r,1-r)	(7,1-r, 1-r)	(5.5,1.5-1.5r,1.5-1.5r)	(5,0.9-0.9r, 0.9-0.9r)	(2.7,0.6-0.6r, 0.6-0.6r)
W4	(7.1,1-r,1-r)	(7.1,1.5-1.5r, 1.5-1.5r)	(7.6,1-r, 1-r)	(3.8,1-r,1-r)	(7.1,1-r,1-r)	(4.4,0.6-0.6r, 0.6-0.6r)
W5	(4.1,0.6-0.6r, 0.6-0.6r)	(3.6,1-r,1-r)	(4.2,1-r, 1-r)	(7.7,1-r,1-r)	(4.9,1-r,1-r)	(5.6,1-r,1-r)
W6	(6,0.9-0.9r, 0.9-0.9r)	(5.8,0.9-0.9r, 0.9-0.9r)	(5.1,1-r, 1-r)	(5.1,1-r,1-r)	(8.9,1.5-1.5r, 1.5-1.5r)	(4.7,1-r,1-r)

Table 6: Row minima subtracted table

	Job-1	Job-2	Job-3	Job-4	Job-5	Job-6
W1	(0,0,0)	(3.4,1.2-1.2r, 1.2-1.2r))	(1.5, 1-r, 1-r)	(0.7,1-r,1-r)	(3.9,1.5-1.5r, 1.5-1.5r)	(2.7, 1-r,1-r)
W2	(0.6, 1-r,1-r)	(3.3, 1-r,1-r)	(0.7, 1-r,1-r)	(0,0,0)	(4.6,1.5-1.5r, 1.5-1.5r)	(1.8,1.5-1.5r, 1.5-1.5r)
W3	(3.5,1.5-1.5r, 1.5-1.5r)	(1.8, 1-r,1-r)	(4.3, 1-r,1-r)	(2.8,1.5-.5r, 1.5-1.5r)	(2.3,1.5-1.5r,1.5-1.5r)	(0,0,0)
W4	(3.3, 1-r,1-r)	(3.3, 1-r,1-r)	(3.8, 1-r,1-r)	(0,0,0)	(3.3, 1-r,1-r)	(0.6, 1-r,1-r)
W5	(0.5, 1-r,1-r)	(0,0,0)	(0.6, 1-r,1-r)	(4.1, 1-r,1-r)	(1.3, 1-r,1-r)	(2, 1-r,1-r)
W6	(1.3, 1-r,1-r)	(1.1, 1-r,1-r)	(0.4, 1-r,1-r)	(0.4, 1-r,1-r)	(4.2, 1-r,1-r)	(0,0,0)

Table 7: Column minima subtracted table

	Job-1	Job-2	Job-3	Job-4	Job-5	Job-6
W1	(0,0,0)	(3.4, 1.2-1.2r, 1.2-1.2r))	(1.1, 1-r, 1-r)	(0.5,0.4-0.4r, 0.4-0.4r)	(2.6, 1.5-1.5r, 1.5-1.5r)	(2.7, 1-r,1-r)
W2	(0.6, 1-r,1-r)	(3.3, 1-r,1-r)	(0.3, 1-r,1-r)	(0,0,0)	(3.3, 1.5-1.5r, 1.5-1.5r)	(1.8, 1.5-1.5r, 1.5-1.5r)
W3	(3.5, 1.5-1.5r, 1.5-1.5r)	(1.8, 1-r,1-r)	(3.9, 1-r,1-r)	(1,1.3-1.3r, 1.3-`1.3r)	(1, 1.5-1.5r, 1.5-1.5r)	(0,0,0)
W4	(3.3, 1-r,1-r)	(3.3, 1-r,1-r)	(3.4, 1-r,1-r)	(0,0,0)	(2, 1-r,1-r)	(0.6, 1-r,1-r)
W5	(0.5, 1-r,1-r)	(0,0,0)	(0.2, 1-r,1-r)	(4.2,0.4-0.4r, 0.4-0.4r)	(0,0,0)	(2, 1-r,1-r)
W6	(1.3, 1-r,1-r)	(1.1, 1-r,1-r)	(0,0,0)	(0.8,0,0)	(2.9, 1-r,1-r)	(0,0,0)

Applying the proposed algorithm, we have

Table 8: Optimum assignment table

	Job-1	Job-2	Job-3	Job-4	Job-5	Job-6
W1	**(0,0,0)**	(2.1, 1.2-1.2r, 1.2-1.2r))	(0.8, 1-r, 1-r)	(0.5,0.4-0.4r, 0.4-0.4r)	(1.3, 1.5-1.5r, 1.5-1.5r)	(2.4, 1-r,1-r)
W2	(0.6, 1-r,1-r)	(2, 1-r,1-r)	**(0,0,0)**	(0,0,0)	(2, 1.5-1.5r, 1.5-1.5r)	(1.5, 1.5-1.5r, 1.5-1.5r)
W3	(3.8, 1.5-1.5r, 1.5-1.5r)	(0.8, 1-r,1-r)	(3.9, 1-r,1-r)	(1.3,1.3-1.3r, 1.3-`1.3r)	**(0,0,0)**	(0,0,0)
W4	(3.3, 1-r,1-r)	(2, 1-r,1-r)	(3.1, 1-r,1-r)	**(0,0,0)**	(0,0,0)	(0.3, 1-r,1-r)
W5	(1.8, 1-r,1-r)	**(0,0,0)**	(1.2, 1-r,1-r)	(5.2,0.4-0.4r, 0.4-0.4r)	(0,0,0)	(3, 1-r,1-r)
W6	(1.6, 1-r,1-r)	(0.1, 1-r,1-r)	(0,0,0)	(1.1,0,0)	(1.9, 1-r,1-r)	**(0,0,0)**

The current assignment is optimum.

The optimum Assignment schedule is $\mathbf{w_1 \rightarrow j_1, w_2 \rightarrow j_3, w_3 \rightarrow j_5, w_4 \rightarrow j_4, w_5 \rightarrow j_2, w_6 \rightarrow j_6}$.

Fuzzy optimal cost = (6,2-2r,2-2r)+(7,2-2r,2-2r)+(3, 1-r,1-r)+(6,2-2r,2-2r)+(6,2-2r,2-2r)

+(8, 2-2r,2-2r) = (36, 2-2r,2-2r) = (34+2r, 36, 38-2r)

Fuzzy optimal time = (4,2-2r,2-2r)+(8,2-r,2-2r)+(15, 3-3r,3-3r)+(6,3-3r,3-3r)+(8, 3-3r,3-3r)

+(7, 2-r,2-2r) = (48, 3-3r,3-3r) = (45+3r,48,51-3r)

Fuzzy optimal quality = (1,1-r,1-r)+(3,2-2r,2-2r)+(5,2-2r,2-2r)+(5,2-2r,2-2r)+(9,2-2r,2-2r)

+(7,2-2r,2-2r = (30, 2-2r,2-2r) = (28+2r,30,32-2r)

5. Comparison of Results

For the same problem Gupta et.al [14] obtained the optimum solution for a particular $r \in [0,1]$ as below:

Table 9: optimum solution by Gupta et.al [14]

r = 0.695652	
Optimum Cost	(24.8, 32, 43.7)
Optimum Time	(38.3, 50, 64.4)
Optimum Quality	(11,20,30.8)

But the proposed method gives vagueness reduced results for different values of $r \in [0,1]$.

Table 10: optimum solution by proposed method

	Fuzzy optimal cost (34+2r, 36, 38-2r)	Fuzzy optimal time (45+3r,48,51-3r)	Fuzzy optimal quality (28+2r,30,32-2r)
r = 0	(34,36,38)	(45,48,51)	(28,30,32)
r = 0.5	(35, 36, 37)	(46.5,48, 49.5)	(29,30,31)
r =1	(36, 36, 36)	(48, 48, 48)	(30,30,30)

6. Conclusion

We have proposed a new approach to solve the fuzzy multi objective assignment problem involving triangular fuzzy numbers without reducing to its classical form. The proposed method gives flexibility to the decision maker to choose his/her suitable r according preference. It is evident from the example that our method provides better solution for the fuzzy multi objective assignment problem.

REFERENCES

[1]. Anita Ravi Tailor and Jayesh M. Dhodiya, 2016, Genetic algorithm based hybrid approach to solve multi-objective assignment problem, International journal of innovative research in science, engineering and technology, **5** (1), 524-535.

[2]. Bao, C. P., Tsai, M. C. and M. I. Tsai, 2007, A new approach to study the multi-objective assignment problem. WHAMPOA- An interdisciplinary journal, **53**, 123-132.

[3]. R. E. Bellman, L. A. Zadeh, 1970, Decision-making in a fuzzy environment, Manage. Sci. **17B**, 141-164.

[4]. P. K. De and Bharti Yadav, 2011, An algorithm to solve multi-objective assignment problem using interactive fuzzy goal programming approach, Int. J. Contemp. Math. Sciences, **6**(34), 1651 – 1662.

[5]. J. Y. Dong and S. P. Wan, 2019, A new method for solving fuzzy multi-objective linear programming problems, Iranian Journal of Fuzzy Systems, **16**(3), 145-159.

[6]. Kagade, K. L., and Bajaj, V. H. 2010, Fuzzy method for solving multi-objective assignment problem with interval cost. Journal of Statistics and Mathematics, **1**, 01-09.

[7]. Kayvan Salehi, 2014, An approach for solving multi-objective assignment problem with interval parameters, Management Science Letters, **4**, 2155–2160.

[8]. T. Leelavathy and K. Ganesan, 2017, A compromise solution to multi objective fuzzy assignment problem, International Journal of Pure and Applied Mathematics, **113**(13), 226 – 235.
[9]. Md. Mahbubur Rahman and Md. Kutub Uddin, 2019, Solving multi-objective assignment problem with decision maker's preferences by using genetic algorithm, Proceedings of the International Conference on Industrial Engineering and Operations Management Bangkok, Thailand, 1153-1162.
[10]. E. Melita Vinoliah and K. Ganesan, 2019, A different approach for the solution of fuzzy multi objective assignment problems, AIP Conference Proceedings **2112**, 020118.
[11]. Ming Ma, Menahem Friedman, Abraham Kandel, 1999, A new fuzzy arithmetic, Fuzzy sets and systems, **108**, 83-90.
[12]. Pankaj Gupta and Mukesh Kumar Mehlawat, 2013, A fuzzy approach to multicriteria assignment problem using exponential membership functions, International Journal of Machine Learning and Cybernet ics, **4**(6), 647-657.
[13]. Gupta P, Mehlawat MK (2014) A new possibilistic programming approach for solving fuzzy multi-objective assignment problem. IEEE Trans Fuzzy Systems, **21**(1),16–34
[14]. Pınar Tapkan, Lale Özbakır and Adil Baykasog, 2013, Solving fuzzy multiple objective generalized assignment problems directly via bees algorithm and fuzzy ranking, Expert Systems with Applications, **40**, 892–898.
[15]. S. Pramanik and P. Biswas, 2011, Multi-objective assignment problem with fuzzy costs for the case military affairs, International Journal of Computer Applications. **30**(10), 7-12.
[16]. S. Pramanik and P. Biswas, 2012, Multi-objective assignment problem with generalized trapezoidal fuzzy numbers, International Journal of Applied Information Systems, **2**(6), 13-20.
[17]. R. Sophia Porchelvi and M. Anitha, 2018, An algorithm to solve multi objective assignment problem using fuzzy programming technique, International Journal of Current Research and Modern Education, **3**(1), 363-367.
[18]. Tailor and Dhodiya, 2016, Genetic algorithm based hybrid approach to solve multi-objective assignment problem, International Journal of Innovative Research in Science, Engineering and Technology, **5**, 524-535.
[19]. Y.L.P.Thorani and N. Ravi Shankar, 2017, Application of Fuzzy Assignment Problem, Advances in Fuzzy Mathematics, **12** (4), 911-939.
[20]. Ventepaka Yadaiah and Haragopal V.V., 2016, Multi objective optimization of time –cost-quality using Hungarian algorithm, American journal of operations research, **2016**(6), 31-35.
[21]. Zadeh L. A. 1965 Fuzzy sets, Information and Control, **8**(3), 338–353.

AUTHOR INDEX